S0-CKV-545

CAMBRIDGE TRACTS IN MATHEMATICS

169 Quantum Stochastic Processes and Noncommutative Geometry

CAMBRIDGE TRACTS IN MATHEMATICS

A complete list of books in the series can be found at
http://www.cambridge.org/series/sSeries.asp?code=CTM

Recent titles include the following:

140. Derivation and Integration. By W. F. PFEFFER
141. Fixed Point Theory and Applications. By R. P. AGARWAL, M. MEEHAN, and D. O'REGAN
142. Harmonic Maps between Riemannian Polyhedra. By J. EELLS and B. FUGLEDE
143. Analysis on Fractals. By J. KIGAMI
144. Torsors and Rational Points. By A. SKOROBOGATOV
145. Isoperimetric Inequalities. By I. CHAVEL
146. Restricted Orbit Equivalence for Actions of Discrete Amenable Groups. By J. KAMMEYER and D. RUDOLPH
147. Floer Homology Groups in Yang–Mills Theory. By S. K. DONALDSON
148. Graph Directed Markov Systems. By D. MAULDIN and M. URBANSKI
149. Cohomology of Vector Bundles and Syzygies. By J. WEYMAN
150. Harmonic Maps, Conservation Laws and Moving Frames. By F. HÉLEIN
151. Frobenius Manifolds and Moduli Spaces for Singularities. By C. HERTLING
152. Permutation Group Algorithms. By A. SERESS
153. Abelian Varieties, Theta Functions and the Fourier Transform. By A. POLISHCHUK
154. Finite Packing and Covering, K. BÖRÖCZKY, JR
155. The Direct Method in Soliton Theory. By R. HIROTA. Edited and translated by A. NAGAI, J. NIMMO, and C. GILSON
156. Harmonic Mappings in the Plane. By P. DUREN
157. Affine Hecke Algebras and Orthogonal Polynomials. By I. G. MACDONALD
158. Quasi-Frobenius Rings. By W. K. NICHOLSON and M. F. YOUSIF
159. The Geometry of Total Curvature. By K. SHIOHAMA, T. SHIOYA, and M. TANAKA
160. Approximation by Algebraic Numbers. By Y. BUGEAUD
161. Equivalence and Duality for Module Categories. By R. R. COLBY, and K. R. FULLER
162. Lévy Processes in Lie Groups. By MING LIAO
163. Linear and Projective Representations of Symmetric Groups. By A. KLESHCHEV
164. The Covering Property Axiom, CPA. K. CIESIELSKI and J. PAWLIKOWSKI
165. Projective Differential Geometry Old and New. By V. OVSIENKO and S. TABACHNIKOV
166. The Lévy Laplacian. By M. N. FELLER
167. Poincaré Duality Algebras, Macaulay's Dual Systems, and Steenrod Operations. By D. M. MEYER and L. SMITH
168. The Cube: A Window to Convex and Discrete Geometry. By CHUANMING ZONG
169. Quantum Stochastic Processes and Noncommutative Geometry. By KALYAN B. SINHA and DEBASHISH GOSWAMI

QUANTUM STOCHASTIC PROCESSES AND NONCOMMUTATIVE GEOMETRY

Kalyan B. Sinha
Indian Statistical Institute
New Delhi, India

Debashish Goswami
Indian Statistical Institute
Kolkata, India

CAMBRIDGE UNIVERSITY PRESS
Cambridge, New York, Melbourne, Madrid, Cape Town, Singapore, São Paulo

Cambridge University Press
The Edinburgh Building, Cambridge CB2 2RU, UK

Published in the United States of America by Cambridge University Press, New York

www.cambridge.org
Information on this title:www.cambridge.org/9780521834506

First published 2007

Printed in the United Kingdom at the University Press, Cambridge

A catalogue record for this publication is available from the British Library

Library of Congress Cataloguing in Publication data

Goswami, Debashish.
Quantum stochastic processes and noncommutative geometry/ Debashish
Goswami, Kalyan B. Sinha. – 1st ed.
p. cm. – (Cambridge tracts in mathematics)
Includes bibliographical references and index.
ISBN-13: 978-0-521-83450-6 (hardback)
1. Stochastic processes. 2. Quantum groups. 3. Noncommutative
differential geometry. 4. Quantum theory. I. Sinha, Kalyan B. II. Title.
III. Series.

QA274.G67 2006
519.2′3 – dc22

2006034088

ISBN-13 978-0-521-83450-6 hardback

Contents

Preface

On the one hand, in almost all the scientific areas, from physical to social sciences, biology to economics, from meteorology to pattern recognition in remote sensing, the theory of classical probability plays a major role and on the other much of our knowledge about the physical world at least is based on the quantum theory [12]. In a way, quantum theory itself is a new kind of theory of probability (in the language of von Neumann and Birkhoff) (see for example [106]) which contains the classical model, and therefore it is natural to extend the other areas of classical probability theory, in particular the theory of Markov processes and stochastic calculus to this quantum model.

There are more than one possible ways (see for example [127]) to construct the above-mentioned extension and in this book we have chosen the one closest to the classical model in spirit, namely that which contains the classical theory as a submodel. This requirement has ruled out any discussion of areas such as free and monotone-probability models. Once we accept this quantum probabilistic model, the 'grand design' that engages us is the 'canonical construction of a $*$-homomorphic flow (satisfying a suitable differential equation) on a given algebra of observables such that the expectation semigroup is precisely the given contractive semigroup of completely positive maps on the said algebra'.

This problem of 'dilation' is here solved completely for the case when the semigroup has a bounded generator, and also for the more general case (of an unbounded generator) with certain additional conditions such as symmetry and/or covariance with respect to a Lie group action. However, a certain amount of space has to be devoted to develop the needed techniques and structures, and the reader is expected to be well equipped with the basics of functional analysis, theory of Hilbert spaces and of operators in them and of probability theory in order to master these.

A beginner with the above-mentioned background may read Chapters 1 to 6 at first and may leave the rest for a second reading. In some places, mathematical assertions have been made without proof wherever we felt that the proof is essentially similar to a detailed proof of an earlier statement or when the verification of the same can be left as an exercise.

Due to lack of space, not all equations have been displayed and long expressions had to be broken at the end of a line, any inconvenience due to this is regretted. The open square symbol denotes the end of a proof. The reference list is far from complete, we have often included only a recent or a representative paper. We apologize for any unintended exclusion of a reference.

It is a pleasure to remember here people who have contributed to the preparation of this book. Professor K. R. Parthasarathy was instrumental in introducing us to the subject and one of us (K. B. S.) has collaborated with him extensively over nearly two decades; without the insights and masterly expositions of him and of Professor P. A. Meyer, the subject may not have reached the stage it is in now. We thank all our friends, collaborators and members of the Q–P club who have helped us directly or indirectly in this endeavor. In particular, we must mention Professors Luigi Accardi, Robin Hudson, V. P. Belavkin, Martin Lindsay, Franco Fagnola, Stephane Attal, Jean-Luc Sauvageot, Burkhard Kümmerer, Hans Maassen, Rajarama Bhat and Dr Arup Pal and Dr Partha Sarathi Chakraborty. We are grateful to the Indian Statistical Institute (both Delhi and Kolkata campuses) for providing the necessary facilities, Indo-French Centre for the Promotion of Advanced Research and DST-DAAD agencies for making many collaborations possible. One of us (D. G.) would like to thank the Alexander von Humboldt Foundation for a postdoctoral fellowship during 2000–01 (and also later visits under its scheme of 'resumption of fellowship'), when part of the work covered by this book was done. We must also thank Dr Lingraj Sahu, who as a graduate student at a critical stage of writing the monograph, helped with introduction of a part of the material and Mr Joydip Jana for help with proofreading. One of the authors (D. G.) dedicates this book to his wife, Gopa and the youngest addition to his family, expected possibly before this book sees the light of the day; and acknowledges with gratitude the constant encouragement and support from his parents, mother-in-law and Amit-da during the writing of the book.

As is often the case in any such enterprise, some important topics (e.g. stop times) have been left out. The responsibility for the choice of topics as well as for any omissions and shortcomings of the text is entirely ours. We can only hope that this monograph will enthuse some researchers and students to solve some of the problems left unsolved.

K. B. Sinha
Debashish Goswami

Notation

$\mathbb{R}$	The set of real numbers
$\mathbb{C}$	The set of complex numbers
$I\!\!N$	The set of natural numbers
$\mathbb{Q}$	The set of rational numbers
$\mathbb{R}_+$	The set of nonnegative real numbers
$\mathbb{Z}$	The set of integers
S^1	The circle group
$\mathbb{T}^n$	The n-torus
C	The set of bounded continuous functions in $L^2(\mathbb{R}_+, k_0)$ or $L^2 \cap L^4_{\mathrm{loc}}(\mathbb{R}_+, k_0)$
$\mathrm{Dom}(T)$	Domain of an operator T
$\mathrm{Ran}(T)$	Range of T
$\mathrm{Ker}(T)$	Kernel of T
$\mathrm{Sp}(A)$	Complex linear span of vectors in the set A
$\dim(V)$	Dimension of the vector space V
$\mathrm{Im}(x)$, $\mathrm{Re}(x)$	Imaginary and real parts of x (complex number or bounded operator)
$h, \mathcal{H}$	Hilbert spaces
$\mathrm{Lin}(V_1, V_2)$	Set of linear maps from V_1 to V_2 (vector spaces).
$\mathcal{B}(\mathcal{H})$	The set of bounded linear operators on a Hilbert space $\mathcal{H}$
$\mathcal{B}(\mathcal{H}, \mathcal{K})$	The set of bounded linear maps from $\mathcal{H}$ to $\mathcal{K}$ (Hilbert spaces)
$\mathcal{B}^{\mathrm{s.a.}}(\mathcal{H})$	The real Banach space of all bounded self-adjoint operators on $\mathcal{H}$
$\mathcal{K}(\mathcal{H})$	The set of compact operators on a Hilbert space $\mathcal{H}$
$\mathcal{B}_1(\mathcal{H})$	The complex Banach space of trace-class operators on a Hilbert space $\mathcal{H}$
$\mathcal{B}_1^{\mathrm{s.a.}}(\mathcal{H})$	The real Banach space of self-adjoint trace-class operators on $\mathcal{H}$
$\mathcal{L}(E, F)$	The set of adjointable maps from E to F (Hilbert modules)
$\mathcal{L}(E)$	The set of adjointable maps on a Hilbert module E

$\mathcal{K}(E)$	The set of compact adjointable maps on a Hilbert module E
$\mathcal{A}'$	The commutant of (C^* or von Neumann algebra) $\mathcal{A}$
$\mathcal{A}_+$	The set of positive elements of C^* or von Neumann algebra $\mathcal{A}$.
$\mathcal{M}(\mathcal{A})$	The multiplier algebra of $\mathcal{A}$
$\Omega, \Omega_{\mathcal{A}}$	The set of all states, of all normal linear functional on a C^*- or von Neumann algebra $\mathcal{A}$
$\otimes_{\text{alg}}$	Algebraic tensor product (between spaces or algebras)
$\Gamma(\mathcal{H})$	The symmetric Fock space over the Hilbert space $\mathcal{H}$
Γ	The symmetric Fock space over $L^2(\mathbb{R}_+, k_0)$ for some Hilbert space k_0
$\Gamma^f(\mathcal{H})$	The free Fock space over $\mathcal{H}$
k_t	The Hilbert space $L^2([0,t], k_0)$ for some Hilbert space k_0
k^t	$L^2((t,\infty), k_0)$
Γ_t	$\Gamma(k_t)$
Γ^t	$\Gamma(k^t)$
$e(f)$	The exponential vector $\oplus_{n=0}^{\infty} \frac{f^{\otimes^n}}{\sqrt{n!}}$
$\Gamma(A)$	The second quantization of A
χ_A	The characteristic function of the set A
f_t	The function $f\chi_{[0,t]}$
f^t	The function $f\chi_{(t,\infty)}$
Θ	The structure matrix
$\mathcal{U}_t$	Time reversal operator in Fock space
$\mathcal{A} >\!\!\lhd_\alpha G$	The crossed product of $\mathcal{A}$ (C^* or von Neumann algebra) by the action α of a group G
$a_R^\dagger(\cdot), a_\delta^\dagger(\cdot)$	The creation integrator processes associated with operator R and map δ respectively
$a_R(\cdot), a_\delta(\cdot)$	The annihilation integrator processes associated with operator R and map δ respectively
$\Lambda_T(\cdot), \Lambda_\sigma(\cdot)$	The number integrator processes associated with operator T and map σ respectively
$\mathcal{I}_{\mathcal{L}}(\cdot)$	The time integrator process associated with map $\mathcal{L}$ on $\mathcal{A}$
L^p_{loc}	The set of all $f \in L^2(\mathbb{R}_+, k_0)$ such that $\int_0^t \|f(s)\|^p ds < \infty$ for every $t \geq 0$
$L^p(\mathcal{A}, \tau)$	The noncommutative L^p spaces associated with trace τ

1
Introduction

The motivations for writing the present monograph are three-fold: firstly from a physical point of view and secondly from two related, but different mathematical angles.

At the present time our mathematical understanding of a conservative quantum mechanical system is reasonably complete, both from the direction of a consistent abstract theory as well as from the one of mathematical theories of applications in many explicit physical systems like atoms, molecules etc. (see for example the books [12] and [108]). However, a nonconservative (open/dissipative) quantum mechanical system does not enjoy a similar status. Over the last seven decades there have been many attempts to make a theory of open quantum systems beginning with Pauli [104]. Some of the typical references are: Van Hove [126], Ford *et al.* [52], along with the mathematical monograph of Davies [35]. The physicists' Master equation (or Langevin equation) was believed to describe the evolution of a nonconservative open quantum (or classical) mechanical system, a mathematical description of which can be found in Feller's book [50].

Physically, one can conceive of an open system as the 'smaller subsystem' of a total ensemble in which the system is in interaction with its 'larger' environment (sometimes called the bath or reservoir). The total ensemble with a very large number of degrees of freedom undergoes (conservative) evolution, obeying the laws of standard quantum mechanics. However, for various reasons, practical or otherwise, it is of interest only to observe the system and not the reservoir, and this 'reduced dynamics' in a certain sense obeys the Master equation (for a more precise description of these, see [35]). Since it is often impossible and impractical to solve the equation of evolution of the total ensemble, it is often meaningful to replace the reservoir by a 'suitable stochastic process' and couple the system with the stochastic process. In the case in which the stochastic process is classical, the total evolution can be described by a suitable stochastic differential equation (for an introduction to this, the reader is referred to [75] and [41]). The standard Langevin equation [52] involving the stochastic process should restore the conservativeness of the total system albeit for almost all paths. However, in many of the models studied by physicists this is not so.

The simplest quantum mechanical system is the so-called harmonic oscillator. However, the (sub-critically) damped harmonic oscillator which has been studied in classical physics since the time of Newton eludes a consistent treatment in conventional quantum mechanics. In the view of the present authors, this happens because the damped harmonic oscillator is a nonconservative, dissipative system and cannot be understood as a flow in a symplectic manifold (classical case) or in a standard Weyl canonical commutation relations (CCR) algebra (quantum case). One possible way to model this is to represent the environment or reservoir (responsible for the friction or the damping term) by an appropriate stochastic process, restore the unitary stochastic evolution of the quantum system and then project back to the 'system space' by 'washing out' the influence of the stochastic process (taking expectation with respect to the stochastic part) to get back the required nonconservative dynamics. This has been studied in [119] and has also been described in some detail in Chapter 7. Thus one can enunciate a philosophy, not too far away from that of the physicists, that given a nonconservative dynamics of a quantum system, one aim is to canonically construct the stochastic process which will represent the environment so that the two together undergo a conservative evolution and the projection to the system space restores exactly the nonconservative evolution. There is a further aim of the physicist, viz. to obtain the stochastic process mentioned above in a suitable approximation from the mechanical descriptions of the particles constituting the reservoir and of their interactions with the observed system. This aspect is not treated in this monograph and the reader is referred to [4], [8] and [35].

There is an exact mathematical counterpart to the picture in physicists' mind as described above. Given a finite probability space $S \equiv \{1, 2, \ldots, n\}$ with probability distribution given by the vector $p \equiv (p_1, p_2, \ldots, p_n)$ on it and a stochastic (or Markov) matrix $(t_{ij})_{ij=1}^n$ such that $t_{ij} \geq 0$, $\sum_{j=1}^n t_{ij} = 1$, one can associate a (discrete) evolution $(Tf)(i) = \sum_{j=1}^n t_{ij} f(j)$ with $f : \{1, 2, \ldots, n\} \to \mathbb{R}$. Then one observes that

(i) T maps positive functions f to positive functions and maps identity function to itself.
(ii) The probability distribution vector p is in one-to-one correspondence with the dual ϕ_p of the algebra of functions on S by $p \mapsto \phi_p$, where

$\phi_p(f) = \sum_{i=1}^{n} p_i f(i)$, and this induces a dual dynamics T^* given by $(T^*\phi_p)(\chi_j) = \sum_{i=1}^{n} p_i t_{ij}$, where χ_k denotes the characteristic function of the singleton set $\{k\}$.

(iii) $T^n, n = 0, 1, \ldots,$ and $T^{*n}, n = 0, 1, 2 \ldots$ provide two discrete (dynamical) semigroups, the second being dual to the first; and clearly T^n for each n satisfies the property (i).

There is a standard construction of a Markov process (in this case Markov chain); see e.g. Feller's book [50]. This procedure extends naturally, beginning with the consideration of the algebra of functions on S as the algebra of $n \times n$ diagonal matrices and $\{T^n\}_{n=0,1,2,\ldots}$ as a positive semigroup on that, to the more general picture considering semigroups (discrete or continuous parameter) on the noncommutative algebra of all $n \times n$ matrices. What is perhaps surprising and is contrary to intuition in classical probability is that a very large class of Markov processes (including Markov chains) can be described by quantum stochastic differential equations in Fock space, again facilitating many computations ([99, 100]).

At this point an important generalization of the class of positive maps on an algebra makes its entrance. From a physical point of view, consider the following scenario. Let $\mathcal{H}$ be the Hilbert space of a localized quantum system A in a box and let there exist another quantum system B with associated Hilbert space $\mathbb{C}^n$. The system B is so far removed from A that there is no interaction between A and B and thus the Hilbert space for the joint system A and B is $\mathcal{H} \otimes \mathbb{C}^n$. Let T_n be the positive linear map which describes an operation on the joint system that does not affect B, given by $T_n(x \otimes y) = T(x) \otimes y$ for $x \in \mathcal{B}(\mathcal{H})$, $y \in \mathcal{B}(\mathbb{C}^n)$ (here $\mathcal{B}(\mathcal{H})$ is the set of all bounded linear operators on the Hilbert space $\mathcal{H}$ defined everywhere) for some positive linear map T on $\mathcal{B}(\mathcal{H})$. It seems reasonable to expect that given a positive linear map T on $\mathcal{B}(\mathcal{H})$, it should be such that for every natural number n, T_n given above should be positive. In such a case, T is said to be completely positive (CP) and such CP maps or semigroups of such maps play a very important role in the description of nonconservative dynamics on quantum systems. It is also useful to note that if the algebra involved is commutative (like the algebra of $n \times n$ diagonal matrices in the first example instead of $\mathcal{B}(\mathcal{H})$ or the whole matrix algebra) positivity and complete positivity are equivalent and that is why complete positivity does not surface in the context of nonconservative evolutions of classical physical systems. A detailed mathematical study of CP maps and of semigroups of CP maps on an algebra is done in Chapters 2 and 3, respectively.

As we had mentioned earlier in the context of a physical subsystem interacting with a reservoir in such a way that the reduced dynamics is governed by a

Master equation, it is natural to assume that the Master equation is just the differential form of a contractive semigroup of CP maps on the algebra describing the subsystem. Now we can turn this into a very interesting (and demanding) mathematical question: does there exist a 'suitable' probabilistic model for (a) the reservoir and for (b) its interaction with the given subsystem such that the expectation of the total evolution with respect to the probabilistic variables give the CP semigroup we started with? This is the general problem of 'dilation of a contractive semigroup of CP maps on a given algebra'. This problem is solved in Chapter 6 in complete generality under the hypotheses that the given semigroup of CP maps is uniformly continuous so that its generator acting on the given algebra is bounded.

There are complete descriptions of the structure of the generator of a uniformly continuous semigroup of CP maps on an algebra in the third chapter. Unfortunately the situation is far from settled for a similar question if the semigroup is only strongly continuous, which is, as is often the case, more interesting from the point of view of applications. However, if we pretend that the generator of the strongly continuous semigroup of CP maps on the algebra formally looks similar to that for the uniformly continuous case, then under certain hypotheses a class of strongly continuous semigroups can be constructed such that its generator coincides with the formal one on suitable domains. This is described in the second section of the same chapter along with an applications to a large class of classical Markov processes and also to the irrational rotation algebra which is a type Π_1 factor von Neumann algebra. More details on these constructions and results on the unital nature of the semigroups, so constructed, can be found in Chebotarev [25]. This chapter ends with an important abstract theorem on noncommutative Dirichlet forms associated with a strongly continuous semigroup of CP maps on a von Neumann algebra equipped with a normal faithful semifinite trace. This result is then used in Chapter 8 to solve the dilation problem for such semigroups.

In order to carry out the program charted out in an earlier paragraph, it is necessary to develop some language and machinery. In Chapter 4, the basic theories of Hilbert C^*- and von Neumann modules and of group actions on them are presented. These ideas are then used to develop an elaborate theory of stochastic integration and quantum Itô formulae in symmetric Fock spaces extending the earlier theory as described in [97]. This language seems to be sufficiently powerful to allow a large class of unbounded operator-valued processes in Fock space to be treated. These methodologies were then used to solve Hudson–Parthasarathy (H–P)-type stochastic quantum differential equations with bounded coefficients (Chapter 5) and with unbounded coefficients (Chapter 7) giving unitary or isometric evolutions in a suitable Hilbert space as solutions. The Evans–Hudson (E–H)-type equation of observable or of an element of an algebra is re-interpreted as an equation on the space of maps on a suitable Fock Hilbert module and for bounded coefficient case, such equations are solved in Chapter 5. This language and associated machinery are important because they allow us to answer in the affirmative the problem of the dilation of a uniformly continuous semigroup of CP maps on an algebra.

Chapter 6 uses the tools of Chapters 4 and 5 to show that given a uniformly continuous semigroup of CP maps on a von Neumann algebra, there exists a quantum probabilistic model in the Fock space such that there is a E–H-type quantum stochastic differential equation describing the stochastic evolution of the observable algebra of the quantum subsystem coupled to the quantum stochastic process in Fock space modeling the reservoir, and such that the expectation gives back the original CP semigroup. This construction is canonical and interestingly gives a quantum stochastic differential equation for the evolution so that further computations for any other observable effects may be facilitated.

The mathematical problem of stochastic dilation of a semigroup of CP maps on a C^*- or von Neumann algebra, uniformly or strongly continuous, with the additional requirement that the dilated map on the algebra satisfies a quantum stochastic differential equation in Fock space and is a $*$-homomorphism on the algebra of observables is the central mathematical problem treated in this book. The property of $*$-homomorphism of such maps is a basic requirement of any quantum theory and the fact that these also satisfy a differential equation makes the family of dilated maps a stochastic flow of $*$-homomorphisms on the algebras. In fact, Chapters 6 and 8 are devoted to the final steps of the solution of this problem, the first for the uniformly continuous semigroup and the second for the strongly continuous one, while the Chapters 2 to 5 and Chapter 7 deal with preliminary materials and develop the machinery needed. This completes our discussions on the central mathematical problem treated here along with its connection to applications, arising from the physics of open quantum systems.

There is a another mathematical direction from which we approach the central mathematical problem of stochastic dilation, viz. that of noncommutative geometry. Chapter 9 should not be and cannot be thought of as an exposition on the rapidly developing subject of noncommutative geometry as created by Alain Connes [28] (the reader may also look at the books [82] and [56]). Instead, after some introduction to basic concepts in differential geometry and elements of noncommutative geometry, three explicit examples are worked out and in each case an appropriate associated stochastic process (classical or quantum) is constructed. Much more study in these areas remains to be done; for example one can investigate whether the nontrivial curvature in the Quantum Heisenberg manifold can be captured in terms of the stochastic processes on it.

We think the spirit of the book is perhaps well-described in the preface by Luigi Accardi in *Probability Towards 2000* [3] and we quote:

> The reason why the interaction of probability with quantum physics is different from the above mentioned ones is that the problem here is not only to apply classical techniques or to extend them to situations which, being even more general, still remain within the same qualitative type of intuition, language and techniques. Furthermore, the formalism of quantum theory, with its complex wave functions and Hilbert spaces, operators instead of random variables, creates a distance between the mathematical model and the physical phenomena which is certainly greater than that of classical physics. For these reasons, these new languages and techniques might be perceived as extraneous by some classical probabilists and researchers in mathematical statistics. However, the developments motivated by quantum theory provide not only powerful theoretical tools to probability, but also some conceptual challenges which can enter into the common education of all mathematicians in the same way as happened for the basic qualitative ideas of non-Euclidean geometries.

2
Preliminaries

In this chapter we shall introduce all the basic materials and preliminary notions needed later on in this book.

2.1 C^* and von Neumann algebras

For the details on the material of this section, the reader may be referred to [125], [40] and [76].

2.1.1 C^*-algebras

An abstract normed $*$-algebra $\mathcal{A}$ is said to be a *pre-C^*-algebra* if it satisfies the C^*-property : $\|x^*x\| = \|x\|^2$. If $\mathcal{A}$ is furthermore complete under the norm topology, one says that $\mathcal{A}$ is a *C^*-algebra*. The famous structure theorem due to Gelfand, Naimark and Segal (GNS) asserts that every abstract C^*-algebra can be embedded as a norm-closed $*$-subalgebra of $\mathcal{B}(\mathcal{H})$ (the set of all bounded linear operators on some Hilbert space $\mathcal{H}$). In view of this, we shall fix a complex Hilbert space $\mathcal{H}$ and consider a concrete C^*-algebra $\mathcal{A}$ inside $\mathcal{B}(\mathcal{H})$. The algebra $\mathcal{A}$ is said to be *unital* or *nonunital* depending on whether it has an identity or not. However, even any nonunital C^*-algebra always has a net (sequence in case the algebra is separable in the norm topology) of *approximate identity*, that is, an nondecreasing net e_μ of positive elements such that $e_\mu a \to a$ for all $a \in \mathcal{A}$. Note that the set of compact operators on an infinite dimensional Hilbert space $\mathcal{H}$, to be denoted by $\mathcal{K}(\mathcal{H})$, is an example of nonunital C^*-algebra.

We now briefly discuss some of the important aspects of C^*-algebra theory. First of all, let us mention the following remarkable characterization of commutative C^*-algebras.

Theorem 2.1.1 *Every commutative C^*-algebra $\mathcal{A}$ is isometrically isomorphic to the C^*-algebra $C_0(X)$ consisting of complex-valued functions on a locally compact Hausdorff space X vanishing at infinity. In case $\mathcal{A}$ is unital, X is compact.*

If $\mathcal{A}$ is nonunital, there is a canonical method of adjoining an identity so that $\mathcal{A}$ is embedded as an ideal in a bigger unital C^*-algebra $\hat{\mathcal{A}}$. In view of this, let us assume $\mathcal{A}$ to be unital for the rest of the subsection, unless otherwise mentioned. For $x \in \mathcal{A}$, the *spectrum* of x, denoted by $\sigma(x)$, is defined as the complement of the set $\{z \in \mathbb{C} : (z1 - x)^{-1} \in \mathcal{A}\}$. It is known that for a self-adjoint element x, $\sigma(x) \subseteq \mathbb{R}$, and moreover, a self-adjoint element x is *positive* (that is, x is of the form y^*y for some y) if and only if $\sigma(x) \subseteq [0, \infty)$. There is a rich functional calculus which enables one to form functions of elements of the C^*-algebra. For any complex function which is holomorphic in some domain containing $\sigma(x)$, one obtains an element $f(x) \in \mathcal{A}$ by the holomorphic functional calculus. Furthermore, for any *normal* element x (that is, $xx^* = x^*x$), there is a continuous functional calculus sending $f \in C(\sigma(x))$ to $f(x) \in \mathcal{A}$ where $f \mapsto f(x)$ is a $*$-isometric isomorphism from $C(\sigma(x))$ onto $C^*(x)$, the sub C^*-algebra of $\mathcal{A}$ generated by x. In particular, for any positive element x, we can form a positive *square root* $\sqrt{x} \in \mathcal{A}$ satisfying $\sqrt{x}^2 = x$. For any element $x \in \mathcal{A}$, we define its *absolute value*, denoted by $|x|$, to be the element $\sqrt{x^*x}$. The *real and imaginary parts* of x, denoted by $\mathrm{Re}(x)$ and $\mathrm{Im}(x)$ respectively, are defined by, $\mathrm{Re}(x) = (x + x^*)/2$, $\mathrm{Im}(x) = (x - x^*)/2i$, so that we have, $x = \mathrm{Re}(x) + i\mathrm{Im}(x)$. For a self-adjoint element x, we define two positive elements x^+ and x^-, called respectively the *positive and negative parts* of x, by setting $x^+ = (x + |x|)/2$, $x^- = (|x| - x)/2$. Clearly, x can be decomposed as $x = x^+ - x^-$ and furthermore $x^+x^- = 0$. A linear functional $\phi : \mathcal{A} \to \mathbb{C}$ is said to be *positive* if $\phi(x^*x) \geq 0$ for all x. It is a useful result that an element $x \in \mathcal{A}$ is positive if and only if $\phi(x) \geq 0$ for every positive functional ϕ on $\mathcal{A}$. It can be shown that the algebraic property of positivity implies the boundedness of ϕ, in particular $\|\phi\| = \phi(1)$. Any positive linear functional ϕ with $\phi(1) = 1$ is called a *state* on $\mathcal{A}$. The set of all states is a convex set which is compact in the weak-$*$ topology, hence it has extreme points, called *pure states*, and the set of states is obtained as the closed convex hull of the pure states. A state ϕ is said to be a *trace* if $\phi(ab) = \phi(ba)$ for all $a, b \in \mathcal{A}$. It is said to be *faithful* if $\phi(x^*x) = 0$ implies $x = 0$. The following result, known as the *GNS construction* for a state, is worthy of mention.

Proposition 2.1.2 *Given a state ϕ on $\mathcal{A}$, there exists a triple (called the GNS triple) $(\mathcal{H}_\phi, \pi_\phi, \xi_\phi)$, consisting of a Hilbert space $\mathcal{H}_\phi$, a $*$-representation π_ϕ of $\mathcal{A}$ into $\mathcal{B}(\mathcal{H}_\phi)$ and a vector $\xi_\phi \in \mathcal{H}_\phi$ which is cyclic in the sense that*

$\{\pi_\phi(x)\xi_\phi : x \in \mathcal{A}\}$ is total in $\mathcal{H}_\phi$, satisfying

$$\phi(x) = \langle \xi_\phi, \pi_\phi(x)\xi_\phi \rangle.$$

Moreover, ϕ is pure if and only if π_ϕ is irreducible.

We shall need to extend the scope of the GNS construction to the case of densely defined positive functionals, at least for *semifinite, faithful, positive traces*, which we discuss now. Let us denote by $\mathcal{A}_+$ the set of positive elements of $\mathcal{A}$. Let $\tau : \mathcal{A}_+ \to [0, \infty]$ be a map satisfying $\tau(a+b) = \tau(a) + \tau(b)$, $\tau(\lambda a) = \lambda\tau(a)$ and $\tau(aa^*) = \tau(a^*a)$ for $a, b \in \mathcal{A}_+, \lambda \in \mathbb{R}_+$. Assume furthermore that $\mathcal{I} \equiv \{a \in \mathcal{A} : \tau(a^*a) < \infty\}$ is norm-dense in $\mathcal{A}$ and $\tau(a^*a) = 0$ implies $a = 0$. Such a map τ is called a semifinite, faithful, positive trace, and it can be uniquely extended to the dense subspace (in fact a both-sided ideal) $\mathcal{I}$ as a linear functional, also denoted by τ. The GNS construction can be generalized to such a trace in the following sense.

Proposition 2.1.3 *There exists a Hilbert space $\mathcal{H}$, a $*$-representation $\pi : \mathcal{A} \to \mathcal{B}(\mathcal{H})$ and a linear map $\eta : \mathcal{I} :\to \mathcal{H}$, such that $\tau(a^*bc) = \langle \eta(a), \pi(b)\eta(c) \rangle$ for all $a, c \in \mathcal{I}, b \in \mathcal{A}$. Furthermore, the range of η is dense in $\mathcal{H}$. Such a triple $(\mathcal{H}, \pi, \eta)$ is unique in the sense that for any other such triple $(\mathcal{H}', \pi', \eta')$, we can find a unitary $\Gamma : \mathcal{H} \to \mathcal{H}'$ such that $\pi'(b) = \Gamma\pi(b)\Gamma^*$ and $\eta'(a) = \Gamma\eta(a)$ for $a \in \mathcal{I}, b \in \mathcal{A}$.*

We shall denote the Hilbert space (unique upto identification) $\mathcal{H}$ in the above proposition by $L^2(\mathcal{A}, \tau)$ or simply $L^2(\tau)$ if $\mathcal{A}$ is understood from the context. It can be shown that $\pi(b)\eta(a) = \eta(ba)$, and thus, if $\mathcal{A}$ is unital and $1 \in \mathcal{I}$, which is equivalent to the boundedness of τ, we have a cyclic vector $\eta(1)$.

For a nonunital C^*-algebra, we say that a positive functional ϕ on $\mathcal{A}$ is a state if $\lim \phi(e_\mu) = 1$ for any approximate identity e_μ of $\mathcal{A}$. A positive element $a \in \mathcal{A}$ is said to be *strictly positive* if $\phi(a)$ is nonzero for every state ϕ on $\mathcal{A}$. It is known that $b \in \mathcal{A}_+$ is strictly positive if and only if $b\mathcal{A}_+ := \{ba, a \in \mathcal{A}_+\}$ is norm-dense in $\mathcal{A}_+$.

We shall conclude the discussion on C^*-algebras with the definition of *multiplier algebra*. For a C^*-algebra $\mathcal{A}$ (possibly nonunital), its multiplier algebra, denoted by $\mathcal{M}(\mathcal{A})$, is defined as the maximal C^*-algebra which contains $\mathcal{A}$ as an *essential two-sided ideal*, that is, $\mathcal{A}$ is an ideal in $\mathcal{M}(\mathcal{A})$ and for $y \in \mathcal{M}(\mathcal{A})$, $ya = 0$ for all $a \in \mathcal{A}$ implies $y = 0$. In case $\mathcal{A}$ is unital, one has $\mathcal{M}(\mathcal{A}) = \mathcal{A}$ and for $\mathcal{A} = C_0(X)$ where X is a noncompact, locally compact Hausdorff space, $\mathcal{M}(\mathcal{A}) = C(\hat{X})$, where $\hat{X}$ denotes the Stone–Čech compactification of X. The norm of $\mathcal{M}(\mathcal{A})$ is given by $\|x\| := \sup_{a \in \mathcal{A}, \|a\| \leq 1}\{\|xa\|, \|ax\|\}$. Furthermore, there is a canonical locally convex topology, called the *strict topology* on $\mathcal{M}(\mathcal{A})$, which is given by the family of seminorms $\{\|.\|_a, a \in \mathcal{A}\}$, where $\|x\|_a := \text{Max}(\|xa\|, \|ax\|)$, for $x \in \mathcal{M}(\mathcal{A})$. We say that an embedding $\mathcal{A} \subseteq \mathcal{B}(\mathcal{H})$ for some Hilbert space $\mathcal{H}$ is *nondegenerate* if for $u \in \mathcal{H}$, $au = 0$ for all $a \in \mathcal{A}$ implies that $u = 0$. It is possible to show by simple arguments that $\mathcal{A} \subseteq \mathcal{B}(\mathcal{H})$ is nondegenerate if and only if $\{au, a \in \mathcal{A}, u \in \mathcal{H}\}$ is total in $\mathcal{H}$. Given a nondegenerate embedding $\mathcal{A} \subseteq \mathcal{B}(\mathcal{H})$, we have that $\mathcal{M}(\mathcal{A}) \cong \{x \in \mathcal{B}(\mathcal{H}) : xa, ax \in \mathcal{A}, \text{ for all } a \in \mathcal{A}\}$.

2.1.2 von Neumann algebras

As a Banach space, $\mathcal{B}(\mathcal{H})$ is equipped with the operator-norm topology, but there are other important and interesting topologies that can be given to it, making it a locally convex (but not normable in general) topological space. The most useful ones are weak, strong, ultra-weak and ultra-strong topologies. However, although $\mathcal{B}(\mathcal{H})$ is complete in each of these topologies, a unital sub C^*-algebra $\mathcal{A}$ of $\mathcal{B}(\mathcal{H})$ need not be so. It can be shown that $\mathcal{A}$ is complete in all of the above four locally convex topologies if and only if it is complete in any one of them, and in such a case $\mathcal{A}$ is said to be a *von Neumann algebra*. Furthermore, the strong (respectively, weak) and ultra-strong (respectively, ultra-weak) topologies coincide on norm-bounded convex subsets of $\mathcal{A}$. It is known that if $\mathcal{H}$ is separable, then any norm-bounded subset of $\mathcal{A}$ is metrizable in each of the ultra-weak and ultra-strong topologies. The natural notion of isomorphism between two von Neumann algebras is an algebraic $*$-isomorphism which is also a homeomorphism of the respective ultra-weak topologies. However, there is a stronger notion, called *spatial isomorphism*. Two von Neumann algebras $\mathcal{A}_1 \subseteq \mathcal{B}(\mathcal{H}_1)$ and $\mathcal{A}_2 \subseteq \mathcal{B}(\mathcal{H}_2)$ are said to be *spatially isomorphic* if there is a unitary operator U from $\mathcal{H}_1$ onto $\mathcal{H}_2$ such that $U^*\mathcal{A}_2U = \mathcal{A}_1$.

The following theorem, known as the *Double commutant theorem* due to von Neumann is of fundamental importance in the study of von Neumann algebras. Note that for any subset $\mathcal{B}$ of $\mathcal{B}(\mathcal{H})$, we denote by $\mathcal{B}'$ the commutant of $\mathcal{B}$, that is, $\mathcal{B}' = \{x \in \mathcal{B}(\mathcal{H}) : xb = bx \text{ for all } b \in \mathcal{B}\}$.

Theorem 2.1.4 *A unital $*$-subalgebra $\mathcal{A} \subseteq \mathcal{B}(\mathcal{H})$ is a von Neumann algebra if and only if $\mathcal{A} = \mathcal{A}''(\equiv (\mathcal{A}')')$.*

For the rest of this subsection, let us denote by $\mathcal{A}$ a unital von Neumann subalgebra of $\mathcal{B}(\mathcal{H})$. $\mathcal{A}$ is said to be *σ-finite* if there does not exist any uncountable family of mutually orthogonal nonzero projections in $\mathcal{A}$.

We say that $\mathcal{A}$ is a *factor* if the center is trivial, that is, $\mathcal{A} \cap \mathcal{A}' = \{\lambda 1, \lambda \in \mathbb{C}\}$. The importance of factors stems from the result (see [40]) that an arbitrary von Neumann algebra can be decomposed in a suitable technical sense as a 'direct integral' of factors. A factor $\mathcal{A}$ is called *hyperfinite* if there is an increasing sequence of finite dimensional $*$-subalgebras, say $\mathcal{A}_n, n = 1, 2, \ldots$, of $\mathcal{A}$ such that the ultra-weak closure of the union of $\mathcal{A}_n$ is the whole of $\mathcal{A}$. It is a remarkable result due to Murray and von Neumann that there is a canonical (unique upto a constant multiple) map $d_{\mathcal{A}}$ from the set of projections of a hyperfinite factor $\mathcal{A}$ into $[0, \infty]$ satisfying certain properties (see [124] for details). This function is called the *dimension function* for $\mathcal{A}$, and depending on the nature of the range of $d_{\mathcal{A}}$, factors are classified into various category, namely type I_n for $n = 1, 2, \ldots, I_\infty, II_1, II_\infty$ and III. Any type I_n factor is isomorphic to the finite dimensional algebra of $n \times n$ matrices. Factors of any other type (that is, $I_\infty, II_1, II_\infty, III$) are infinite dimensional. There is a unique hyperfinite type I_∞ factor, which is isomorphic with $\mathcal{B}(\mathcal{H})$ for a separable infinite dimensional Hilbert space $\mathcal{H}$, and the range of $d_{\mathcal{A}}$ for this factor is a countable infinite set (constant multiple of $\{0, 1, \ldots\} \cup \{\infty\}$). Similarly, there is a unique (up to isomorphism) hyperfinite type II_1 factor, say $\mathcal{R}$, and this is characterized by the property of having the range of the dimension function equal to an interval of the form $[0, c]$ for some finite positive number c. The hyperfinite II_∞ factor (also unique upto isomorphism) has the set $[0, \infty]$ as the range of its dimension function. However, the case of type III factors is much more delicate as the range of dimension function is a trivial set $\{0, \infty\}$ for any such factor. Nevertheless, there is a complete classification of hyperfinite type III factors due to the work of Alain Connes. We shall not discuss that here.

A von Neumann algebra $\mathcal{A}$ has enough projections and unitaries, in the sense that $\mathcal{A}$ is the strong closure of the $*$-subalgebra generated by all projections (respectively unitaries) in $\mathcal{A}$. Furthermore, any $x \in \mathcal{A}$ can be decomposed into its real and imaginary parts, each of which is self-adjoint and is in $\mathcal{A}$, and if $E_y(\cdot)$ denotes the family of spectral measures of a self-adjoint element $y \in \mathcal{A}$, then $E_y(\Delta) \in \mathcal{A}$ for all Borel set Δ. We remark that this fact is not true for a general C^*-algebra which is not a von Neumann algebra. A state ϕ on $\mathcal{A}$ is said to be *normal* if $\phi(x_\alpha)$ increases to $\phi(x)$ whenever x_α increases to x for a net $\{x_\alpha\}$ of positive elements from $\mathcal{A}$. More generally, we call a linear map $\Phi : \mathcal{A} \to \mathcal{B}$ (where $\mathcal{B}$ is a von Neumann algebra) positive if it takes positive elements of $\mathcal{A}$ to positive elements of $\mathcal{B}$ and Φ is called normal if whenever x_α increases to x for a net $\{x_\alpha\}$ of positive elements from $\mathcal{A}$, one has that $\Phi(x_\alpha)$ increases to $\Phi(x)$ in $\mathcal{B}$. It is known that a positive linear map is normal if and only if it is continuous with respect to the ultra-weak topology mentioned earlier. In view of this fact, we shall say that a bounded linear map between two von Neumann algebras is normal if it is continuous with respect to the respective two ultra-weak topologies. Normal states, and more generally normal positive linear maps (in particular, normal $*$-homomorphisms) play a major role in the study of von Neumann algebras. The following result describes the structure of a normal state.

Proposition 2.1.5 *A state ϕ on $\mathcal{A}$ is normal if and only if there is a trace-class operator ρ on $\mathcal{H}$ such that $\phi(x) = \mathrm{tr}(\rho x)$ for all $x \in \mathcal{A}$.*

An abelian von Neumann algebra $\mathcal{A}$ is said to be *maximal abelian* if $\mathcal{A}'$ is also abelian. The structure of abelian von Neumann algebras is described by the following proposition, which can be thought of as a von-Neumann algebra analogue of Gelfand's theorem about abelian C^*-algebras.

Proposition 2.1.6 *An abelian von Neumann algebra acting on a separable Hilbert space is $*$-isomorphic with $L^\infty(\Omega, \mathcal{F}, \mu)$ for some measure space $(\Omega, \mathcal{F}, \mu)$. Moreover, if the von Neumann algebra is maximal abelian, the above isomorphism can be chosen to be a spatial isomorphism.*

In view of this result, the theory of von Neumann algebras can be looked upon as a noncommutative measure or probability theory. Many of the well-known theorems, such as the Radon–Nikodym theorem, Martingale convergence theorem have their appropriate generalizations in the set-up of von Neumann algebras.

A remarkable property of von Neumann algebras is the beautiful and particularly simple structure of its normal $*$-homomorphisms. This plays a canonical role in a major portion of the present work. There are three basic and natural ways in which a normal $*$-homomorphism π of $\mathcal{A}$ can arise.

(i) *Reduction:* $\pi(x) = PxP$, where P is a projection. It can be shown that (see Lemma 4.5.2 for a proof) P necessarily belongs to $\mathcal{A}'$.
(ii) *Dilation:* $\pi(x) = x \otimes 1_k$ for some Hilbert space k.
(iii) *Unitary conjugation:* $\pi(x) = \Gamma^* x \Gamma$ where Γ is a unitary in $\mathcal{B}(\mathcal{H})$.

The following result asserts that every normal $*$-homomorphisms of $\mathcal{A}$ is a composition of the above three types.

Proposition 2.1.7 *Given a normal $*$-homomorphism $\pi : \mathcal{A}(\subseteq \mathcal{B}(\mathcal{H})) \to \mathcal{B}(\mathcal{K})$ for some Hilbert space $\mathcal{K}$, there exists a pair (Γ, k) where k is a Hilbert space and Γ is a partial isometry from $\mathcal{K}$ to $\mathcal{H} \otimes k$ such that $\pi(x) = \Gamma^*(x \otimes 1_k)\Gamma$, and the projection $\Gamma\Gamma^*$ commutes with $x \otimes 1_k$ for all $x \in \mathcal{A}$. Moreover, if π is unital, Γ is an isometry. In case $\mathcal{H}$ is separable, one can choose k to be separable as well.*

Corollary 2.1.8 *Any normal unital $*$-homomorphism $\pi : \mathcal{B}(\mathcal{H}) \to \mathcal{B}(\mathcal{K})$ (where $\mathcal{K}$ is a Hilbert space) is of the form $\pi(x) = U^*(x \otimes 1_k)U$, where k is a Hilbert space (which can be chosen to be separable if $\mathcal{H}$ is so) such that $\mathcal{K}$ can be identified with $\mathcal{H} \otimes k$ and U is a unitary in $\mathcal{H} \otimes k$.*

We say that a closed, densely defined linear operator B on $\mathcal{H}$, with the polar decomposition $B = V|B|$ is *affiliated* to the von Neumann algebra $\mathcal{A}$ if $V \in \mathcal{A}$

and $f(|B|) \in \mathcal{A}$ for any bounded measurable function $f : \mathbb{R}_+ \to \mathbb{R}$. We have the following characterization of the property of being affiliated to $\mathcal{A}$:

Lemma 2.1.9 *A possibly unbounded, closed, densely defined operator B acting on a Hilbert space $\mathcal{H}$ is affiliated to a von Neumann algebra $\mathcal{A}$ if and only if for every $a' \in \mathcal{A}'$ and $u \in \mathcal{D}(B)$, we have that $a'u \in \mathcal{D}(B)$ and $Ba'u = a'Bu$. Moreover, if B and C are two closed, densely defined operators affiliated to $\mathcal{A}$, such that BC is also a closed, densely defined operator, then BC is affiliated to $\mathcal{A}$.*

We now discuss the concept of the *enveloping von Neumann algebra* and the *predual* of a von Neumann algebra. Given a unital C^*-algebra $\mathcal{B}$, denote the set of all states by Ω. For $\phi \in \Omega$, we denote by $(\mathcal{H}_\phi, \pi_\phi, \xi_\phi)$ the associated GNS triple. Let $\mathcal{H} = \oplus_{\phi \in \Omega} \mathcal{H}_\phi$ (to be called the *universal enveloping GNS space* and $\pi = \oplus_{\phi \in \Omega} \pi_\phi$. We call π to be the *universal representation* of $\mathcal{B}$ and the weak closure of $\pi(\mathcal{B})$, that is, $\pi(\mathcal{B})''$ in $\mathcal{B}(\mathcal{H})$, is known as the universal enveloping von Neumann algebra of $\mathcal{B}$. We denote it by $\tilde{\mathcal{B}}$. Indeed, it has the following universal property.

Proposition 2.1.10 *Given any $*$-homomorphism ρ of $\mathcal{B}$ in some Hilbert space $\mathcal{K}$, there exists a unique normal $*$-homomorphism $\tilde{\rho} : \tilde{\mathcal{B}} \to \mathcal{B}(\mathcal{K})$ such that $\tilde{\rho} \circ \pi = \rho$, where π is the universal representation mentioned before. Moreover, the image of $\tilde{\rho}(\tilde{\mathcal{B}})$ is the weak closure of $\rho(\mathcal{B})$ in $\mathcal{B}(\mathcal{K})$.*

For a nonunital C^*-algebra, we perform the similar construction after adjoining a unit to it and get similar universal property as described above.

For a von Neumann algebra $\mathcal{A} \subseteq \mathcal{B}(\mathcal{H})$, there is a Banach space $\mathcal{A}_*$, called the predual of $\mathcal{A}$, such that the Banach dual of $\mathcal{A}_*$ coincides with $\mathcal{A}$ with norm topology, whereas the weak-$*$ topology coincides with the ultra-weak topology of $\mathcal{A}$. In fact, Sakai showed that a von Neumann algebra can be characterized in the class of C^*-algebras by this property of having predual. Let us give an explicit description of the predual. For a real linear space we shall consider as its dual the space of all real linear functionals on it. We denote by $\mathcal{B}_1(\mathcal{H})$ and $\mathcal{B}_2(\mathcal{H})$ the set of all trace-class operators and of all Hilbert–Schmidt operators on $\mathcal{H}$ respectively. Let $\mathcal{B}^{\mathrm{s.a.}}(\mathcal{H})$ and $\mathcal{B}_1^{\mathrm{s.a.}}(\mathcal{H})$ stand for the real linear spaces of all bounded self-adjoint operators and all trace-class self-adjoint operators on $\mathcal{H}$ respectively. For a von Neumann algebra $\mathcal{A}$ contained in $\mathcal{B}(\mathcal{H})$, we denote by $\mathcal{A}_h$ the subset of all self-adjoint elements in $\mathcal{A}$. Let $\mathcal{A}_{h,*}$ be the predual of $\mathcal{A}_h$. We define an equivalence relation $\sim$ on $\mathcal{B}_1(\mathcal{H})$ by saying $\rho_1 \sim \rho_2$ if and only if $\mathrm{tr}(\rho_1 x) = \mathrm{tr}(\rho_2 x)$ for all $x \in \mathcal{A}$. We denote by $\mathcal{A}^\perp$ the closed subspace $\{\rho \in \mathcal{B}_1(\mathcal{H}) : \rho \sim 0\}$. For $\rho \in \mathcal{B}_1(\mathcal{H})$, we denote by $[\rho]$ its equivalence class with respect to $\sim$, and $|||[\rho]||| = \inf_{\eta \sim \rho} ||\eta||_1$, where $||.||_1$ denotes trace-class norm. By $\mathcal{A}_h^{\perp}$ we shall denote the set of all self-adjoint elements in $\mathcal{A}^{\perp}$. Clearly, $\mathcal{A}_h^{\perp}$ is a closed subspace of the real Banach space $\mathcal{B}_1^{\text{s.a.}}(\mathcal{H})$ and hence one can consider the quotient space $\mathcal{B}_1^{\text{s.a.}}(\mathcal{H})/\mathcal{A}_h^{\perp}$. For $\rho \in \mathcal{B}_1^{\text{s.a.}}(\mathcal{H})$, let us denote by $[\rho]_h$ the equivalence class corresponding to ρ in the above quotient. It is not difficult to observe that the quotient norm of $[\rho]_h$, say $|||[\rho]_h||$, coincides with $|||[\rho]|||$ defined earlier. To see this, it is enough to note that whenever $\eta \sim \rho$ and ρ is self-adjoint, then $\eta^* \sim \rho$, and thus $(\eta+\eta^*)/2 \in [\rho]_h$ and $||(\eta+\eta^*)/2||_1 \leq ||\eta||_1$. This implies that $|||[\rho]_h|| \leq |||[\rho]|||$ and hence they are equal. We now describe the structure of $\mathcal{A}_*$ and $\mathcal{A}_{h,*}$ as follows.

Proposition 2.1.11 *(i) We have the isometric isomorphism* $\mathcal{A}_* \cong \mathcal{B}_1(\mathcal{H})/\mathcal{A}^{\perp} \cong \Omega_{\mathcal{A}}$, *where* $\Omega_{\mathcal{A}}$ *denotes the space of all normal complex linear bounded functionals on* $\mathcal{A}$. *The canonical identification between* $\mathcal{A}$ *and* $(\mathcal{B}_1(\mathcal{H})/\mathcal{A}^{\perp})^*$ *is given by,* $\mathcal{A} \ni x \mapsto \zeta_x \in (\mathcal{B}_1(\mathcal{H})/\mathcal{A}^{\perp})^*$ *where* $\zeta_x([\rho]) \equiv \text{tr}(\rho x)$. *Moreover, an element* $[\rho]$ *of* $\mathcal{B}_1(\mathcal{H})/\mathcal{A}^{\perp}$ *is canonically associated with* $\phi_{[\rho]}$ *in* $\Omega_{\mathcal{A}}$ *where* $\phi_{[\rho]}(x) \equiv \text{tr}(\rho x)$, $x \in \mathcal{A}$.

(ii) Furthermore, we have $\mathcal{A}_{h,*} \cong \mathcal{B}_1^{\text{s.a.}}(\mathcal{H})/\mathcal{A}_h^{\perp} \cong \Omega_{\mathcal{A}_h}$, *where* $\Omega_{\mathcal{A}_h}$ *denotes the space of all real linear normal bounded functionals on* $\mathcal{A}_h$. *The identification between* $[\rho]_h$ *and its counterpart* $\varphi_{[\rho]_h}$ *(say) in* $\Omega_{\mathcal{A}_h}$ *is given by,* $\varphi_{[\rho]_h} = \text{tr}(\rho x)$, $x \in \mathcal{A}_h$.

Proof:
Part (i) is the exercise (0.4.6) in page 14 of [124] and the proof is omitted. We prove (ii) as an application of (i). Let us consider $\psi : \mathcal{A}_h \to (\mathcal{B}_1^{\text{s.a.}}(\mathcal{H})/\mathcal{A}_h^{\perp})^*$ defined by, $\psi(x)([\rho]_h) = \text{tr}(\rho x)$, $\rho \in \mathcal{B}_1^{\text{s.a.}}(\mathcal{H})/\mathcal{A}_h^{\perp}$, $x \in \mathcal{A}_h$; which is clearly well-defined, linear and one-to-one. To see the surjectivity of ψ, it is enough to note that given $\vartheta \in (\mathcal{B}_1^{\text{s.a.}}(\mathcal{H})/\mathcal{A}_h^{\perp})^*$, we can extend it to $\hat{\vartheta} \in (\mathcal{B}_1(\mathcal{H})/\mathcal{A}^{\perp})^*$ by defining $\hat{\vartheta}([\rho]) = \vartheta([\text{Re}(\rho)]_h) + i\vartheta([\text{Im}(\rho)]_h)$ and by (i), there is an $x \in \mathcal{A}$ such that $\hat{\vartheta}([\rho]) = \text{tr}(\rho x)$ for all ρ. Thus, $\vartheta([\rho]_h) = \hat{\vartheta}([\rho]) = \text{tr}(\rho\text{Re}(x)) + i\ \text{tr}(\rho\text{Im}(x))$ for $\rho \in \mathcal{B}_1^{\text{s.a.}}(\mathcal{H})$; and since $\vartheta([\rho]_h)$ is real, we must have that $\vartheta([\rho]_h) = \text{tr}(\rho\text{Re}(x)) = \psi(\text{Re}(x))([\rho]_h)$. Now we observe that for a positive $\rho \in \mathcal{B}_1(\mathcal{H})$, $\varphi_{[\rho]}$ (as defined in the statement of (i)) is a positive linear normal functional on $\mathcal{A}$, and hence $|||[\rho]_h|| = |||[\rho]||| = ||\varphi_{[\rho]}|| = \varphi_{[\rho]}(1) = \text{tr}(\rho)$. We have $|\psi(x)([\rho]_h)| \leq ||x||||[\rho]_h||$, hence $||\psi(x)|| \leq ||x||$. On the other hand, for any self-adjoint x, $||x|| = \sup_{u\in h,||u||=1}|\langle u, xu\rangle| = \sup_{u\in h,||u||=1}|\text{tr}(\rho_u x)| = \sup_{u\in h,||u||=1}|\psi(x)([\rho_u]_h)| \leq \sup_{u\in h,||u||=1}||\psi(x)|| \cdot |||[\rho_u]_h|| = ||\psi(x)||$ (where $\langle\cdot,\cdot\rangle$ denotes the inner product in h and ρ_u denotes the rank-one operator $|u><u|$). Thus, $||x|| \leq ||\psi(x)||$ also, proving that $||x|| = ||\psi(x)||$.

The assertion that $[\rho]_h \mapsto \varphi_{[\rho]_h}$ is an isometric isomorphism is a straightforward consequence of (i), after noting that any $\alpha \in \Omega_{\mathcal{A}_h}$ can be extended to $\hat{\alpha}$ defined by $\hat{\alpha}(x) = \alpha(\text{Re}(x)) + i\alpha(\text{Im}(x))$ for $x \in \mathcal{A}$ and it is straightforward to prove that $||\hat{\alpha}|| = ||\alpha||$. In fact, $\hat{\varphi}_{[\rho]_h} = \varphi_{[\rho]}$, and $||[\rho]_h|| = ||[\rho]||$, which completes the proof by invoking (*i*). □

Let us fix some more notational convention which will be useful later on. For Hilbert spaces $\mathcal{H}_i$ and von Neumann algebras $\mathcal{A}_i$ contained in $\mathcal{B}(\mathcal{H}_i)$ $(i = 1, 2)$, and a linear map $T : \mathcal{B}_1^{\text{s.a.}}(\mathcal{H}_i) \to \mathcal{B}_1^{\text{s.a.}}(\mathcal{H}_j)$ $(i, j \in (1, 2))$, we shall say that T induces a map $\widetilde{T} : \mathcal{B}_1^{\text{s.a.}}(\mathcal{H}_i)/\mathcal{A}_{i\,h}^{\perp} \to \mathcal{B}_1^{\text{s.a.}}(\mathcal{H}_j)/\mathcal{A}_{j\,h}^{\perp}$ if $T\mathcal{A}_{i\,h}^{\perp} \subseteq \mathcal{A}_{j\,h}^{\perp}$; and in such a case we define $\widetilde{T}$ by $\widetilde{T}([\rho]_h) = [T\rho]_h$. Upto a canonical identification, $\widetilde{T}$ will give rise to a map from $\Omega_{\mathcal{A}_{i\,h}}$ to $\Omega_{\mathcal{A}_{j\,h}}$ which we shall denote by the same notation as $\widetilde{T}$. For $[\rho]_h \in \mathcal{B}_1^{\text{s.a.}}(\mathcal{H})$ and $x \in \mathcal{A}_h$, $[\rho]_h \mapsto \text{tr}([\rho]_h x) \equiv \text{tr}(\rho x)$ is a class-map. We say $[\rho]_h$ to be positive if $\text{tr}([\rho]_h x) \geq 0$ for all positive $x \in \mathcal{A}_h$. It has already been noted in the proof of the previous proposition that for positive ρ, $[\rho]_h$ is also positive, and $||[\rho]_h|| = \text{tr}(\rho) = ||\rho||_1$. Conversely, if $[\rho]_h$ is positive, then the associated functional $\varphi_{[\rho]}$ in $\Omega_{\mathcal{A}}$ is positive and hence there exists a positive ρ_0 such that $\text{tr}(\rho x) \equiv \varphi_{[\rho]}(x) = \text{tr}(\rho_0 x)$ for all $x \in \mathcal{A}$; which in particular implies $[\rho]_h = [\rho_0]_h$. Thus, $||[\rho]_h|| = ||[\rho_0]_h|| = \text{tr}(\rho_0) = \text{tr}(\rho)$. This observation will be useful and let us summarize it in the following manner.

Lemma 2.1.12 *$[\rho]_h$ is positive if and only if* there exists *a positive $\rho_0 \in [\rho]_h$; and in such a case, $||[\rho]_h|| = \text{tr}(\rho) = \text{tr}(\rho_0)$.*

We also have the following lemma.

Lemma 2.1.13 *For two positive elements $[\rho]_h$ and $[\sigma]_h \in \mathcal{B}_1^{\text{s.a.}}(\mathcal{H})/\mathcal{A}_h^{\perp}$, $||[\rho]_h + [\sigma]_h|| = ||[\rho]_h|| + ||[\sigma]_h||$.*

Proof:
It is immediate from the previous lemma since $[\rho + \sigma]_h = [\rho]_h + [\sigma]_h$, which is clearly positive; and thus $||[\rho + \sigma]_h|| = \text{tr}(\rho + \sigma)$. □

Let us mention without proof a noncommutative analogue of the well-known Hahn decomposition for signed measures. We refer to [125] (Theorem 4.2 (ii) of page 140) for the proof.

Proposition 2.1.14 *Given $\rho \in \mathcal{B}_1^{\text{s.a.}}(\mathcal{H})$, it is possible to find two positive elements ρ_1, ρ_2 (not necessarily the positive and negative parts of ρ) in $\mathcal{B}_1^{\text{s.a.}}(\mathcal{H})$ such that*

$$[\rho]_h = [\rho_1]_h - [\rho_2]_h, \quad \|[\rho]_h\| = \|[\rho_1]_h\| + \|[\rho_2]_h\|.$$

Let us conclude our preliminary discussion on von Neumann algebras with a brief account of densely defined normal traces. We have already introduced the concept of a semifinite, faithful, positive trace on a C^*-algebra. However, for von Neumann algebras, the natural topology is the ultra-weak topology, and thus we take the following definition of a normal, semifinite, faithful, positive trace on a von Neumann algebra.

Definition 2.1.15 Let $\mathcal{A}$ be a von Neumann algebra and $\mathcal{A}_+$ be the set of its positive elements. A map $\tau : \mathcal{A}_+ \to [0, \infty]$ is called a *normal, semifinite, faithful, positive trace* on $\mathcal{A}$ if:

(i) $\tau(a + b) = \tau(a) + \tau(b)$, $\tau(\lambda a) = \lambda\tau(a)$ and $\tau(a^*a) = \tau(aa^*)$ for all $a, b \in \mathcal{A}_+, \lambda \in \mathbb{R}_+$;
(ii) $\{a \in \mathcal{A} : \tau(a^*a) < \infty\}$ is dense in $\mathcal{A}$ with respect to the ultra-weak topology;
(iii) $\tau(a^*a) = 0$ if and only if $a = 0$; and
(iv) given any nondecreasing net a_α of positive elements of $\mathcal{A}$, such that $a_\alpha \uparrow a$, we have $\tau(a) = \sup_\alpha \tau(a_\alpha)$.

There exists an analogue of Proposition 2.1.3 for a normal semifinite faithful positive trace τ on $\mathcal{A}$, with the additional feature of π being normal. The associated Hilbert space is denoted by $L^2(\mathcal{A}, \tau)$ or simply $L^2(\tau)$. Taking analogy from classical measure theory, one denotes $\mathcal{A}$ by $L^\infty(\mathcal{A}, \tau)$ or $L^\infty(\tau)$ for short. In fact, this analogy can be carried much farther, leading to a beautiful theory of noncommutative L^p spaces associated with τ, denoted by $L^p(\mathcal{A}, \tau)$ (or simply $L^p(\tau)$) for $1 \leq p \leq \infty$, which share many of the properties of their classical counterparts. We, however, do not go into the details of it, and refer the reader to [96] and [115]. We remark that in particular an exact analogue of the classical Hölder inequality is valid for such noncommutative L^p spaces, which we shall need later on.

2.1.3 Group actions on C^* and von Neumann algebras

Let $\mathcal{A}$ be a unital C^*-algebra, Aut($\mathcal{A}$) be the group of automorphisms of $\mathcal{A}$ and G be a locally compact group such that $g \mapsto \alpha_g$ is a strongly continuous action, taking values in Aut($\mathcal{A}$) (strong continuity refers to the norm-continuity of the map $G \ni g \mapsto \alpha_g(x) \in \mathcal{A}$ for each fixed $x \in \mathcal{A}$). Let μ be a fixed *left Haar measure* on G, and $\delta(g)$ be the corresponding modular function. We define the following multiplication $\bullet$ and involution $*$ on the set $C_c(G, \mathcal{A})$ of compactly supported continuous functions from G to $\mathcal{A}$:

$$(f_1 \bullet f_2)(g) = \int f_1(h)\alpha_h(f_2(h^{-1}g))\mu(dh),$$

$$(f^*)(g) = \delta(g)^{-1}\alpha_g(f(g^{-1})^*);$$

for $f_1, f_2, f \in C_c(G, \mathcal{A})$, $g \in G$. Given a $*$-representation π of $\mathcal{A}$ in $\mathcal{B}(\mathcal{H})$, where $\mathcal{H}$ is a Hilbert space, and a strongly continuous unitary representation

$g \mapsto U_g$ of G in $\mathcal{H}$ (where strong continuity means that for each vector $\xi \in \mathcal{H}$, $g \mapsto U_g\xi$ is continuous), we say that (π, U) is a *covariant representation* of $\mathcal{A}$ (with respect to the given action α) if $\pi(\alpha_g(x)) = U_g \pi(x) U_g^*$ for all $x \in \mathcal{A}$, $g \in G$. For any covariant representation (π, U), we define a $*$-homomorphism $\tilde{\pi}$ of $C_c(G, \mathcal{A})$ (with the $*$-algebraic structure defined above) to $\mathcal{B}(\mathcal{H})$ as follows:

$$\tilde{\pi}(f) = \int \pi(f(g)) U_g \mu(dg), \quad f \in C_c(G, \mathcal{A}).$$

Definition 2.1.16 The *crossed product* $\mathcal{A} >\!\!\lhd_\alpha G$ (α may be omitted if understood from the context) is defined to be the completion of $C_c(G, \mathcal{A})$ with respect to the norm given by $\|f\| := \sup\{\|\tilde{\pi}(f)\|\}$, where the supremum is taken over all possible covariant representation (π, U).

Clearly, the above crossed product is a C^*-algebra, with the extension of the multiplication and involution of $C_c(G, \mathcal{A})$ by continuity. One has the following proposition.

Proposition 2.1.17 *Given any $*$-representation ρ of $\mathcal{A} > \lhd_\alpha G$ in $\mathcal{B}(\mathcal{K})$ for some Hilbert space $\mathcal{K}$, there is a covariant representation (π, U) of $\mathcal{A}$ in the same Hilbert space such that $\rho = \tilde{\pi}$.*

Now, let us fix some faithful imbedding of $\mathcal{A}$ in $\mathcal{B}(\mathcal{H})$ for some Hilbert space $\mathcal{H}$ and let $\tilde{\mathcal{H}} = L^2(G, \mu) \otimes \mathcal{H}$. There is a canonical covariant representation (π, U) of $\mathcal{A}$ in $\tilde{\mathcal{H}}$ which is defined on the dense set $C_c(G, \mathcal{H}) \subset \tilde{\mathcal{H}}$ by the following:

$$(\pi(a)\xi)(g_1) = \alpha_{g_1^{-1}}(a)\xi(g_1), \; (U_g\xi)(g_1) = \xi(g^{-1}g_1),$$

for $g, g_1 \in G, a \in \mathcal{A}, \xi \in C_c(G, \mathcal{H})$. This induces a $*$-representation $\tilde{\pi}$ of $\mathcal{A} >\!\!\lhd_\alpha G$ into $\mathcal{B}(\tilde{\mathcal{H}})$, and we define the image C^*-algebra $\tilde{\pi}(\mathcal{A} >\!\!\lhd_\alpha G)$ to be the reduced crossed product and denote it by $\mathcal{A} >\!\!\lhd_{\alpha,r} G$. This algebra actually does not depend on which $\mathcal{H}$ we choose to embed $\mathcal{A}$, and it is isomorphic with the crossed product $\mathcal{A} >\!\!\lhd_\alpha G$ (called sometimes the full crossed product) whenever G is *amenable* (that is, there is a left-invariant state, called a left-invariant mean, on $L^\infty(G)$). For the details of this topic, including the results on the duality of crossed product, the reader is referred to [28] and [105] and the references cited in those books.

We say that the action α of G on a unital C^*-algebra $\mathcal{A}$ is *ergodic* if $\alpha_g(a) = a$ for all $g \in G$ implies a is a constant multiple of 1. There is a considerable amount of literature devoted to the study of ergodic action of compact groups, and we shall quote one interesting result which will be useful for us.

Proposition 2.1.18 *Let G be a compact group acting ergodically on a unital C^*-algebra $\mathcal{A}$. Then there is a set of elements $\{t_{ij}^{\pi}, \pi \in \hat{G}, i = 1, \ldots, n_{\pi}, j = 1, \ldots, m_{\pi}\}$ of $\mathcal{A}$, where $\hat{G}$ is the set of irreducible representation of G, n_{π} is the dimension of the representation space of the irreducible representation π, $m_{\pi} \leq n_{\pi}$ is a natural number, such that the followings hold.*

(i) There is a unique faithful G-invariant state τ on $\mathcal{A}$, which is in fact a trace.
(ii) The linear span of $\{t_{ij}^{\pi}\}$ is norm-dense in $\mathcal{A}$.
(iii) $\{t_{ij}^{\pi}\}$ is an orthonormal basis of $\mathcal{H} := L^2(\mathcal{A}, \tau)$.
(iv) The action α_g of G coincides with the irreducible representation π on the n_{π}-dimensional vector space $\mathcal{H}_j^{\pi}$ spanned by $\{t_{ij}^{\pi}, i = 1, \ldots, n_{\pi}\}$ for each fixed j and π.
(v) $\sum_{i=1,\ldots,n_{\pi}} (t_{ij}^{\pi})^* t_{ik}^{\pi} = \delta_{jk} n_{\pi} 1$, where δ_{jk} denotes the Kronecker delta symbol. Thus, in particular, $\|t_{ij}^{\pi}\| \leq \sqrt{n_{\pi}}$ for all π, i, j.

The proof can be obtained by combining the results of [9], [67] and [116].

Let us now move on to the von Neumann algebra case. If $\mathcal{A} \subseteq \mathcal{B}(\mathcal{H})$ is a von Neumann algebra, and $g \mapsto U_g$ is a strongly continuous unitary representation in $\mathcal{H}$ of a locally compact group G, then we can define an action α_g of G on $\mathcal{A}$ by $\alpha_g(x) = U_g x U_g^*$, and we define the *crossed product von Neumann algebra* $\mathcal{A} >\!\!\lhd_{\alpha} G$ to be the weak closure of the algebraic linear span of the operators $(a \otimes 1), a \in \mathcal{A}$ and $U_g \otimes L_g, g \in G$ in $\mathcal{B}(\mathcal{H} \otimes L^2(G))$ (where L_g denotes the left regular representation of G in $L^2(G)$). In fact, this von Neumann algebra is spatially isomorphic with the more conventional definition of the crossed product, namely the weak closure of the algebraic linear span of the operators $(1 \otimes L_g), g \in G$ and $\tilde{\alpha}(a), a \in \mathcal{A}$, where $(\tilde{\alpha}(a)\xi)(g) = \alpha_g^{-1}(a)\xi(g)$, for $\xi \in \mathcal{H} \otimes L^2(G) \equiv L^2(G, \mathcal{H})$. If the group G is amenable, it can be shown that given any normal covariant representation (π, V) (a normal covariant representation is defined essentially in the same way as in case of C^*-algebras, only with the additional assumption of normality of π) in some Hilbert space $\mathcal{K}$, there is a normal representation $\tilde{\pi}$ of the crossed product von Neumann algebra in $\mathcal{K}$ satisfying $\tilde{\pi}(a \otimes 1) = \pi(a)$, $\tilde{\pi}(U_g \otimes L_g) = V_g$.

It is interesting and important to know when $\mathcal{A} >\!\!\lhd_{\alpha} G$ is isomorphic with the von Neumann subalgebra $\{a, U_g : a \in \mathcal{A}, g \in G\}''$ of $\mathcal{B}(\mathcal{H})$. Let us mention one general result (due to Haga, [64]), in this direction.

Proposition 2.1.19 *Let us denote by $\pi : \mathcal{A} \to \mathcal{A} >\!\!\lhd_{\alpha} G$ the representation $\pi(a) = a \otimes 1$, and let $\pi(\mathcal{A})'$ be the commutant of the image of π in $\mathcal{B}(\mathcal{H} \otimes L^2(G))$. If we have $(\mathcal{A} >\!\!\lhd_{\alpha} G) \cap (\pi(\mathcal{A}))' \subseteq \pi(\mathcal{A})$, then $\mathcal{A} >\!\!\lhd_{\alpha} G$ is isomorphic with the weak closure of $\mathcal{A}$ and $\{U_g, g \in G\}$ in $\mathcal{B}(\mathcal{H})$.*

Let now $\mathcal{A}_0$ be a unital C^*-algebra equipped with an ergodic action of a compact group G, and let $\mathcal{H} = L^2(\mathcal{A}_0, \tau)$, where τ is the unique invariant faithful trace described in Proposition 8.1.18. Denoting by $\bar{\mathcal{A}}$ the weak closure of $\mathcal{A}_0$ in $\mathcal{B}(\mathcal{H})$, we have the following result.

Theorem 2.1.20 *Let U_g be the unitary in $\mathcal{H}$ induced by the action of G, that is, on the dense set $\mathcal{A}_0 \subseteq \mathcal{H}$, $U_g(a) := \alpha_g(a)$, where α_g denotes the G-action on $\mathcal{A}_0$. Denoting also by α the extended action $g \mapsto U_g.U_g^*$ on $\bar{\mathcal{A}}$, we have that $\bar{\mathcal{A}} >\!\!\lhd_\alpha G$ is isomorphic with the von Neumann algebra generated by $\bar{\mathcal{A}}$ and $U_g, g \in G$.*

Proof:
We shall use the notation of the Proposition 8.1.18, replacing $\mathcal{A}$ by $\mathcal{A}_0$. The crossed product von Neumann algebra $\mathcal{C} := \bar{\mathcal{A}} >\!\!\lhd G$ is by definition the von Neumann algebra generated by $\{(t_{ij}^\pi \otimes 1), \pi, i, j; (U_g \otimes L_g), g \in G\}$ in $L^2(\tau) \otimes L^2(G)$, where L_g is the left regular representation of G in $L^2(G)$. Let ρ be the normal $*$-homomorphism from $\mathcal{C}$ onto $\{\bar{\mathcal{A}}, U_g, g \in G\}'' \subseteq \mathcal{B}(L^2(\tau))$ which satisfies $\rho(t_{ij}^\pi \otimes 1) = t_{ij}^\pi$ and $\rho(U_g \otimes L_g) = U_g$. We have to show that this is an isomorphism, that is, the kernel of ρ is trivial. Clearly, the set of elements of the form $\sum c_{\pi ij} t_{ij}^\pi U_{g_{\pi ij}}$ (finitely many terms), with $c_{\pi ij} \in C$; $g_{\pi ij} \in G$ is dense with respect to the strong-operator topology in $\{\bar{\mathcal{A}}, U_g, g \in G\}''$. Similarly, the set of elements of the form $\sum c_{\pi ij}(t_{ij}^\pi \otimes 1)(U_{g_{\pi ij}} \otimes L g_{\pi ij})$ (finitely many terms) will be strongly dense in $\mathcal{C}$. Now, let $\mathcal{I} \equiv \{X \in \mathcal{C} : \rho(X) = 0\}$. We need to show that $\mathcal{I} = \{0\}$. Let $X \in \mathcal{I}$ and let $X_p = \sum c_{\pi ij}^{(p)}(t_{ij}^\pi \otimes 1)(U_{g_{\pi ij}^{(p)}} \otimes L g_{\pi ij}^{(p)})$ be a net (indexed by p) of elements from the above dense algebra such that X_p converges strongly to X. Hence we have, $\sum |c_{\pi ij}^{(p)}|^2 = \|\rho(X_p)(1)\|^2 \to 0$. This implies that for any $\phi \in L^2(G)$, $\|X_p(1 \otimes \phi)\|^2 = \sum |c_{\pi ij}^{(p)}|^2 \|L_{g_{\pi ij}^{(p)}} \phi\|^2 \leq \|\phi\|^2 \sum |c_{\pi ij}^{(p)}|^2 \to 0$, which proves that $X(1 \otimes \phi) = 0$ for every $X \in \mathcal{I}$. But since $\mathcal{I}$ is an ideal in $\mathcal{C}$, this shows that for $a \in \mathcal{A}$, $X(a \otimes \phi) = (X(a \otimes 1))(1 \otimes \phi) = 0$, and by the fact that $\{a \otimes \phi : a \in \mathcal{A}, \phi \in L^2(G)\}$ is total in $\mathcal{H} \otimes L^2(G)$ we conclude that $\mathcal{I} = \{0\}$. □

From this follows the immediate corollary.

Corollary 2.1.21 *Suppose that G is a compact group acting transitively on a compact Hausdorff topological space X and let μ be a G-invariant Borel measure on X, and let $\mathcal{H} = L^2(X, \mu)$, $\mathcal{A} = L^\infty(X, \mu) \subseteq \mathcal{B}(\mathcal{H})$. Let $U_g, g \in G$ be the unitary operator on $\mathcal{H}$ induced by the G-action on X. Then $\mathcal{A} >\!\!\lhd_\alpha G$ is isomorphic with the von Neumann algebra generated by $\{\mathcal{A}, U_g, g \in G\}$.*

2.2 Completely positive maps

Let us consider two unital $*$-algebras $\mathcal{A}$ and $\mathcal{B}$ and a linear map $T : \mathcal{A} \to \mathcal{B}$. Recall that T is said to be positive if it takes positive elements of $\mathcal{A}$ to positive elements of $\mathcal{B}$. It is clear that a positive map is 'real' in the sense that it takes a self-adjoint element into a self-adjoint element. Given any such positive map T, it is natural to consider $T_n \equiv T \otimes \mathrm{Id} : \mathcal{A} \otimes \mathcal{M}_n \to \mathcal{B} \otimes \mathcal{M}_n$ where $\mathcal{M}_n$ denotes the algebra of $n \times n$ complex matrices. A natural question which arises is the following:
is T_n a positive map for each n ?
The answer to this question is negative, as the following simple example illustrates.

Example 2.2.1 *Let $\mathcal{A}$ be $\mathcal{M}_2$, the algebra of 2×2 complex matrices, and $T : \mathcal{A} \to \mathcal{A}$ be the map given by $T(X) = X'$, where $'$ denotes transpose. That is, (i, j)th element of $T(X)$ is the (j, i)th element of X. Clearly, $T(X^*X) = \overline{X}^*\overline{X} \geq 0$, where the (i, j)th element of $\overline{X}$ is the complex conjugate of (i, j)th element of X. Hence T is positive. We claim that T is not 2-positive. Take $X_1 = \begin{pmatrix} 1 & 0 \\ i & 0 \end{pmatrix}$, $X_2 = \begin{pmatrix} 0 & 0 \\ 1 & 1 \end{pmatrix}$. Consider the element of $\mathcal{M}_2 \otimes \mathcal{M}_2$ given by the block matrix of the form $X = \begin{pmatrix} X_1^*X_1 & X_1^*X_2 \\ X_2^*X_1 & X_2^*X_1 \end{pmatrix}$. Clearly X is positive in $\mathcal{M}_2 \otimes \mathcal{M}_2 \cong \mathcal{M}_4$. But by a simple computation it can be verified that $T(X)$ is the 4×4 matrix*

$$\begin{pmatrix} 2 & 0 & -i & 0 \\ 0 & 0 & -i & 0 \\ i & i & 1 & 1 \\ 0 & 0 & 1 & 1 \end{pmatrix},$$

which is not positive since its determinant $= -2 < 0$.

We say that T is *n-positive* if T_k is positive for all $k \leq n$, but not necessarily for $k = n + 1$. T is said to be *completely positive* (CP for short) if it is n-positive for each n. The role of positivity in classical probability is played by complete positivity in the quantum theory. Let us now formulate the notion of complete positivity in a slightly different but convenient language, namely that of positive definite kernels. For this purpose, we first need a few definitions and facts. For a set X and a Hilbert space $\mathcal{H}$, a map $K : X \times X \to \mathcal{B}(\mathcal{H})$ is called a *kernel*. The set of all kernels is a vector space, denoted by $K(X; \mathcal{H})$.

Definition 2.2.2 A kernel K in $K(X; \mathcal{H})$ is said to be *positive definite* if for each positive integer n and each choice of vectors $u_1, \ldots, u_n$ in $\mathcal{H}$ and elements $x_1, \ldots, x_n \in X$, one has $\sum_{i,j=1}^{n} \langle K(x_i, x_j)u_j, u_i \rangle \geq 0$.

Definition 2.2.3 Kolmogorov decomposition
Let $K \in K(X;\mathcal{H})$. Let $\mathcal{H}_V$ be a Hilbert space and $V : X \to \mathcal{B}(\mathcal{H},\mathcal{H}_V)$ be a map such that $K(x,y) = V(x)^*V(y)$ for all $x, y \in X$. Then $(V,\mathcal{H}_V)$ is said to be a *Kolmogorov decomposition* of K. It is said to be minimal if the set $\{V(x)u : x \in X, u \in \mathcal{H}\}$ is total in $\mathcal{H}_V$. Two Kolmogorov decompositions $(V,\mathcal{H}_V)$ and $(V',\mathcal{H}_{V'})$ are said to be equivalent if there exists a unitary $U : \mathcal{H}_V \to \mathcal{H}_{V'}$ such that $V'(x) = UV(x)$ for all $x \in X$.

Let us now prove that any positive definite kernel admits a canonical minimal Kolmogorov decomposition. Let $F_0 = F_0(X;\mathcal{H})$ denote the vector space of $\mathcal{H}$-valued functions on X having finite support and let $F = F(X;\mathcal{H})$ denote the vector space of all $\mathcal{H}$-valued functions on X. We identify F with a subspace of the algebraic dual F_0' of F_0 by defining for $p \in F$ the functional $\langle p, .\rangle$ on F_0 given by

$$\langle p, f\rangle = \sum_{x\in X}\langle p(x), f(x)\rangle$$

for $f \in F_0$; where the summation is actually over a finite set since f has finite support. Given $K \in K(X;\mathcal{H})$ we define an associated operator $\tilde{K} : F_0(X;\mathcal{H}) \to F(X;\mathcal{H})$ by

$$(\tilde{K}f)(x) = \sum_{y\in X} K(x,y)f(y).$$

Then it is possible to verify that K is positive if and only if $\langle \tilde{K}f, f\rangle \geq 0$ for all $f \in F_0(X;\mathcal{H})$.

Lemma 2.2.4 *Let V be a vector space, V' be its algebraic dual, with the pairing $V' \times V \to \mathbb{C}$ written as $v', v \mapsto \langle v', v\rangle$. Let $A : V \to V'$ be a linear map such that $\langle Av, v\rangle \geq 0$ for all $v \in V$. Then there exists a well defined inner product on the image space AV given by, $\langle Av_1, Av_2\rangle = \langle Av_1, v_2\rangle$.*

Proof:
The sesquilinear form $v_1, v_2 \mapsto a(v_1, v_2) \equiv \langle Av_1, v_2\rangle$ is nonnegative, so that by Schwarz's inequality one obtains

$$|\langle Av_1, v_2\rangle|^2 \leq \langle Av_1, v_1\rangle\langle Av_2, v_2\rangle.$$

It follows that the set $V_A \equiv \{v \in V : \langle Av, v\rangle = 0\}$ coincides with Ker A and the natural projection $\pi : V \to V/\text{Ker } A$ carries the form $a(\cdot,\cdot)$ into an inner product $\langle ., .\rangle_A$ on $V/\text{Ker } A$ given by, $\langle \pi(v_1), \pi(v_2)\rangle_A = a(v_1, v_2)$. The vector space isomorphism $A' : V/\text{Ker } A \to AV$ given by $A'\pi = A$ carries the inner product $\langle ., .\rangle_A$ into an inner product $\langle ., .\rangle$ on AV given by, $\langle Av_1, Av_2\rangle = \langle A'\pi(v_1), A'\pi(v_2)\rangle = \langle \pi(v_1), \pi(v_2)\rangle_A = \langle Av_1, v_2\rangle$. □

We now prove the existence and uniqueness for a minimal Kolmogorov decomposition in a few steps.

Theorem 2.2.5 *Given a positive definite kernel $K \in K(X; \mathcal{H})$, there exists a unique Hilbert space $\mathcal{R}(K)$ of $\mathcal{H}$-valued functions on X such that:*

(i) $\mathcal{R}(K)$ is the closed linear span of $\{K(\cdot, x)u; x \in X, u \in \mathcal{H}\}$;
(ii) $\langle f(x), u\rangle = \langle f, K(\cdot, x)u\rangle$ for all $f \in \mathcal{R}(K), x \in X$ and $u \in \mathcal{H}$.

Proof:
Since K is positive definite, the associated operator $\tilde{K}$ satisfies the hypothesis of Lemma 2.2.4, so that we obtain an inner product $\langle ., .\rangle$ on $\tilde{K}F_0$ and let us denote by $\overline{\tilde{K}F_0}$ the completion of $\tilde{K}F_0$ with respect to the norm inherited from this inner product, and identify $\tilde{K}F_0$ with a dense subset of $\overline{\tilde{K}F_0}$. For each $x \in X$ and $u \in \mathcal{H}$, define the function u_x in F_0 by setting $u_x(y) = u$ if $y = x$ and 0 otherwise. Clearly, $(\tilde{K}u_x)(y) = K(y, x)u$. Define K_x on $\mathcal{H}$ by setting $K_x u = \tilde{K}u_x$ for all $x \in X, u \in \mathcal{H}$. Then $\|K_x u\| \leq \|K(x, x)\|^{\frac{1}{2}}.\|u\|$; and hence K_x is a bounded linear map. A straightforward calculation shows that on $\tilde{K}F_0$ we have $K_x^* f = f(x)$. The mapping from $\overline{\tilde{K}F_0}$ into the space of all $\mathcal{H}$-valued functions on X which sends f into the function $x \mapsto K_x^* f$ is linear, injective and compatible with the identification of $\tilde{K}F_0$ with a dense subset of $\overline{\tilde{K}F_0}$. Thus we regard $\overline{\tilde{K}F_0}$ as a Hilbert space $\mathcal{R}(K)$ consisting of $\mathcal{H}$-valued functions on X. We have already proved that $\mathcal{R}(K)$ satisfies (i) and (ii). Uniqueness of $\mathcal{R}(K)$ is proved by standard arguments. □

Definition 2.2.6 $\mathcal{R}(K)$ in the above theorem is called the *reproducing kernel Hilbert space* for K.

Now we can prove the following theorem.

Theorem 2.2.7 *A kernel K is positive definite if and only if it admits a minimal Kolmogorov decomposition. Moreover, any two minimal Kolmogorov decompositions for the same kernel are equivalent.*

Proof:
The 'if part' of the first statement is trivial; for the 'only if' part we take $\mathcal{H}_V = \mathcal{R}(K)$ and $V(x) = K_x : \mathcal{H} \to \mathcal{R}(K)$ as in the proof of Lemma 2.2.4, noting that $(V, \mathcal{H}_V)$ is minimal by (i) of that result. To prove the second part of the present theorem, let us assume that $(V_1, \mathcal{H}_{V_1})$ and $(V_2, \mathcal{H}_{V_2})$ are two minimal Kolmogorov decompositions for the same kernel K. Define a unitary $U : \mathcal{H}_{V_1} \to \mathcal{H}_{V_2}$ by setting $U(V_1(x)u) = V_2(x)u$ and extend it by linearity and density to the whole of $\mathcal{H}_{V_1}$. It is clear that U is well defined and unitary. □

Let us now come back to complete positivity and deduce the fundamental theorem of Stinespring. Let us fix a Hilbert space $\mathcal{H}$ and a unital $*$-subalgebra $\mathcal{A}$ of $\mathcal{B}(\mathcal{H})$.

Theorem 2.2.8 *Let T be a linear map from $\mathcal{A}$ into $\mathcal{B}(\mathcal{H})$. Define an associated kernel $K_T : \mathcal{A} \times \mathcal{A} \to \mathcal{B}(\mathcal{H})$ given by, $K_T(x, y) = T(x^* y)$ for $x, y \in \mathcal{A}$. Then T is completely positive (CP) if and only if K_T is positive definite.*

Proof:
For $u_1, \ldots, u_n \in h$ and $x_1, \ldots, x_n \in \mathcal{A}$, $\sum_{i,j} \langle K_T(x_i, x_j) u_j, u_i \rangle = \langle T_n(X) \tilde{u}, \tilde{u} \rangle$, where X denotes the element in $\mathcal{A} \otimes \mathcal{M}_n \equiv \mathcal{M}_n(\mathcal{A})$ given by the $\mathcal{A}$-valued $n \times n$ matrix $((x_i^* x_j))$, T_n denotes the map $T \otimes I_{\mathcal{M}_n}$ and $\tilde{u}$ denotes the vector $u_1 \oplus u_2 \oplus \cdots \oplus u_n$ in $h \oplus \cdots \oplus h$. Since by our choice $X \geq 0$ as an element of $\mathcal{A} \otimes \mathcal{M}_n$, it is clear that positivity of T_n for each n is equivalent to the positive definiteness of K_T. □

Theorem 2.2.9 (Stinespring's theorem)
A linear map $T : \mathcal{A} \to \mathcal{B}(\mathcal{H})$ is CP if and only if there is a triple $(\mathcal{K}, \pi, V)$ consisting of a Hilbert space $\mathcal{K}$, a unital $*$-homomorphism $\pi : \mathcal{A} \to \mathcal{B}(\mathcal{K})$ and $V \in \mathcal{B}(\mathcal{H}, \mathcal{K})$ such that $T(x) = V^* \pi(x) V$ for all $x \in \mathcal{A}$, and $\{\pi(x) V u : u \in h, x \in \mathcal{A}\}$ is total in $\mathcal{K}$. Such a triple, to be called the 'Stinespring triple' associated with T, is unique in the sense that if $(\mathcal{K}', \pi', V')$ is another such triple, then there is a unitary operator $\Gamma : \mathcal{K} \to \mathcal{K}'$ such that $\pi'(x) = \Gamma \pi(x) \Gamma^*$ and $V' = \Gamma V$. Furthermore, if $\mathcal{A}$ is a von Neumann algebra and T is normal, π can be chosen to be normal.

Proof:
Let $(\lambda, \mathcal{K})$ be the minimal Kolmogorov decomposition for the kernel K_T defined in the statement of Theorem 2.2.8. For $x \in \mathcal{A}$, define a map $\pi(x)$ on the linear span of vectors $\lambda(y)u$ by setting $\pi(x)(\lambda(y)u) = \lambda(xy)u$, and by extending linearly. The complete positivity of T enables us to verify that indeed $\pi(x)$ is well defined and one has

$$\left\| \pi(x) \left(\sum_{i=1}^{n} \lambda(x_i) u_i \right) \right\|^2 \leq \|x\|^2 \left\| \sum_{i=1}^{n} \lambda(x_i) u_i \right\|^2$$

for any finite collection $x_1, \ldots, x_n$ of elements of $\mathcal{A}$ and $u_1, \ldots, u_n$ in $\mathcal{H}$. Thus, $\pi(x)$ extends as a bounded linear map on the whole of $\mathcal{K}$ and it is also clear that $\pi : \mathcal{A} \to \mathcal{B}(\mathcal{K})$ is a $*$-homomorphism. To complete the proof of the existence part, we choose $V = \lambda(1)$ and note that $T(x) = \lambda(1)^* \lambda(x) = V^* \pi(x) V$. The proof of uniqueness is straightforward and omitted.

In case $\mathcal{A}$ is a von Neumann algebra and T is normal, let us prove the normality of π. Let $0 \leq x_\alpha \uparrow x$ where x_α is a net of elements in $\mathcal{A}$ and $x \in \mathcal{A}$. Note that normality of T implies its ultra-weak continuity, which coincides with the weak continuity on norm-bounded convex sets. Thus, for $y, z \in \mathcal{A},\ u, v \in \mathcal{H}$, we have

$$\langle \pi(x_\alpha)(\lambda(y)u), \lambda(z)v\rangle = \langle u, T(y^* x_\alpha z)v\rangle$$

$$\to \langle u, T(y^* x z)v\rangle = \langle \pi(x)(\lambda(y)u), \lambda(z)v\rangle.$$

Here, we have used the fact that x_α is a bounded net and hence so is the net $y^* x_\alpha z$. Thus, for any vector ξ which is a finite linear combination of vectors of the form $\lambda(y)u,\ y \in \mathcal{A},\ u \in \mathcal{H}$, we have that $\langle \pi(x_\alpha)\xi, \xi\rangle$ converges to $\langle \pi(x)\xi, \xi\rangle$. Since the net $\pi(x_\alpha)$ is norm-bounded, this holds for all ξ in $\mathcal{K}$. Hence π is normal. □

Corollary 2.2.10 *Given any normal CP map $T : \mathcal{B}(\mathcal{H}) \to \mathcal{B}(\mathcal{H})$, with $\mathcal{H}$ separable, there exist elements $R_n, n = 1, 2, \ldots$ of $\mathcal{B}(\mathcal{H})$ such that T has the form*

$$T(x) = \sum_n R_n^* x R_n,$$

where the sum converges strongly.

Proof:
The proof is a simple application of the Theorem 2.2.9 combined with the result about the structure of normal $*$-homomorphism of $\mathcal{B}(\mathcal{H})$, which implies that the Hilbert space $\mathcal{K}$ obtained by the Theorem 2.2.9 is of the form $\mathcal{K} = \mathcal{H} \otimes k$ for some separable k, and π in that theorem can be taken to be $\pi(x) = U^*(x \otimes 1_k)U$, where U is a unitary. We choose $R = UV$, where V is as in Theorem 2.2.9, and also choose orthonormal bases $\{f_i\}$, $\{e_n\}$ of $\mathcal{H}$ and k respectively. By defining R_n to be the unique bounded operator satisfying $\langle u, R_n v\rangle = \langle (u \otimes e_n), Rv\rangle$, $(u, v \in \mathcal{H})$, we have for $x \in \mathcal{B}(\mathcal{H})$,

$$\begin{aligned}
&\langle u, T(x)v\rangle \\
&= \langle Ru, (x \otimes 1)Rv\rangle \\
&= \sum_{i,n} \langle Ru, f_i \otimes e_n\rangle \langle f_i \otimes e_n, (x \otimes 1)Rv\rangle \\
&= \sum_{i,n} \langle R_n u, f_i\rangle \langle f_i, x R_n v\rangle \\
&= \sum_n \langle R_n u, x R_n v\rangle.
\end{aligned}$$

Thus, $\sum_n R_n^* x R_n$ converges weakly. To show strong convergence as claimed, let us first note that $\sum_n R_n^* R_n$ is weakly convergent to the bounded operator $R^* R$. Clearly, for any $n \geq m \geq 1$, $\sum_{k=m}^n R_k^* R_k \leq R^* R$, hence $\| \sum_m^n R_k^* R_k \|$

$\leq \|R^* R\| = \|R\|^2$. Now, we observe that by Cauchy–Schwarz inequality

$$\begin{aligned}
&\left|\left\langle u, \left(\sum_m^n R_k^* x R_k\right) v\right\rangle\right| \\
&\leq \left(\sum_m^n \|R_k u\|^2\right)^{\frac{1}{2}} \left(\sum_m^n \|x R_k v\|^2\right)^{\frac{1}{2}} \\
&= \left(\sum_m^n \langle u, R_k^* R_k u\rangle\right)^{\frac{1}{2}} \left(\sum_m^n \langle v, R_k^* x^* x R_k v\rangle\right)^{\frac{1}{2}} \\
&\leq \|u\| \|R\| \left(\sum_m^n \langle v, R_k^* x^* x R_k v\rangle\right)^{\frac{1}{2}}.
\end{aligned}$$

Taking supremum over u with $\|u\| = 1$, we get

$$\left\|\left(\sum_{k=m}^n R_k^* x R_k\right) v\right\| \leq \|R\| \left(\sum_m^n \langle v, R_k^* x^* x R_k v\rangle\right)^{\frac{1}{2}},$$

which goes to 0 as $m, n \to \infty$, since $\sum_k R_k^* x^* x R_k$ is weakly convergent.

□

Let us prove here a version of Cauchy–Schwarz inequality for CP maps.

Corollary 2.2.11 *Let $T : \mathcal{A} \to \mathcal{B}(\mathcal{H})$ be a CP map as in the Theorem 2.2.9. Then we have the following operator inequality:*

$$T(a)^* T(a) \leq \|T(1)\| T(a^* a) \text{ for all} a \in \mathcal{A}.$$

Proof:
Let $(\mathcal{K}, \pi, V)$ be the triple obtained by the Theorem 2.2.9. Since $T(1) = V^* V$, it is clear that $\|T(1)\| = \|V\|^2 = \|V V^*\|$. Now, for $a \in \mathcal{A}$, we have

$$\begin{aligned}
&T(a)^* T(a) \\
&= V^* \pi(a) * V V^* \pi(a) V \\
&\leq \|V V^*\| V^* \pi(a)^* \pi(a) V \\
&= \|T(1)\| V^* \pi(a^* a) V \\
&= \|T(1)\| T(a^* a).
\end{aligned}$$

□

We now prove a result (see also [45]) which shows that the distinction between positivity and complete positivity appears only for noncommutative algebras.

Theorem 2.2.12 *If $\mathcal{A}$ is a commutative C^*-algebra and $\mathcal{B}$ is any C^*-algebra, then any positive map from $\mathcal{A}$ to $\mathcal{B}$ is automatically CP. Similar statement holds for any positive map from $\mathcal{B}$ to $\mathcal{A}$.*

Proof:
Without loss of generality we assume that $\mathcal{A}$ and $\mathcal{B}$ are unital, and thus $\mathcal{A}$ is of the form $C(X)$ for some compact Hausdorff space X. We also assume that $\mathcal{B} \subseteq \mathcal{B}(\mathcal{H})$ for some Hilbert space $\mathcal{H}$. Let T be a positive map from $\mathcal{A}$ to $\mathcal{B}$. Since for any fixed $u, v \in \mathcal{H}$, the map $C(X) \in f \mapsto \langle u, T(f)v\rangle \in \mathbb{C}$ is a continuous linear functional, by the well-known Riesz' representation theorem there is a regular Borel signed measure $\mu_{(u,v)}$ on X such that $\langle u, T(f)v\rangle = \int f d\mu_{(u,v)}$. Clearly, the positivity of T implies that $\mu_{(u,u)}$ is nonnegative measure for any u, and furthermore, for any $u_1, \ldots, u_n \in \mathcal{H}$ and any measurable subset Δ of X, the $n \times n$ matrix $((\mu_{(u_i,u_j)}(\Delta)))_{ij}$ is positive definite, which in particular implies that $|\mu_{(u_i,u_j)}(\Delta)| \leq \mu(\Delta) := \sum_{k=1}^{n} \mu_{(u_k,u_k)}(\Delta)$. Thus, each $\mu_{(u_i,u_j)}$ is absolutely continuous with respect to μ, and if g_{ij} denotes the associated Radon–Nikodym derivative $d\mu_{(u_i,u_j)}/d\mu$, then the positive definiteness of $((\mu_{(u_i,u_j)}(\Delta)))_{ij}$ for each measurable Δ implies that g_{ij} can be chosen so that $((g_{ij}(\cdot)))_{ij}$ is μ-a.e. positive definite. Now, let f_{ij} be elements of $C(X)$ such that $((f_{ij}(x)))_{ij=1,\ldots,n}$ is positive definite for all x. Since the Schur product of two positive definite matrices with complex entries is again positive definite, we have that $((f_{ij}(\cdot)g_{ij}(\cdot)))$ is μ-a.e. positive definite, hence the positive definiteness of $((\langle u_i, T(f_{ij})u_j\rangle)) = ((\int f_{ij}g_{ij}d\mu))$ follows. This is enough to prove the complete positivity of T.

Now, let $S : \mathcal{B} \to C(X)$ be a positive map. We want to show that S is CP. For each $x \in X$, the map $T_x : \mathcal{B} \to \mathbb{C}$ defined by $T_x(b) = T(b)(x)$ is clearly a positive linear functional. So, by the GNS Theorem there is a Hilbert space $\mathcal{H}_x$, a vector $\xi_x \in \mathcal{H}_x$ and a $*$-homomorphism $\pi_x : \mathcal{B} \to \mathcal{B}(\mathcal{H}_x)$ such that $T_x(b) = \langle \xi_x, \pi_x(b)\xi_x\rangle$ for all $b \in \mathcal{B}$. From this it follows that T_x is CP for each x. Hence for $A \equiv ((a_{ij})) \in M_n(\mathbb{C}) \otimes \mathcal{B}$, $((T(a_{ij})(x)))$ is positive definite for each x, which means that $((T(a_{ij})))$ is a positive element of $M_n(\mathbb{C}) \otimes C(X) \equiv C(X, M_n(\mathbb{C}))$. This completes the proof. □

2.3 Semigroups of linear maps on locally convex spaces

2.3.1 Some general theory

In this brief subsection, we mention a few standard and useful results from the theory of *one-parameter semigroups* of continuous linear maps acting on a locally convex topological vector space. For a more elaborate account and proofs, the reader may be referred to [130] and other references cited in appropriate places. Let X be a locally convex, sequentially complete, linear topological space and $(T_t)_{t \geq 0}$ be a one-parameter family of continuous linear operators from X to itself satisfying $T_t T_s = T_{t+s}$, $T_0 = I$, and $\lim_{t \to t_0} T_t x = T_{t_0} x$ for all $x \in X$ and $t_0 \geq 0$. Such a family is called a one-parameter semigroup of *class* C_0 (or strongly continuous) of operators on X. The family T_t is called equi-continuous if given any continuous seminorm p on X, there exists a continuous seminorm q on X such that $p(T_t x) \leq q(x)$ for all $t \geq 0$ and $x \in X$. If X is Banach space and $t \mapsto T_t$ is continuous in map-norm, that is, $\lim_{t \to t_0} \|T_t - T_{t_0}\| = 0$, then we say that T_t is uniformly continuous or norm-continuous. It can be proved that T_t is uniformly continuous if and only if there exists $L \in \mathcal{B}(X)$ such that $T_t = e^{tL}$ for all t. Given an equi-continuous semigroup of class C_0 on X, we define a linear operator A on X, called the *generator* of T_t, with the domain $\mathrm{Dom}(A) = \{x \in X : \lim_{t \to 0+} (T_t x - x)/t \text{ exists}\}$, given by $Ax = \lim_{t \to 0+} (T_t x - x)/t$ for $x \in \mathrm{Dom}(A)$. It is a remarkable fact that $\mathrm{Dom}(A)$ is dense and A is closed. Furthermore, for any $x \in \mathrm{Dom}(A)$, $T_t x$ also belongs to $\mathrm{Dom}(A)$ for all $t \geq 0$ and $AT_t x = T_t Ax$. The following theorem due to Hille and Yosida characterizes generators of equi-continuous semigroups of class C_0.

Theorem 2.3.1 *A closed linear operator A on X with dense domain is the generator of an equi-continuous semigroup of class C_0 if and only if for every positive integer n, $(nI - A)^{-1}$ exists as a bounded operator and the family of operators $\{(I - n^{-1}A)^{-m}\}_{n=1,2,\ldots;\ m=0,1,2,\ldots}$ is equi-continuous.*

Specializing to Banach spaces, the above equi-continuity translates into the existence of a positive constant C satisfying $\|(I - n^{-1}A)^{-m}\| \leq C$ for all n, m; and furthermore that A is the generator of a contraction semigroup (that is, each T_t is a contraction) if and only if $\|(I - n^{-1}A)^{-1}\| \leq 1$ for all $n = 1, 2, \ldots$.

We now mention a useful characterization of the domain of the generator of a contraction semigroup acting on the dual of a Banach space.

Proposition 2.3.2 *Let X be a Banach space and $(T_t)_{t \geq 0}$ be a C_0-semigroup of contractions on the dual space X^* of X, and assume furthermore that each T_t is weak-$*$ continuous. Then the domain of the generator, say $\mathcal{L}$, of $(T_t)_{t \geq 0}$ (with respect to the norm-topology of X^*) is given by*

$$\mathrm{Dom}(\mathcal{L}) = \{\rho \in X^* : \mathrm{liminf}_{t \downarrow 0}\, t^{-1} \|T_t(\rho) - \rho)\| < \infty\}.$$

Convergence of semigroups is an important problem in the theory of semigroups and in this context, we state the following result.

Proposition 2.3.3 *Let X be a locally convex, sequentially complete, complex linear space and for each $n = 1, 2, \ldots$ let $(T_t^{(n)})_{t \geq 0}$ be an equi-continuous semigroup of class C_0 with generator A_n. Assume furthermore the following.*

(i) *For any continuous seminorm* p *on* X, *there exists a continuous seminorm* q *on* X *such that* $p(T_t^{(n)}x) \leq q(x)$ *for all* $t \geq 0, n = 1, 2, \ldots$ *and* $x \in X$.

(ii) *For some* λ_0 *with* $\mathrm{Re}(\lambda_0) > 0$, *there exists an invertible operator* $J : X \to X$ *such that the range of* J *is dense in* X *and* $J(x) = \lim_{n\to\infty}(\lambda_0 - A_n)^{-1}(x)$ $x \in X$. *Then* $(\lambda_0 - J^{-1})$ *is the generator of an equi-continuous semigroup* $(T_t)_{t\geq 0}$ *of class* C_0 *satisfying* $T_t(x) = \lim_{n\to\infty} T_t^{(n)}(x) x \in X$ *and the above convergence is uniform over compact subintervals of* $[0, \infty)$.

In case X is a Banach space and $T_t^{(n)}$, T_t are contraction semigroups, it is not difficult to see that the following are equivalent.

(a) $T_t^{(n)}(x)$ converges (uniformly on compacts) to $T_t(x)$.

(b) $(\lambda - A_n)^{-1}(x) \to (\lambda - A)^{-1}(x)$ for all x as $n \to \infty$ for some λ with $\mathrm{Re}(\lambda) > 0$, where A_n and A denote the generators of $T_t^{(n)}$ and T_t, respectively.

(c) $(\lambda - A_n)^{-1}(x) \to (\lambda - A)^{-1}(x)$ for all x as $n \to \infty$ uniformly in λ over compact subsets of the right half plane $\{\lambda : \mathrm{Re}(\lambda) > 0\}$.

We shall note here the following useful results due to Chernoff.

Proposition 2.3.4 *Let* X *be a Banach space,* $\mathcal{D}$ *be a dense subspace of* X, *and* $C_n, n = 1, 2, \ldots, C$ *be generators of* C_0*-contraction semigroups* $T_t^{(n)}, T_t$ *(respectively) on* X *such that* $\mathcal{D}$ *is contained in the domain of* C_n *for each* n *and also of* C. *Furthermore, we assume that the closure of the restriction of* C *on* $\mathcal{D}$ *coincides with* C, *and* $C_n u \to Cu$ for all $u \in \mathcal{D}$. *Then* $T_t^{(n)} \to T_t$ *in the strong topology, uniformly on compact subsets of* $[0, \infty)$.

Proposition 2.3.5 *Let* X *be a Banach space and* $F : [0, \infty) \to \mathcal{B}(X)$ *be a strongly continuous map such that* $F(t)$ *is a contraction for every* $t \geq 0$ *and* $F(0) = I$. *Suppose also that the closure of the strong derivative* $F'(0)$ *generates a* C_0*-contraction semigroup* T_t. *Then* $F(\frac{t}{n})^n \to T_t$ *in the strong topology, uniformly on compact subsets of* $[0, \infty)$ *as* $n \to \infty$.

For the proof of the above propositions we refer the reader to [27].

For a locally convex, sequentially complete, linear topological space X, let us denote by X^* its dual, viewed naturally as a locally convex space. Given an equi-continuous semigroup T_t of class C_0 on X, it is natural to consider the dual semigroup T_t^* on X^*. T_t^* will be equi-continuous but not in general of class C_0. It is of class C_0 when X^* is also sequentially complete.

2.3.2 Some results on perturbation of semigroups

In this subsection we state without proofs some important and well-known results from the theory of perturbation of semigroups. We refer the reader for more details to the book by Kato [77]. We let X be a Banach space throughout the discussion that follows.

Definition 2.3.6 Given two (possibly unbounded) linear operators T, A on X, we say that A is *T-bounded* (or, *relatively bounded* with respect to T) if $\mathrm{Dom}(T) \subseteq \mathrm{Dom}(A)$ and there are nonnegative constants a, b such that $\|Au\| \leq a\|u\| + b\|Tu\|$ for all $u \in \mathrm{Dom}(T)$. The infimum over all choices of b as above is defined to be the *relative bound* of A with respect to T.

Proposition 2.3.7 *If $\mathcal{H}$ is a Hilbert space, T is a self-adjoint operator on $\mathcal{H}$ and A is a symmetric operator which is T-bounded with the relative bound strictly less than* 1, *then $T + A$ is also self-adjoint.*

Now, let us recall some notations from [77]. For positive constants M, β, we shall denote by $\mathcal{G}(M, \beta)$ the set of all linear operators A on X such that A generates a C_0-semigroup T_t satisfying $\|T_t\| \leq Me^{\beta t}$ for all t. For β real and $\omega > 0$, we shall denote by $\mathcal{H}(\omega, \beta)$ the set of all densely defined closed linear operators A on X with the property that for every $\epsilon > 0$, there exists a positive constant M_ϵ such that for all complex number ξ with $\mathrm{Re}(\xi) > 0$ and $|arg(\xi)| \leq \frac{\pi}{2} + \omega - \epsilon$, the operator $(A - \beta - \xi)$ has a bounded inverse and $\|(A - \beta - \xi)^{-1}\| \leq M_\epsilon/|\xi|$. The semigroups generated by elements of $\mathcal{H}(\omega, \beta)$ for some ω, β are called holomorphic semigroups. With this notation and terminology, we have the following results.

Proposition 2.3.8 *Let $A, B \in \mathcal{G}(1, 0)$ and B is A-bounded with the relative bound less than $\frac{1}{2}$. Then $A + B \in \mathcal{G}(1, 0)$. Furthermore, in case X is a Hilbert space, the same result holds under the weaker condition that the relative bound is less than* 1.

Proposition 2.3.9 *Let $A, B \in \mathcal{G}(1, 0)$ and* $\mathrm{Dom}(A) \cap \mathrm{Dom}(B)$ *be dense in X. Suppose furthermore that $A + B$ is closable and for sufficiently large real ξ, $A + B - \xi$ has a dense range. Then the closure of $A + B$ is in $\mathcal{G}(1, 0)$.*

Proposition 2.3.10 *Given $A \in \mathcal{H}(\omega, \beta)$ and $\epsilon > 0$, there are positive constants γ, δ such that whenever B is A-bounded and $\|Bu\| \leq a\|u\| + b\|Au\|$* for all $u \in \mathrm{Dom}(A)$, *with $a < \delta, b < \delta$, then we have $A + B \in \mathcal{H}(\omega - \epsilon, \gamma)$.*

2.4 Fock spaces and Weyl operators

In this brief section, we recall some well-known facts about Fock spaces. For a Hilbert space $\mathcal{H}$ and positive integer n, let $\mathcal{H}_n \equiv \mathcal{H}^{\otimes^n}$ denote the n-fold tensor product of $\mathcal{H}$, and $\mathcal{H}_0$ denote the one-dimensional Hilbert space $\mathbb{C}$. The free Fock space $\Gamma^f(\mathcal{H})$ is defined as

$$\Gamma^f(\mathcal{H}) = \oplus_{n=0}^{\infty} \mathcal{H}_n.$$

The distinguished vector $\Omega \equiv 1 \oplus 0 \oplus 0 \oplus \ldots$ is called the vacuum. For two Hilbert spaces $\mathcal{H}, \mathcal{K}$ and a contraction $T : \mathcal{H} \to \mathcal{K}$, we denote by T_n the n-fold tensor product of T and set $T_0 = I$. Let us define $\Gamma^f(T) \equiv \oplus_{n=0}^{\infty} T_n : \Gamma^f(\mathcal{H}) \to \Gamma^f(\mathcal{K})$. Then, it is possible to verify the following.

Lemma 2.4.1 *Γ^f is a functor on the category whose objects are Hilbert spaces and morphisms are contractions, that is, $\Gamma^f(ST) = \Gamma^f(S)\Gamma^f(T)$, $\Gamma^f(I) = I$. Furthermore, $\Gamma^f(0)$ is the projection on the Fock vacuum vector and $\Gamma^f(T^*) = (\Gamma^f(T))^*$.*

The proof of the lemma is straightforward and hence omitted.

Let us now discuss symmetric and antisymmetric Fock spaces. Let $\mathcal{H}_n^s$ and $\mathcal{H}_n^a$ denote respectively the symmetric and antisymmetric n-fold tensor products of $\mathcal{H}$ for any positive integer n, and $\mathcal{H}_0^s = \mathcal{H}_0^a = \mathcal{H}_0$. Then the symmetric (or Boson) and antisymmetric (or Fermion) Fock spaces over $\mathcal{H}$, denoted respectively by $\Gamma^s(\mathcal{H})$ and $\Gamma^a(\mathcal{H})$, are defined as

$$\Gamma^s(\mathcal{H}) = \oplus_{n=0}^{\infty} \mathcal{H}_n^s,$$

$$\Gamma^a(\mathcal{H}) = \oplus_{n=0}^{\infty} \mathcal{H}_n^a.$$

We shall be mostly concerned with the symmetric Fock spaces in the present work, and hence for simplicity of notation, we shall use the notation $\Gamma(\mathcal{H})$ for the symmetric Fock space. Let us mention the basic factorization property of $\Gamma(\mathcal{H})$.

Theorem 2.4.2 *Consider the map $\mathcal{H} \ni u \mapsto e(u) \in \Gamma(\mathcal{H})$ given by $e(u) = \oplus_{n=0}^{\infty} (n!)^{-\frac{1}{2}} u^{\otimes^n}$, where $u^{\otimes^n}$ is the n-fold tensor product of u for positive n and $u^{\otimes^0} = 1$. Then the map $e(.)$ is the minimal Kolmogorov decomposition for the positive definite kernel $\mathcal{H} \times \mathcal{H} \mapsto \mathbb{C}$ given by $u, v \mapsto \exp(\langle u, v \rangle)$. Furthermore, $\{e(u) : u \in \mathcal{H}\}$ is a linearly independent total set of vectors in $\Gamma(\mathcal{H})$.*

Proof:

That $e(.)$ is a Kolmogorov decomposition for the above-mentioned kernel is verified by the relation $\langle e(u), e(v) \rangle = \exp(\langle u, v \rangle)$. Furthermore, the relation

$$\frac{d^n}{dt^n} e(tu)|_{t=0} = (n!)^{-\frac{1}{2}} u^{\otimes^n}$$

shows that for every $u \in \mathcal{H}, u^{\otimes^n}$ belongs to the closed linear span of $e(u)$. Since the vectors of the form $u^{\otimes^n}$ where n varies over $\{0, 1, 2, \ldots\}$ are total in $\Gamma(\mathcal{H})$, the assertion about minimality follows. To prove the linear independence, suppose that $u_1, u_2, \ldots, u_n$ are distinct vectors in $\mathcal{H}$ and $z_1, \ldots, z_n$ are complex numbers such that $\sum_{j=1}^{n} z_j e(u_j) = 0$. Then we have, for all $t \in \mathbb{R}, \sum_{j=1}^{n} z_j \exp(t\langle u_j, v\rangle) = 0$ for all $v \in \mathcal{H}$. Since $u_1, u_2, \ldots, u_n$ are distinct, there exists $v \in \mathcal{H}$ such that the scalars $\langle u_j, v\rangle$ are distinct and hence the functions $\{e^{t\langle u_j, v\rangle}\}$ are linearly independent, which implies that $z_j = 0$ for all j. □

Corollary 2.4.3 *For any dense subset S of $\mathcal{H}$, the set $\{e(u) : u \in S\}$ is total in $\Gamma(\mathcal{H})$.*

The proof is straightforward and we refer the reader to [100] (Corollary 19.5, page 127).

Corollary 2.4.4 *There is a natural identification of $\Gamma(\mathcal{H} \oplus \mathcal{K})$ with $\Gamma(\mathcal{H}) \otimes \Gamma(\mathcal{K})$ under which $e(u \oplus v) \mapsto e(u) \otimes e(v)$.*

Proof:
The proof is a straightforward consequence of the minimality of the Kolmogorov decomposition mentioned in the Theorem 2.4.2. □

Let us conclude this section by mentioning the second quantization and Weyl operators. For a contraction C on $\mathcal{H}$, we define the *second quantization* $\Gamma(C)$ on $\Gamma(\mathcal{H})$ by

$$\Gamma(C)e(u) = e(Cu).$$

We observe the following.

Lemma 2.4.5 *$\Gamma(C)$ admits an extension to $\Gamma(\mathcal{H})$ and the extension, denoted also by $\Gamma(C)$, is a contraction. Moreover, if C is isometry (respectively unitary), then so is $\Gamma(C)$.*

Proof:
For $\alpha_i \in \mathbb{C}$, vectors $u_i \in \mathcal{H}$ $(i = 1, \cdots, n)$, we have

$$\begin{aligned}
&\left\| \Gamma(C)\left(\sum_{i=1}^{n} \alpha_i \, e(u_i)\right)\right\|^2 \\
&= \sum_{i,j=1}^{n} \bar{\alpha}_i \alpha_j \, \exp(\langle u_i, C^*Cu_j\rangle) \\
&= \sum_{m=0}^{\infty} \frac{1}{m!} \sum_{i,j=1}^{n} \bar{\alpha}_i \alpha_j \, \langle u_i, C^*Cu_j\rangle^m
\end{aligned}$$

$$= \sum_{m=0}^{\infty} \frac{1}{m!} \sum_{i,j=1}^{n} \bar{\alpha}_i \alpha_j \, \langle u_i^{\otimes^m}, (C^*C)^{\otimes^m} u_j^{\otimes^m} \rangle,$$

$$= \sum_{m=0}^{\infty} \frac{1}{m!} \langle \tilde{u}_m, (C^*C)^{\otimes^m} \tilde{u}_m \rangle,$$

where we have denoted the m-fold tensor product of a vector or an operator by the symbol $\otimes^m$ (with $v^{\otimes^0} := 1 \in \mathbb{C}$, $(C^*C)^{\otimes^0} := I$), and by $\tilde{u}_m$ the vector $\sum_{i=1}^n \alpha_i u_i^{\otimes^m}$. Since $(C^*C)^{\otimes^m}$ is a positive contraction for every m, we have

$$\begin{aligned} &\sum_{m=0}^{\infty} \frac{1}{m!} \langle \tilde{u}_m, (C^*C)^{\otimes^m} \tilde{u}_m \rangle \\ &\leq \sum_{m=0}^{\infty} \frac{1}{m!} \langle \tilde{u}_m, \tilde{u}_m \rangle \\ &= \sum_{m=0}^{\infty} \frac{1}{m!} \sum_{i,j=1}^{n} \bar{\alpha}_i \alpha_j \, \langle u_i, u_j \rangle^m \\ &= \sum_{i,j=1}^{n} \bar{\alpha}_i \alpha_j \, \exp(\langle u_i, u_j \rangle) \\ &= \left\| \sum_{i,j=1}^{n} \alpha_i \, e(u_i) \right\|^2 . \end{aligned}$$

This completes the proof that $\Gamma(C)$ extends to a contraction, since the linear span of exponential vectors is a dense subset of $\Gamma(\mathcal{H})$. It is straightforward to see that $\Gamma(C)$ is an isometry (respectively unitary) whenever C is so. □

For $u \in \mathcal{H}$ and unitary operator U in $\mathcal{H}$, we define the Weyl operators $W(u, U)$ by setting

$$W(u, U)e(v) = \exp\left(-\frac{1}{2}\|u\|^2 - \langle u, Uv \rangle\right) e(u + Uv).$$

It is known that the von Neumann algebra generated by the family $\{W(u, I) : u \in S\}$ is the whole of $\mathcal{B}(\Gamma(\mathcal{H}))$ whenever S is a dense subspace of $\mathcal{H}$. One can also verify the following properties of the Weyl operators:

(i) the correspondence $(u, U) \mapsto W(u, U)$ is strongly continuous; and
(ii) they satisfy the Weyl commutation rule:

$$W(u_1, U_1)W(u_2, U_2) = \exp(-i \, Im\langle u_1, U_1 u_2 \rangle) W(u_1 + U_1 u_2, U_1 U_2).$$

We refer the reader to [100] for a further detailed description of Fock spaces and related topics with applications to quantum probability.

3

Quantum dynamical semigroups

Let us now restrict ourselves to the case when the general locally convex space X is replaced by a C^* or a von Neumann algebra $\mathcal{A}$, and study the implications of the complete positivity of a semigroup T_t acting on it.

Definition 3.0.1 A *quantum dynamical semigroup (Q.D.S)* on a C^*-algebra $\mathcal{A}$ is a contractive semigroup T_t of class C_0 such that each T_t is a completely positive map from $\mathcal{A}$ to itself. T_t is said to be *conservative* if $T_t(1) = 1$ for all $t \geq 0$.

3.1 Generators of uniformly continuous quantum dynamical semigroups: the theorems of Lindblad and Christensen–Evans

For a uniformly continuous semigroup on a von Neumann algebra $\mathcal{A} \subseteq \mathcal{B}(h)$, we have the following result.

Lemma 3.1.1 *Let $T_t = e^{t\mathcal{L}}$ be a uniformly continuous contractive semigroup acting on $\mathcal{A}$ with $\mathcal{L}$ as the generator. Then T_t is normal for each t if and only if $\mathcal{L}$ is ultra-strongly (and hence ultra-weakly) continuous on any norm-bounded subset of $\mathcal{A}$.*

Proof:
Let us first note that $\mathcal{L}$ is norm-bounded. If $\mathcal{L}$ is ultra-strongly continuous on bounded sets, then clearly $e^{t\mathcal{L}}$ is ultra-strongly continuous on bounded sets for each t, and hence normal. For the converse, first note that for any $t \geq 0$ and $x \in \mathcal{A}$, we have

$$\|(T_t(x) - x)\| \leq \int_0^t \|T_s(\mathcal{L}(x))\| ds \leq \|\mathcal{L}\| \|x\| t.$$

Hence it is not difficult to see that

$$\begin{aligned}&\left\|\mathcal{L}(x) - \frac{T_t(x) - x}{t}\right\| \\ &= \left\|\frac{1}{t}\int_0^t \{\mathcal{L}(x) - T_s(\mathcal{L}(x))\}ds\right\| \\ &\leq \frac{1}{t}\int_0^t \|\mathcal{L}\|\|\mathcal{L}(x)\|sds \\ &\leq \|\mathcal{L}\|^2\|x\|\frac{t}{2}.\end{aligned}$$

Now suppose that x_α is a net of elements in $\mathcal{A}$ such that x_α strongly converges to $x \in \mathcal{A}$ and there exists positive constant M such that $\|x_\alpha\| \leq M$ for all α. Fix $u \in h$ and $\epsilon > 0$. Choose t_0 small enough so that $\|\mathcal{L}\|^2 M\|u\|t_0 < \epsilon$. Clearly,

$$\begin{aligned}&\|\mathcal{L}(x_\alpha - x)u\| \\ &\leq \frac{1}{2}\epsilon + \left\|\left\{\frac{T_{t_0}(x_\alpha - x) - (x_\alpha - x)}{t_0}\right\}u\right\|,\end{aligned}$$

which proves that $\mathcal{L}(x_\alpha - x)u \to 0$ since T_t is strongly continuous on bounded sets. This completes the proof, since on bounded sets ultra-strong and strong topologies coincide. □

In view of the above result, we shall make the following definition.

Definition 3.1.2 We define a *quantum dynamical semigroup (Q.D.S.) on a von Neumann algebra* to be a semigroup T_t of completely positive, contractive maps on the von Neumann algebra such that T_t is normal for each t. In case when T_t is uniformly continuous, its norm-bounded generator is ultra-strongly (hence ultra-weakly) continuous on bounded sets.

It is to be noted that although each T_t acts on an algebra, the domain of the generator need not be an algebra, nor it may contain any $*$-subalgebra which is sufficiently large in any reasonable sense. This is a fundamental difficulty in translating the complete positivity of the semigroup into some property of its generator. However, when T_t is uniformly continuous, its generator is defined as a bounded map on the whole of $\mathcal{A}$, facilitating the analysis of complete positivity. Thus, for the sake of convenience, we assume for the rest of the present subsection that T_t is uniformly continuous. To understand the implication of complete positivity, let us first note some definitions and results.

Definition 3.1.3 Let $\mathcal{A}$ and $\mathcal{B}$ be two C^*-algebras such that the former is a subalgebra of the latter, and $\mathcal{L} : \mathcal{A} \to \mathcal{B}$ be a bounded linear map with the property that $\mathcal{L}$ is real, that is, $\mathcal{L}(x^*) = \mathcal{L}(x)^*$ for all $x \in \mathcal{A}$. We call $\mathcal{L}$

conditionally completely positive (CCP) if

$$\sum_{i,j=1}^{n} b_i^* \mathcal{L}(a_i^* a_j) b_j \geq 0$$

for all $a_1, \ldots, a_n$ in $\mathcal{A}$ and $b_1, \ldots, b_n$ in $\mathcal{B}$ satisfying $\sum_{i,j=1}^{n} a_i b_i = 0$.

We first give a few characterizations of conditional complete positivity in the language of positive definite kernels. We state the result without proof, which can be found in [45] (page 70–71, Lemma 14.5).

Lemma 3.1.4 *The following are equivalent.*

(i) *For each $a \in \mathcal{A}$, the kernel $\mathcal{A} \times \mathcal{A} \ni (b, c) \mapsto K_a(b, c) \equiv \mathcal{L}(b^* a^* ac) + b^* \mathcal{L}(a^* a)c - \mathcal{L}(b^* a^* a)c - b^* \mathcal{L}(a^* ac)$ is positive definite.*
(ii) *The kernel $(\mathcal{A} \times \mathcal{A}) \times (\mathcal{A} \times \mathcal{A}) \ni (b_1, b_2), (c_1, c_2) \mapsto \mathcal{L}(b_1^* b_2^* c_2 c_1) + b_1^* \mathcal{L}(b_2^* c_2) c_1 - \mathcal{L}(b_1^* b_2^* c_2) c_1 - b_1^* \mathcal{L}(b_2^* c_2 c_1)$ is positive definite.*
(iii) *$\mathcal{L}$ is CCP.*

Lemma 3.1.5 *Let $\mathcal{L}$ be a bounded linear real map from a unital C^*-algebra $\mathcal{A}$ to itself such that $e^{t\mathcal{L}}$ is a contraction for each $t \geq 0$. Then the following are equivalent:*

(i) *$e^{t\mathcal{L}}$ is positive for all positive t.*
(ii) *$(\lambda - \mathcal{L})^{-1}$ is positive for all sufficiently large positive λ.*
(iii) *For $y, a \in \mathcal{A}$ with the property that $ya = 0$, one has $a^* \mathcal{L}(y^* y) a \geq 0$.*

Proof:
(i) $\Rightarrow$ *(iii)*:
If y, a satisfies the hypothesis of (iii), we have that $(a^* e^{t\mathcal{L}}(y^* y)a - a^* y^* ya)/t = (a^* e^{t\mathcal{L}}(y^* y)a)/t \geq 0$ for all positive t, and hence taking limit as $t \to 0+$, we obtain $a^* \mathcal{L}(y^* y)a \geq 0$.
(iii) $\Rightarrow$ *(ii)*:
Let λ be greater than $\|\mathcal{L}\|$. Then $(\lambda - \mathcal{L})^{-1}$ exists, or equivalently, $\mathrm{Ran}(\lambda - \mathcal{L}) = \mathcal{A}$. Thus there exists an x in $\mathcal{A}$ such that $(\lambda - \mathcal{L})(x) \geq 0$. Since $\mathcal{L}$ is real, one has $\mathrm{Im}((\lambda - \mathcal{L})(x)) = (\lambda - \mathcal{L})(\mathrm{Im}(x))$, and positivity of $(\lambda - \mathcal{L})(x)$ implies in particular that $\mathrm{Im}((\lambda - \mathcal{L})(x)) = 0$. Hence we have $(\lambda - \mathcal{L})(x) = (\lambda - \mathcal{L})(\mathrm{Re}(x))$, and thus we may assume without loss of generality that x is self-adjoint. We want to show that x is positive. Let $x = x^+ - x^-$, with x^+ and x^- positive and $x^+ x^- = 0$. Then, by *(iii)*, we have $x^- \mathcal{L}(x^+) x^- \geq 0$, so that $0 \leq x^- [x - \lambda^{-1} \mathcal{L}(x)] x^- = -(x^-)^3 - \lambda^{-1} x^- \mathcal{L}(x^+) x^- + \lambda^{-1} x^- \mathcal{L}(x^-) x^-$. Thus $0 \leq (x^-)^3 \leq \lambda^{-1} x^- \mathcal{L}(x^-) x^-$, and hence $\|x^-\|^3 \leq \lambda^{-1} \|\mathcal{L}\| \|x^-\|^3$, which implies $\|x^-\| = 0$, since $\|\mathcal{L}\| < \lambda$.

(ii) $\Rightarrow$ *(i)*:
This follows from the identity $e^{t\mathcal{L}} = \lim_{n\to\infty}(1 - \frac{t}{n}\mathcal{L})^{-n}$. □

As a simple application of the above lemma, we obtain the following useful result.

Theorem 3.1.6 *A bounded linear adjoint-preserving map $\mathcal{L}$ from a unital C^*-algebra $\mathcal{A}$ to itself is CCP if and only if $e^{t\mathcal{L}}$ is CP for all positive t.*

Proof:
It is enough to observe that $\mathcal{L}$ is CCP if and only if $(\mathcal{L} \otimes I) : \mathcal{A} \otimes \mathcal{M}_n \to \mathcal{A} \otimes \mathcal{M}_n$ satisfies the hypothesis of (iii) in Lemma 3.1.5 with $\mathcal{A}$ replaced by $\mathcal{A} \otimes \mathcal{M}_n$. □

We shall now prove a structure theorem for normal CCP maps acting on a von Neumann algebra. For this, we first quote a result without proof (see [24] for a proof.).

Proposition 3.1.7 *Let $\mathcal{A}$ be a von Neumann subalgebra of $\mathcal{B}(h)$ for some Hilbert space h and $W : \mathcal{A} \to \mathcal{B}(h)$ be a derivation, that is, $W(ab) = W(a)b + aW(b)$. Then there exists an operator $T \in \mathcal{B}(h)$ such that $W(a) = Ta - aT$ for all $a \in \mathcal{A}$.*

As an application of this result, the canonical structure theorem for normal CCP maps is established.

Theorem 3.1.8 *(***Christensen–Evans***)*
Let $(T_t)_{t\geq 0}$ be a uniformly continuous quantum dynamical semigroup (Q.D.S.) on a unital von Neumann algebra $\mathcal{A} \subseteq \mathcal{B}(h)$ with $\mathcal{L}$ as its ultra-weakly continuous generator. Then there is a quintuple $(\rho, \mathcal{K}, \alpha, H, R)$ where ρ is a unital normal $*$-representation of $\mathcal{A}$ in a Hilbert space $\mathcal{K}$ and a ρ-derivation $\alpha : \mathcal{A} \to \mathcal{B}(\mathcal{K})$ (that is, $\alpha(xy) = \alpha(x)y + \rho(x)\alpha(y)$) such that the set $\mathcal{D} \equiv \{\alpha(x)u : x \in \mathcal{A}, u \in h\}$ is total in $\mathcal{K}$, H is a self-adjoint element of $\mathcal{A}$, and $R \in \mathcal{B}(h, \mathcal{K})$ such that $\alpha(x) = Rx - \rho(x)R$, and $\mathcal{L}(x) = R^*\rho(x)R - \frac{1}{2}(R^*R - \mathcal{L}(1))x - \frac{1}{2}x(R^*R - \mathcal{L}(1)) + i[H, x]$ for all $x \in \mathcal{A}$. Furthermore, $\mathcal{L}$ satisfies the following algebraic identity, called the cocycle property (or cocycle relation) with α as coboundary.

$$\mathcal{L}(x^*y) - \mathcal{L}(x^*)y - x^*\mathcal{L}(y) + x^*\mathcal{L}(1)y = \alpha(x)^*\alpha(y).$$

Moreover, R can be chosen from the ultra-weak closure of $\mathrm{sp}\{\alpha(x)y : x, y \in \mathcal{A}\}$ *and hence in particular $R^*\rho(x)R \in \mathcal{A}$.*

Proof:
We briefly sketch only the main ideas behind the proof and refer the reader to [45] and [24] for details.

Consider the trilinear map D on $\mathcal{A} \times \mathcal{A} \times \mathcal{A}$ defined by, $D(x, y, z) = \mathcal{L}(xyz) + x\mathcal{L}(y)z - \mathcal{L}(xy)z - x\mathcal{L}(yz)$. It can be verified that the kernel (a_1, a_2), $(b_1, b_2) \mapsto D(a_1^*, a_2^* b_2, b_1)$ is positive definite. By Theorem 2.2.7, we obtain a Hilbert space $\mathcal{K}$ and $\lambda : \mathcal{A} \times \mathcal{A} \to \mathcal{B}(h, \mathcal{K})$ such that $(\lambda, \mathcal{K})$ is the minimal Kolmogorov decomposition for the above kernel. As in the proof of Theorem 2.2.9, it is possible to verify that $\rho : \mathcal{A} \to \mathcal{B}(\mathcal{K})$ defined by $\rho(x)(\lambda(a, b)u) = \lambda(a, xb)u$ extends to a normal $*$-homomorphism of $\mathcal{A}$. The proof of normality is similar to that of the representation π in Theorem 2.2.9, using the ultra-weak continuity of $\mathcal{L}$ on norm-bounded sets. Denote by $\alpha(x)$ the operator $\lambda(x, 1) \in \mathcal{B}(h, \mathcal{K})$. Then, it can be verified that $\lambda(x^*, y^*)^*[\alpha(ab) - \rho(a)\alpha(b) - \alpha(a)b] = D(x, y, ab) - D(x, ya, b) - D(x, y, a)b = 0$. By minimality of $(\lambda, \mathcal{K})$ we conclude that $\alpha(ab) = \rho(a)\alpha(b) + \alpha(a)b$, that is, α is a ρ-derivation. Now, to obtain R, consider the faithful normal representation π: $\mathcal{A} \to \mathcal{B}(h \oplus \mathcal{K})$ given by, $\pi(a) = \begin{pmatrix} a & 0 \\ 0 & \rho(a) \end{pmatrix}$. Let $W : \pi(\mathcal{A}) \to \mathcal{B}(h \oplus \mathcal{K})$ given by, $W(\pi(a)) = \begin{pmatrix} 0 & 0 \\ \alpha(a) & 0 \end{pmatrix}$. Then $W(\pi(a)\pi(b)) = W(\pi(a))\pi(b) + \pi(a)W(\pi(b))$. By the Proposition 3.1.7, there exists $T \in \mathcal{B}(h \oplus \mathcal{K})$ such that $W(\pi(a)) = T\pi(a) - \pi(a)T$. Writing $T = \begin{pmatrix} P & Q \\ R & S \end{pmatrix}$ with respect to the canonical decomposition of $\mathcal{B}(h \oplus \mathcal{K})$, we obtain $\alpha(a) = Ra - \rho(a)R$. Consider the map $\Psi(x) \equiv \mathcal{L}(x) - R^*\rho(x)R$. A simple algebraic calculation will show that $x \mapsto \Psi(x) - \frac{1}{2}(\Psi(1)x + x\Psi(1))$ is a derivation, and hence by the Proposition 3.1.7 and also using the fact that Ψ is adjoint-preserving, we obtain a self-adjoint H in $\mathcal{B}(h)$ such that $\Psi(x) - \frac{1}{2}(\Psi(1)x + x\Psi(1)) = i[H, x]$. The proof that R can be chosen from the ultra-weak closure of $\mathrm{Sp}\{\alpha(x)y : x, y \in \mathcal{A}\}$ and H can be chosen from $\mathcal{A}$ is omitted; referring the reader to the original paper by Christensen and Evans [24].

To complete the proof, we need to verify that $R^*\rho(x)R \in \mathcal{A}$. Since R belongs to the ultra-weak closure of right $\mathcal{A}$-linear span of elements of the form $\alpha(a)$ $(a \in \mathcal{A})$ and $\mathcal{A}$ is closed under the ultra-weak topology, it suffices to show that $\alpha(a)^*\rho(b)\alpha(c) \in \mathcal{A}$ for $a, b, c \in \mathcal{A}$. To this end, we observe that due to the cocycle property of $\mathcal{L}$, $\alpha(x)^*\alpha(y) \in \mathcal{A}$ whenever $x, y \in \mathcal{A}$. Thus,

$$\begin{aligned} \alpha(a)^*\rho(b)\alpha(c) &= \alpha(a)^*(\alpha(bc) - \alpha(b)c) \\ &= \alpha(a)^*\alpha(bc) - \alpha(a)^*\alpha(b)c \ \in \mathcal{A}. \end{aligned}$$

□

Theorem 3.1.9 *The generator $\mathcal{L}$ of a uniformly continuous Q.D.S. on the von Neumann algebra $\mathcal{B}(h)$ (where h is separable) can be written as*

$$\mathcal{L}(x) = \sum_{n=1}^{\infty} R_n^* x R_n + Gx + xG^*,$$

for $x \in \mathcal{B}(h)$; where $R_n (n = 1, 2, \ldots)$ and G are elements of $\mathcal{B}(h)$, such that $-\mathrm{Re}(G)$ is a nonnegative operator, and the sum on the right-hand side above converges strongly.

Proof:
By Theorem 3.1.8 we obtain a Hilbert space $\mathcal{K}$, a normal unital representation $\rho : \mathcal{B}(h) \to \mathcal{B}(\mathcal{K})$ and elements $V \in \mathcal{B}(h, \mathcal{K})$, $H \in \mathcal{B}^{\text{s.a.}}(h)$ such that $\mathcal{L}(x) = V^*\rho(x)V - \frac{1}{2}V^*Vx - \frac{1}{2}xV^*V + i[H, x]$. Furthermore, by the Corollary 2.1.8 it follows that there exists a separable Hilbert space k_0 and a unitary operator $U \in \mathcal{B}(h \otimes k_0)$ such that $\mathcal{K} = h \otimes k_0$ and $\rho(x) = U^*(x \otimes 1)U$. We take $R = UV$ and $G = -\frac{1}{2}R^*R + iH$. Fixing any orthonormal basis $\{e_n\}$ of k_0, we define R_n as the unique bounded operator on h satisfying $\langle u, R_n v\rangle = \langle u \otimes e_n, (Rv)\rangle$, and the rest of the proof will be similar to the proof of the Corollary 2.2.10 of Chapter 2. □

Remark 3.1.10 *The above theorem also applies to the generator of a uniformly continuous Q.D.S. acting on a C^*-algebra $\mathcal{A}$, with the only essential modification that $R^*\rho(x)R$ and H will belong to the ultra-weak closure of $\mathcal{A}$ instead of $\mathcal{A}$ itself.*

Remark 3.1.11 *From the structure obtained by the above Theorem 3.1.8 it is clear (since ρ is normal) that $\mathcal{L}$ is normal, that is, ultra-weakly continuous.*

Let us complete this section by introducing a few notations which will be useful later. For a unital C^*-algebra $\mathcal{A} \subseteq \mathcal{B}(h)$, a representation $(\pi, \mathcal{H})$ of $\mathcal{A}$, and operators $R \in \mathcal{B}(h, \mathcal{H})$ and $H \in \mathcal{B}(h)$ we write $\delta_{R,\pi}$ and $\mathcal{L}_{R,\pi,H}$ for the operators given by

$$\delta_{R,\pi}(a) = Ra - \pi(a)R, \quad \mathcal{L}_{R,\pi,H}(a) = R^*\pi(a)R - \tfrac{1}{2}\{R^*R, a\} + i[a, H],$$

where $\{b, c\} \equiv bc + cb$. Thus $\delta_{R,\pi} : \mathcal{A} \to \mathcal{B}(h, \mathcal{H})$ is a π-derivation, that is,

$$\delta_{R,\pi}(ab) = \delta_{R,\pi}(a)b + \pi(a)\delta_{R,\pi}(b) \ \forall a, b \in \mathcal{A},$$

and $\mathcal{L}_{R,\pi,H} : \mathcal{A} \to \mathcal{B}(h)$ satisfies

$$\partial\mathcal{L}_{R,\pi,H}(a, b) = \delta_{R,\pi}(a)^*\delta_{R,\pi}(b) - a^*R^*(1 - \pi(1))Rb, \tag{3.1}$$

where, given $\eta : \mathcal{A} \to \mathcal{B}(h)$, the map $\partial\eta : \mathcal{A} \times \mathcal{A} \to \mathcal{B}(h)$ is defined by

$$\partial\eta(a, b) = \eta(a^*b) - a^*\eta(b) - \eta(a^*)b + a^*\eta(1)b.$$

We quote the following result without proof (which can be found in [24]), and this we shall use later.

Lemma 3.1.12 *Let $(\eta, \rho, \mathcal{H}, \delta)$ consist of a map $\eta \in \mathcal{B}(\mathcal{A})$, a $*$-representation $(\rho, \mathcal{H})$ of $\mathcal{A}$ and a ρ-derivation $\delta : \mathcal{A} \to \mathcal{B}(h, \mathcal{H})$ satisfying $\partial\eta(a, b) = \delta(a)^*\delta(b)$ and $\delta(1) = 0$. Then there is an operator $R \in \mathcal{B}(h, \mathcal{H})$ which lies in the ultra-weak closure of* $\mathrm{sp}\{\delta(a)b : a, b \in \mathcal{A}\}$ *and an element $H \in \mathcal{A}''$ such that*

$$\delta(\cdot) = \delta_{R,\rho}(\cdot) \text{ and } \eta(\cdot) = \mathcal{L}_{R,\rho,H}(\cdot) + \tfrac{1}{2}\{\eta(1), \cdot\}.$$

If η is real in the sense that $\eta(x^*) = \eta(x)^*$, then H may be chosen so that $H = H^*$.

3.2 The case of strongly continuous quantum dynamical semigroups

There is no complete characterization of the generator of a general strongly continuous Q.D.S. In this section, we shall outline some of the structure theorems known in this direction. These partial results are aimed at expressing the generator of the Q.D.S. (under certain assumptions) formally as $R^*\rho(x)R + Gx + xG^*$ for unbounded operators R and G, with suitable domains. Thus, first of all it will be useful to discuss the technical details of how to construct a Q.D.S. from a given formal generator expressed in terms of unbounded operators.

3.2.1 Construction of quantum dynamical semigroups from form-generators

Suppose that h and $\mathcal{K}$ are two Hilbert spaces, $\mathcal{A} \subseteq \mathcal{B}(h)$ is a von Neumann algebra, $\pi : \mathcal{A} \to \mathcal{B}(\mathcal{K})$ is a normal, unital, $*$-representation; $(P_t)_{t\geq 0}$ is a C_0-contraction semigroup on h; and $R : h \to \mathcal{K}$ is a closed, densely defined, linear (possibly unbounded) map. Formally we introduce a map $\mathcal{L}$ by, $\mathcal{L}(x) = R^*\pi(x)R + xG + G^*x$, $x \in \mathcal{A}$; where G is the generator of $(P_t)_{t\geq 0}$. Let us make the following assumptions on G and R.

(Ai) G is affiliated to $\mathcal{A}$ and $R^*\pi(x)R$ is affiliated to $\mathcal{A}$, for all $x \in \mathcal{A}$.

(Aii) $\mathrm{Dom}(G) \subseteq \mathrm{Dom}(R)$ and for all $u, v \in \mathrm{Dom}(G)$, $\langle Ru, Rv\rangle + \langle u, Gv\rangle + \langle Gu, v\rangle = 0$, where it may be recalled that $\mathrm{Dom}(A)$ denotes the domain of a linear map A.

Remark 3.2.1 *(i) Note that for a uniformly continuous quantum dynamical semigroup, its generator is given by Christensen–Evans form obtained in [24]: $\mathcal{L}(x) = R^*\pi(x)R + xG + G^*x$, where $R \in \mathcal{B}(h, \mathcal{K})$ such that $R^*\pi(x)R$ and G are in $\mathcal{A}$ for $x \in \mathcal{A}$. Thus the assumption* **(Ai)** *is a natural generalization of the Christensen–Evans form to the case where*

$\mathcal{L}$ is unbounded. Note also that **(Aii)** *corresponds to formal statement that* $\mathcal{L}(1) = 0$. *It is also clear that when* $\mathcal{A} = \mathcal{B}(h)$, **(Ai)** *is trivially satisfied.*

(ii) However, it is to be noted that the assumption **(Ai)** *does not cover the case of the heat semigroup (see Section 9.2).*

(iii) Note that **(Aii)** *is equivalent to the following*
(Aii′) : $(1 - G^*)^{-1}R^*R(1 - G)^{-1} + (1 - G^*)^{-1}G(1 - G)^{-1} + (1 - G^*)^{-1}G^*(1 - G)^{-1} = 0.$

We recall the discussion on the predual of a von Neumann algebra in Chapter 2. Let us now consider $\pi(\mathcal{A})$ as a von Neumann algebra in $\mathcal{B}(\mathcal{K})$ and define a map π_* from $\pi(\mathcal{A}_h)_*$ to $\mathcal{A}_{h,*}$ by, $(\pi_*\psi)(a) = \psi(\pi(a))$, $a \in \mathcal{A}_h$, and $\psi \in \Omega_{\pi(\mathcal{A}_h)} \equiv (\pi(\mathcal{A}_h))_*$. Upto canonical identification, π_* can be viewed as a map from $\mathcal{B}_1^{s.a.}(\mathcal{K})/(\pi(\mathcal{A}_h))^{\perp}$ to $\mathcal{B}_1^{\text{s.a.}}(h)/\mathcal{A}_h^{\perp}$; and we shall not notationally distinguish between these two views. It is clear that $tr((\pi_*[\eta]_h)x) = tr([\eta]_h\pi(x))$ for all $x \in \mathcal{A}_h$, $\eta \in \mathcal{B}_1^{s.a.}(\mathcal{K})$.

Lemma 3.2.2 $||\pi_*|| \le 1$, *π_* is positive, and the dual of π_* is the restriction of π to $\mathcal{A}_h$.*

Proof:
For $\psi \in \Omega_{\pi(\mathcal{A}_h)}$, we have $||\pi_*\psi|| = \sup_{a\in\mathcal{A}_h,||a||\le 1}|(\pi_*\psi)(a)| = \sup_{a\in\mathcal{A}_h,||a||\le 1}|\psi(\pi(a))| \le \sup_{b\in\pi(\mathcal{A}_h),||b||\le 1}|\psi(b)|$ (since $||\pi(c)|| \le ||c||$ for any $*$-representation π) $= ||\psi||$.

The map π_* is positive because for any positive functional $\psi \in \Omega_{\pi(\mathcal{A}_h)}$, $(\pi_*\psi)(a^*a) = \psi(\pi(a)^*\pi(a)) \ge 0$ for all $a \in \mathcal{A}$. It can be seen that the dual of π_* is the restriction of π to $\mathcal{A}_h$. □

We shall adopt the convention of denoting C_0-contraction semigroups on predual of a von Neumann algebra by a suffix $*$, for example, $(C_{*,t})_{t\ge 0}$. We observe that the dual of $C_{*,t}$, to be denoted by C_t, is a contractive, ultra-weakly continuous map on the von Neumann algebra, and in fact $(C_t)_{t\ge 0}$ is a C_0-semigroup on the von Neumann algebra. Now, let us define $S_{*,t} : \mathcal{B}_1^{\text{s.a.}}(h) \to \mathcal{B}_1^{\text{s.a.}}(h)$ by, $S_{*,t}(\rho) = P_t\rho P_t^*$ $(t \ge 0)$. It is immediate that $(S_{*,t})_{t\ge 0}$ is a positive, C_0-contraction semigroup.

Lemma 3.2.3 *For all $\lambda > 0$, $(\lambda - G)^{-1} \in \mathcal{A}$. Moreover,* for all $x \in \mathcal{A}_h$, $P_t^*xP_t \in \mathcal{A}_h$ for all $t \ge 0$ *and* $(1 - G^*)^{-1}R^*\pi(x)R(1 - G)^{-1} \in \mathcal{A}_h$.

Proof:
It follows from **(Ai)** and Lemma 2.1.9 that for $\lambda > 0$, $a' \in \mathcal{A}'$ and $u \in h$, one has, $(\lambda - G)a'(\lambda - G)^{-1}u = a'(\lambda - G)(\lambda - G)^{-1}u = a'u$. Thus, $(\lambda - G)a'(\lambda - G)^{-1} = a'$, that is, $a'(\lambda - G)^{-1} = (\lambda - G)^{-1}a'$, for all $a' \in \mathcal{A}'$. So we have $(\lambda - G)^{-1} \in \mathcal{A}'' = \mathcal{A}$, since $(\lambda - G)^{-1}$ is bounded. This also implies that

$P_t = \text{s} - \lim_{n\to\infty}((1 - \frac{tG}{n})^{-1})^n \in \mathcal{A}$. Thus, for any x in $\mathcal{A}_h$, $P_t^* x P_t$ is a self-adjoint element in $\mathcal{A}$, that is, belongs to $\mathcal{A}_h$.

Now, let $x \in \mathcal{A}_h$. Let us consider the polar decomposition of R as, $R = U|R|$, where $|R|$ is a positive operator affiliated to $\mathcal{A}$ (because $|R| = (R^*R)^{1/2}$, $R^*R = R^*\pi(1)R$ is affiliated to $\mathcal{A}$) and U is a partial isometry with the closure of range of $|R|$ as the initial space and the closure of range of R as the final space. We note that $|R|(1-G)^{-1}$ is bounded (as $\text{Ran}((1-G)^{-1}) \subseteq \text{Dom}(R) = \text{Dom}(|R|)$) and by Lemma 2.1.9 (second part), it is affiliated to $\mathcal{A}$, hence belongs to $\mathcal{A}$. Thus, for any bounded continuous function f from $[0, \infty)$ to $\mathbb{R}$, $f(|R|)$ belongs to $\mathcal{A}$, as $|R|$ is affiliated to $\mathcal{A}$. In particular, $n(1+n|R|)^{-1} \in \mathcal{A}$ for any positive integer n. So, $n^2(1+n|R|)^{-1}R^*\pi(x)R(1+n|R|)^{-1} = (n(1+n|R|)^{-1}|R|)U^*\pi(x)U(n|R|(1+n|R|)^{-1}) \in \mathcal{A}$ (since it is clearly bounded and is affiliated to $\mathcal{A}$ by Lemma 2.1.9). But $n|R|(1+n|R|)^{-1} \to 1$ strongly as $n \to \infty$; hence

$$U^*\pi(x)U = \text{s} - \lim_{n\to\infty} n^2(1+n|R|)^{-1}R^*\pi(x)R(1+n|R|)^{-1} \in \mathcal{A}.$$

Thus, $(1-G^*)^{-1}R^*\pi(x)R(1-G)^{-1} = (|R|(1-G)^{-1})^*(U^*\pi(x)U)(|R|(1-G)^{-1}) \in \mathcal{A}$, as $|R|(1-G)^{-1} \in \mathcal{A}$. Furthermore, if $x \in \mathcal{A}_h$, $(1-G^*)^{-1}R^*\pi(x)$ $R(1-G)^{-1}$ is self-adjoint and hence belongs to $\mathcal{A}_h$. □

Lemma 3.2.4 *For $t \geq 0$, $S_{*,t}$ induces a linear map $\widetilde{S_{*,t}} : \mathcal{A}_{h,*} \to \mathcal{A}_{h,*}$, and $(\widetilde{S_{*,t}})_{t\geq 0}$ is a positive, C_0-contraction semigroup on $\mathcal{A}_{h,*}$ (positivity means that $\widetilde{S_{*,t}}([\rho]_h)$ is a positive element whenever $[\rho]_h$ is so in $\mathcal{A}_{h,*}$).*

Proof:
If we take $\rho_1, \rho_2 \in \mathcal{B}_1^{\text{s.a.}}(h)$ such that $\rho_1 \sim \rho_2$, then for any $x \in \mathcal{A}_h$ and $t \geq 0$, $tr(S_{*,t}(\rho_1)x) = tr(P_t\rho_1 P_t^* x) = tr(\rho_1 P_t^* x P_t) = tr(\rho_2 P_t^* x P_t)$ (by Lemma 3.2.3) $= tr(S_{*,t}(\rho_2)x)$. Thus, $S_{*,t}(\rho_1) \sim S_{*,t}(\rho_2)$, which proves that $S_{*,t}$ induces a map $\widetilde{S_{*,t}}$. Semigroup property and positivity of $(\widetilde{S_{*,t}})_{t\geq 0}$ are immediate, whereas strong continuity follows from the fact that $(S_{*,t})_{t\geq 0}$ is strongly continuous and $||\widetilde{S_{*,t}}([\rho]_h) - [\rho]_h|| \leq ||S_{*,t}(\rho) - \rho||_1$ for all $\rho \in \mathcal{B}_1^{\text{s.a.}}(h)$, $t \geq 0$. □

Let us denote the generator of $(S_{*,t})_{t\geq 0}$ by Z. Since each $S_{*,t}$ induces an operator on $\mathcal{A}_{h,*}$, Z will do so. It is not difficult to see that the generator of $(\widetilde{S_{*,t}})_{t\geq 0}$ will be a closed extension of the map induced by Z. We, by slight abuse of notation, denote by $\widetilde{Z}$ the generator of $(\widetilde{S_{*,t}})_{t\geq 0}$. Let us define φ: $\mathcal{B}_1^{\text{s.a.}}(h) \to \mathcal{B}_1^{\text{s.a.}}(h)$ by $\varphi(\rho) = (1-G)^{-1}\rho(1-G^*)^{-1}$. Since $(1-G)^{-1}$ and $(1-G^*)^{-1}$ belong to $\mathcal{A}$, φ will induce a map $\widetilde{\varphi}$ from $\mathcal{A}_{h,*}$ to $\mathcal{A}_{h,*}$, which can be proven in a way similar to the proof of Lemma 3.2.4. Let us denote by $\mathcal{D}$ and $\widetilde{\mathcal{D}}$ the ranges of φ and $\widetilde{\varphi}$ respectively.

Lemma 3.2.5 *$\mathcal{D}$ and $\widetilde{\mathcal{D}}$ are dense in $\mathcal{B}_1^{\text{s.a.}}(h)$ and $\mathcal{A}_{h,*}$ respectively. Moreover, they are cores for Z and $\widetilde{Z}$ respectively.*

Proof:
Since $\text{Dom}(G)$ is dense in h, the real linear span of rank-one operators of the form $|u><u|$, $u \in \text{Dom}(G)$ is dense in $\mathcal{B}_1^{\text{s.a.}}(h)$. But for $u \in \text{Dom}(G) = \text{Ran}((1-G)^{-1})$, $|u><u|$ is clearly in $\mathcal{D}$, which proves the density of $\mathcal{D}$ in $\mathcal{B}_1^{\text{s.a.}}(h)$. The core property of $\mathcal{D}$ follows because each $S_{*,t}$ leaves $\mathcal{D}$ invariant (by Theorem 1.9 in [37]). The assertions about $\widetilde{\mathcal{D}}$ follow similarly, only thing to note is that $||[\rho]_h - [\sigma]_h|| \leq ||\rho - \sigma||_1$ for all $\rho, \sigma \in \mathcal{B}_1^{\text{s.a.}}(h)$. □

Lemma 3.2.6 *Given $[\rho]_h \in \tilde{\mathcal{D}}$ and $\epsilon > 0$, we can get positive elements $[\rho_1]_h$ and $[\rho_2]_h$ in $\tilde{\mathcal{D}}$ such that $[\rho]_h = [\rho_1]_h - [\rho_2]_h$ and $||[\rho_1]_h|| + ||[\rho_2]_h|| \leq ||[\rho]_h|| + \epsilon$.*

Proof:
Let us consider a general $[\rho]_h \in \widetilde{\mathcal{D}}$ such that $\rho = \varphi(\sigma)$, $\sigma \in \mathcal{B}_1^{\text{s.a.}}(h)$. Given $\epsilon > 0$, we can choose sufficiently large n such that $||[\sigma_n]_h - [\rho]_h|| \leq \epsilon$, where $\sigma_n = (1 - G/n)(1-G)^{-1}\sigma((1-G/n)(1-G)^{-1})^* = (1-G/n)\varphi(\sigma)(1 - G^*/n) = (1-G/n)\rho(1-G^*/n)$. By the noncommutative Hahn decomposition (Proposition 2.1.14 of Chapter 2), we can find two positive elements σ_n^+, σ_n^- such that $||[\sigma_n]_h|| = ||[\sigma_n^+]_h|| + ||[\sigma_n^-]_h||$ and $[\sigma_n]_h = [\sigma_n^+]_h - [\sigma_n^-]_h$. Take $\rho_1 = (1-G/n)^{-1}\sigma_n^+(1-G^*/n)^{-1}$, $\rho_2 = (1-G/n)^{-1}\sigma_n^-(1-G^*/n)^{-1}$; and observe that

$$
\begin{aligned}
&||[\rho_1]_h|| + ||[\rho_2]_h|| \\
&= ||(1-G/n)^{-1}\sigma_n^+(1-G^*/n)^{-1}||_1 + ||(1-G/n)^{-1}\sigma_n^-(1-G^*/n)^{-1}||_1 \\
&\quad \text{(as } \rho_1, \rho_2 \text{ are positive)} \\
&\leq ||\sigma_n^+||_1 + ||\sigma_n^-||_1 \\
&= ||[\sigma_n^+]_h|| + ||[\sigma_n^-]_h|| \\
&= ||[\sigma_n]_h|| \leq ||[\rho]_h|| + \epsilon.
\end{aligned}
$$

Now, $[\rho_1]_h - [\rho_2]_h = [(1-G/n)^{-1}(\sigma_n^+ - \sigma_n^-)(1-G^*/n)^{-1}]_h = [(1-G/n)^{-1}\sigma_n(1-G^*/n)^{-1}]_h = [\rho]_h$, because $(\sigma_n^+ - \sigma_n^-) \sim \sigma_n$ implies $(1-G/n)^{-1}(\sigma_n^+ - \sigma_n^-)(1-G^*/n)^{-1} \sim (1-G/n)^{-1}\sigma_n(1-G^*/n)^{-1}$.

□

Let us now define $J : \mathcal{D} \to \mathcal{A}_{h,*}$ by $J(\varphi(\rho)) = \pi_*([R(1-G)^{-1}\rho(1-G^*)^{-1}R^*]_h)$, $\rho \in \mathcal{B}_1^{\text{s.a.}}(h)$; where $(1-G^*)^{-1}R^*$ is to be interpreted as the bounded operator $(R(1-G)^{-1})^*$. Next two lemmas give some useful properties of the map $\widetilde{J}$ induced by J.

Lemma 3.2.7 *The map $\widetilde{J} : \widetilde{\mathcal{D}} \to \mathcal{A}_{h,*}$ given by, $\widetilde{J}(\widetilde{\varphi}([\sigma]_h)) = J(\varphi(\sigma))$ for $\sigma \in \mathcal{B}_1^{\text{s.a.}}(h)$ is well-defined and linear.*

Proof:
It is enough to prove that whenever $\varphi(\sigma) \sim 0$, we must have $J(\sigma) = [0]_h$. Given $\varphi(\sigma) \sim 0$, we first show that $\sigma \sim 0$. For $n = 1, 2, \ldots$, let us denote by G_n the operator $(1 - G/n)^{-1}$. Clearly, each G_n is in $\mathcal{A}$, and $G_n \to 1$ strongly. So, $(1 - GG_n)(1 - G)^{-1} \to 1$ strongly. This implies that $(1 - GG_n)(1 - G)^{-1}\sigma(1 - G^*)^{-1}(1 - GG_n)^* x$ converges to σx in trace-norm as n tends to ∞, for any $x \in \mathcal{A}_h$, and hence $\text{tr}(\sigma x)$
$= \lim_{n\to\infty}\text{tr}((1 - GG_n)(1 - G)^{-1}\sigma(1 - G^*)^{-1}(1 - GG_n)^* x)$
$= \lim_{n\to\infty}\text{tr}(\varphi(\sigma).(1 - GG_n)^* x(1 - GG_n)) = 0$ for all $x \in \mathcal{A}_h$; where in the last step we have noted that $(1 - GG_n) \in \mathcal{A}$, which implies $(1 - GG_n)^* x(1 - GG_n) \in \mathcal{A}_h$. Hence, $\sigma \sim 0$.

Now, for all $x \in \mathcal{A}_h$, we have

$$\begin{aligned}
&\text{tr}(J(\sigma x)) \\
&= \text{tr}(R(1 - G)^{-1}\sigma(1 - G^*)^{-1}R^*\pi(x)) \\
&= \text{tr}(\sigma(1 - G^*)^{-1}R^*\pi(x)R(1 - G)^{-1}) \\
&= 0, \ \text{ since } (1 - G^*)^{-1}R^*\pi(x)R(1 - G)^{-1} \in \mathcal{A}_h.
\end{aligned}$$

This completes the proof. □

Lemma 3.2.8 *$\widetilde{J}$ is positive, and* $\text{tr}(\widetilde{J}([\rho]_h) + \widetilde{Z}([\rho]_h)) = 0$ for all $[\rho]_h \in \widetilde{\mathcal{D}}$.

Proof:
The positivity of $\widetilde{J}$ follows from the positivity of π_*. It also follows that for $\rho \in \mathcal{D}$, $Z(\rho) = G\rho + \rho G^*$; hence for $[\rho]_h \in \widetilde{\mathcal{D}}$, $\text{tr}(\widetilde{Z}([\rho]_h)) = \text{tr}(Z(\rho)) = \text{tr}(G\rho + \rho G^*)$. Now for $\sigma \in \mathcal{B}_1^{\text{s.a.}}(h)$, we have by **(Aii′)** that

$$\begin{aligned}
&\text{tr}(\widetilde{J}(\widetilde{\varphi}([\sigma]_h))) \\
&= \text{tr}(J(\varphi(\sigma))) \\
&= \text{tr}(R(1 - G)^{-1}\sigma(1 - G^*)^{-1}R^*\pi(1)) \\
&= \text{tr}(\sigma(1 - G^*)^{-1}R^*R(1 - G)^{-1}) \\
&= -\text{tr}(\sigma(1 - G^*)^{-1}G(1 - G)^{-1} + \sigma(1 - G^*)^{-1}G^*(1 - G)^{-1}) \\
&= -\text{tr}(G\varphi(\sigma) + \varphi(\sigma)G^*) \\
&= -\text{tr}(\widetilde{Z}(\widetilde{\varphi}([\sigma]_h))),
\end{aligned}$$

which completes the proof. □

For $\lambda > 0$, $(\lambda - Z)^{-1}$ can be expressed as $\int_0^\infty e^{-\lambda t} S_{*,t}\,dt$, hence $(\lambda - Z)^{-1}$ leaves $\mathcal{D}$ invariant and is positive. Similar statements are valid about $(\lambda - \widetilde{Z})^{-1}$.

Let us define $\widetilde{B}(\lambda) : \widetilde{\mathcal{D}} \to \mathcal{A}_{h,*}$ by, $\widetilde{B}(\lambda) = \widetilde{J}(\lambda - \widetilde{Z})^{-1}$.

Lemma 3.2.9 *$\widetilde{B}(\lambda)$ extends to a positive linear contractive map from $\mathcal{A}_{h,*}$ to $\mathcal{A}_{h,*}$, which we denote by the same notation.*

Proof:
For positive $[\rho]_h \in \widetilde{\mathcal{D}}$, with $\rho \in \mathcal{D}$, we have that $||\widetilde{B}(\lambda)([\rho]_h)|| = \text{tr}(\widetilde{B}(\lambda)([\rho]_h)) = \text{tr}(\widetilde{J}(\lambda - \widetilde{Z})^{-1}([\rho]_h)) = -\text{tr}(\widetilde{Z}(\lambda - \widetilde{Z})^{-1}([\rho]_h)) = \text{tr}(\rho) - \lambda\ \text{tr}((\lambda - \widetilde{Z})^{-1}([\rho]_h)) \leq \text{tr}(\rho) = ||[\rho]_h||$. For an arbitrary $[\rho]_h$ in $\tilde{\mathcal{D}}$ and any positive number ϵ, we choose two positive elements $[\rho_1]_h$ and $[\rho_2]_h$ satisfying the conclusions of Lemma 3.2.6. Therefore, $||\widetilde{B}(\lambda)([\rho]_h)|| \leq ||\widetilde{B}(\lambda)([\rho_1]_h)|| + ||\widetilde{B}(\lambda)([\rho_2]_h)|| \leq ||[\rho_1]_h|| + ||[\rho_2]_h|| \leq ||[\rho]_h|| + \epsilon$. This proves that $||\widetilde{B}(\lambda)([\rho]_h)|| \leq ||[\rho]_h||$ for all $[\rho]_h \in \widetilde{\mathcal{D}}$, and we complete the proof by the density of $\widetilde{\mathcal{D}}$ in $\mathcal{A}_{h,*}$. □

Lemma 3.2.10 *$\widetilde{J}$ extends to $\widetilde{J}'$ on* $\text{Dom}(\widetilde{Z})$ *such that* $\text{tr}(\widetilde{Z}([\rho]_h) + \widetilde{J}'([\rho]_h)) = 0$ for all $[\rho]_h \in \text{Dom}(\widetilde{Z})$.

Proof:
It is enough to take $\widetilde{J}'([\rho]_h) = \widetilde{B}(1)(1 - \widetilde{Z})([\rho]_h)$, for $[\rho]_h \in \text{Dom}(\widetilde{Z})$, and to note that $\widetilde{\mathcal{D}}$ is a core for $\widetilde{Z}$. □

By an abuse of notation, we shall continue to denote $\widetilde{J}'$ by $\widetilde{J}$.

Theorem 3.2.11 *For $0 \leq r < 1$, define $\widetilde{G_r} = \widetilde{Z} + r\widetilde{J}$. Then $\widetilde{G_r}$ generates a positive, C_0 contraction semigroup, say $(\widetilde{T_{*,t}^{(r)}})_{t \geq 0}$, on $\mathcal{A}_{h,*}$; and $(\lambda - \widetilde{G_r})^{-1} = (\lambda - \widetilde{Z})^{-1} \sum_{n=0}^{\infty} r^n \widetilde{B}(\lambda)^n$ for $\lambda > 0$.*

Proof:
Fix $\lambda > 0$, $r \in [0, 1)$. We have, $(\lambda - \widetilde{G_r}) = (1 - r\widetilde{J}(\lambda - \widetilde{Z})^{-1})(\lambda - \widetilde{Z}) = (1 - r\widetilde{B}(\lambda))(\lambda - \widetilde{Z})$. Since $||r\widetilde{B}(\lambda)|| < 1$, we can use the Neumann series to get $(1 - r\widetilde{B}(\lambda))^{-1} = \sum_{n=0}^{\infty} r^n \widetilde{B}(\lambda)^n$, which shows in particular that $(\lambda - \widetilde{G_r})^{-1} = (\lambda - \widetilde{Z})^{-1} \sum_{n=0}^{\infty} r^n \widetilde{B}(\lambda)^n$ exists as a bounded operator. Now, for any positive $[\rho]_h \in \mathcal{A}_{h,*}$, let us denote by $[\sigma]_h$ the element $(\lambda - \widetilde{G_r})^{-1}([\rho]_h)$. Clearly, $(\lambda - \widetilde{G_r})^{-1}$ is positive since $(\lambda - \widetilde{Z})^{-1}$ and $\widetilde{B}(\lambda)$ are so; and hence $[\sigma]_h$ is positive. Thus,

$$
\begin{aligned}
\|[\rho]_h\| &= \text{tr}([\rho]_h) \\
&= \lambda \text{tr}([\sigma]_h) + (1 - r)\text{tr}(\widetilde{J}([\sigma]_h)) \ (\text{as } \text{tr}((\widetilde{J} + \widetilde{Z})([\sigma]_h)) = 0) \\
&\geq \lambda\ \text{tr}([\sigma]_h) = \lambda\ ||[\sigma]_h||.
\end{aligned}
$$

So, $||(\lambda - \widetilde{G_r})^{-1}([\rho]_h)|| \leq ||[\rho]_h||/\lambda$ for all positive $[\rho]_h$ in $\mathcal{A}_{h,*}$ and hence for all $[\rho]_h$ in $\mathcal{A}_{h,*}$, since by Proposition 2.1.14 we can decompose any $[\rho]_h$ as $[\rho]_h = [\rho_1]_h - [\rho_2]_h$ with $||[\rho]_h|| = ||[\rho_1]_h|| + ||[\rho_2]_h||$, where $[\rho_1]_h$ and $[\rho_2]_h$ are positive. We complete the proof of the theorem by appealing to the Hille–Yosida theorem, and also noting that positivity of $(\lambda - \widetilde{G_r})^{-1}$ implies positivity of $\widetilde{T_{*,t}^{(r)}}$. □

Theorem 3.2.12 *As $r \uparrow 1$ (that is, as r increases to 1), $\widetilde{T_{*,t}}^{(r)} \uparrow \widetilde{T_{*,t}}^{(\min)}$, where $(\widetilde{T_{*,t}}^{(\min)})_{t\geq 0}$ is a positive, C_0 contraction semigroup on $\mathcal{A}_{h,*}$; and the above convergence is strong and uniform for t in compact subsets of $[0,\infty)$.*

Proof:
We see that for positive element $[\rho]_h \in \mathcal{A}_{h,*}$, $\widetilde{T_{*,t}}^{(r)}([\rho]_h) \geq \widetilde{T_{*,t}}^{(s)}([\rho]_h) \geq 0$ for r, s such that $1 > r \geq s \geq 0$. This follows from the series expansion of $(\lambda - \widetilde{G_r})^{-1}$ as in the preceding theorem and from the fact that $\widetilde{T_{*,t}}^{(r)} = s - \lim_{n\to\infty}(n/t)^n(n/t - \widetilde{G_r})^{-n}$. Moreover, $\mathrm{tr}(\widetilde{T_{*,t}}^{(r)}([\rho]_h)) \leq ||[\rho]_h||$, since $||\widetilde{T_{*,t}}^{(r)}|| \leq 1$. So $(\mathrm{tr}(\widetilde{T_{*,t}}^{(r)}([\rho]_h)))_{r\in(0,1)}$ is an increasing bounded net of positive numbers and hence it converges to some finite positive limit as $r \uparrow 1$. For $r \geq s$ we have

$$||\widetilde{T_{*,t}}^{(r)}([\rho]_h) - \widetilde{T_{*,t}}^{(s)}([\rho]_h)|| = \mathrm{tr}(\widetilde{T_{*,t}}^{(r)}([\rho]_h) - \widetilde{T_{*,t}}^{(s)}([\rho]_h)) \to 0 \text{ as } r, s \to 1,$$

and hence $\widetilde{T_{*,t}}^{(r)}([\rho]_h)$ converges in the norm of $\mathcal{A}_{h,*}$ to say $\widetilde{T_{*,t}}^{(\min)}([\rho]_h)$ as $r \uparrow 1$. We extend $\widetilde{T_{*,t}}^{(\min)}$ to the whole of $\mathcal{A}_{h,*}$ by linearity using Proposition 2.1.14, and observe that this extension will be a contractive, positive linear map from $\mathcal{A}_{h,*}$ to itself, since each $\widetilde{T_{*,t}}^{(r)}$ has the same properties. The semigroup property of $(\widetilde{T_{*,t}}^{(\min)})_{t\geq 0}$ follows since it is the strong limit of a net of contractive semigroups.

We now show that the convergence is uniform over compacts in t, which will also prove strong continuity of $(\widetilde{T_{*,t}}^{(\min)})_{t\geq 0}$. Suppose that this is not true. Then, there exist positive $[\rho]_h \in \mathcal{A}_{h,*}$, positive number ϵ_0, and sequences $(r_n)_{n=1,2,\ldots}$, $(t_n)_{n=1,2,\ldots}$ such that

$$0 \leq r_n \uparrow 1, \;\; 0 \leq t_n \to t_0$$

for some $t_0 \geq 0$; and

$$||\widetilde{T_{*,t_n}}^{(r_n)}([\rho]_h) - \widetilde{T_{*,t_n}}^{(\min)}([\rho]_h)|| \geq \epsilon_0$$

for all positive integer n. Since $\widetilde{T_{*,t_n}}^{(r_n)}([\rho]_h)$ and $(-\widetilde{T_{*,t_n}}^{(r_n)}([\rho]_h) + \widetilde{T_{*,t_n}}^{(\min)}([\rho]_h))$ are positive, we have by Lemma 2.1.13 that

$$||\widetilde{T_{*,t_n}}^{(\min)}([\rho]_h)|| = ||\widetilde{T_{*,t_n}}^{(r_n)}([\rho]_h)|| + ||\widetilde{T_{*,t_n}}^{(\min)}([\rho]_h) - \widetilde{T_{*,t_n}}^{(r_n)}([\rho]_h)||,$$

which implies,

$$||\widetilde{T_{*,t_n}}^{(\min)}([\rho]_h)|| - ||\widetilde{T_{*,t_n}}^{(r_n)}([\rho]_h)|| \geq \epsilon_0.$$

Thus for all $m \leq n$,

$$||\widetilde{T_{*,t_n}}^{(r_m)}([\rho]_h)|| \leq ||\widetilde{T_{*,t_n}}^{(r_n)}([\rho]_h)|| \leq ||\widetilde{T_{*,t_n}}^{(\min)}([\rho]_h)|| - \epsilon_0.$$

Keeping m fixed and letting n tend to ∞, we obtain that

$$||\widetilde{T_{*,t_0}}^{(r_m)}([\rho]_h)|| \leq ||\widetilde{T_{*,t_0}}^{(\min)}([\rho]_h)|| - \epsilon_0;$$

and then by letting m tend to ∞,

$$||\widetilde{T_{*,t_0}}^{(\min)}([\rho]_h)|| \leq ||\widetilde{T_{*,t_0}}^{(\min)}([\rho]_h)|| - \epsilon_0,$$

which is clearly a contradiction. □

Theorem 3.2.13 *The generator of $(\widetilde{T_{*,t}}^{(\min)})_{t\geq 0}$, say $\widetilde{A}$, is an extension of $\widetilde{Z}+\widetilde{J}$; and we have the following minimality property.*
Whenever $(\widetilde{T_{*,t}}')_{t\geq 0}$ is a positive, C_0 contraction semigroup on $\mathcal{A}_{h,*}$ whose generator (say $\widetilde{A}'$) extends $\widetilde{Z}+\widetilde{J}$, we must have $\widetilde{T_{*,t}}' \geq \widetilde{T_{*,t}}^{(\min)}$ for all $t \geq 0$.

Proof:
The first part of the theorem follows from the fact that $\widetilde{G_r}([\rho]_h) \to (\widetilde{Z}+\widetilde{J})([\rho]_h)$ as $r \uparrow 1$, for all $[\rho]_h \in \mathrm{Dom}(\widetilde{Z})$. For minimality, it is required to observe that for $\lambda > 0$, $(\lambda - \widetilde{A}')^{-1} - (\lambda - \widetilde{G_r})^{-1} = (\lambda - \widetilde{A}')^{-1}(\widetilde{A}' - \widetilde{G_r})(\lambda - \widetilde{G_r})^{-1} \geq 0$, since the restriction of $\widetilde{A}'$ to the range of $(\lambda - \widetilde{G_r})^{-1}$, i. e. to $\mathrm{Dom}(\widetilde{G_r})$, is the same as $\widetilde{Z}+\widetilde{J} \geq \widetilde{G_r}$, and $(\lambda-\widetilde{A}')^{-1}$, $(\lambda-\widetilde{G_r})^{-1}$ are positive. We complete the proof by noting that $\widetilde{T_{*,t}}' = \mathrm{s}-\lim_{n\to\infty}(n/t)^n(n/t-\widetilde{A}')^{-n} \geq \mathrm{s}-\lim_{n\to\infty}(n/t)^n(n/t-\widetilde{G_r})^{-n} = \widetilde{T_{*,t}}^{(r)}$ for all r, t. □

Lemma 3.2.14 *For $\lambda > 0$ and nonnegative integer n, define $R_n(\lambda) := (\lambda - \widetilde{Z})^{-1}\sum_{k=0}^{n}(\widetilde{B(\lambda)})^k$. Then $R_n(\lambda) \to R(\lambda) \equiv (\lambda - \widetilde{A})^{-1}$ strongly as $n \to \infty$.*

Proof:
For $0 \leq r < 1$, let $R_n^{(r)}(\lambda) := (\lambda - \widetilde{Z})^{-1}\sum_{k=0}^{n} r^k(\widetilde{B(\lambda)})^k$. Clearly, $R_n^{(r)}(\lambda) \uparrow (\lambda - \widetilde{Z})^{-1}\sum_{k=0}^{\infty} r^k(\widetilde{B(\lambda)})^k = (\lambda - \widetilde{G_r})^{-1}$ as $n \to \infty$. Furthermore, it follows from Theorem 3.2.12 that $(\lambda - \widetilde{G_r})^{-1} \uparrow (\lambda - \widetilde{A})^{-1} = R(\lambda)$ as $r \uparrow 1$, and hence

$$R_n^{(r)}(\lambda) \leq (\lambda - \widetilde{G_r})^{-1} \leq R(\lambda). \tag{3.2}$$

Letting $r \uparrow 1$, $R_n(\lambda) \leq R(\lambda)$. This, combined with the fact that $R_n(\lambda)$ is nondecreasing in n, allows us to conclude that there is an operator, say $R'(\lambda)$, such that $R_n(\lambda) \uparrow R'(\lambda)$ as $n \to \infty$. Clearly, $R'(\lambda) \leq R(\lambda)$. On the other hand, $R_n^{(r)}(\lambda) \leq R'(\lambda)$ for all r, n, and $(\lambda - \widetilde{G_r})^{-1} = \lim_{n\to\infty} R_n^{(r)}(\lambda) \leq R'(\lambda)$ for all r. Now, taking limit $r \uparrow 1$, we get $R(\lambda) \leq R'(\lambda)$, which completes the proof of the lemma. □

We shall now write down a number of equivalent criteria for the conservativity of the minimal semigroup we have constructed. For $\lambda > 0$ denote the dual of the map $\widetilde{B(\lambda)}$ by Q_λ. The following lemma gives a computable expression for Q_λ.

Lemma 3.2.15 *We have the following:*

$$\langle u, Q_\lambda(x)v\rangle = \int_0^\infty e^{-\lambda t}\langle RP_t u, \pi(x)RP_t v\rangle dt, \tag{3.3}$$

for $u, v \in \mathrm{Dom}(G)$ *and* $x \in \mathcal{A}$. *Furthermore,* $||Q_\lambda(x)|| \leq ||x||$ *for all* $x \in \mathcal{A}$.

Proof:
Take $x \in \mathcal{A}$ and $[\rho] \in \tilde{\mathcal{D}}$, where $\rho = (1-G)^{-1}\eta(1-G^*)^{-1}$ for some $\eta \in \mathcal{B}_1(h)$. We have

$$\begin{aligned}&(\lambda - \tilde{Z})^{-1}([\rho])\\ &= \int_0^\infty e^{-\lambda t} P_t(1-G)^{-1}[\eta](1-G^*)^{-1}P_t^* dt\\ &= [(1-G)^{-1}y(1-G^*)^{-1}],\end{aligned}$$

where $y = \int_0^\infty e^{-\lambda t} P_t \eta P_t^* dt$, and we have used the fact that P_t and G commute with each other. It now follows from the definition of $\widetilde{B(\lambda)} = \tilde{J}(\lambda - \tilde{Z})^{-1}$ that

$$\widetilde{B(\lambda)}([\rho]) = \pi_*([R(1-G)^{-1}y(R(1-G)^{-1})^*]).$$

Thus,

$$\begin{aligned}&\mathrm{tr}(Q_\lambda(x)\rho)\\ &= \mathrm{tr}(\pi(x)R(1-G)^{-1}y(1-G^*)^{-1}R^*)\\ &= \mathrm{tr}\left(\int_0^\infty e^{-\lambda t}\pi(x)R(1-G)^{-1}P_t\eta P_t^*(1-G^*)^{-1}R^* dt\right)\\ &= \mathrm{tr}\left(\int_0^\infty e^{-\lambda t}\pi(x)RP_t(1-G)^{-1}\eta(1-G^*)^{-1}P_t^* R^* dt\right)\\ &= \mathrm{tr}\left(\int_0^\infty e^{-\lambda t}\pi(x)RP_t\rho P_t^* R^* dt\right).\end{aligned}$$

Taking $\rho = |v><u|$, with $u, v \in \mathrm{Dom}(G)$, we get the formula (3.3). The second conclusion follows by using **(Aii)**, observing that for $u \in Dom(G)$, $\|RP_t u\|^2 = -\frac{d}{dt}\|P_t u\|^2$, and by Schwartz' inequality and an integration by parts. □

Theorem 3.2.16 *The following are equivalent.*

(i) The dual semigroup $(\widetilde{T}_t)_{t\geq 0}$ *defined by* $\widetilde{T}_t = (\widetilde{T_{*,t}}^{(\min)})^* : \mathcal{A}_h \to \mathcal{A}_h$, *is conservative; that is,* $\widetilde{T}_t(1) = 1$ *for all* $t \geq 0$.
(ii) $\mathrm{tr}(\widetilde{T_{*,t}}^{(\min)}([\rho]_h)) = \mathrm{tr}(\rho)$ *for all* $\rho \in \mathcal{B}_1^{\mathrm{s.a.}}(h)$ *and for all* $t \geq 0$.
(iii) $\widetilde{B}(\lambda)^n \to 0$ *strongly as* $n \to \infty$, *for any* $\lambda > 0$.
(iv) $(\lambda - \tilde{J} - \tilde{Z})(\widetilde{\varphi}(\mathcal{A}_{h,*}))$ *is dense in* $\mathcal{A}_{h,*}$, *for any* $\lambda > 0$.

(v) $\{x \in \mathcal{A} : \langle Ru, \pi(x)Rv\rangle + \langle Gu, xv\rangle + \langle u, xGv\rangle = \lambda\langle u, xv\rangle$ *for all* $u, v \in \mathrm{Dom}(G)\} = 0$ *for some* $\lambda > 0$.

(vi) *For any* $\lambda > 0$, $Q_\lambda^n(1) \to 0$ *strongly as* $n \to \infty$, *where* $Q_\lambda : \mathcal{A} \to \mathcal{A}$ *is the dual of* $\tilde{B}(\lambda)$.

(vii) *For any* $\lambda > 0$, $Q_\lambda^n(1) \to 0$ *weakly as* $n \to \infty$.

Proof:
The statements (i) and (ii) are equivalent by duality. To show the equivalence of (ii) and (iii), we first note that for positive $\rho \in \mathcal{B}_1^{\text{s.a.}}(h)$,

$$\|R(\lambda)([\rho]_h)\| = \mathrm{tr}(R(\lambda)([\rho]_h)) = \int_0^\infty e^{-\lambda t}\,\mathrm{tr}(\widetilde{T_{*,t}}^{(\min)}([\rho]_h))dt = \int_0^\infty e^{-\lambda t}\|\widetilde{T_{*,t}}^{(\min)}([\rho]_h)\|dt,$$

which is a consequence of the fact that $R(\lambda)([\rho]_h),\ \widetilde{T_{*,t}}^{(\min)}([\rho]_h) \geq 0$. Furthermore, it is possible to derive the following identity from the definitions of the maps involved:

$$[\rho]_h = (\lambda - \widetilde{Z} - \widetilde{J})(R_n(\lambda)([\rho]_h)) + \widetilde{B}(\lambda)^{n+1}([\rho]_h). \tag{3.4}$$

Since $\mathrm{tr}\left((\widetilde{Z} - \widetilde{J})(R_n(\lambda)([\rho]_h))\right) = 0$ by Lemma 3.2.10, and $R_n(\lambda), \widetilde{B}(\lambda) \geq 0$, we have

$$\|[\rho]_h\| = \lambda\,\|R_n(\lambda)([\rho]_h)\| + \|\widetilde{B}(\lambda)^{n+1}([\rho]_h)\|$$

for all $\rho \geq 0$. This implies that $\lim_{n\to\infty}\|\widetilde{B}(\lambda)^n([\rho]_h)\|$ exists and is equal to

$$\begin{aligned}&\|[\rho]_h\| - \lambda \lim_{n\to\infty}\|R_n(\lambda)([\rho]_h)\|\\ &= \|[\rho]_h\| - \lambda\,\|R(\lambda)([\rho]_h)\|\\ &= \lambda\int_0^\infty e^{-\lambda t}\left(\|[\rho]_h\| - \|\widetilde{T_{*,t}}^{(\min)}([\rho]_h)\|\right)dt.\end{aligned}$$

Thus, it is obvious that (ii) implies $\lim_{n\to\infty}\widetilde{B}(\lambda)^n([\rho]_h) = 0$ for all $\rho \geq 0$, hence for all $\rho \in \mathcal{B}_1(h)$. On the other hand, if (iii) holds, we conclude by using the continuity of $t \mapsto \|\widetilde{T_{*,t}}^{(\min)}([\rho]_h)\|$ and the nonnegativity of $\|[\rho]_h\| - \|\widetilde{T_{*,t}}^{(\min)}([\rho]_h)\|$ for all $\rho \geq 0$ that $\|[\rho]_h\| = \|\widetilde{T_{*,t}}^{(\min)}([\rho]_h)\|$, which proves (ii).

Next, we show $(iii) \Leftrightarrow (iv)$. To this end, note that by the identity (3.4), (iii) is equivalent to $(iii)'$: $\lim_{n\to\infty}(\lambda - \widetilde{Z} - \widetilde{J})(R_n(\lambda)([\rho]_h)) = [\rho]_h$ for all $[\rho]_h \geq 0$. So, we need to prove $(iii)' \Leftrightarrow (iv)$. Since $\widetilde{\mathcal{D}} = \mathrm{Ran}(\widetilde{\varphi})$ is a core for $\widetilde{Z}$ (see Lemma 3.2.5), given $[\rho]_h \in \mathrm{Dom}(\widetilde{Z})$, we can choose a sequence $[\rho_n]_h \in \widetilde{\mathcal{D}}$ such that $[\rho_n]_h \to [\rho]_h$ and $\widetilde{Z}([\rho_n]_h) \to \widetilde{Z}([\rho]_h)$ as $n \to \infty$.

However, by definition $\widetilde{J} = \widetilde{B}(1)(1-\widetilde{Z})$, and $\widetilde{B}(1)$ is contractive, so it follows that $\widetilde{J}([\rho_n]_h) \to \widetilde{J}([\rho]_h)$ also, that is,

$$\lim_{n\to\infty}(\lambda - \widetilde{Z} - \widetilde{J})([\rho_n]_h) = (\lambda - \widetilde{Z} - \widetilde{J})([\rho]_h).$$

This shows that $\widetilde{\mathcal{D}}$ is a core for $\lambda - \widetilde{Z} - \widetilde{J}$, and in particular, *(iv)* holds if and only if *(iv)'* : $\overline{\mathrm{Ran}(\lambda - \widetilde{Z} - \widetilde{J})} = \mathcal{A}_{h,*}$. It is obvious that *(iii)'* implies *(iv)'*, hence also *(iv)*. To see the converse, consider the sequence of maps $C_n(\lambda) := \frac{1}{n+1}\sum_{k=0}^{n}\widetilde{B}(\lambda)^k$. Clearly, $\|C_n(\lambda)\| \leq 1$ for all n. Moreover, $C_n(\lambda)(I-\widetilde{B}(\lambda)) = \frac{1}{n+1}(I-\widetilde{B}(\lambda)^{n+1})$, which shows that $C_n(\lambda)(I-\widetilde{B}(\lambda))([\rho]_h) \to 0$ as $n \to \infty$, for $[\rho]_h \in \mathcal{A}_{h,*}$. We claim that $\mathrm{Ran}(I - \widetilde{B}(\lambda))$ is dense in $\mathcal{A}_{h,*}$. Indeed, as $(I - \widetilde{B}(\lambda))$ is bounded and $\mathrm{Ran}(\lambda - \widetilde{Z})$ is dense, we have, by using *(iv)'*,

$$\begin{aligned}
&\overline{\mathrm{Ran}(I - \widetilde{B}(\lambda))}\\
&= \overline{\mathrm{Ran}(I - \widetilde{B}(\lambda)(\lambda - \widetilde{Z}))}\\
&= \overline{\mathrm{Ran}(\lambda - \widetilde{Z} - \widetilde{J})}\\
&= \mathcal{A}_{h,*}.
\end{aligned}$$

Since we have already shown that $C_n(\lambda)([\sigma]_h) \to 0$ as $n \to \infty$ for $[\sigma]_h$ belonging to $\mathrm{Ran}(I - \widetilde{B}(\lambda))$, and $\|C_n(\lambda)\| \leq 1$ for all n, it follows that $\lim_{n\to\infty} C_n(\lambda)([\rho]_h) = 0$ for all $[\rho]_h \in \mathcal{A}_{h,*}$. On the other hand, as $\widetilde{B}(\lambda)$ is a contractive positive map, we have $\|\widetilde{B}(\lambda)^m\| \leq \|\widetilde{B}(\lambda)^n\|$ whenever $m \geq n$. This, together with the obvious positivity of $C_n(\lambda)$, imply that for positive $[\rho]_h$,

$$\|\widetilde{B}(\lambda)^n([\rho]_h)\| \leq \frac{1}{n+1}\sum_{k=0}^{n}\|\widetilde{B}(\lambda)^k([\rho]_h)\| = \|C_n(\lambda)([\rho]_h)\| \to 0,$$

which proves *(iii)*.

Now, we come to the proof of *(iv)* $\Leftrightarrow$ *(v)*. Suppose first that *(iv)* holds, and let x be an element of $\mathcal{A}$ such that

$$\langle Ru, \pi(x)Rv\rangle + \langle Gu, xv\rangle + \langle u, xGv\rangle = \lambda\langle u, xv\rangle \text{ for all } u, v \in \mathrm{Dom}(G). \tag{3.5}$$

As $\mathrm{Dom}(G) = \mathrm{Ran}((1 - G)^{-1})$, (3.5) is equivalent to

$$\begin{aligned}
&\langle R(1-G)^{-1}u, \pi(x)R(1-G)^{-1}v\rangle + \langle G(1-G)^{-1}u, x(1-G)^{-1}v\rangle\\
&\quad + \langle (1-G)^{-1}u, xG(1-G)^{-1}v\rangle = \lambda\langle u, xv\rangle \text{ for all } u, v \in h,
\end{aligned}$$

which can also be written as

$$\mathrm{tr}\,((J + Z)(\varphi(|v><u|))x) = \lambda\,\mathrm{tr}(x|v><u|) \quad \text{for all } u, v \in h,$$

or,

$$\mathrm{tr}\left((\widetilde{J}+\widetilde{Z})(\widetilde{\varphi}([|v><u|]_h))x\right) = \lambda\, \mathrm{tr}(x[|v><u|]_h) \quad \text{for all } u, v \in h.$$

Thus, $\mathrm{tr}\left(x(\lambda - \widetilde{Z} - \widetilde{J})(\widetilde{\varphi}([\sigma]_h))\right) = 0$ for all $\sigma \in \mathcal{B}_1^{\text{s.a.}}(h)$, and as $\mathrm{Ran}(\lambda - \widetilde{Z} - \widetilde{J})$ is dense by *(iv)*, we conclude $x = 0$ by the Hahn–Banach theorem. The converse is also straightforward to see.

Let us now prove the equivalence of *(iii)* and *(vii)*, by noting that $\tilde{B}(\lambda)$ is a positive map, so for all nonnegative elements $\rho \in \mathcal{A}_*$, we have $\|\tilde{B}(\lambda)^n([\rho])\| = \mathrm{tr}(\tilde{B}(\lambda)^n([\rho])) = \mathrm{tr}(Q_\lambda^n(1)\rho)$. Thus, *(iii)* implies that $\lim_{n\to\infty} \mathrm{tr}(Q_\lambda^n(1)\rho) = 0$ for any nonnegative ρ, and hence for all $\rho \in \mathcal{A}_*$, since a general element of $\mathcal{A}_*$ can be expressed as a linear combination of four nonnegative elements of $\mathcal{A}_*$. In particular, this implies that for $u, v \in h$, $\langle u, Q_\lambda^n(1)v\rangle = \mathrm{tr}(Q_\lambda^n(1)|v><u|)$ converges to 0. Conversely, if $Q_\lambda^n(1) \to 0$ weakly, it follows that $\|\tilde{B}(\lambda)^n([\rho])\|_1 \to 0$ for all nonnegative ρ, and hence for any $\rho \in \mathcal{A}_*$. Note that here we have used the well-known result that whenever $A_n \to 0$ weakly as $n \to \infty$ and B is a trace-class operator, we must have $\mathrm{tr}(A_n B) \to 0$.

Finally, let us argue that *(vi)* and *(vii)* are equivalent. It is trivial that *(vi)* implies *(vii)*, so let us prove the other direction. Assume that $Q_\lambda^n(1) \to 0$ weakly. It can be seen that Q_λ is a contractive positive map on $\mathcal{A}$, so $B_n := Q_\lambda^n(1)$ is a sequence of positive contractions on h. Thus, for any $u \in h$, $\|B_n u\|^2 = \langle u, B_n^2 u\rangle \leq \|B_n\|\langle u, B_n u\rangle \leq \langle u, B_n u\rangle \to 0$ as $n \to \infty$. This completes the proof. □

Remark 3.2.17 *We can replace* $\mathrm{Dom}(G)$ *in (v) of the statement of the above theorem by any subspace* $\mathcal{D}_0$ *which is a core for* G*. That is,* $\widetilde{T}_t$ *is conservative if and only if for some* $\lambda > 0$*,*

$$\{x \in \mathcal{A} : \langle Ru, \pi(x)Rv\rangle + \langle Gu, xv\rangle + \langle u, xGv\rangle = \lambda\langle u, xv\rangle \text{ for all } u, v \in \mathcal{D}_0\} = \{0\}.$$

This follows from the boundedness of $R(1-G)^{-1}$*. Indeed, for* $u, v \in \mathrm{Dom}(G)$*, we can choose sequences* $\{u_n\}, \{v_n\}$ *of vectors from* $\mathcal{D}_0$*, such that* u_n, v_n, Gu_n, Gv_n *converge to* u, v, Gu, Gv *respectively as* $n \to \infty$*. Clearly we have that* $\lim_{n\to\infty} Ru_n = \lim_{n\to\infty} R(1-G)^{-1}(u_n - Gu_n) = R(1-G)^{-1}(u - Gu) = Ru$*, and similarly* $\lim_{n\to\infty} Rv_n = Rv$*. Thus* $\langle Ru, \pi(x)Rv\rangle + \langle Gu, xv\rangle + \langle u, xGv\rangle = \lambda\langle u, xv\rangle$ *holds for all* $u, v \in \mathcal{D}_0$ *if and only if it holds for all* $u, v \in \mathrm{Dom}(G)$*, which proves the claim made in this remark.*

For a linear map V from $\mathcal{A}_{h,*}$ to $\mathcal{A}_{h,*}$, we consider its canonical extension $\widehat{V}$ as a map from $\mathcal{A}_*$ to $\mathcal{A}_*$ defined by, $\widehat{V}([\rho]) = V([\mathrm{Re}(\rho)]_h) + i.V([\mathrm{Im}(\rho)]_h)$. We shall say that V is completely positive if the dual of $\widehat{V}$, say $\widehat{V}^*$, is completely positive as a map from $\mathcal{A}$ to $\mathcal{A}$, that is, for any $x_1, x_2, \ldots, x_n; y_1, y_2, \ldots, y_n$ in

$\mathcal{A}$, $\sum_{i,j=1}^{n} x_i^* \widehat{V}^*(y_i^* y_j)x_j$ is positive. This is again equivalent to the following:

$$\sum_{i,j=1}^{n} \operatorname{tr}(x_i^* \widehat{V}^*(y_i^* y_j)x_j\rho) \geq 0 \text{ for all positive } \rho \in \mathcal{B}_1^{\text{s.a.}}(h),$$

which can be re-written as

$$\sum_{i,j=1}^{n} \operatorname{tr}(y_j \widehat{V}([x_j\rho x_i^*])y_i^*) \geq 0 \text{ for all positive } \rho \in \mathcal{B}_1^{\text{s.a.}}(h).$$

It may be noted that complete positivity is preserved under taking finite sum, composition and strong limit of operators. Now we prove in the following theorem that the minimal semigroup constructed in the present section is not only positive but also completely positive.

Theorem 3.2.18 *The semigroup* $\widetilde{T_{*,t}}^{(\min)}$ *is completely positive for all* $t \geq 0$.

Proof:
It is enough to prove that for any $r \in [0, 1)$ and $t \geq 0$, $\widetilde{T_{*,t}}^{(r)}$ is completely positive. For this, it suffices to verify that $(\lambda - \widetilde{G_r})^{-1}$ is completely positive, because $\widetilde{T_{*,t}}^{(r)}$ is strong limit of $(n/t)^n(n/t - \widetilde{G_r})^{-n}$ as $n \to \infty$. But, $(\lambda - \widetilde{G_r})^{-1} = (\lambda - \widetilde{Z})^{-1}\sum_{j=0}^{\infty} r^j \widetilde{B}(\lambda)^j$, and so it is enough to check that $(\lambda - \widetilde{Z})^{-1}$ and $\widetilde{B}(\lambda)$ are completely positive.

Now, for $x_1, \ldots, x_n; y_1, \ldots, y_n \in \mathcal{A}$ and positive $\rho \in \mathcal{B}_1^{\text{s.a.}}(h)$, we have

$$\begin{aligned}
&\sum_{i,j=1}^{n} \operatorname{tr}(y_j \widehat{(\lambda - \widetilde{Z})^{-1}}([x_j\rho x_i^*])y_i^*) \\
&= \int_0^\infty e^{-\lambda t} \sum_{i,j=1}^{n} \operatorname{tr}(y_j P_t x_j \rho x_i^* P_t^* y_i^*) dt \\
&= \int_0^\infty e^{-\lambda t} \operatorname{tr}(\eta_t \eta_t^*) dt \text{ (where } \eta_t = \sum_{i=1}^{n} y_i P_t x_i \rho^{1/2}) \ \geq 0.
\end{aligned}$$

We are left to show that $\sum_{i,j=1}^{n} \operatorname{tr}(y_j \widehat{\widetilde{B}(\lambda)}([x_j\rho x_i^*])y_i^*) \geq 0$. By the density of Dom$(G)$ in h, we choose sequences $\rho_i^{(k)} \in \mathcal{B}_2(h)$, $i = 1, 2, \ldots, n$ such that $(1 - G)^{-1}\rho_i^{(k)} \to x_i\rho^{1/2}$ in Hilbert–Schmidt norm as $k \to \infty$, for each i. Recalling the definition of $\widetilde{B}(\lambda)$ and proceeding as in the previous paragraph, we get

$$\begin{aligned}
&\sum_{i,j=1}^{n} \operatorname{tr}(y_j \widehat{\widetilde{B}(\lambda)}([(1 - G)^{-1}\rho_j^{(k)}\rho_i^{(k)*}(1 - G^*)^{-1}])y_i^*) \\
&= \sum_{i,j=1}^{n} \operatorname{tr}(\widehat{\widetilde{B}(\lambda)}([(1 - G)^{-1}\rho_j^{(k)}\rho_i^{(k)*}(1 - G^*)^{-1}])y_i^* y_j)
\end{aligned}$$

$$= \int_0^\infty e^{-\lambda t} \sum_{i,j=1}^n \mathrm{tr}(R(1-G)^{-1} P_t \rho_j^{(k)} \rho_i^{(k)*} P_t^*(R(1-G)^{-1})^* \pi(y_i^* y_j))dt$$

$$= \int_0^\infty e^{-\lambda t} \sum_{i,j=1}^n \mathrm{tr}(\pi(y_j) R(1-G)^{-1} P_t \rho_j^{(k)} \rho_i^{(k)*} P_t^*(R(1-G)^{-1})^* \pi(y_i)^*)dt$$

$$= \int_0^\infty e^{\lambda t} \mathrm{tr}(\zeta_t^{(k)} \zeta_t^{(k)*})dt \ \ (\text{where } \zeta_t^{(k)} := \sum_{i=1}^n \pi(y_i) R(1-G)^{-1} P_t \rho_i^{(k)}),$$

which is clearly positive, and which also converges to $\sum_{i,j=1}^n \mathrm{tr}(y_j \widehat{\widetilde{B}(\lambda)} [(x_j \rho x_i^*)] y_i)$. Therefore, by the discussion preceding this theorem, the result follows. □

Let us give a few examples of the construction discussed in this subsection.

Example 1

Let us consider a σ-finite measure space $(\Omega, \mathcal{B}, \mu)$ and denote by $L^p(\mu)$ and $L^p(\mu)_{\mathbb{R}}$ the space of all measurable complex-valued (real-valued, respectively) functions on Ω with finite L^p-norm, for $p \geq 1$. We denote by h the Hilbert space $L^2(\mu)$ and by $\mathcal{A}$ the abelian von Neumann algebra $L^\infty(\mu)$, to be identified as multiplication operators on h. It is well-known that $\mathcal{A}_{h,*} \cong L^1(\mu)_{\mathbb{R}}$. Our aim is to construct the minimal semigroup when the generator is given by

$$(\mathcal{L}(\varphi) f)(\omega) := \left\{ \int a(\omega, z)(\varphi(z) - \varphi(\omega)) \mu(dz) \right\} f(\omega),$$

whenever the right-hand side exists ($\varphi \in \mathcal{A}$ and $f \in h$). Its formal predual is given by, $(A\psi)(x) = \int_\Omega a(y, x)\psi(y)\mu(dy) - (\int_\Omega a(x, y)\mu(dy))\psi(x)$, whenever the right-hand side makes sense for $\psi \in L^1(\mu)$. We assume that $a : \Omega \times \Omega \to [0, \infty)$ is measurable and $\int a(x, y)\mu(dy)$ is finite for almost all x. This is the obvious generalization of classical semigroups studied by Feller [50] and Kato [78] where μ was chosen to be the counting measure on the set of positive integers.

Let us consider the Hilbert space $h \otimes k$, where $h = L^2(\mu(d\omega))$, $k = L^2(\mu(dz))$ and $\mathcal{K} = h \otimes k \cong L^2(\mu(d\omega) \otimes \mu(dz))$, where $\mu \otimes \mu$ denotes the measure-theoretic product of two copies of μ. Let $\langle \cdot, \cdot \rangle_h$ and $\langle \cdot, \cdot \rangle_{\mathcal{K}}$ denote the inner products in h and $\mathcal{K}$ respectively, whereas $\| \ \|_h$ and $\| \ \|_{\mathcal{K}}$ are the respective norms. Define $\pi : \mathcal{A} \to \mathcal{B}(\mathcal{K})$ by

$$(\pi(\varphi)F)(\omega, z) = \varphi(z)F(\omega, z); \ z, w \in \Omega, \ F \in \mathcal{K}.$$

It is straightforward to verify that π is a normal, $*$-representation. Note here that the representation chosen is *not* $(\hat{\pi}(\phi)F)(\omega, z) = \phi(\omega)F(\omega, z)$. This change in the choice of representation has deeper implication at the level of stochastic dilation as was observed in [99], (see also [93]) where it was shown that this necessitates introduction of a unitary operator-valued number process. Now, define $R : h \to \mathcal{K}$ as follows.

Let Dom(R) be the set of all f in h such that $\int\int a(\omega, z)|f(\omega)|^2 \mu(d\omega)\mu(dz)$ is finite; and $(Rf)(\omega, z) = \sqrt{a(\omega, z)} f(\omega)$ for $f \in$ Dom(R).

Theorem 3.2.19 *The operator R is densely defined and closed; and* Dom(R^*) *contains the set $\mathcal{D}$ of all $G \in \mathcal{K}$ such that $\int \sqrt{a(\cdot, z)} G(\cdot, z)\mu(dz)$ is in $L^2(\mu)$. For $G \in \mathcal{D}$, $(R^*G)(\omega) = \int \sqrt{a(\omega, z)} G(\omega, z)\mu(dz)$; and the set $\hat{\mathcal{D}} \equiv \{f \in$ Dom$(R) : Rf \in \mathcal{D}\}$ is a core for $R^*\pi(\varphi)R$ for any positive $\varphi \in \mathcal{A}$. Moreover, if we put an additional restriction, namely,* $\sup_{\omega \in \Delta} \int_\Delta a(\omega, z)\mu(dz) < \infty$ *for all measurable set Δ having finite μ-measure, then* Dom(R^*) *and $\mathcal{D}$ coincide.*

Proof:
Let us denote by $g(\omega)$ the function $\int a(\omega, z)\mu(dz)$. Clearly, as $\mu\{\omega : g(\omega) = \infty\} = 0$, the linear span of elements of the form $f\chi_{\{\omega : g(\omega) \le n\}}$, where n is any positive number and $f \in L^2(\mu)$ is dense in $L^2(\mu)$, where χ_B denotes indicator of B. But $f\chi_{\{\omega:g(\omega)\le n\}} \in$ Dom(R) for $f \in L^2(\mu), n \ge 1$; which proves that R is densely defined. To see that R is closed, suppose that a sequence f_n in Dom(R) converges (in $|| \ ||_h$) to $f \in L^2(\mu)$ and Rf_n converges in $|| \ ||_\mathcal{K}$ to $\psi \in \mathcal{K}$. Then, we can choose a subsequence $(n_k)_{k=1,2,...}$ such that $f_{n_k} \to f$ a.e. (μ) and $Rf_{n_k} \to \psi$ a.e. $(\mu \otimes \mu)$. But, for all $z \in \Omega$, $(Rf_{n_k})(\omega, z) = \sqrt{a(\omega, z)} f_{n_k}(\omega)$ $\to \sqrt{a(\omega, z)} f(\omega)$ for almost all ω; and hence $\psi(\omega, z) = \sqrt{a(\omega, z)} f(\omega)$ a.e. $(\mu \otimes \mu)$. Since $\psi \in L^2(\mu \otimes \mu)$, it is clear that $f \in$ Dom(R), which proves that R is closed.

We now want to show first that $\mathcal{D} \subseteq$ Dom(R^*). Suppose $G \in \mathcal{D}$. It is a simple observation that, if $\int \sqrt{a(\omega, z)} G(\omega, z)\mu(dz)$ is denoted by $\gamma(\omega)$, then $\gamma \in L^2(\mu)$ by hypothesis and for all $f \in$ Dom(R),

$$\begin{aligned}
\langle f, \gamma \rangle_h &= \int\int \bar{f}(\omega)\sqrt{a(\omega, z)} G(\omega, z)\mu(dz)\mu(d\omega) \\
&= \int\int \bar{f}(\omega)\sqrt{a(\omega, z)} G(\omega, z)\mu(d\omega)\mu(dz) \\
&= \langle Rf, G \rangle_\mathcal{K},
\end{aligned}$$

where the interchange of the order of integration is justified because $\int\int \sqrt{a(\omega, z)} |\bar{f}(\omega)G(\omega, z)|\mu(dz)\mu(d\omega) \le ||Rf||_\mathcal{K} ||G||_\mathcal{K} < \infty$, by Cauchy–Schwarz inequality. This proves $G \in$ Dom(R^*) and $(R^*G)(\omega) = \int \sqrt{a(\omega, z)}$ $G(\omega, z)\mu(dz)$.

To show that $\hat{\mathcal{D}}$ is a core for $R^*\pi(\varphi)R$ for positive $\varphi \in \mathcal{A}$, we consider the semigroup $(L_t)_{t \ge 0}$ where $L_t = e^{-tR^*\pi(\varphi)R}$. It is enough to show that for each t, $L_t\hat{\mathcal{D}} \subseteq \hat{\mathcal{D}}$, which can be verified by straightforward arguments.

Now, assume also that $\sup_{\omega \in \Delta} \int_\Delta a(\omega, z)\mu(dz) < \infty$ for all Δ with finite μ-measure. Let $G \in$ Dom(R^*). Then, there exists $\varphi \in h$ such that

$\langle Rf, G\rangle_{\mathcal{K}} = \langle f, \varphi\rangle_h$ for all $f \in \mathrm{Dom}(R)$. But $\langle Rf, G\rangle_{\mathcal{K}} = \int \bar{f}(\omega)p(\omega)\mu(d\omega)$, as we have already computed (where $p(\omega) = \int \sqrt{a(\omega, z)}G(\omega, z)\mu(dz)$). Thus, for all $f \in \mathrm{Dom}(R)$, we have $\int \bar{f}(\omega)(p(\omega) - \varphi(\omega))\mu(d\omega) = 0$.

Partitioning Ω into disjoint sets of finite μ-measure, say $\{\Omega_n\}_{(n=1,2,\ldots)}$ and choosing f to be $\chi_{\Omega_n \cap \{\omega:\varphi(\omega)-p(\omega)\geq\epsilon\}}$ for $\epsilon > 0$, we can deduce from the above identity that $\mu(\{\omega : \varphi(\omega) - p(\omega) \geq \epsilon;\ \omega \in \Omega_n\}) = 0$. Note that this argument requires $\chi_{\Omega_n \cap \{\omega:\ \varphi(\omega)-p(\omega)\geq\epsilon\}}$ to belong to $\mathrm{Dom}(R)$, which is a consequence of the assumption that $\sup_{\omega\in\Delta} \int_\Delta a(\omega, z)\mu(dz) < \infty$ for all Δ with $\mu(\Delta) < \infty$. Similarly, one obtains that $\mu(\{\omega : \varphi(\omega) - p(\omega) \leq -\epsilon;\ \omega \in \Omega_n\}) = 0$ for all n and positive ϵ. Thus, $\varphi = p$ a.e. (μ), and hence $p \in L^2(\mu)$, proving $G \in \mathcal{D}$. This completes the proof. □

In order to incorporate the present example into the framework of the general theory developed here, one only has to identify the generator G with $-\frac{1}{2}R^*R$, which is a multiplication operator by the measurable function $-\frac{1}{2}\int a(\cdot, z)\mu(dz)$ and hence is affiliated to $\mathcal{A}$. Similarly, $R^*\pi(\varphi)R$ can be seen to be a multiplication operator by the measurable function $\int a(\cdot, z)\varphi(z)\mu(dz)$ and hence is affiliated to $\mathcal{A}$. Thus, for positive $\varphi \in L^\infty(\mu)$ and $f \in \hat{\mathcal{D}}$, it is not difficult to verify that

$$\begin{aligned}
&((R^*\pi(\varphi)R - \tfrac{1}{2}R^*R\varphi - \tfrac{1}{2}\varphi R^*R)f)(\omega)\\
&= ((R^*\pi(\varphi)R - \varphi R^*R)f)(\omega)\\
&= (\textstyle\int a(\omega, z)(\varphi(z) - \varphi(\omega))\mu(dz))f(\omega)\\
&= (\mathcal{L}(\varphi)f)(\omega).
\end{aligned}$$

If we make the assumption that $\sup_{\omega\in\Delta} \int_\Delta a(\omega, z)\mu(dz) < \infty$ for all Δ with finite μ-measure, then the above identity holds for all f in the domain of R^*R, which is the same as the domain of $R^*\pi(\varphi)R - \varphi R^*R$ for any $\varphi \in \mathcal{A}$ in this case.

Example 2

Consider another example where the above theory works on a von Neumann algebra which is a type II_1 factor. Note that such a von Neumann algebra is neither commutative nor of the form $\mathcal{B}(\mathcal{H})$.

We fix some irrational number θ and consider $h = L^2(\mathbb{R})$ and the C^*-algebra $\mathcal{A}_\theta$ generated by the unitaries U and V where $(Uf)(s) = f(s+1)$, $(Vf)(s) = e^{2\pi i s\theta}f(s)$. In this case U and V obey the commutation relation $UV = e^{2\pi i\theta}VU$, It is known ([28]) that the double commutant $\mathcal{A}_\theta''$ is a type II_1 factor in $\mathcal{B}(h)$. We refer the reader to Chapter 9 for some more details about the C^*-algebra $\mathcal{A}_\theta$ (called the *irrational rotation algebra* or the *noncommutative torus*). Let us consider a canonical derivation δ with the domain $\mathcal{D}$ consisting of all polynomials in U and V, and given by, $\delta(U) = U$, $\delta(V) = 0$.

We observe that $\delta(X) = [S, X]$, for $X \in \mathcal{D}$, where $(Sf)(s) = -sf(s)$ for all $f \in L^2(\mathbb{R})$ such that $sf(s)$ is also in $L^2(\mathbb{R})$. We note that S is affiliated to $\mathcal{A}''_\theta$ since for each integer n, $\mathcal{A}_\theta$ contains multiplication by $e^{2\pi i s n\theta} = V^n$ and since θ is irrational, multiplication by $e^{i\alpha s}$ belongs to $\mathcal{A}''_\theta$ for any $\alpha \in \mathbb{R}$. Thus, $S^*\pi(x)S = S^*xS$ is affiliated to $\mathcal{A}''_\theta$ for $x \in \mathcal{A}_\theta$. We now take the formal expression $\mathcal{L}(X) = [S, [S, X]] = S^*XS - \frac{1}{2}S^*SX - \frac{1}{2}XS^*S$, for $X \in \mathcal{D}$, since $S^* = S$. It is now easy to see that $\mathcal{L}(U^m V^n) = -m^2 U^m V^n$.

3.2.2 Structure theorem for a class of strongly continuous quantum dynamical semigroups on $\mathcal{B}(h)$

In this subsection, we shall discuss some of the results obtained by E. B. Davies in [36]. Under certain assumptions, the Lindblad-form was derived there for the generator (to be understood in a suitable sense, taking into account the domain-questions) of a strongly continuous Q.D.S. acting on $\mathcal{B}(h)$. We shall denote by $\mathcal{K}(h)$ the C^*-algebra of compact operators on h. We need a few lemmas before we present the structure theorem of Davies.

Lemma 3.2.20 *Let $(T_{*,t})_{t\geq 0}$ be a C_0-contraction semigroup on $\mathcal{B}_1(h)$ with the generator denoted by W, such that its dual semigroup $(T_t)_{t\geq 0}$ is a strongly continuous (with respect to the ultra-weak topology) Q.D.S. on $\mathcal{B}(h)$, where h is a separable Hilbert space. Assume that (T_t) has a stationary pure state; that is, there is a unit vector Ω of h such that $\langle \Omega, T_t(X)\Omega\rangle = \langle \Omega, X\Omega\rangle$ for all $t > 0$, $X \in \mathcal{B}(h)$. Then there exists a C_0-semigroup $(C_t)_{t\geq 0}$ of contractions on h satisfying $C_t(X\Omega) = T_t(X)\Omega$ for $t \geq 0$, $X \in \mathcal{B}(h)$.*

Proof:
We define C_t as follows. Clearly, the closed subspace spanned by $\mathcal{K}_0 \equiv \{X\Omega : X \in \mathcal{B}(h)\}$ is the whole of h. Define C_t on $\mathcal{K}_0$ by

$$C_t(X\Omega) = (T_t(X))\Omega.$$

Clearly, C_t keeps $\mathcal{K}_0$ invariant. It can be seen that C_t is well-defined and is contractive on $\mathcal{K}_0$, so extends to h as a contraction. Indeed, T_t is a contractive and completely positive map, hence $T_t(X)^*T_t(X) \leq T_t(X^*X)$. Thus, $\|T_t(X)\Omega\|^2 = \langle \Omega, T_t(X)^*T_t(X)\Omega\rangle \leq \langle \Omega, T_t(X^*X)\Omega\rangle = \langle \Omega, X^*X\Omega\rangle = \|X\Omega\|^2$, which shows that C_t is well-defined and contractive on $\mathcal{K}_0$. This defines C_t as a contraction on the whole of h. Furthermore, $C_{t+s}(X\Omega) = (C_t \circ C_s)(X\Omega)$ follows clearly from the definition of C_t and the semigroup property of (T_t). The weak-continuity (which is equivalent to strong continuity for semigroups) of (C_t) is a consequence of that of (T_t). This completes the proof of the lemma. □

Let Y denote the generator of the C_0 semigroup (C_t), so that Y^* is the generator of (C_t^*). Define the following C_0-semigroup on $\mathcal{B}_1(h)$:

$$S_{*,t}(\rho) := C_t^* \rho C_t,$$

for $t \geq 0$.

Lemma 3.2.21 *The vector Ω belongs to the domain of the operators Y and Y^* with $Y\Omega = Y^*\Omega = 0$.*

Proof:
For $t > 0$, $\langle \Omega, C_t \Omega \rangle = \langle \Omega, T_t(1)\Omega \rangle = \langle \Omega, \Omega \rangle = 1$. As C_t is a contraction, it follows that $C_t \Omega = \Omega$ for all $t > 0$. So, $\frac{d}{dt} C_t \Omega$ exists and is equal to 0, that is, Ω belongs to the domain of Y with $Y\Omega = 0$. A similar argument (replacing C_t by C_t^*) proves that Ω belongs to the domain of Y^* with $Y^*\Omega = 0$. □

We need to investigate the special case of uniformly continuous semigroup at this point.

Lemma 3.2.22 *If $(T_{*,t})_{t \geq 0}$ in Lemma 3.2.20 is uniformly continuous Q.D.S. then the operators Y and Y^* are bounded. Furthermore,*

$$\| T_{*,t}(\phi(\rho)) - S_{*,t}(\phi(\rho)) \| \leq 4t \|\rho\|$$

for all $\rho \in \mathcal{B}_1(h)$, where $\phi(\rho) = (Y^* - 1)^{-1} \rho (Y - 1)^{-1}$. Moreover, for positive $\rho \in \mathcal{B}_1(h)$, we have $T_{*,t}(\rho) \geq S_{*,t}(\rho) \geq 0$.

Proof:
It is obvious that the uniform continuity of $T_{*,t}$ implies the same of the dual semigroup T_t. Denote the bounded generator of T_t by $\mathcal{L}$. Clearly, for $\xi, \eta \in h$, we have $\langle \xi, \mathcal{L}(X)\eta \rangle = \mathrm{tr}(XW(|\eta >< \xi|))$. Since the operator Y is closed, it is sufficient, in order to prove the boundedness of Y (and hence of Y^*), to show that the domain of Y is the whole of h. For any $\xi \in h$, we write it as $\xi = X\Omega$, where $X = |\xi >< \Omega|$. Thus $C_t(\xi) = T_t(X)\Omega$. Since the domain of $\mathcal{L}$ is the whole of $\mathcal{B}(h)$ by our assumption, it follows that $\lim_{t \to 0+}(C_t(\xi) - \xi)/t = \lim_{t \to 0+}((T_t(X) - X)/t)\Omega$ exists and is equal to $\mathcal{L}(X)\Omega$. Thus, $Y(X\Omega) = \mathcal{L}(X)\Omega$,that is, $Y(\xi) = \mathcal{L}(|\xi >< \Omega|)\Omega$, so that Y is defined everywhere Therefore Y is bounded.

We now claim that $\mathcal{L}$ can be written as

$$\mathcal{L}(X) = \sum_{n=1}^{\infty} R_n^* X R_n + YX + XY^*$$

for some countable collection of operators $R_n \in \mathcal{B}(h)$ such that $R_n \Omega = 0$ for all n and the infinite sum in the above expression converges strongly. Define a map $Q : \mathcal{B}(h) \to \mathcal{B}(h)$ by $Q(X) = YX + XY^*$. We shall prove that $\mathcal{L} - Q$ is completely positive. To this end, recall the trilinear map D given by

$$D(x, y, z) = \mathcal{L}(xyz) + x\mathcal{L}(y)z - \mathcal{L}(xy)z - x\mathcal{L}(yz)$$

for $x, y, z \in \mathcal{B}(h)$, introduced in the proof of Theorem 3.1.8, and note that D has been shown to be positive definite in a suitable sense. From this, it follows that for $X_0, X_1 \in \mathcal{B}(h)$,

$$\mathcal{L}(X_0^* X_1^* X_1 X_0) + X_0^* \mathcal{L}(X_1^* X_1) X_0 \geq \mathcal{L}(X_0^* X_1^* X_1) X_0 + X_0^* \mathcal{L}(X_1^* X_1 X_0).$$

Since $\langle \Omega, \mathcal{L}(X)\Omega \rangle = \frac{d}{dt}|_0 \langle \Omega, T_t(X)\Omega \rangle = \frac{d}{dt}|_0 \langle \Omega, X\Omega \rangle = 0$, we have that

$$\begin{aligned}
&\langle X_0\Omega, \mathcal{L}(X_1^* X_1) X_0 \Omega \rangle \\
&\geq \langle \Omega, \mathcal{L}(X_0^* X_1^* X_1) X_0 \Omega \rangle + \langle X_0 \Omega, \mathcal{L}(X_1^* X_1 X_0)\Omega \rangle \\
&= \langle Y(X_1^* X_1 X_0 \Omega), X_0 \Omega \rangle + \langle X_0 \Omega, Y(X_1^* X_1 X_0 \Omega) \rangle \\
&= \langle X_0 \Omega, (Q(X_1^* X_1)) X_0 \Omega \rangle.
\end{aligned}$$

As the set $\{X_0\Omega; X_0 \in \mathcal{B}(h)\}$ is dense in h, we get the operator inequality $\mathcal{L}(X_1^* X_1) \geq Q(X_1^* X_1)$, which proves that $\mathcal{L} - Q$ is a positive map. By similar arguments, replacing h by $h \otimes \mathbb{C}^N$ (N positive integer), and $\mathcal{L}$ and Q by their lifting to $\mathcal{B}(h \otimes \mathbb{C}^N)$, we can establish the complete positivity of the map $\mathcal{L} - Q$. The normality of $\mathcal{L} - Q$ is an immediate consequence of the fact that both $\mathcal{L}$ and Q are clearly so.

So, by the Corollary 2.2.10, there are $R_n, n = 1, 2, \ldots$ such that $\mathcal{L}(X) - Q(X) = \sum_n R_n^* X R_n$. Note that $\langle \Omega, \mathcal{L}(1)\Omega \rangle = 0$ and $\langle \Omega, Q(1)\Omega \rangle = \langle Y\Omega, \Omega \rangle + \langle \Omega, Y\Omega \rangle = 0$. Thus, $\sum_n \langle \Omega, R_n^* R_n \Omega \rangle = 0$, implying $R_n \Omega = 0$ for all n. This completes the proof of the claim.

From the relation $\langle \xi, \mathcal{L}(X)\eta \rangle = \mathrm{tr}(XW(|\eta><\xi|))$ for $\xi, \eta \in h$, it can be seen that W has the following form:

$$W(\rho) = \sum_n R_n \rho R_n^* + \rho Y + Y^* \rho.$$

Now, since $T_t(1)$ is contractive positive operator for each t, $T_t(1) - 1 \leq 0$, hence $\mathcal{L}(1) \leq 0$ as operator. This implies the operator inequality

$$0 \leq \sum_n R_n^* R_n \leq -(Y + Y^*),$$

and hence the norm inequality

$$\left\| (Y-1)^{-1} \left(\sum_n R_n^* R_n \right) (Y^* - 1)^{-1} \right\| \leq \| (Y-1)^{-1} (Y + Y^*)(Y^* - 1)^{-1} \|.$$

Moreover, we have

$$\|(Y-1)^{-1}\| \leq \int_0^\infty \|e^{-t}C_t\|dt \leq 1, \text{ and } \|(Y-1)^{-1}Y\| \leq \|1+(Y-1)^{-1}Y\| \leq 2.$$

It follows that $\|(Y-1)^{-1}(Y+Y^*)(Y^*-1)^{-1}\| \leq 4$. Consider the linear map $J_1 = J \circ \phi$ on $\mathcal{B}_1(h)$ where J is given by $J(\rho) = \sum_n R_n \rho R_n^*$, and note that J_1 is positive in the sense that $J_1(\rho) \geq 0$ for $\rho \geq 0$. Furthermore, its dual $J_1^* : \mathcal{B}(h) \to \mathcal{B}(h)$ is nothing but the completely positive map $X \mapsto (Y-1)^{-1}\left(\sum_n R_n^* X R_n\right)(Y^*-1)^{-1}$. So,

$$\|J_1\| = \|J_1^*\| = \|J_1^*(1)\| = \left\|(Y-1)^{-1}\left(\sum_n R_n^* R_n\right)(Y^*-1)^{-1}\right\| \leq 4.$$

Note that S_t and ϕ commute for each t, since C_t^* commutes with its generator Y^* and C_t with Y. Now, using the perturbation expansion

$$T_{*,t}(\rho) = S_{*,t}(\rho) + \int_0^t T_{*,t-s} \circ J \circ S_s(\rho)ds,$$

we see that $T_{*,t}(\rho) \geq S_{*,t}(\rho)$ for positive ρ, and furthermore,

$$\begin{aligned}
&\|T_{*,t}(\phi(\rho)) - S_{*,t}(\phi(\rho))\|_1 \\
&\leq \int_0^t \|T_{*,t-s} \circ J \circ S_{*,s}(\phi(\rho))\|_1 ds \\
&= \int_0^t \|T_{*,t-s} \circ J_1 \circ S_{*,s}(\rho))\|_1 ds \leq 4t\|\rho\|_1,
\end{aligned}$$

where we have used the facts that the maps ϕ and S_s commute, $T_{*,s}$, $S_{*,s}$ are contractive and $\|J_1\| \leq 4$. □

Before we state and prove the main theorem, we need one more lemma, which can be viewed as a 'predual' version of the Corollary 2.2.10.

Lemma 3.2.23 *Let $\Psi : \mathcal{B}_1(h) \to \mathcal{B}_1(h)$ be a linear map which is positive in the sense that $\Psi(\rho) \geq 0$ whenever $\rho \geq 0$. Then the dual of Ψ, say Ψ^*, which is a linear map from $\mathcal{B}(h)$ to $\mathcal{B}(h)$, is positive and normal. If Ψ is completely positive in the sense that $(\Psi \otimes I) : \mathcal{B}_1(h) \otimes M_n \equiv \mathcal{B}_1(h \otimes \mathbb{C}^n) \to \mathcal{B}_1(h \otimes \mathbb{C}^n)$ is positive for all $n = 1, 2, \ldots$, Ψ^* is completely positive and normal; and we can find $C_1, C_2, \ldots$ in $\mathcal{B}(h)$ such that $\Psi(\rho) = \sum_n C_n \rho C_n^*$, where the sum converges weakly.*

Proof:
For a positive $X \in \mathcal{B}(h)$, $\xi \in h$, we have $\langle \xi, \Psi^*(X)\xi \rangle = \mathrm{tr}(\Psi^*(X)|\xi><\xi|) = \mathrm{tr}(X\Psi(|\xi><\xi|)) \geq 0$, since $\Psi(|\xi><\xi|)$ is a positive operator by the assumption of positivity of Ψ. This proves that Ψ^* is positive. Similarly

the complete positivity of Ψ^* can be shown in case Ψ is completely positive. To show the normality of Ψ^*, we take a net X_α of positive operators such that $X_\alpha \uparrow X$ for some X. To show that $\Psi^*(X_\alpha) \to \Psi^*(X)$ weakly, it is enough to note that for $\xi, \eta \in h$, $\langle \xi, \Psi^*(X_\alpha)\eta \rangle = \mathrm{tr}(X_\alpha \Psi(|\xi >< \eta|)) \to \mathrm{tr}(X\Psi(|\xi >< \eta|))$, as $\Psi(|\xi >< \eta|) \in \mathcal{B}_1(h)$ and $X_\alpha \to X$ weakly.

Applying the Corollary 2.2.10 to the normal completely positive map Ψ^*, we can find $C_n, n = 1, 2, \ldots$, such that $\Psi^*(X) = \sum_n C_n^* X C_n, X \in \mathcal{B}(h)$, from which it follows that $\Psi(\rho) = \sum_n C_n \rho C_n^*$ for $\rho \in \mathcal{B}_1(h)$. □

We are now in a position to prove the main theorem.

Theorem 3.2.24 *Let $(T_{*,t})_{t \geq 0}$ be a C_0-contraction semigroup on $\mathcal{B}_1(h)$, such that its dual semigroup $(T_t)_{t \geq 0}$ is a strongly continuous (with respect to the ultra-weak topology) Q.D.S. on $\mathcal{B}(h)$, where h is a separable Hilbert space. Assume that (T_t) satisfies the following conditions.*
(i) There is a unit vector Ω of h such that $\langle \Omega, T_t(X)\Omega \rangle = \langle \Omega, X\Omega \rangle$ for all $t > 0$, $X \in \mathcal{B}(h)$; and (ii) $T_t(\mathcal{K}(h)) \subseteq \mathcal{K}(h)$ for all $t > 0$, where $\mathcal{K}(h)$ denotes the set of compact operators on h.
Then there is a dense subspace $\mathcal{D}$ containing Ω, linear (densely defined, closable) operators B_n, Y, with $\mathcal{D} \subseteq \mathrm{Dom}(B_n), \mathrm{Dom}(Y)$, for $n = 1, 2, \ldots$, satisfying

$$Y\Omega = Y^*\Omega = B_n\Omega = 0,$$

$$\langle Y\xi, \xi \rangle + \langle \xi, Y\xi \rangle + \sum_n \|B_n\xi\|^2 \leq 0 \quad \text{for all } \xi \in \mathrm{Dom}(Y),$$

and the generator of $(T_{*,t})$, say W, is of the form

$$W(\rho) = Y^*\rho + \rho Y + \sum_n B_n \rho B_n^*,$$

for all ρ in a suitable dense domain μ, and where the sum in the above formula converges weakly.

Proof:
We retain the notation of the Lemma 3.2.20 and the discussion following it. Thus, Y is taken to be the generator of the semigroup (C_t) as before.

By *(ii)*, it is possible to view the semigroup $T_{*,t}$ on the Banach space $\mathcal{B}_1(h) = (\mathcal{K}(h))^*$ as the dual of the restriction of T_t to $\mathcal{K}(h)$. Furthermore, each $T_{*,t}$ is continuous in the weak-$*$ topology of $(\mathcal{K}(h))^*$, as the duality between $\mathcal{B}_1(h)$ and $\mathcal{K}(h)$ is given by the restriction of the duality between $\mathcal{B}_1(h)$ and $\mathcal{B}(h)$, and $T_t = T_{*,t}^*$ is weak* continuous as a map on $\mathcal{B}(h)$. Thus, by the Proposition 2.3.2, the domain of the generator of $(T_{*,t})$, say W, is given by

$$\mathrm{Dom}(W) = \left\{ \rho \in \mathcal{B}_1(h) : \mathrm{liminf}_{t \to 0+} \frac{\|T_{*,t}(\rho) - \rho\|_1}{t} < \infty \right\}.$$

Define a family of semigroups $(T^{\epsilon}_{*,t})$ with bounded generators W_{ϵ} indexed by $\epsilon > 0$ where $W_{\epsilon} = (T_{*,\epsilon} - I)/\epsilon$. Note that $W_{\epsilon}(\rho) \to W\rho$ as $\epsilon \to 0+$, for all ρ in the domain of W, so in particular $(W_{\epsilon} - W)(I - W)^{-1}(\rho) \to 0$ as $\epsilon \to 0+$ for all $\rho \in \mathcal{B}_1(h)$. Furthermore, since $T_{*,\epsilon}$ is contractive operator, we have $\|\exp(tW_{\epsilon})\| \leq e^{-\frac{t}{\epsilon}}\|\exp(\frac{t}{\epsilon}T_{*,\epsilon})\| \leq e^{-\frac{t}{\epsilon}(1-\|T_{*,\epsilon}\|)} \leq 1$. Thus, for each $\epsilon > 0$, $(T^{\epsilon}_{*,t})_{t\geq 0}$ is a C_0 semigroup of contractions on $\mathcal{B}_1(h)$. Moreover, for any $\rho \in \mathcal{B}_1(h)$ and $\lambda > 1$,

$$
\begin{aligned}
&\|(\lambda - W_{\epsilon})^{-1}(\rho) - (\lambda - W)^{-1}(\rho)\|_1 \\
&= \|(\lambda - W_{\epsilon})^{-1}(W_{\epsilon} - W)(\lambda - W)^{-1}(\rho)\|_1 \\
&\leq \frac{1}{\lambda}\|(W_{\epsilon} - W)(\lambda - W)^{-1}(\rho)\|_1 \\
&\to 0 \text{ as } \epsilon \to 0+ .
\end{aligned}
$$

This shows, by Proposition 2.3.3 and the related discussion in Chapter 2, that $T^{\epsilon}_{*,t}\rho \to T_{*,t}\rho$ in the norm of $\mathcal{B}_1(h)$ for all $t \geq 0$ and $\rho \in \mathcal{B}_1(h)$,as $\epsilon \to 0+$.

It is clear that the dual of $T^{\epsilon}_{*,t}$, say T^{ϵ}_t, is nothing but the uniformly continuous Q.D.S. with the generator $\mathcal{L}_{\epsilon} := (T_{\epsilon} - I)/\epsilon$. From the series expansion $T^{\epsilon}_t = e^{-t\epsilon^{-1}}\sum_{n\geq 0}(n!)^{-1}(\frac{t}{\epsilon})^n T_{\epsilon n}$ it follows that $\langle\Omega, T^{\epsilon}_t(X)\Omega\rangle = \langle\Omega, X\Omega\rangle$ for all $t \geq 0$ and $X \in \mathcal{B}(h)$. Thus, C^{ϵ}_t defined by $C^{\epsilon}_t(X\Omega) = T^{\epsilon}_t(X)\Omega$ is a C_0 semigroup of contractions, and in fact, $C^{\epsilon}_t = \exp(-\frac{t}{\epsilon}(1 - C_{\epsilon}))$ Set $S^{\epsilon}_{*,t}(\rho) = (C^{\epsilon}_t)^*\rho C^{\epsilon}_t$. it can be shown, by very similar arguments as in the case of $T^{\epsilon}_{*,t}$, that $S^{\epsilon}_{*,t}(\rho) \to S_{*,t}(\rho)$ as $\epsilon \to 0+$, for all $t \geq 0$ and $\rho \in \mathcal{B}_1(h)$.

Since by the Lemma 3.2.22 $T^{\epsilon}_{*,t}(\rho) \geq S^{\epsilon}_{*,t}(\rho)$ for $\rho \geq 0$; and $\|T^{\epsilon}_{*,t}(\rho) - S^{\epsilon}_{*,t}(\rho)\|_1 \leq 4\|\rho\|_1$ for all $\rho \in \mathcal{B}_1(h)$ and for all $\epsilon > 0$, it follows that $T_{*,t}(\rho) \geq S_{*,t}(\rho)$ for $\rho \geq 0$; and $\|T_{*,t}(\rho) - S_{*,t}(\rho)\|_1 \leq 4\|\rho\|_1$ for all $\rho \in \mathcal{B}_1(h)$.

For ρ in the domain of the generator of $S_{*,t}$, say W_S, we have $\text{liminf}_{t\to 0+}(\|S_{*,t}(\rho) - \rho\|_1)/t < \infty$, hence $\text{liminf}_{t\to 0+}(\|T_{*,t}(\rho) - \rho\|_1)/t$ is finite also. Thus, the domain of W_S is contained in the domain of W. Now, it is clear that the range of ϕ, that is, $\{(Y^* - 1)^{-1}\rho(Y - 1)^{-1} : \rho \in \mathcal{B}_1(h)\}$ is contained in the domain of the generator W_S of $S_{*,t}$ (hence it is in $\text{Dom}(W)$ as well), and $W_S(\phi(\rho)) = Y^*(Y^* - 1)^{-1}\rho(Y - 1)^{-1} + (Y^* - 1)^{-1}\rho(Y - 1)^{-1}Y$. We take $\mathcal{D}$ to be the dense subspace $\text{Ran}(Y^* - 1)^{-1} = \text{Dom}(Y^*)$, and μ to be $\text{Ran}(\phi)$. Clearly, given any rank one operator $|\xi >< \eta|$, we can choose $\xi_n, \eta_n, n = 1, 2, \cdots$ in $\mathcal{D}$ such that $\xi_n \to \xi$, $\eta_n \to \eta$ as $n \to \infty$; so that $\rho_n := |\xi_n >< \eta_n|$ converges to $|\xi >< \eta|$ in the norm of $\mathcal{B}_1(h)$. Writing $\xi_n = (Y^* - 1)^{-1}\xi'_n$, $\eta_n = (Y^* - 1)^{-1}\eta'_n$, for some ξ'_n, η'_n in h, it is not difficult to see that $\rho_n = \phi(|\xi'_n >< \eta'_n|) \in \mathcal{C}$. This proves that μ is dense in $\mathcal{B}_1(h)$.

We now define a linear map $\Psi : \mathcal{B}_1(h) \to \mathcal{B}_1(h)$ by $\Psi(\rho) := W(\phi(\rho)) - W_S(\phi(\rho))$. We claim that Ψ is completely positive in the sense of the

Lemma 3.2.23. Let us prove the positivity only, as the complete positivity can be deduced by very similar arguments, replacing h by $h \otimes \mathbb{C}^n$. Note that $\phi(\rho) \geq 0$ for $\rho \geq 0$, and furthermore we have

$$\begin{aligned}
&\Psi(\rho) \\
&= \lim_{t\to 0+} \frac{T_{*,t}(\phi(\rho)) - \phi(\rho)}{t} - \lim_{t\to 0+} \frac{S_{*,t}(\phi(\rho)) - \phi(\rho)}{t} \\
&= \lim_{t\to 0+} \frac{T_{*,t}(\phi(\rho)) - S_{*,t}(\phi(\rho))}{t} \\
&\geq 0,
\end{aligned}$$

since for each t, $T_{*,t} - S_{*,t}$ is a positive map as noted before.

By Lemma 3.2.23, we can find $D_n, n = 1, 2, \ldots$ in $\mathcal{B}(h)$ such that $\Psi(\rho) = \sum_n D_n \rho D_n^*$, where the sum converges weakly. We take $B_n := D_n(Y^* - 1)$ (so that the domain of B_n is same as that of Y^*), and observe that for $\rho \in \mathcal{M}$, say of the form $\rho = (Y^* - 1)^{-1}\eta(Y - 1)^{-1}$ for some $\eta \in \mathcal{B}_1(h)$,

$$\begin{aligned}
&W(\rho) \\
&= (W - W_S)(\phi(\eta)) + W_S(\phi(\eta)) \\
&= \sum_n D_n \eta D_n^* + Y^*\phi(\eta) + \phi(\eta)Y \\
&= \sum_n B_n(Y^* - 1)^{-1}\eta(Y - 1)^{-1}B_n^* + Y^*\rho + \rho Y \\
&= \sum_n B_n \rho B_n^* + Y^*\rho + \rho Y.
\end{aligned}$$

□

3.2.3 Structure of generator of symmetric quantum dynamical semigroups

In this subsection we shall study Q.D.S. on a von Neumann algebra which are symmetric with respect to a normal faithful semifinite trace on the von Neumann algebra and introduce the concept of noncommutative Dirichlet forms. But before we do that a brief survey of the classical Dirichlet form will be useful. Let Ω be an open region in $\mathbb{R}^n$ with connected components $\Omega_1, \ldots$. The *Dirichlet Laplacian* Δ_D is the unique self-adjoint operator in $L^2(\Omega, dx)$ associated with the *classical Dirichlet form* $\mathcal{E}$, which is the closure of the quadratic form

$$\mathcal{E}_0(f, g) = \int_\Omega \overline{\nabla f}(x)\nabla g(x)dx \equiv \sum_{j=1}^n \int_\Omega \frac{\partial \bar{f}(x)}{\partial x_j}\frac{\partial g(x)}{\partial x_j}dx$$

for $f, g \in C_0^\infty(\Omega)$. From the definition, it is evident that the $\mathrm{Dom}(\mathcal{E}) = H_0^1(\Omega)$, the completion of $C_0^\infty(\Omega)$ in the Sobolev norm of $H^1(\Omega)$. Also, $-\Delta_D$

is precisely the Friedrichs extension and $-\Delta_D = A^*A$, where A is the closure of the gradient operator on $C_0^\infty(\Omega)$. If we define the maps J and $P : L^2(\Omega) \to L^2(\Omega)$ by setting $(Jf)(x) = \bar{f}(x)$ and $(Pf)(x) = \mathrm{Min}(1, \mathrm{Max}(0, f(x)))$, and if we set $\mathcal{E}(f) = \mathcal{E}(f, f)$ for $f \in \mathrm{Dom}(\mathcal{E})$, then $Jf, Pf \in \mathrm{Dom}(\mathcal{E})$, $\mathcal{E}(Jf) = \mathcal{E}(f)$, and $\mathcal{E}(Pf) \leq \mathcal{E}(f)$.

The Dirichlet Laplacian has many more interesting properties. If Ω has 'nice' boundary, $-\Delta_D$ has a core containing of functions, which are C^∞ upto the boundary, and vanish on the boundary, and it is this property that associates this Laplacian with classical Dirichlet boundary value problems of potential theory (see for example [110] for details on this). Furthermore, the properties of the maps J and P with respect to the associated closed Dirichlet form has far-reaching implications in the theory of probability, namely, (a) such a form can be extended to the cases where the Lebesgue measure in $L^2(\Omega)$ is replaced by a larger class of measures μ and then one gets an associated self-adjoint operator H_μ in $L^2(\Omega, \mu)$; (b) with every closed Dirichlet from $\mathcal{E}$ on $L^2(\Omega, \mu)$ satisfying the properties of the maps J and P described earlier, the associated self-adjoint operator $H_\mathcal{E}$ generates a Markov semigroup in $L^2(\Omega, \mu)$: $\{e^{-tH_\mathcal{E}}\}_{t\geq 0}$ is a contractive C_0-semigroup such that for $f \in L^2(\Omega, \mu)$ with $0 \leq f \leq 1$, $0 \leq e^{-tH_\mathcal{E}} f \leq 1$ for all $t \geq 0$. For details on these aspects, the reader is referred to [54] and it is property (b) that is used in the rest of this subsection for the definition of a noncommutative Dirichlet form.

Let us now introduce the concept of noncommutative Dirichlet forms and discuss elements of the theory of such forms. Let $\mathcal{A}$ be a C^*-algebra and τ be a densely defined, faithful, lower semicontinuous and semifinite trace on $\mathcal{A}$. Let $\overline{\mathcal{A}}$ be the von Neumann algebra $\pi_\tau(\mathcal{A})''$ in $\mathcal{B}(L^2(\mathcal{A}, \tau))$, where π_τ denotes the GNS representation for τ. Clearly, τ extends as a normal faithful semifinite trace on $\overline{\mathcal{A}}$. Let $\mathcal{A}_\tau$ be the dense subset of $\mathcal{A}$ (which can be shown to be a dense $*$-ideal by using Cauchy–Schwarz inequality) consisting of elements a such that $\tau(a^*a) < \infty$. Consider the antilinear isometry J on the Hilbert space $L^2 \equiv L^2(\mathcal{A}, \tau)$ which is given by $J(a) = a^*$ for $a \in \mathcal{A}_\tau$. We denote the subspace of elements a in $L^2(\mathcal{A}, \tau)$ satisfying $J(a) = a$ by L_h^2. Such elements will be called *hermitian* or *real* elements. Furthermore, let P be the projection from $L^2(\mathcal{A}, \tau)$ onto the subspace obtained by taking the L^2-closure of $\{a \in \overline{\mathcal{A}} : 0 \leq a \leq 1\}$. Now, we are ready to define a Dirichlet form.

Definition 3.2.25 A *Dirichlet form* on $L^2(\mathcal{A}, \tau)$ is a closed, densely defined, positive quadratic form $(\mathcal{E}, \mathrm{Dom}(\mathcal{E}))$ where $\mathrm{Dom}(\mathcal{E})$ denotes the domain of the form, and the following conditions are satisfied:

(i) $J(a) \in \mathrm{Dom}(\mathcal{E})$ for all $a \in \mathrm{Dom}(\mathcal{E})$ with $\mathcal{E}(J(a)) = \mathcal{E}(a)$; and
(ii) $Pa \in \mathrm{Dom}(\mathcal{E})$ with $\mathcal{E}(Pa) \leq \mathcal{E}(a)$ for all $a \in \mathrm{Dom}(\mathcal{E}) \cap L_h^2$.

For every positive integer n, the quadratic form $(\mathcal{E}_n, \mathrm{Dom}(\mathcal{E}_n))$ is defined as follows:

$$\mathrm{Dom}(\mathcal{E}_n) := M_n(\mathrm{Dom}(\mathcal{E})) \equiv \mathrm{Dom}(\mathcal{E}) \otimes_{\mathrm{alg}} M_n(\mathbb{C}) \subseteq L^2(\mathcal{A} \otimes M_n, \tau \otimes \mathrm{tr}_n),$$

where tr_n is the normalized trace on the matrix algebra $M_n = M_n(\mathbb{C})$), and

$$\mathcal{E}_n(((a_{ij}))) := \sum_{ij=1}^{n} \mathcal{E}(a_{ij}) \text{ for } ((a_{ij}))_{ij=1}^n \in M_n(\mathrm{Dom}(\mathcal{E})).$$

Definition 3.2.26 A Dirichlet form $(\mathcal{E}, \mathrm{Dom}(\mathcal{E}))$ is said to be *completely Dirichlet* if for every $n \geq 1$, $(\mathcal{E}_n, \mathrm{Dom}(\mathcal{E}_n))$ is a Dirichlet form. A completely Dirichlet form is called a *C^*-Dirichlet form* if $\mathcal{A} \cap \mathrm{Dom}(\mathcal{E})$ is dense (in the C^*-norm) in $\mathcal{A}$ and is also a form-core for $(\mathcal{E}, \mathrm{Dom}(\mathcal{E}))$.

The following result, which we quote without proof, is due to Davies and Lindsay [38].

Proposition 3.2.27 *If $(\mathcal{E}, \mathrm{Dom}(\mathcal{E}))$ is a completely Dirichlet form on $L^2(\mathcal{A}, \tau)$ then $\overline{\mathcal{A}} \cap \mathrm{Dom}(\mathcal{E})$ is a $*$-subalgebra of $\overline{\mathcal{A}}$. Furthermore, if $(\mathcal{E}, \mathrm{Dom}(\mathcal{E}))$ is a C^*-Dirichlet form, then $\mathcal{A} \cap \mathrm{Dom}(\mathcal{E})$ is a norm-dense $*$-subalgebra of $\mathcal{A}$, to be called the 'Dirichlet algebra' associated with the form $\mathcal{E}$.*

Let Ψ be a contractive CP map on $\mathcal{A}$, which is also *symmetric*, that is, $\tau(\Psi(a)b) = \tau(a\Psi(b))$ for all $a, b \geq 0$ in $\mathcal{A}$. Let us denote $L^p(\mathcal{A}, \tau)$ by simply L^p, and the L^p-norm of an element a by $\|a\|_p$. We also denote by L^∞ the universal enveloping von Neumann algebra of $\mathcal{A}$ and denote by $\|\cdot\|_\infty$ the norm on this von Neumann algebra, which extends the C^*-norm on $\mathcal{A}$. It follows from Theorem 2.2.9 and Proposition 2.1.10 that Ψ can be extended as a normal, contractive, CP map on L^∞. For $a \in \mathcal{A} \cap L^2$, we have by using the symmetry of Ψ and Corollary 2.2.11 that

$$\begin{aligned} \tau(\Psi(a)^*\Psi(a)) &\leq \tau(\Psi(a^*a)) \\ = \tau(a^*a\Psi(1)) &= \tau(a\Psi(1)a^*) \leq \tau(aa^*) = \tau(a^*a). \end{aligned}$$

Thus, Ψ maps $\mathcal{A} \cap L^2$ (which is dense in L^2 in the L^2-norm) into itself and is a contraction in the norm of L^2, hence extends to L^2 as a contraction. We denote this L^2-extension also by Ψ.

Lemma 3.2.28 *Let $(T_t)_{t\geq 0}$ be a Q.D.S. on $\mathcal{A}$, such that each T_t is symmetric with respect to τ. Then the L^2-extensions of T_t, as discussed above, form a C_0-semigroup of positive contractions on the Hilbert space L^2.*

Proof:
It is clear that the L^2-extensions form a semigroup, so it suffices to prove that T_t is a positive map on L^2 and $\|(T_t - I)u\|_2 \to 0$ as $t \to 0+$ for $u \in L^2$. Now, for $a \in \mathcal{A} \cap L^2$, we have

$$\begin{aligned} &\tau(T_t(a)^* a) \\ &= \tau(T_{\frac{t}{2}}(a)^* T_{\frac{t}{2}}(a)) \\ &\geq 0, \end{aligned}$$

where we have used the fact that $T_{\frac{t}{2}}$ is symmetric. This shows the positivity of T_t, since $\mathcal{A} \cap L^2$ is dense in L^2.

To complete the proof of the lemma, we need to show $\lim_{t \to 0+} \|(T_t - I)a\|_2 = 0$ for $a \in \mathcal{A} \cap L^1$, as T_t is L^2-contractive and $\mathcal{A} \cap L^1$ is dense in L^2. Fix $a \in \mathcal{A} \cap L^1$. It is straightforward to see that $|\tau(T_t(a)^* a) - \tau(a^* a)| \leq \|T_t(a) - a\|_\infty \|a\|_1 \to 0$ as $t \to 0+$. Thus, we have

$$\begin{aligned} &\|(T_t a - a)\|_2^2 \\ &= \tau(T_t(a^*) T_t(a)) + \tau(a^* a) - 2\tau(T_t(a^*) a) \\ &= \tau(T_{2t}(a^*) a) + \tau(a^* a) - 2\tau(T_t(a^*) a) \\ &\to 0 \text{ as } t \to 0+. \end{aligned}$$

□

The following well-known result (see [10] and [32] for details) describes how a canonical Dirichlet form is obtained from a symmetric Q.D.S.

Proposition 3.2.29 *Let $(T_t)_{t \geq 0}$ be a Q.D.S. on $\mathcal{A}$ with the generator $\mathcal{L}$. Assume furthermore that T_t is symmetric (with respect to τ) in the sense discussed before. Let us view T_t as a C_0-semigroup of positive contractions on $L^2(\mathcal{A}, \tau)$ as in the Lemma 3.2.28, and let $\mathcal{L}_2$ be the negative operator which generates this semigroup of contractions. Consider the following quadratic form $\mathcal{E}$ on $L^2(\mathcal{A}, \tau)$:*

$$\mathrm{Dom}(\mathcal{E}) = \mathrm{Dom}((-\mathcal{L}_2)^{\frac{1}{2}}), \quad \mathcal{E}(a) = \|(-\mathcal{L}_2)^{\frac{1}{2}}(a)\|_2^2, \ a \in \mathrm{Dom}(\mathcal{E}).$$

Then $(\mathcal{E}, \mathrm{Dom}(\mathcal{E}))$ is a C^*-Dirichlet form. Conversely, any C^*-Dirichlet form on $\mathcal{A}$ can be obtained in this way from some symmetric Q.D.S.

We call $\mathcal{E}$ constructed from the Q.D.S. (T_t) as in the above proposition the Dirichlet form associated with (T_t). The following result, most of which is contained in [32], provides all the necessary ingredients for obtaining a Christensen–Evans type structure for the generator of (T_t).

Theorem 3.2.30 *Let $(\mathcal{E}, \mathrm{Dom}(\mathcal{E}))$ be the C^*-Dirichlet form associated with a symmetric Q.D.S. (T_t) on $\mathcal{A}$. Let $\mathcal{B} = \mathcal{A} \cap \mathrm{Dom}(\mathcal{E})$ be the associated Dirichlet algebra. Assume furthermore that there exists a norm-dense $*$-subalgebra $\mathcal{A}_0 \subseteq \mathrm{Dom}(\mathcal{L}) \cap \mathrm{Dom}(\mathcal{L}_2)$ of $\mathcal{A}$ which is a core for the norm-generator $\mathcal{L}$ of (T_t) on one hand, and a form-core for $\mathcal{E}$ on the other. Suppose also that $\mathcal{L}(\mathcal{A}_0) \subseteq \mathcal{A}_0$. Then the following conditions hold.*

(i) There is a canonical Hilbert space $\mathcal{K}$ equipped with an $\mathcal{A}$-$\mathcal{A}$ bimodule structure, in which the right action is denoted by $(a, \xi) \mapsto \xi a, \xi \in \mathcal{K}, a \in \mathcal{A}$ and the left action by $(a, \xi) \mapsto \pi(a)\xi, \xi \in \mathcal{K}, a \in \mathcal{A}$.

(ii) There is a densely defined closable linear map δ_0 from $\mathcal{A}$ into $\mathcal{K}$ such that $\mathcal{B} = \mathrm{Dom}(\delta_0)$, and δ_0 is a bimodule derivation, that is, $\delta_0(ab) = \delta_0(a)b + \pi(a)\delta_0(b)$ for all $a, b \in \mathcal{B}$.

(iii) For $a \in \mathcal{A}_0$, $b \in \mathcal{B}$, $\|\delta_0(a)b\|_{\mathcal{K}} \leq C_a \|b\|_2$, where $\|.\|_{\mathcal{K}}$ denotes the Hilbert space norm of $\mathcal{K}$, and C_a is a constant depending only on a. Thus, for any fixed $a \in \mathcal{A}_0$, the map $\mathcal{B} \ni b \mapsto \sqrt{2}\delta_0(a)b \in \mathcal{K}$ extends to a unique bounded linear map between the Hilbert spaces $L^2(\mathcal{A}, \tau)$ and $\mathcal{K}$, and this bounded map will be denoted by $\delta(a)$.

(iv) The triple $(\mathcal{L}, \pi, \delta)$ satisfy the following cocycle relation:

$$\partial\mathcal{L}(a, b, c) \equiv \delta(a)^*\pi(b)\delta(c) = \mathcal{L}(a^*bc) - \mathcal{L}(a^*b)c - a^*\mathcal{L}(bc) + a^*\mathcal{L}(b)c,$$

for $a, b, c \in \mathcal{A}_0$.

(v) $\mathcal{K}$ is the closed linear span of $\{\delta(a)b : a, b \in \mathcal{A}_0\}$.

(vi) π extends to a normal $*$-homomorphism on $\bar{\mathcal{A}}$.

Proof:
The proof of (i), (ii) and (iii) can be found in [32]. We give a brief account of it here. We begin with the observation that for any CP map Ψ from a C^*-algebra $\mathcal{C} \subseteq \mathcal{B}(\mathcal{H}_0)$ to itself, the map $\beta := \Psi - I$ is CCP. This can be proved by writing the CP map Ψ in the form $\Psi(x) = V^*\pi(x)V$ by the Theorem 2.2.9, where π is a $*$-representation of $\mathcal{C}$ in $\mathcal{B}(\mathcal{H})$ for some Hilbert space $\mathcal{H}$ and $V \in \mathcal{B}(\mathcal{H}_0, \mathcal{H})$. It follows by a straightforward calculation that for $a, b, c \in \mathcal{C}$, the kernel $K_a(b, c) := \beta(b^*a^*ac) + b^*\beta(a^*a)c - \beta(b^*a^*a)c - b^*\beta(a^*ac)$ can be written as

$$K_a(b, c) = (Vb - \pi(b)V)^*\pi(a^*a)(Vc - \pi(c)V),$$

from which it is immediate that K_a is a positive definite kernel, hence β is CCP.

Now, for $\epsilon > 0$, consider the operator $(I - \epsilon\mathcal{L})^{-1} = \int_0^\infty e^{-t} T_{\epsilon t} dt$. This can be viewed as a contractive operator on each of the spaces $\mathcal{A}$, $\overline{\mathcal{A}}$ $(= \pi_\tau(\mathcal{A})'')$ as well as $L^2(\mathcal{A}, \tau)$. In fact, it is clearly a CP map on $\overline{\mathcal{A}} \supseteq \mathcal{A}$. Thus,

$\Psi_\epsilon := \frac{1}{\epsilon}((I - \epsilon\mathcal{L})^{-1} - I) = -\mathcal{L}(I - \epsilon\mathcal{L})^{-1}$ is a bounded CCP map, hence $K_\epsilon : \overline{\mathcal{A}} \times \overline{\mathcal{A}} \to \overline{\mathcal{A}}$ defined below is positive definite:

$$K_\epsilon(a, b) = \Psi_\epsilon(a^* b) + a^* \Psi_\epsilon(1) a - \Psi_\epsilon(a^*) b - a^* \Psi_\epsilon(b).$$

Since $0 \leq T_t(1) \leq 1$, it is obvious that $\Psi_\epsilon(1) \leq 0$, hence the kernel $K'_\epsilon(a, b) := K_\epsilon(a, b) - a^* \Psi_\epsilon(1) b$ is also positive definite. Note that for $a, b \in \mathcal{B} \subseteq \mathrm{Dom}(\mathcal{L}_2)$,

$$\begin{aligned} &\tau(a^* \Psi_\epsilon(b)) \\ &= \langle a, (-\mathcal{L}_2)(1 - \epsilon\mathcal{L}_2)^{-1} b \rangle \\ &= \langle (-\mathcal{L}_2)^{\frac{1}{2}}(a), (-\mathcal{L}_2)^{\frac{1}{2}}(1 - \epsilon\mathcal{L}_2)^{-1}(b) \rangle \end{aligned}$$

which converges to

$$\langle (-\mathcal{L}_2)^{\frac{1}{2}}(a), (-\mathcal{L}_2)^{\frac{1}{2}}(b) \rangle = \langle (-\mathcal{L}_2)^{\frac{1}{2}}(a), (-\mathcal{L}_2)^{\frac{1}{2}}(b) \rangle, \text{ as } \epsilon \to 0+.$$

This allows us to define the following positive definite sesquilinear form η on $\mathcal{B} \otimes_{\mathrm{alg}} \mathcal{B}$:

$$\eta(a \otimes b, c \otimes d) = \lim_{\epsilon \to 0+} \frac{1}{2} \tau(b^* K'_\epsilon(a, c) d).$$

We construct a Hilbert space $\mathcal{K}$ from $\mathcal{B} \otimes_{\mathrm{alg}} \mathcal{B}$ by taking quotient by the subspace $\{b : \eta(x, x) = 0\}$ and then completing the quotient. Let us denote the inner product and norm on $\mathcal{K}$ by $\langle \cdot, \cdot \rangle_{\mathcal{K}}$ and $\| \cdot \|_{\mathcal{K}}$ respectively. Let $\mathcal{K}_0$ denote the $\mathbb{C}$-linear span of elements of the form $[b \otimes c]$, $b, c \in \mathcal{B}$, where $[b \otimes c]$ denotes the equivalence class of $(b \otimes c)$ in $\mathcal{K}$ after taking quotient and completion mentioned before. It is obvious that $\mathcal{K}_0$ is dense in $\mathcal{K}$. To define the left and right $\mathcal{A}$-module structure on $\mathcal{K}$, we proceed as follows. For $a, b, c \in \mathcal{B}$, define

$$[b \otimes c].a = [b \otimes ca],$$

$$a.[b \otimes c] = [ab \otimes c] - [a \otimes bc],$$

and then extend on $\mathcal{K}_0$ by linearity. By a simple calculation (see pages 8–11 of [32] for details), we can derive the following inequalities for $a \in \mathcal{B}, \xi \in \mathcal{K}_0$:

$$\|a.\xi\|^2_{\mathcal{K}} \leq \|a\|^2_\infty \|\xi\|^2_{\mathcal{K}}, \quad \|\xi.a\|^2_{\mathcal{K}} \leq \|a\|^2_\infty \|\xi\|^2_{\mathcal{K}}.$$

By the density of $\mathcal{B}$ in $\mathcal{A}$, we can define left and right actions by an arbitrary element of $\mathcal{A}$. This proves (*i*).

For $a \in \mathcal{B}$, we define $\delta_0(a) \in \mathcal{K}$ as the unique element satisfying

$$\langle \delta_0(a), [b \otimes c] \rangle_{\mathcal{K}} = \frac{1}{2} \left(\tilde{\mathcal{E}}(a, bc) + \tilde{\mathcal{E}}(b^*, ca^*) - \tilde{\mathcal{E}}(b^* a, c) \right),$$

where $b, c \in \mathcal{B}$ and $\tilde{\mathcal{E}}(x, y) := \langle (-\mathcal{L}_2)^{\frac{1}{2}}(x), (-\mathcal{L}_2)^{\frac{1}{2}}(y) \rangle$ for all $x, y \in \mathrm{Dom}(\mathcal{E})$. We refer to [32] for the estimates needed for proving the existence and uniqueness of the vector $\delta_0(a)$ defined above, and also the bimodule derivation property.

To prove (*iii*), let us first note that by Proposition 4.4 (*ii*) of [32] (page 13), we have

$$\delta_0(a)b = [a \otimes b],$$

where $a, b \in \mathcal{B}$. Now, for $a, b \in \mathcal{B}$,

$$\begin{aligned}
&\|[a \otimes b]\|_{\mathcal{K}}^2 \\
&= \lim_{\epsilon \to 0+} \frac{1}{2} \tau(b^* K'_\epsilon(a, a)b) \\
&\leq \limsup_{\epsilon \to 0+} \frac{1}{2} \|K'_\epsilon(a, a)\|_\infty \tau(b^* b) \\
&= \limsup_{\epsilon \to 0+} \frac{1}{2} \|K'_\epsilon(a, a)\|_\infty \|b\|_2^2.
\end{aligned}$$

Choose $a \in \mathcal{A}_0 \subseteq \mathrm{Dom}(\mathcal{L})$. Since $\mathcal{A}_0$ is a $*$-algebra, we have $a^*, a^*a \in \mathcal{A}_0$, and it is clear that $K'_\epsilon(a, a) \to -\mathcal{L}(a^*a) + \mathcal{L}(a^*)a + a^*\mathcal{L}(a)$ as $\epsilon \to 0+$, hence $c_a := \frac{1}{2} \lim_{\to 0+} \|K'_\epsilon(a, a)\|_\infty$ exists and is finite. This proves (*iii*).

We verify (*iv*) by straightforward calculations, which we omit. To prove (*v*), we first recall from [32] that $\mathcal{K}$ can be taken to be the closed linear span of the vectors of the form $\delta_0(a)b, a, b \in \mathcal{B}$. Now, by Lemma 3.3 of [32], $\|\delta_0(a)b\|_{\mathcal{K}}^2 \leq \|b\|_\infty^2 \mathcal{E}(a)$. Since $\mathcal{A}_0$ is on the one hand norm-dense in $\mathcal{A}$ and also forms the core for $\mathcal{E}$ on the other, (*v*) follows.

Let us now prove (*vi*). It is enough to prove that whenever we have a Cauchy net $a_\mu \in \mathcal{A}_0$ in the weak topology, then $\langle \xi, \pi(a_\mu)\xi \rangle$ is also Cauchy for any fixed ξ belonging to the dense subspace of $\mathcal{K}$ spanned by the vectors of the form $\delta(b)c, b, c \in \mathcal{A}_0$. But it is clear that for this, it suffices to show that $a \mapsto \langle \delta(b)b', \pi(a)\delta(b)b' \rangle_{\mathcal{K}}$ is weakly continuous. Now, by the symmetry of $\mathcal{L}$ and the trace property of τ, we have that for $a \in \mathcal{A}_0$,

$$\begin{aligned}
\langle \delta(b)b', \pi(a)\delta(b)b' \rangle_{\mathcal{K}} &= \langle b, ab\mathcal{L}(b'b'^*) \rangle - \langle b, a\mathcal{L}(bb'b'^*) \rangle \\
&\quad - \langle \mathcal{L}(bb'b'^*), ab \rangle + \langle \mathcal{L}(bb'(bb')^*), a \rangle.
\end{aligned}$$

The first three terms in the right-hand side are clearly weakly continuous in a, so we have to concentrate only on the last term, which is of the form $\tau(\mathcal{L}(xx^*)a)$ for $x \in \mathcal{A}_0$. Now, we have

$$\tau(\mathcal{L}(xx^*)a) = \tau(\mathcal{L}(x)x^*a) + \tau(x\mathcal{L}(x^*)a) + \tau(\delta(x^*)^*\delta(x^*)a),$$

and since $\mathcal{L}(\mathcal{A}_0) \subseteq \mathcal{A}_0$, the first two terms in the right hand side of the above expression are weakly continuous in a, so we are left with the term $\tau(\delta(x^*)^*\delta(x^*)a)$. Let us choose an approximate identity e_n of the C^*-algebra

$\mathcal{A}$ such that each e_n belongs to $\mathcal{A}_\tau$ (this is clearly possible, since $\mathcal{A}_\tau$ is a norm-dense $*$-ideal, and for $z \in \mathcal{A}_\tau$, one has that $|z| \in \mathcal{A}_\tau$). By normality of τ, $\tau(\delta(x^*)^*\delta(x^*)) = \sup_n \tau(e_n\delta(x^*)^*\delta(x^*)e_n) = 2\sup_n \|\delta_0(x^*)e_n\|^2_{\mathcal{K}} \leq 2\sup_n \|e_n\|^2_\infty \mathcal{E}(x^*) < \infty$, since $\|e_n\|_\infty \leq 1$ and $x^* \in \mathcal{A}_0 \subseteq \mathrm{Dom}(\mathcal{E})$. Thus, $\delta(x^*)^*\delta(x^*) = y^2$ for some $y \in \mathcal{A}_\tau$, hence $\tau(\delta(x^*)^*\delta(x^*)a) = \tau(yay)$, which proves the required weak continuity. □

Now we obtain a Christensen–Evans type structure of the generator $\mathcal{L}$.

Theorem 3.2.31 *In the notation of Theorem 3.2.30, let $R : L^2 \to \mathcal{K}$ be defined as follows:*

$$\mathrm{Dom}(R) = \mathcal{A}_0, \quad Rx \equiv \sqrt{2}\delta_0(x).$$

Then R has a densely defined adjoint R^*, whose domain contains the linear span of the vectors $\delta(x)y$, $x, y \in \mathcal{A}_0$ and

$$R^*(\delta(x)y) = x\mathcal{L}(y) - \mathcal{L}(x)y - \mathcal{L}(xy).$$

We denote the closure of R by the same notation R. For $x, y \in \mathcal{A}_0$,

$$(R^*\pi(x)R - \frac{1}{2}R^*Rx - \frac{1}{2}xR^*R)(y) = \mathcal{L}(x)y.$$

Furthermore,

$$\delta(x)y = (Rx - \pi(x)R)(y), x, y \in \mathcal{A}_0,$$
$$\mathcal{L}_2 = -\frac{1}{2}R^*R.$$

Proof:
For $x, y, z \in \mathcal{A}_0$, we observe by using the symmetry of $\mathcal{L}$ that

$$\begin{aligned}
&\langle \delta(x)y, Rz\rangle \\
&= 2\langle \delta_0(x)y, \delta_0(z)\rangle \\
&= \tau(y^*\mathcal{L}(x^*z) - y^*\mathcal{L}(x^*)z - y^*x^*\mathcal{L}(z)) \\
&= \tau(\mathcal{L}(y^*)x^*z - (\mathcal{L}(x)y)^*z - \mathcal{L}(xy)^*z) \\
&= \langle \{x\mathcal{L}(y) - \mathcal{L}(x)y - \mathcal{L}(xy)\}, z\rangle.
\end{aligned}$$

This suffices for the proof of the statements regarding R^*. It can be verified by a straightforward computation that $(R^*\pi(x)R - \frac{1}{2}R^*Rx - \frac{1}{2}xR^*R)(y) = \mathcal{L}(x)y$ holds for $x, y \in \mathcal{A}_0$. The remaining statements are also verified in a straightforward manner. □

3.2.4 A class of quantum dynamical semigroups on uniformly hyperfinite (U.H.F.) C^*-algebra

In this subsection we shall describe a physically interesting class of conservative Q.D.S. constructed by T. Matsui in [88] on the U.H.F. C^*-algebras of class N^∞. The reader is also referred to [63] for some of the details.

Let $\mathcal{A}$ be the uniformly hyperfinite (U.H.F.) C^*-algebra generated as the C^*-completion of infinite tensor product $\bigotimes_{j \in \mathbb{Z}^d} M_N(\mathbb{C})$, where N and d are two fixed positive integers. This is a particular type of approximately finite dimensional (AF) C^*-algebras discussed in, for example, [33]. Let us give a concrete description of this C^*-algebra here. Choose and fix a unit vector ξ_0 in $\mathbb{C}^N$ and consider the Hilbert space h obtained as the countable infinite tensor product of copies of $\mathbb{C}^N$, with ξ_0 as the stabilizing vector, as discussed below. Let J denote the set of all sequences (v_j) indexed by $j \in \mathbb{Z}^d$, with $v_j \in \mathbb{C}^N$ for all j, and $v_j = \xi_0$ for all but finitely many values of j. Let h_0 be the space of formal $\mathbb{C}$-linear combinations of elements in J, equipped with the semiinner product $\langle \cdot, \cdot \rangle$ defined on the elements of J by setting

$$\langle (v_j), (u_j) \rangle := \prod_j \langle v_j, u_j \rangle,$$

and extending on h_0 by linearity. Note that the above infinite product is finite because for all but finitely many j, $u_j = v_j = \xi_0$, thus $\langle v_j, u_j \rangle = 1$ for all but finitely many j's. We denote by h the Hilbert space obtained by taking quotient by the subspace $\{x \in h_0 \mid \langle x, x \rangle = 0\}$ and then completing the quotient. For $x \in M_N(\mathbb{C})$ and fixed $j \in \mathbb{Z}^d$, let $x^{(j)}$ be the unique bounded operator given by its action on the dense subspace h_0 as follows:

$$x^{(j)}(v_k) := (v'_k),$$

where $v'_k = v_k$ for $k \neq j$, $v'_j = x v_j$. Consider the $*$-subalgebra $\mathcal{A}_{\text{loc}}$ of $\mathcal{B}(h)$ generated by the operators $\{x^{(j)},\ x \in M_N(\mathbb{C}),\ j \in \mathbb{Z}^d\}$, and let $\mathcal{A}$ be its norm-completion with respect to the norm of $\mathcal{B}(h)$. It is well-known (see [33]) that there is a unique normalized trace on $\mathcal{A}$, which we shall denote by tr.

It is clear that for any finite subset $\Lambda = \{j_1, \ldots, j_k\}$ of $\mathbb{Z}^d$, the tensor product $\bigotimes_{j \in \Lambda} M_N(\mathbb{C})$ can be embedded as a $*$-subalgebra $\mathcal{A}_\Lambda$ of $\mathcal{A}_{\text{loc}}$ in a canonical way, by identifying an element of the form $\bigotimes_{j \in \Lambda} x_j$ (where $x_j \in M_N(\mathbb{C})$) with $a = x_{j_1}^{(j_1)} \cdots x_{j_k}^{(j_k)}$ in $\mathcal{A}_{\text{loc}}$. We shall also use the notation $x_{j_1}^{(j_1)} \otimes \cdots \otimes x_{j_k}^{(j_k)}$ to denote the above element a. Such elements will be called simple tensor elements of $\mathcal{A}$ and the jth component of a, denoted by $a_{(j)}$, is defined to be x_{j_l} if $j = j_l$ for some l, and 1_{M_N} otherwise. Thus, for $x \in M_N(\mathbb{C})$ and $j \in \mathbb{Z}^d$, we have $x^{(j)}_{(j)} = x$ and $x^{(j)}_{(k)} = 1$ for $k \neq j$. The support of a, denoted by $\text{supp}(a)$, is defined to be the set $\{j \in \mathbb{Z}^d : a_{(j)} \neq 1\}$. For a general element $a \in \mathcal{A}$ such that $a = \sum_{n=1}^{\infty} c_n a_n$ with a_n being simple tensor elements in $\mathcal{A}$ and c_n being complex coefficients, we define $\text{supp}(a) := \cup_{n \geq 1} \text{supp}(a_n)$ and we set $|a|$ = cardinality of $\text{supp}(a)$. Thus, $\mathcal{A}_\Lambda$ is the $*$-subalgebra generated by elements of $\mathcal{A}$ with support contained in Λ, and $\mathcal{A}_{\text{loc}}$ consists of elements with finite support. When $\Lambda = \{k\}$, we write $\mathcal{A}_k$ instead of $\mathcal{A}_{\{k\}}$. For $k \in \mathbb{Z}^d$, the translation τ_k on

$\mathcal{A}$ is an automorphism determined by $\tau_k(x^{(j)}) := x^{(j+k)}$ for all $x \in M_N(\mathbb{C})$ and $j \in \mathbb{Z}^d$. Thus, we get an action τ of the infinite discrete group $\mathbb{Z}^d$ on $\mathcal{A}$. For $x \in \mathcal{A}$ we denote $\tau_k(x)$ by x_k.

We also need another dense subset of $\mathcal{A}$, which is in a sense like the first Sobolev space in $\mathcal{A}$. For this, we need to note that $M_N(\mathbb{C})$ is spanned by a pair of noncommutative representatives $\{U, V\}$ of $\mathbb{Z}_N = \{0, 1, \ldots, N-1\}$ such that $U^N = V^N = 1$ and $UV = \omega VU$, where $\omega \in \mathbb{C}$ is the primitive Nth root of unity. These U, V can be chosen to be the $N \times N$ circulant matrices. In particular for $N = 2$, a possible choice is given by $U = \sigma_x$ and $V = \sigma_z$, where σ_x and σ_z denote the Pauli-spin matrices. Let G be the cyclic group $\mathbb{Z}_N \times \mathbb{Z}_N$. For $g = (\alpha, \beta) \in G$, its inverse is $-g = (-\alpha, -\beta)$. Now for $j \in \mathbb{Z}^d$ and $g = (\alpha, \beta) \in G$, we set $W_{j,g} = U^{(j)^\alpha} V^{(j)^\beta} \in \mathcal{A}_{\mathrm{loc}}$ and an automorphism $\pi_{j,g}$ of $\mathcal{A}_{\mathrm{loc}}$, given by $\pi_{j,g}(x) = W_{j,g}\, x\, W^*_{j,g}$. Now we define

$$\sigma_{j,g}(x) = \pi_{j,g}(x) - x, \quad \text{for all } x \in \mathcal{A}, \text{ and } \|x\|_1 = \sum_{j,g} \|\sigma_{j,g}(x)\|.$$

Let $\mathcal{C}^1(\mathcal{A}) = \{x \in \mathcal{A}: \|x\|_1 < \infty\}$. It is not difficult to see that $\|x^*\|_1 = \|\tau_j(x)\|_1 = \|x\|_1$ and since $\mathcal{C}^1(\mathcal{A})$ contains the dense $*$-subalgebra $\mathcal{A}_{\mathrm{loc}}$, $\mathcal{C}^1(\mathcal{A})$ is a dense τ invariant $*$-subalgebra of $\mathcal{A}$.

Let $\mathcal{G} := \prod_{j \in \mathbb{Z}^d} G$ be the infinite direct product of the finite group G at each lattice site. Thus each $g \in \mathcal{G}$ has jth component $g_{(j)} = (\alpha_j, \beta_j) \in G$. For $g \in \mathcal{G}$ we define its support by $\mathrm{supp}(g) = \{j \in \mathbb{Z}^d : g_{(j)} \neq (0,0)\}$ and $|g| =$ cardinality of $\mathrm{supp}(g)$. Let us consider the projective unitary representation of $\mathcal{G}$, given by $\mathcal{G} \ni g \mapsto U_g = \prod_{j \in \mathbb{Z}^d} W_{j, g^{(j)}} \in \mathcal{A}$.

For a given completely positive map T on $\mathcal{A}$, we formally define the Lindbladian

$$\mathcal{L} = \sum_{k \in \mathbb{Z}^d} \mathcal{L}_k,$$

where

$$\mathcal{L}_k(x) = \tau_k \mathcal{L}_0(\tau_{-k} x), \text{ for all } x \in \mathcal{A},$$

with

$$\mathcal{L}_0(x) = -\frac{1}{2}\{T(1), x\} + T(x), \tag{3.6}$$

and $\{A, B\} := AB + BA$.

In particular we consider the Lindbladian $\mathcal{L}$ for the completely positive map

$$T(x) = \sum_{l=0}^{\infty} a(l)^* x a(l), \quad \text{for all } x \in \mathcal{A},$$

associated with a sequence of elements $\{a(l)\}$ in $\mathcal{A}$, with $a(l) = \sum_{g \in G} c_{l,g} U_g$ such that $\sum_{l=0}^{\infty} \sum_{g \in G} |c_{l,g}|\, |g|^2 < \infty$.

Before we prove that the above formal generator can indeed be given a rigorous meaning, we need a technical lemma.

Lemma 3.2.32 *Let $\lambda > 0$, and x be a self-adjoint element in $\mathcal{C}^1(\mathcal{A})$. Then there exists a bounded operator Γ on $l^1(\mathbb{Z}^d \times G)$ such that*

$$(\lambda - \Gamma)\|\sigma_{\cdot}(x)\|(j, g') \leq \|\sigma_{j,g'}((\lambda - \mathcal{L})x)\|.$$

In fact Γ is an infinite positive matrix which can be expressed as a sum of two matrices:

$$\Gamma = \Gamma^{(0)} + \Gamma^{(1)},$$

where

$$\Gamma^{(0)} f\,(j, g') := 2 \sum_{g \in \mathcal{G}} |c_g| \sum_{k \in \mathrm{supp}(g)} \;\sum_{l: l+k \in \mathrm{supp}(g)} \;\sum_{g'' \in G} f(j - k + l, g'')$$

and

$$\Gamma^{(1)} f\,(j, g') := 2 \sum_{k \in \mathbb{Z}^d} \left\{ \sum_{\mathcal{G} \ni g : j-k \in \mathrm{supp}(g)} |c_g| \right\} \sum_{\nu \in \mathcal{G}} |c_\nu| \sum_{l \in \mathrm{supp}(\nu)} \;\sum_{g'' \in G} f(l + k, g'')$$

for $f \in l^1(\mathbb{Z}^d \times G)$.

Proof:
For simplicity, we shall only consider the case when there is just one $a(l)$, say r. That is, we assume that T is of the form, $T(x) = r^* x r$ for some $r \in \mathcal{A}$, given by, $r = \sum c_g U_g$, $\sum |c_g| \, |g|^2 < \infty$. The general case can be treated in a very similar way.

For a fixed $\lambda > 0$, and a self-adjoint element x in $\mathcal{C}^1(\mathcal{A})$, let us write $y := (\lambda - \mathcal{L})x$. For $(j, g') \in \mathbb{Z}^d \times G$, we have,

$$\sigma_{j,g'}(x) = \frac{1}{\lambda}\{\sigma_{j,g'}(\sigma_{j,g'}(y) + \mathcal{L}(x))\}.$$

Now it follows from the definition that

$$\begin{aligned}
&\sigma_{j,g'}(\mathcal{L}(x)) \\
&= \frac{1}{2} \sum_{k \in \mathbb{Z}^d} \{\pi_{j,g'}([r_k^*, x] r_k) - [r_k^*, x] r_k + \pi_{j,g'}(r_k^*[x, r_k]) - r_k^*[x, r_k]\} \\
&= \frac{1}{2} \sum_{k \in \mathbb{Z}^d} \{A_k(\sigma_{j,g'}(x)) + [\sigma_{j,g'}(r_k^*), x] \pi_{j,g'}(r_k) \pi_{j,g'}(r_k^*)[x, \sigma_{j,g'}(r_k)] \\
&\quad + [r_k^*, x] \sigma_{j,g'}(r_k) + \sigma_{j,g'}(r_k^*)[x, r_k]\}, \qquad (3.7)
\end{aligned}$$

where

$$A_k(x) = [\pi_{j,g'}(r_k^*), x]\pi_{j,g'}(r_k) + \pi_{j,g'}(r_k^*)[x, \pi_{j,g'}(r_k)].$$

It is clear that for each k, A_k is a bounded CCP map and $A_k(1) = 0$. Thus A_k is the generator of a contractive CP semigroup, say $(T_t^{(k)})_{t\geq 0}$. As x is self-adjoint, so is $\sigma_{j,g'}(x)$, so we can find a state ψ on $\mathcal{A}$ such that

$$|\psi(\sigma_{j,g'}(x))| = \|\sigma_{j,g'}(x)\|.$$

First let us assume,

$$\psi(\sigma_{j,g'}(x)) = \|\sigma_{j,g'}(x)\|. \tag{3.8}$$

Since $T_t^{(k)}$ is contractive and positive, we have

$$\begin{aligned}
&\psi(T_t^{(k)}(\sigma_{j,g'}(x))) \\
&\leq |\psi(T_t^{(k)}(\sigma_{j,g'}(x)))| \leq \|T_t^{(k)}(\sigma_{j,g'}(x))\| \\
&\leq \|\sigma_{j,g'}(x)\| = \psi(T_0^{(k)}(\sigma_{j,g'}(x))).
\end{aligned}$$

So,

$$\psi(A_k(\sigma_{j,g'}(x))) \leq 0. \tag{3.9}$$

Now evaluating the state ψ on $\sigma_{j,g'}(x)$ we have by (3.8),

$$\|\sigma_{j,g'}(x)\| = \frac{1}{\lambda}\{\psi(\sigma_{j,g'}(y)) + \psi(\sigma_{j,g'}(\mathcal{L}(x)))\}.$$

By (3.7) and (3.9), this gives

$$\begin{aligned}
\|\sigma_{j,g'}(x)\| &\leq \frac{1}{\lambda}\{\psi(\sigma_{j,g'}(y)) + \frac{1}{2}\sum_{k\in\mathbb{Z}^d}\{\psi([\sigma_{j,g'}(r_k^*), x]\pi_{j,g'}(r_k)) \\
&\quad +\psi(\pi_{j,g'}(r_k^*)[x, \sigma_{j,g'}(r_k)]) + \psi([r_k^*, x]\sigma_{j,g'}(r_k)) + \psi(\sigma_{j,g'}(r_k^*)[x, r_k])\} \\
&\leq \frac{1}{\lambda}\|\sigma_{j,g'}(y)\| + \frac{1}{2\lambda}\sum_{k\in\mathbb{Z}^d}\{\|r\|\,\|[\sigma_{j,g'}(r_k^*), x]\| + \|r\|\,\|[\sigma_{j,g'}(r_k), x]\| \\
&\quad +\|[r_k^*, x]\|\,\|\sigma_{j,g'}(r_k)\| + +\|\sigma_{j,g'}(r_k^*)\|\,\|[r_k, x]\|\}.
\end{aligned} \tag{3.10}$$

If $\psi(\sigma_{j,g'}(x)) = -\|\sigma_{j,g'}(x)\|$, replacing x by $-x$, same argument as above gives the inequality (3.10).

Now in order to estimate the second term of (3.10), first let us consider, for $g \in \mathcal{G} : j \in \text{supp}(g)$

$$\begin{aligned}
\|[\sigma_{j,g'}(U_g), x]\| &= \|[U^{(j)^\alpha} V^{(j)^\beta} U_g V^{(j)^{-\beta}} U^{(j)^{-\alpha}} - U_g, x]\| \\
&= \|[(w^{\alpha(\beta_j-\beta)-\beta(\alpha_j-\alpha)} - 1)U_g, x]\| \leq 2\|[U_g, x]\|
\end{aligned}$$

So we have

$$\begin{aligned}
&\|[\sigma_{j,g'}(r_k), x]\| \\
&= \sum_{k\in\mathbb{Z}^d} \sum_{g: j-k\in \mathrm{supp}(g)} |c_g| \|[\sigma_{j,g'}(\tau_k U_g), x]\| \\
&\leq 2 \sum_{k\in\mathbb{Z}^d} \sum_{g: j-k\in \mathrm{supp}(g)} |c_g| \|[\tau_k U_g, x]\| \\
&\leq 2 \sum_{k\in\mathbb{Z}^d} \sum_{g: j-k\in \mathrm{supp}(g)} |c_g| \sum_{l: l-k\in \mathrm{supp}(g)} \|\sigma_{l,g^{(l-k)}}(x)\| \\
&\leq 2 \sum_{k\in\mathbb{Z}^d} \sum_{g: j-k\in \mathrm{supp}(g)} |c_g| \sum_{l: l-k\in \mathrm{supp}(g)} \sum_{g''\in G} \|\sigma_{l,g''}(x)\| \\
&\leq 2 \sum_{k\in\mathbb{Z}^d} \sum_{g: j+k\in \mathrm{supp}(g)} |c_g| \sum_{l: l+k\in supp(g)} \sum_{g''\in G} \|\sigma_{l,g''}(x)\| \\
&\leq 2 \sum_{g\in\mathcal{G}} |c_g| \sum_{k\in \mathrm{supp}(g)-j} \sum_{l: l+k\in \mathrm{supp}(g)} \sum_{g''\in G} \|\sigma_{l,g''}(x)\| \\
&\leq 2 \sum_{g\in\mathcal{G}} |c_g| \sum_{k\in \mathrm{supp}(g)} \sum_{l: l+k\in \mathrm{supp}(g)} \sum_{g''\in G} \|\sigma_{j-k+l,g''}(x)\|.
\end{aligned}$$

Thus,

$$\frac{1}{2} \sum_{k\in\mathbb{Z}^d} \{\|r\| \, \|[\sigma_{j,g'}(r_k^*), x]\| + \|r\| \, \|[\sigma_{j,g'}(r_k), x]\|\} \leq \Gamma^{(0)}(\|\sigma_{\cdot}(x)\|)(j, g').$$

A similar estimate gives

$$\|\sigma_{j,g'}(r_k^*)\| \, \|[r_k, x]\| \leq \left\{ 2 \sum_{g: j-k\in \mathrm{supp}(g)} |c_g| \right\} \sum_{\nu\in\mathcal{G}} |c_\nu| \sum_{l\in \mathrm{supp}(\nu)} \sum_{g''\in G} \|\sigma_{l+k,g''}(x)\|.$$

So, we have

$$\frac{1}{2} \sum_{k\in\mathbb{Z}^d} \|\sigma_{j,g'}(r_k^*)\| \, \|[r_k, x]\| + \|[r_k^*, x]\| \, \|\sigma_{j,g'}(r_k)\| \leq \Gamma^{(1)} \|\sigma_{\cdot}(x)\|(j, g').$$

A simple estimate enables us to conclude that

$$\|\Gamma(f)\|_{l_1} \leq N^2 \left\{ \sum_{g\in\mathcal{G}} |c_g| \, |g|^2 \right\} \left\{ 4\|r\| + \sum_{g\in\mathcal{G}} |c_g| \, |g|^2 \right\} \|f\|_{l_1}.$$

□

Remark 3.2.33 *This lemma implies that for* $\lambda > \|\Gamma\|_{l^1(\mathbb{Z}^d\times G)}$, $\Gamma - \lambda$ *is invertible and*

$$\|\sigma_{j,g'}(x)\| \leq (\Gamma - \lambda)^{-1} \|\sigma_{\cdot}((\mathcal{L} - \lambda)x)\|(j, g'). \tag{3.11}$$

We are now in a position to state and prove the following theorem contained in [88].

Theorem 3.2.34 *(i) The map $\mathcal{L}$ formally defined above is well defined on $\mathcal{C}^1(\mathcal{A})$. (ii) The closure $\hat{\mathcal{L}}$ of $\mathcal{L}|_{\mathcal{C}^1(\mathcal{A})}$ is the generator of a conservative Q.D.S. $(T_t)_{t\geq 0}$ on $\mathcal{A}$.*

Proof:
For simplicity let us consider the Lindbladian $\mathcal{L}$ associated with an element $r = \sum_{g\in\mathcal{G}} c_g U_g \in \mathcal{A}$ such that $|r|_2 := \sum_{g\in\mathcal{G}} |c_g|\, |g|^2 < \infty$. Here $\mathcal{L}$ takes the form

$$\mathcal{L}(x) = \sum_{k\in\mathbb{Z}^d} \mathcal{L}_k(x)$$

with

$$\mathcal{L}_k(x) = \frac{1}{2}\{[r_k^*, x]r_k + r_k^*[x, r_k]\}, \text{ for all } k \in \mathbb{Z}^d.$$

Denoting these two bounded derivations $\left[r_k^*, .\right]$ and $[., r_k]$ on $\mathcal{A}$ by $\delta_k^\dagger$ and δ_k respectively, $\mathcal{L}(x) = \frac{1}{2}\sum_{k\in\mathbb{Z}^d} \delta_k^\dagger(x) r_k + r_k^*\delta_k(x)$.
In order to prove (i), for $x \in \mathcal{C}^1(\mathcal{A})$, let us estimate the norm of $\mathcal{L}(x)$:

$$\begin{aligned}
\|\mathcal{L}(x)\| &\leq \frac{1}{2}\sum_{k\in\mathbb{Z}^d} \|\delta_k^\dagger(x)r_k + r_k^*\delta_k(x)\| \\
&\leq \frac{\|r\|}{2}\sum_{k\in\mathbb{Z}^d} \|\delta_k^\dagger(x)\| + \|\delta_k(x)\| \\
&\leq \frac{\|r\|}{2}\sum_{k\in\mathbb{Z}^d}\sum_{g\in\mathcal{G}} |c_g|\|[\tau_k U_g, x]\| + \|[\tau_k U_g^*, x]\|.
\end{aligned}$$

Since we have

$$\|[U_g, x]\| = \|[\prod_{j\in\mathbb{Z}^d} W_{j,g(j)}, x]\| \leq \sum_{j\in\text{supp}(g)} \|[W_{j,g(j)}, x]\| = \sum_{j\in\text{supp}(g)} \|\sigma_{j,g(j)}(x)\|,$$

it follows that

$$\begin{aligned}
\|\mathcal{L}(x)\| &\leq \frac{\|r\|}{2}\sum_{k\in\mathbb{Z}^d}\sum_{g\in\mathcal{G}} |c_g| \sum_{j\in\text{supp}(g)+k} \|\sigma_{j,g(j-k)}(x)\| + \|\sigma_{j,-g(j-k)}(x)\| \\
&\leq \|r\|\sum_{k\in\mathbb{Z}^d}\sum_{g\in\mathcal{G}} |c_g| \sum_{j\in\text{supp}(g)+k}\sum_{g'\in G} \|\sigma_{j,g'}(x)\| \\
&\leq \|r\|\sum_{g\in\mathcal{G}} |c_g||g| \sum_{k\in\mathbb{Z}^d}\sum_{g'\in G} \|\sigma_{j,g'}(x)\| \\
&\leq \|r\|\sum_{g\in\mathcal{G}} |c_g|\; |g|\; ||x||_1 \\
&\leq |r|_2{}^2 ||x||_1.
\end{aligned}$$

Now let us come to the proof of (*ii*). For each $n \geq 1$, we set $\mathcal{L}^{(n)} = \sum_{|k|\leq n} \mathcal{L}_k$. It is clear that $\mathcal{L}^{(n)}$ is a bounded CCP map on $\mathcal{A}$. So $\mathcal{L}^{(n)}$ is the generator of contractive Q.D.S. $(T_t^{(n)})$ on $\mathcal{A}$ and for $\lambda \geq 0$, $\|(\mathcal{L}^{(n)} - \lambda)(x)\| \geq \lambda\|x\|$, for all $x \in \mathcal{A}$ and $\|(\mathcal{L}^{(n)} - \lambda)^{-1}\| \leq \frac{1}{\lambda}$ for $\lambda > 0$.

For $\lambda > \|\Gamma\|_{l^1}$, in order to show that $\mathrm{Ran}(\hat{\mathcal{L}} - \lambda)$ is dense in $(\mathcal{A}, \|\cdot\|)$, we consider the following. Let y be a self-adjoint element in $\mathcal{C}^1(\mathcal{A})$. Since $(\mathcal{L}^{(n)} - \lambda)$ is invertible for every n, we can choose self-adjoint $x_n \in \mathcal{A}$ so that $(\mathcal{L}^{(n)} - \lambda)(x_n) = y$. A careful look at the arguments for proving (3.11) reveals that it holds also for $\mathcal{L}^{(n)}$ and thus we have,

$$\|\sigma_{\alpha'}(x_n)\| \leq \sum_{\alpha} \{(\lambda - \Gamma)^{-1}\}_{\alpha',\alpha} \|\sigma_\alpha(y)\|. \tag{3.12}$$

Summing over α', it follows that

$$\|(x_n)\|_1 \leq \|(\lambda - \Gamma)^{-1}\| \ \|y\|_1 < \infty$$

and hence $x_n \in \mathcal{C}^1(\mathcal{A})$. Now setting $y_n = (\mathcal{L} - \lambda)(x_n)$, we have

$$\|y_n - y\| = \|(\mathcal{L} - \mathcal{L}^{(n)})x_n\| = \sum_{|k|>n} \mathcal{L}_k(x_n).$$

The above quantity is clearly dominated by

$$\|r\| \sum_{|k|>n} \sum_{g\in\mathcal{G}} |c_g| \sum_{g_1\in G} \sum_{j\in \mathrm{supp}(g)+k} \|\sigma_{j,g_1}(x_n)\|$$

$$\leq \|r\| \sum_{|k|>n} \sum_{g\in\mathcal{G}} |c_g| \{ \sum_{g_1\in G} \sum_{j\in \mathrm{supp}(g)+k} (\lambda - \Gamma)^{-1}\{\|\sigma_{\cdot}(y)\|\}(j, g_1)\} \text{ (by (3.12))},$$

which goes to 0 as n tends to ∞ since

$$\begin{aligned}
&\sum_{|k|>1} \sum_{g\in\mathcal{G}} |c_g| \sum_{g_1\in G} \sum_{j\in \mathrm{supp}(g)+k} (\lambda - \Gamma)^{-1}\{\|\sigma_{\cdot}(y)\|\}(j, g_1) \\
&\leq \sum_{g\in\mathcal{G}} |c_g| \, |g| \quad \sum_{j\in\mathbb{Z}^d} \sum_{g_1\in G} (\lambda - \Gamma)^{-1}\{\|\sigma_{\cdot}(y)\|\}(j, g_1) \\
&\leq \sum_{g\in\mathcal{G}} |c_g| \, |g| \, \|(\lambda - \Gamma)^{-1}\| \ \|y\|_1 < \infty.
\end{aligned}$$

This shows that y_n converges to y. For an arbitrary element $y \in \mathcal{C}^1(\mathcal{A})$, we apply the above line of reasoning to the real and imaginary parts of y and conclude that $\mathrm{Ran}(\hat{\mathcal{L}} - \lambda)$ as well as $(\mathcal{L} - \lambda)(\mathcal{C}^1(\mathcal{A}))$ are dense in $\mathcal{C}^1(\mathcal{A})$ and therefore they are also dense in $\mathcal{A}$.

Now for $y = (\mathcal{L} - \lambda)(x)$ in the dense $*$-subalgebra $(\mathcal{L} - \lambda)(\mathcal{C}^1(\mathcal{A}))$, let us consider

$$\begin{aligned}
&\|(\mathcal{L}^{(n)} - \lambda)^{-1}(y) - (\hat{\mathcal{L}} - \lambda)^{-1}(y)\| \\
&= \|(\mathcal{L}^{(n)} - \lambda)^{-1}(\hat{\mathcal{L}} - \mathcal{L}^{(n)})(\hat{\mathcal{L}} - \lambda)^{-1}(y)\| \\
&\leq \left\| (\mathcal{L}^{(n)} - \lambda)^{-1} \right\| \left\| \sum_{|k|>n} \mathcal{L}_k(x) \right\| \\
&\leq \frac{1}{\lambda} \left\| \sum_{|k|>n} \mathcal{L}_k(x) \right\|
\end{aligned}$$

which tends to 0 as n tends to ∞. Thus

$$\|(\hat{\mathcal{L}} - \lambda)^{-1}(y)\| \leq \frac{1}{\lambda}\|y\|$$

and by the Hille–Yosida Theorem $\hat{\mathcal{L}}$ generate a contractive semigroup, say (T_t). By Proposition 2.3.3, it follows that the contractive CP semigroup $T_t^{(n)}$ converges to T_t strongly as n tends to ∞, which proves that T_t is contractive and CP.

Finally, the semigroup T_t satisfies

$$T_t(x) = x + \int_0^t T_s(\hat{\mathcal{L}}(x))ds, \quad \text{for all } x \in \mathrm{Dom}(\hat{\mathcal{L}}).$$

Since $1 \in \mathcal{C}^1(\mathcal{A})$ and $\hat{\mathcal{L}}(1) = \mathcal{L}(1) = 0$, it follows that $T_t(1) = 1$, for all $t \geq 0$. □

We conclude this subsection with a brief discussion on ergodicity of the semigroups constructed here following [88]. We say that T_t is *ergodic* if there exists an invariant state ψ satisfying

$$\|T_t(x) - \psi(x)1\| \to 0 \text{ as } t \to \infty, \quad \text{for all } x \in \mathcal{A}. \tag{3.13}$$

In [88], T. Matsui has discussed some criteria for ergodicity of the Q.D.S. T_t. Some examples of such semigroups associated with partial states on the U.H.F. algebra and their perturbation are given.

For a state ϕ on $M_N(\mathbb{C})$ and $k \in \mathbb{Z}^d$, the partial state ϕ_k on $\mathcal{A}$ is a CP map determined by $\phi_k(x) = \phi(x_{(k)})x_{\{k\}^c}$, for $x = x_{(k)}x_{\{k\}^c}$, where $x_{(k)} \in \mathcal{A}_k$ and $x_{\{k\}^c} \in \mathcal{A}_{\{k\}^c}$. We can find a natural number $N' \leq N^2$ and elements $\{L^{(m)} : m = 1, 2, \ldots, N'\}$ in $M_N(\mathbb{C})$ such that

$$\phi(x) = \sum_{m=1}^{N'} L^{(m)*} x L^{(m)}, \quad \text{for all } x \in M_N(\mathbb{C}) \text{ and } \sum_{m=1}^{N'} L^{(m)*} L^{(m)} = 1.$$

For $m = 1, \ldots, N'$, let us consider the element $L_0^{(m)} \in \mathcal{A}_0$ with the zero-th component being $L^{(m)}$. Now for $k \in \mathbb{Z}^d$, writing $L_k^{(m)} = \tau_k(L_0^{(m)})$, the partial state ϕ_k is given by

$$\phi_k(x) = \sum_{m=1}^{N'} L_k^{(m)^*} x L_k^{(m)}, \quad \text{for all } x \in \mathcal{A}.$$

Here the Linbladian $\mathcal{L}^\phi$ corresponding to the partial state ϕ_0 is formally given by

$$\mathcal{L}^\phi(x) = \sum_{k \in \mathbb{Z}^d} \mathcal{L}_k^\phi(x),$$

where

$$\mathcal{L}_k^\phi(x) = \phi_k(x) - x = \frac{1}{2} \sum_{m=1}^{N'} [L_k^{(m)^*}, x] L_k^{(m)} + L_k^{(m)^*}[x, L_k^{(m)}].$$

It follows from Theorem 3.2.34 that $\mathcal{L}^\phi$ is defined on $\mathcal{C}^1(\mathcal{A})$. Moreover, the closure $\hat{\mathcal{L}}^\phi$ of $\mathcal{L}^\phi/_{\mathcal{C}^1(\mathcal{A})}$ generates a conservative Q.D.S. T_t^ϕ on $\mathcal{A}$ given by

$$T_t^\phi\left(\prod_{k \in \Lambda} x_{(k)}\right) = \prod_{k \in \Lambda} \{\phi(x_{(k)}) + e^{-t}(x_{(k)} - \phi(x_{(k)}))\}.$$

Here we note that the map Φ defined by

$$\Phi\left(\prod_{k \in \Lambda} x_{(k)}\right) = \lim_{t \to \infty} T_t^\phi\left(\prod_{k \in \Lambda} x_{(k)}\right) = \prod_{k \in \Lambda} \phi(x_{(k)})$$

extends as a state on $\mathcal{A}$ which is an invariant state for the ergodic Q.D.S. T_t^ϕ.

For any real number c, let us consider the perturbation

$$\mathcal{L}^{(c)}(x) = \mathcal{L}^\phi(x) + c\mathcal{L}(x), \text{ for all } x \in \mathcal{C}^1(\mathcal{A}).$$

It is clear that $\mathcal{L}^{(c)}$ is the Linbladian associated with the CP map

$$T(x) = \sum_{m=1}^{N'} L_k^{(m)^*} x L_k^{(m)} + c \sum_{l=0}^{\infty} a_l^* x a_l, \text{ for all } x \in \mathcal{A}$$

and by Theorem 3.2.34 it follows that the closure $\hat{\mathcal{L}^{(c)}}$ of $\mathcal{L}^{(c)}/\mathcal{C}^1(\mathcal{A})$ generate a Q.D.S. $T_t^{(c)}$. Moreover, the following can be shown.

Proposition 3.2.35 *There exists a constant c_0 such that for $0 \leq c \leq c_0$ the above Q.D.S. $T_t^{(c)}$ is ergodic with the invariant state $\Phi^{(c)}$ and we have*

$$\|T_t^{(c)}(x)\|_1 \leq 2e^{-(1-\frac{c}{c_0})t}\|x\|_1, \tag{3.14}$$

$$\|T_t^{(c)}(x) - \Phi^{(c)}(x)1\| \leq \frac{4}{N^2}e^{-(1-\frac{c}{c_0})t}\|x\|_1, \quad \text{for all } x \in \mathcal{C}^1(\mathcal{A}).$$

We omit the proof of this result and refer the reader to [88].

Remark 3.2.36 *The invariant state $\Phi^{(c)}$ corresponding to the ergodic Q.D.S. $T_t^{(c)}$ is given by*

$$\Phi^{(c)}(x) = \Phi(x) + c\int_0^\infty \Phi(\mathcal{L}(T_t^{(c)}(x)))dt, \quad \text{for all } x \in \mathcal{C}^1(\mathcal{A}).$$

4
Hilbert modules

In this chapter we introduce the idea of *Hilbert modules* and briefly discuss some useful results on them. For a more detailed account on this subject, the reader is referred to [81], [90], [98] (and [122] for von Neumann modules).

4.1 Hilbert C^*-modules

A Hilbert space is a complex vector space equipped with a complex-valued inner product. A natural generalization of this is the concept of Hilbert module, which has become quite an important tool of analysis and mathematical physics in recent times.

Definition 4.1.1 Given a $*$-subalgebra $\mathcal{A} \subseteq \mathcal{B}(h)$ (where h is a Hilbert space), a *semi-Hilbert $\mathcal{A}$-module* E is a right $\mathcal{A}$-module equipped with a sesquilinear map $\langle .,. \rangle : E \times E \rightarrow \mathcal{A}$ satisfying $\langle x, y\rangle^* = \langle y, x\rangle$, $\langle x, ya\rangle = \langle x, y\rangle a$ and $\langle x, x\rangle \geq 0$ for $x, y \in E$ and $a \in \mathcal{A}$. A semi-Hilbert module E is called a *pre-Hilbert module* if $\langle x, x\rangle = 0$ if and only if $x = 0$; and it is called a *Hilbert C^*-module* if furthermore $\mathcal{A}$ is a C^*-algebra and E is complete in the norm $x \mapsto \|\langle x, x\rangle\|^{\frac{1}{2}}$ where $\|.\|$ the C^*-norm of $\mathcal{A}$.

It is clear that any semi-Hilbert $\mathcal{A}$-module can be made into a Hilbert module in a canonical way: first quotienting it by the ideal $\{x : \langle x, x\rangle = 0\}$ and then completing the quotient.

Let us assume that $\mathcal{A}$ is a C^*-algebra. The $\mathcal{A}$-valued inner product $\langle .,. \rangle$ of a Hilbert module shares some of the important properties of usual complex-valued inner product of a Hilbert space, such as the Cauchy–Schwartz inequality, which we prove now.

Theorem 4.1.2 *Let $\mathcal{A}$ be a C^*-algebra and E be a pre-Hilbert $\mathcal{A}$-module. Then, for $x, y \in E$, we have the following inequality in the sense of C^*-algebra:*

$$0 \leq \langle x, y \rangle \langle y, x \rangle \leq \langle x, x \rangle \| \langle y, y \rangle \|.$$

Proof:
Let $a = \langle x, x \rangle$, $b = \langle y, y \rangle$ and $c = \langle x, y \rangle$. It is enough to prove that for every positive linear functional ϕ on $\mathcal{A}$,

$$\phi(cc^*) \leq \|b\| \phi(a).$$

Fix any positive linear functional ϕ on $\mathcal{A}$. Since the inequality $\phi(cc^*) \leq \|b\|\phi(a)$ is clearly valid in case $\phi(cc^*) = 0$, let us assume without loss of generality that $\phi(cc^*) > 0$. For any real number λ, we have from the property of the inner product $\langle \cdot, \cdot \rangle$ that $\langle x - y\lambda c^*, x - y\lambda c^* \rangle \geq 0$ in $\mathcal{A}$, which on simplification becomes $a - 2\lambda cc^* + \lambda^2 cbc^* \geq 0$. But $0 \leq cbc^* \leq \|b\| cc^*$ since b is a nonnegative element of $\mathcal{A}$, and hence we get that $a - 2\lambda cc^* + \lambda^2 \|b\| cc^* \geq 0$. After applying ϕ on this C^*-algebraic inequality, we conclude that $\phi(a) - 2\lambda\phi(cc^*) + \lambda^2 \|b\| \phi(cc^*) \geq 0$ for all real values of λ, which is equivalent to the discriminant of the expression in the left-hand side being nonpositive, that is, $4\phi(cc^*)^2 - 4\phi(a)\|b\|\phi(cc^*) \leq 0$. Since $\phi(cc^*)$ is strictly positive by our assumption, we get the desired inequality $\phi(cc^*) \leq \|b\|\phi(a)$. □

From this C^*-algebraic Cauchy–Schwartz inequality, it is possible to deduce a weaker version of this inequality by taking the norm on both sides, namely, $\|\langle x, y \rangle\|^2 \leq \|\langle x, x \rangle\| \|\langle y, y \rangle\|$, or in other words, $\|\langle x, y \rangle\|^2 \leq \|x\|^2 \|y\|^2$, since the norm on E is defined by $\|x\| := \|\langle x, x \rangle\|^{\frac{1}{2}}$.

However, some of the crucial properties of Hilbert spaces do not extend to general Hilbert modules: the most remarkable ones are the projection theorem and self-duality. We say that a vector-subspace F of a Hilbert $\mathcal{A}$-module E is a *submodule* if $xa \in F$ for all $x \in F, a \in \mathcal{A}$. A norm-closed submodule F of E is said to be *orthocomplemented*, or simply *complemented* if there is a direct sum decomposition of the form $E = F \oplus G$, where G is a closed submodule with $\langle f, g \rangle = 0$, for all $f \in F, g \in G$. It is remarkable that, in contrast to the Hilbert space case, closed submodules of a general Hilbert module need not be complemented. To give a very simple example of this, let us take $\mathcal{A}$ to be the commutative C^*-algebra $C[0, 1]$ and take E to be the same as $\mathcal{A}$ as a Banach space, with its own right action and the inner product given by $\langle x, y \rangle := x^* y, x, y \in \mathcal{A}$. Consider the closed subspace $F = C_0[0, 1] = \{f \in C[0, 1] : f(0) = f(1) = 0\}$. It can be observed that any element f in E such that $\langle g, f \rangle = 0$ for all $g \in F$ must be the zero function, and therefore the existence of a decomposition of the form $E = F \oplus G$, with

$\langle f_1, f_2 \rangle = 0$ for all $f_1 \in F, f_2 \in G$, implies $G = \{0\}$, hence $F = E$. But clearly F is a strict subset of E: for example the identity of $\mathcal{A}$ is not in F.

Furthermore, the Banach space of all $\mathcal{A}$-valued, $\mathcal{A}$-linear, bounded maps on a Hilbert $\mathcal{A}$-module E may not be isometrically anti-isomorphic to E, in contrast to the Riesz' theorem for complex Hilbert space. These unpleasant features make the study of Hilbert modules considerably difficult and challenging. For example, a bounded $\mathcal{A}$-linear map from one Hilbert $\mathcal{A}$-module to another may not have an adjoint. For this reason, the role played by the set of bounded linear maps between Hilbert spaces is taken over by the set of *adjointable* linear maps. To be more precise, let us make the following definition.

Definition 4.1.3 Let E and F be two Hilbert $\mathcal{A}$-modules. We say that a $\mathbb{C}$-linear map L from E to F is *adjointable* if there exists a $\mathbb{C}$-linear map L^* from F to E such that $\langle L(x), y \rangle = \langle x, L^*(y) \rangle$ for all $x \in E, \ y \in F$. We call L^* the adjoint of L. The set of all adjointable maps from E to F is denoted by $\mathcal{L}(E, F)$. In case $E = F$, we write $\mathcal{L}(E)$ for $\mathcal{L}(E, E)$.

The first important fact to be noted is that an adjointable map is automatically $\mathcal{A}$-linear and norm-bounded, which we prove below.

Theorem 4.1.4 *Let E, F be Hilbert $\mathcal{A}$-modules and $L \in \mathcal{L}(E, F)$. Then we have the following conditions.*

(i) Both L and its adjoint L^* are $\mathcal{A}$-linear.
(ii) L is a norm-bounded map from E to F, viewed as Banach spaces. Similarly, L^* is bounded.
(iii) L^* is the unique $\mathbb{C}$-linear map satisfying $\langle L(x), y \rangle = \langle x, L^*(y) \rangle$ for all $x \in E$,
$y \in F$. Furthermore, $(L^*)^* = L$.
(iv) If E_1, E_2, E_3 are Hilbert $\mathcal{A}$-modules and $L_1 \in \mathcal{L}(E_1, E_2)$, $L_2 \in \mathcal{L}(E_2, E_3)$, then $L_2 L_1 := L_2 \circ L_1 \in \mathcal{L}(E_1, E_3)$ with $(L_2 L_1)^* = L_1^* L_2^*$.

Proof:
(i) Fix $x \in E, a \in \mathcal{A}$. We have that for any $y \in F$, $\langle y, L(xa) \rangle = \langle L^*(y), xa \rangle = \langle L^*(y), x \rangle a = \langle y, L(x) \rangle a = \langle y, L(x)a \rangle$. Since $\langle y, z \rangle = 0$ for all $y \in F$ implies $z = 0$, we conclude that $L(xa) = L(x)a$. Similarly the $\mathcal{A}$-linearity of L^* can be proved.
(ii) By the well-known closed graph theorem, it is enough to show that L is a closed map from the Banach space E to the Banach space F. Let $x_n, n = 1, 2, \ldots$ be a sequence of elements in E with $x_n \to x \in E$, and also assume that $L(x_n) \to y \in F$. For any $z \in F$, we have that $\langle z, (y - Lx) \rangle = \langle z, y \rangle - \langle L^*(z), x \rangle = \lim_{n \to \infty}(\langle z, L(x_n) \rangle - \langle L^*(z), x_n \rangle) == 0$, which proves that $y = L(x)$. This shows that L is closed, hence bounded. Similar arguments can

be given to prove the boundedness of L^*. We omit the proof of (*iii*) and (*iv*) since they are very straightforward. □

We say that an element $L \in \mathcal{L}(E, F)$ is an *isometry* if $\langle Lx, Ly \rangle = \langle x, y \rangle$ for all $x, y \in E$. L is said to be *unitary* if L is isometry and its range is the whole of F.

We observe that for $x \in E$, $\|x\| = \sup_{z \in E, \|z\| \leq 1} \|\langle x, z \rangle\|$. The supremum in the right-hand side is in fact attained by taking $z = x/\|x\|$, when x is not 0, and $z = 0$ otherwise. We define a norm on $\mathcal{L}(E, F)$ by $\|L\| := \sup_{x \in E, \|x\| \leq 1} \|L(x)\|$. We have already proven that for $L \in \mathcal{L}(E, F)$, $\|L\|$ is indeed finite. Clearly, $\mathcal{L}(E)$ is a C^*-algebra with this norm. We also observe the following useful fact.

Lemma 4.1.5 *An element L of the C^*-algebra $\mathcal{L}(E)$ is nonnegative, that is, of the form $L = K^*K$ for some $K \in \mathcal{L}(E)$, if and only if $\langle x, L(x) \rangle \geq 0 \in \mathcal{A}$ for all $x \in E$.*

Proof:
If $L = K^*K$, it is obvious that $\langle x, L(x) \rangle \geq 0$ for all x. To show the converse, let us assume that $\langle x, L(x) \rangle \geq 0$ for all $x \in E$. So, in particular, $\langle x, L(x) \rangle = \langle x, L(x) \rangle^* = \langle L(x), x \rangle = \langle x, L^*(x) \rangle$. By usual polarization argument, it follows that $\langle x, L(y) \rangle = \langle x, L^*(y) \rangle$ for all $x, y \in E$, hence $L = L^*$, that is, L is a self-adjoint element of the C^*-algebra $\mathcal{L}(E)$. By the general theory of C^*-algebras as discussed in Chapter 2, we can write $L = L_+ - L_-$, where L_+, L_- are nonnegative elements of $\mathcal{L}(E)$ with $L_+L_- = L_-L_+ = 0$. From the assumption that $\langle x, L(x) \rangle \geq 0$ we have that $0 \leq \langle x, L_-(x) \rangle \leq \langle x, L_+(x) \rangle$ for all x. Choosing $x = L_-(y)$ for $y \in E$, we have that $0 \leq \langle y, L_-^3(y) \rangle \leq \langle L_-(y), L_+L_-(y) \rangle = 0$. By polarization we conclude that $L_-^3 = 0$, and as L_- is a nonnegative element of the C^*-algebra $\mathcal{L}(E)$, it follows that $L_- = 0$, so that $L = L_+$, which is a nonnegative element. □

Fix two Hilbert $\mathcal{A}$-modules E and F. For $t \in \mathcal{L}(E, F)$ and $x \in E$, it can be proved by straightforward arguments that $\langle tx, tx \rangle = \langle x, t^*t(x) \rangle \leq \|t\|^2 \langle x, x \rangle$, where $\|t\|$ denotes the map-norm of t. Indeed, as t^*t is a nonnegative element of the unital C^*-algebra $\mathcal{L}(E)$, we have that $\|t\|^2 I - t^*t$ is nonnegative, hence by the Lemma 4.1.5, $\|t\|^2 \langle x, x \rangle \geq \langle x, t^*t(x) \rangle$. The topology on $\mathcal{L}(E, F)$ given by the family of seminorms $\{\|.\|_x, \|.\|_y : x \in E, y \in F\}$ where $\|t\|_x = \|\langle tx, tx \rangle^{\frac{1}{2}}\|$ and $\|t\|_y = \|\langle t^*y, t^*y \rangle^{\frac{1}{2}}\|$, is known as the *strict topology*. For $x \in E$, $y \in F$, we denote by $|y >< x| \equiv \theta_{x,y}$ the element of $\mathcal{L}(E, F)$ defined by $\theta_{x,y}(z) = y\langle x, z \rangle$ $(z \in E)$. The $\mathcal{L}(E)$-norm-closed subset generated by $\mathcal{A}$-linear span of $\{\theta_{x,y} : x \in E, y \in F\}$ is called the set of *compact operators* and denoted by $\mathcal{K}(E, F)$, and we denote $\mathcal{K}(E, E)$ by $\mathcal{K}(E)$. It should be noted that these objects need not be compact in the sense of compact operators between two Banach spaces, though this abuse of terminology has become standard. It is known that $\mathcal{K}(E, F)$ is dense in $\mathcal{L}(E, F)$ in the strict topology. It is clear that $\mathcal{K}(E)$ is a C^*-algebra which is dense in the C^*-algebra $\mathcal{L}(E)$ with respect to the strict topology.

At this point, we shall prove the following result which is going to be useful in the sequel.

Theorem 4.1.6 *Given a $*$-subalgebra $\mathcal{A} \subseteq \mathcal{B}(h)$, and a pre-Hilbert $\mathcal{A}$-module E, there is a Hilbert space $\mathcal{H}$ and an $\mathcal{A}$-linear isometric embedding $\Gamma : E \to \mathcal{B}(h, \mathcal{H})$, such that $\mathcal{H}$ is the closed linear span of elements of the form $\Gamma(e)u$ for $e \in E, u \in h$. Furthermore, the choice of the pair $(\mathcal{H}, \Gamma)$ is unique in the following sense: if there is any other pair $(\mathcal{H}', \Gamma')$ with the above property, then there is a unitary operator $U : \mathcal{H} \to \mathcal{H}'$ such that $U\Gamma = \Gamma'$.*

Proof:
We let $\mathcal{H}_0 \equiv E + h$ be the algebraic complex linear span of all formal symbols (e, u) where $e \in E, u \in h$, and define a complex-valued semi-inner product $\langle \cdot, \cdot \rangle$ on $\mathcal{H}_0$ given by

$$\langle (e, u), (f, v) \rangle := \langle u, \langle e, f \rangle_{\mathcal{A}} v \rangle_h$$

(where $\langle \cdot, \cdot \rangle_h$ is the complex-valued inner-product of h, $\langle \cdot, \cdot \rangle_{\mathcal{A}}$ is the $\mathcal{A}$-valued inner product of E: we shall however not use these suffixes in our later discussion any more), and then linearly extend it on the whole of $\mathcal{H}_0$, and note that this form is positive definite. We then take the quotient of $\mathcal{H}_0$ by the subspace $\{X \in \mathcal{H}_0 : \langle X, X \rangle = 0\}$, and complete the quotient space in the norm coming from $\langle \cdot, \cdot \rangle$ to get a Hilbert space $\mathcal{H}$. Let us denote by $[X]$ the equivalence class in $\mathcal{H}$ corresponding to an element $X \in \mathcal{H}_0$. By construction the set of all such $[X]$ is dense in $\mathcal{H}$. Now, let $\Gamma : E \to \mathcal{B}(h, \mathcal{H})$ be defined by

$$\Gamma(e)(u) := [(e, u)],$$

for $e \in E, u \in h$. To verify that Γ is indeed $\mathcal{A}$-linear, we need to verify that

$$\langle (\Gamma(ea) - \Gamma(e)a)u, (\Gamma(ea) - \Gamma(e)a)u \rangle = 0$$

for $e \in E, a \in \mathcal{A}, u \in h$. Indeed,

$$\begin{aligned}
&\langle (\Gamma(ea) - \Gamma(e)a)u, (\Gamma(ea) - \Gamma(e)a)u \rangle \\
&= \langle [ea, u], [ea, u] \rangle - \langle [e, au], [ea, u] \rangle - \langle [ea, u], [e, au] \rangle + \langle [e, au], [e, au] \rangle \\
&= \langle u, \langle ea, ea \rangle u \rangle - \langle au, \langle e, ea \rangle u \rangle - \langle u, \langle ea, e \rangle au \rangle + \langle au, \langle e, e \rangle au \rangle \\
&= \langle u, a^* \langle e, e \rangle au \rangle - \langle u, a^* \langle e, e \rangle au \rangle - \langle u, a^* \langle e, e \rangle au \rangle + \langle u, a^* \langle e, e \rangle au \rangle \\
&= 0.
\end{aligned}$$

Clearly, for any fixed $e \in E$, $\Gamma(e)$ is a bounded operator, as we have $\|\Gamma(e)u\|^2 = \langle u, \langle e, e\rangle u\rangle \leq \|e\|^2\|u\|^2$. Furthermore, it is clear that $\Gamma(e)^* \in \mathcal{B}(\mathcal{H}, h)$ given by $\Gamma(e)^*([(f, v)]) = \langle e, f\rangle v$ for $f \in E, v \in h$, and from this it follows that $\langle u, \Gamma(e)^*\Gamma(f)v\rangle = \langle u, \langle e, f\rangle v\rangle$, that is, the inner product between $\Gamma(e)$ and $\Gamma(f)$ as elements of $\mathcal{B}(h, \mathcal{H})$, namely $\Gamma(e)^*\Gamma(f)$, coincides with the $\mathcal{A}$-valued inner product $\langle e, f\rangle$ of E. In other words, Γ is indeed an isometry, and thus clearly continuous and injective. The fact that $\mathcal{H}$ is the closed linear span of elements of the form $\Gamma(e)u$, $e \in E, u \in h$, is obvious from our construction. Finally, the uniqueness of $(\mathcal{H}, \Gamma)$ also follows by straightforward arguments. □

Remark 4.1.7 *It is a simple observation from the proof of the above theorem that if $\mathcal{A}$ is a C^*-algebra and E is a Hilbert C^* $\mathcal{A}$-module, then the image of E under Γ in $\mathcal{B}(h, \mathcal{H})$ will be closed in the norm topology inherited from $\mathcal{B}(h, \mathcal{H})$.*

Let us now prove a few useful results. For a C^*-algebra $\mathcal{A}$ and a Hilbert $\mathcal{A}$-module E, we denote by $\langle E, E\rangle$ the norm-closed $\mathbb{C}$-linear span of elements of the form $\langle e, f\rangle, e, f \in E$. Since $\langle e, f\rangle^* = \langle f, e\rangle$, and $\langle e, f\rangle a = \langle e, fa\rangle \in \langle E, E\rangle$ for all $a \in \mathcal{A}$, it is clear that $\langle E, E\rangle$ is a norm-closed two sided ideal in $\mathcal{A}$. In particular, $\langle E, E\rangle$ is a C^*-algebra itself.

Lemma 4.1.8 *The complex linear span of $\{eb : e \in E, b \in \langle E, E\rangle\}$ is norm-dense in E.*

Proof:
Since $\langle E, E\rangle$ is a C^*-algebra, take an approximate identity p_ν indexed by some net, such that $\sup_\nu \|p_\nu\| \leq 1$. Fix any $e \in E$ and observe that $\langle (ep_\nu - e), (ep_\nu - e)\rangle = p_\nu\langle e, e\rangle p_\nu + \langle e, e\rangle - \langle e, e\rangle p_\nu - p_\nu\langle e, e\rangle$. Since $\lim_\nu p_\nu\langle e, e\rangle = \langle e, e\rangle = \lim_\nu\langle e, e\rangle p_\nu$, it follows that $ep_\nu \to e$. □

Recall the definition of multiplier algebra $\mathcal{M}(\mathcal{A})$ of a C^*-algebra $\mathcal{A}$ discussed in Chapter 2. One has (see [81]) the following result.

Theorem 4.1.9 *The multiplier algebra $\mathcal{M}(\mathcal{K}(E))$ of $\mathcal{K}(E)$, the compact adjointable operators on E, is isomorphic with $\mathcal{L}(E)$ for any Hilbert module E.*

Proof:
By the Theorem 4.1.6, let us assume that E is a submodule of $\mathcal{B}(h, \mathcal{H})$ for some Hilbert spaces $h, \mathcal{H}$, where $\mathcal{A} \subseteq \mathcal{B}(h)$ and $\{\xi u, \xi \in E, u \in h\}$ is total in $\mathcal{H}$. We shall first identify $\mathcal{L}(E)$ with a suitable subalgebra of $\mathcal{B}(\mathcal{H})$. Define a map $\Pi : \mathcal{L}(E) \to \mathcal{B}(\mathcal{H})$ by

$$\Pi(L)(\xi u) := L(\xi)u,$$

for $\xi \in E$, $u \in h$ and $L \in \mathcal{L}(E)$; and then extend it $\mathbb{C}$-linearly on the dense set spanned by vectors of the form ξu with ξ, u as above. We need to verify that this is well defined. Take $\Phi = \sum_{i=1}^{n} \xi_i u_i \in \mathcal{H}$, where $\xi_i \in E$ and $u_i \in h$, n integer. Considering L as an element of $\mathcal{B}(\mathcal{H})$ and using the Hilbert space inequality $< \Phi, L^*L\Phi > \leq \|L\|^2\|\Phi\|^2$, we see that $\|\Pi(L)\Phi\|^2 = \sum_{i,j} < u_i, < \xi_i, L^*L(\xi_j) > u_j > = < \Phi, L^*L\Phi > \leq \|L\|^2\|\Phi\|^2$. From this, it follows that $\Pi(L)$ is well defined and it extends to a bounded operator on $\mathcal{H}$ with the norm less than or equal to the norm of L. Moreover, for $\xi, \eta \in E$ and $u, v \in h$, we have that $\langle \xi u, \Pi(L)(\eta v)\rangle = \langle \xi u, L(\eta)v\rangle = \langle u, \langle \xi, L(\eta)\rangle v\rangle = \langle u, \langle L^*(\xi), \eta\rangle v\rangle = \langle \Pi(L^*)(\xi u), \eta v\rangle$. By the totality of the vectors of the form ξu, ξ, u as above, we conclude that $\Pi(L)^* = \Pi(L^*)$. Thus, Π is a $*$-homomorphism from the C^*-algebra $\mathcal{L}(E)$ into the C^*-algebra $\mathcal{B}(\mathcal{H})$. Furthermore, $\Pi(L) = 0$ for some $L \in \mathcal{L}(E)$ implies that $L(\xi)u = 0$ for all $\xi \in E$ and $u \in h$, which in turn implies that $L(\xi) = 0$ for all $\xi \in E$, so $L = 0$. Thus, Π is a one-to-one map, and we have $\mathcal{L}(E) \cong \mathrm{Ran}(\Pi) \subseteq \mathcal{B}(\mathcal{H})$. We claim that in fact, $\mathrm{Ran}(\Pi) = \{B \in \mathcal{B}(\mathcal{H}) : B^*\xi, B\xi \in E, \text{ for all } \xi \in E\}$. Note that since an element ξ of E is also an element of $\mathcal{B}(h, \mathcal{H})$, $B\xi$ just means the composition of $B \in \mathcal{B}(\mathcal{H})$ with ξ, so that $B\xi \in \mathcal{B}(h, \mathcal{H})$, for any $B \in \mathcal{B}(\mathcal{H})$, although $B\xi$ in general may not be in E. To verify the claim, first observe that for any $L \in \mathcal{L}(E)$, $\xi \in E$, we have that $\Pi(L)\xi = L(\xi) \in E$. Similarly, $\Pi(L)^*\xi = \Pi(L^*)\xi \in E$. Conversely, take $B \in \mathcal{B}(\mathcal{H})$ such that $B\xi, B^*\xi \in E$, for all $\xi \in E$. Define L from E to E by setting $L(\xi) := B\xi$, so that we have $\Pi(L) = B$. We need to check that L is indeed $\mathcal{A}$-linear and adjointable. For $a \in \mathcal{A}$, $L(\xi a) = B\xi a = (B\xi)a = L(\xi)a$, which proves $\mathcal{A}$-linearity. Furthermore, L^* is given by $L^*(\xi) := B^*\xi$. This proves the claim.

Next, we note that the restriction of Π to the sub C^*-algebra $\mathcal{K}(E)$ of $\mathcal{L}(E)$ is nondegenerate. Indeed, for $\xi, \eta, \theta \in E$ and $u \in h$, we have that $\Pi(|\xi >< \eta|)(\theta u) = \xi\langle \eta, \theta\rangle u$, and since by Lemma 4.1.8, linear span of $\{\xi\langle \eta, \theta\rangle, \xi, \eta, \theta \in E\}$ is norm-dense in E, it follows that the complex linear span of elements of the form $\Pi(|\xi >< \eta|)X$, where $X \in \mathcal{H}$ is dense in $\mathcal{H}$. This proves that $\Pi|_{\mathcal{K}(E)}$ is nondegenerate and therefore, $\mathcal{M}(\mathcal{K}(E)) \cong \{B \in \mathcal{B}(\mathcal{H}) : B\Pi(K), \Pi(K)B \in \Pi(\mathcal{K}(E)), \text{ for all } K \in \mathcal{K}(E)\}$. Now, suppose that $B \in \mathcal{B}(\mathcal{H})$ is such that $B\Pi(K), \Pi(K)B \in \Pi(\mathcal{K}(E))$ for all $K \in \mathcal{K}(E)$, or equivalently $B\Pi(K), B^*\Pi(K) \in \Pi(\mathcal{K}(E))$ for all $K \in \mathcal{K}(E)$. In particular, choosing $K = |\xi >< \eta|, \xi, \eta \in E$, we see that $B\xi\langle \eta, \theta\rangle \in E$ and $B^*\xi\langle \eta, \theta\rangle \in E$ for all $\xi, \eta, \theta \in E$. By applying Lemma 4.1.8, we get that $BE \subseteq E$, $B^*E \subseteq E$, that is, $B \in \Pi(\mathcal{L}(E))$. Conversely, since $\mathcal{K}(E)$ is an ideal in $\mathcal{L}(E)$, it follows that for any $B = \Pi(L)$ for $L \in \mathcal{L}(E)$, we have $B\Pi(K) = \Pi(LK) \in \Pi(\mathcal{K}(E))$ for all $K \in \mathcal{K}(E)$. Similarly, $B^*\Pi(\mathcal{K}(E)) \subseteq \Pi(\mathcal{K}(E))$. Thus we have proved that $\mathcal{M}(\mathcal{K}(E)) \cong \Pi(\mathcal{L}(E)) \cong \mathcal{L}(E)$, since Π is one-to-one. $\square$

Let us now give a few concrete examples of Hilbert modules. For any Hilbert space $\mathcal{H}$ and C^*-algebra $\mathcal{A}$, one may consider the algebraic tensor product $\mathcal{A} \otimes_{\text{alg}} \mathcal{H}$ as a pre-Hilbert module by putting an inner product given by

$$\left\langle \sum_i x_i \otimes \eta_i, \sum_j x'_j \otimes \eta'_j \right\rangle = \sum_{i,j} x_i^* x'_j \langle \eta_i, \eta'_j \rangle,$$

which is clearly a candidate for inner product. The right $\mathcal{A}$-module structure is defined by taking $(x \otimes \eta)a := (xa \otimes \eta)$ and extending linearly. The completion of this pre-Hilbert module under the norm inherited from the above inner product is denoted by $\mathcal{H}_{\mathcal{A}}$ or $\mathcal{A} \otimes_{C^*} \mathcal{H}$ (sometimes the suffix C^* may be dropped if understood from the context). In case $\mathcal{H}$ is separable and infinite dimensional, $\mathcal{H}_{\mathcal{A}}$ can be described as a set as the collection of all sequences $(x_k)_{k=1}^{\infty}$ with $x_k \in \mathcal{A}$, such that the series $\sum x_k^* x_k$ is convergent in the norm of $\mathcal{A}$. If $\mathcal{A} = C[0,1]$, $\mathcal{H}_{\mathcal{A}}$ can be described as the set $C([0,1], \mathcal{H}) := \{f : [0,1] \to \mathcal{H} : [0,1] \ni t \mapsto \|f(t)\|_{\mathcal{H}}$ continuous$\}$, where $\|.\|_{\mathcal{H}}$ denotes the norm of the Hilbert space $\mathcal{H}$. For $f, g \in C([0,1], \mathcal{H})$, the $C[0,1]$-valued inner product is given by $\langle f, g\rangle(t) := \langle f(t), g(t)\rangle_{\mathcal{H}}$, with $\langle \cdot, \cdot \rangle_{\mathcal{H}}$ denoting the ($\mathbb{C}$-valued) inner product of $\mathcal{H}$.

It turns out that the relatively simple Hilbert modules of the form $\mathcal{H}_{\mathcal{A}}$ are a kind of universal objects, as the following theorem due to Kasparov asserts.

Theorem 4.1.10 (Kasparov's stabilization theorem)
Let E be a countably generated Hilbert $\mathcal{A}$-module, that is, there is a countable set $B = \{y_1, y_2, \ldots\}$ in E such that the norm closure of the right $\mathcal{A}$-linear span of B is the whole of E. Then there exists a unitary element t in $\mathcal{L}(E \oplus \mathcal{H}_{\mathcal{A}}, \mathcal{H}_{\mathcal{A}})$, where $\mathcal{H}_{\mathcal{A}}$ is as above with $\mathcal{H}$ a separable infinite dimensional Hilbert space. In other words, $E \oplus \mathcal{H}_{\mathcal{A}}$ is isomorphic as a Hilbert module with $\mathcal{H}_{\mathcal{A}}$, and in particular E is embedded (by the map $t|_E$) in $\mathcal{H}_{\mathcal{A}}$ as a complemented closed submodule.

Proof:
We briefly sketch the main steps of the proof, following Mingo and Phillips [90]. Without loss of generality one can assume $\mathcal{A}$ to be unital; the nonunital case can be taken care of by the standard unitization procedure. Let $\{\eta_i\}$ be a countable normalized set of generators of E with each generator repeated infinitely often. Let $\{e_i\}$ be an orthonormal basis of $\mathcal{H}$, and let $\xi_i \equiv 1 \otimes e_i \in \mathcal{H}_{\mathcal{A}}$. Consider the map $T \in \mathcal{L}(\mathcal{H}_{\mathcal{A}}, E \oplus \mathcal{H}_{\mathcal{A}})$ given by

$$T(\xi_i) = 2^{-i} \eta_i \oplus 4^{-i} \xi_i,$$

and extending by $\mathcal{A}$-linearity and continuity. Clearly, as each η_i is repeated infinitely often, one has for any fixed i, $\eta_i \oplus 2^{-k} \xi_k \in \text{Ran}(T)$ for infinitely many k, and thus $\eta_i \oplus 0$ is in the closure of $\mathrm{Ran}(T)$ for every i. Hence, $0 \oplus \xi_k = 4^k T(\xi_k) - 2^k(\eta_k \oplus 0) \in \overline{\mathrm{Ran}(T)}$ too for each k, which proves that the closure of $\mathrm{Ran}(T)$ contains $\eta_i \oplus \xi_k$ for all i, k, hence must be the whole of $E \oplus \mathcal{H}_{\mathcal{A}}$.

We shall now argue that $\mathrm{Ran}(T^*T)$ is dense in $\mathcal{H}_{\mathcal{A}}$. First of all, it is straightforward to see that $T \in \mathcal{K}(\mathcal{H}_{\mathcal{A}}, E \oplus \mathcal{H}_{\mathcal{A}})$, and hence $T^*T \in \mathcal{K}(\mathcal{H}_{\mathcal{A}})$. Furthermore, it can be verified that T^*T can be written in the following infinite matrix form, with respect to the basis $\{\xi_i\}_i$ of $\mathcal{H}_{\mathcal{A}}$:

$$T^*T = K_1 + K_2,$$

where

$$K_1 = diag(4^{-4i})_{i=1,2,\ldots}, K_2 = ((4^{-i-j}\langle \eta_i, \eta_j \rangle))_{i,j=1,2,\ldots}.$$

Clearly K_2 is nonnegative element of $\mathcal{K}(\mathcal{H}_{\mathcal{A}})$, so $T^*T \geq K_1$. However, from the matrix representation of K_1 it is obvious that K_1 , and hence T^*T is strictly positive element of $\mathcal{K}(\mathcal{H}_{\mathcal{A}})$ in the sense of page 9, that is, for any positive functional ρ on the C^*-algebra $\mathcal{K}(\mathcal{H}_{\mathcal{A}})$ such that $\lim_\mu \rho(e_\mu) = 1$ for any approximate identity e_μ, we have $\rho(T^*T)$ to be strictly positive. By the result mentioned in page 9, $(T^*T)\mathcal{K}(\mathcal{H}_{\mathcal{A}})$ is dense in $\mathcal{K}(\mathcal{H}_{\mathcal{A}})$, and since the linear span of the set $\{Ax, A \in \mathcal{K}(\mathcal{H}_{\mathcal{A}}), x \in \mathcal{H}_{\mathcal{A}}\}$ is dense in $\mathcal{H}_{\mathcal{A}}$, it follows that $\mathrm{Ran}(T^*T)$ is dense in $\mathcal{H}_{\mathcal{A}}$, hence so is $\mathrm{Ran}(|T|)$. We now define an $\mathcal{A}$-linear map $V : \mathcal{H}_{\mathcal{A}} \to E \oplus \mathcal{H}_{\mathcal{A}}$ by $V(|T|\xi) := T\xi$, which is clearly seen to be an isometry, and furthermore, since $\mathrm{Ran}(|T|)$ and $\mathrm{Ran}(T)$ are dense in $\mathcal{H}_{\mathcal{A}}$ and $E \oplus \mathcal{H}_{\mathcal{A}}$ respectively, V has a unique continuous extension, which is a unitary isomorphism between the Hilbert modules $\mathcal{H}_{\mathcal{A}}$ and $E \oplus \mathcal{H}_{\mathcal{A}}$. Thus the choice $t = V^*$ does the job in the statement of the present theorem. □

Let us mention an important consequence of the above theorem, which will be useful in Chapter 5.

Theorem 4.1.11 *Let $\mathcal{H}$ be a separable Hilbert space. Then, for any C^*-algebra $\mathcal{A}$, we have, $\mathcal{L}(\mathcal{A} \otimes_{C^*} \mathcal{H}) \cong \mathcal{M}(\mathcal{A} \otimes \mathcal{K}(\mathcal{H}))$. Thus, in the notation of Theorem 4.1.10, for every $s \in \mathcal{L}(E)$, $tst^* \in \mathcal{L}(\mathcal{A} \otimes_{C^*} \mathcal{H}) \cong \mathcal{M}(\mathcal{A} \otimes \mathcal{K}(\mathcal{H}))$; hence $\mathcal{L}(E)$ can be embedded as a sub C^*-algebra in $\mathcal{M}(\mathcal{A} \otimes \mathcal{K}(\mathcal{H}))$.*

We shall conclude our preliminary discussion on Hilbert C^*-modules with the following unification of Stinespring and GNS constructions in the framework of Hilbert modules.

Theorem 4.1.12 (KSGNS construction)
Let $\mathcal{A}$, $\mathcal{B}$ be C^*-algebras, F be a Hilbert $\mathcal{B}$-module and $\rho : \mathcal{A} \to \mathcal{L}(F)$ be continuous with respect to the strict topologies on $\mathcal{A}$ and $\mathcal{L}(F)$. Furthermore, assume that ρ is completely positive. Then we have the following.

(i) *There exist a Hilbert $\mathcal{B}$-module F_ρ, $*$-homomorphism $\pi_\rho : \mathcal{A} \to \mathcal{L}(F_\rho)$ and an element v_ρ in $\mathcal{L}(F, F_\rho)$ such that $\rho(a) = v_\rho^* \pi_\rho(a) v_\rho$ for all $a \in \mathcal{A}$ and the $\mathcal{B}$-linear span of $\{\pi_\rho(a)v_\rho f : a \in \mathcal{A}, f \in F\}$ is dense in F_ρ.*

(ii) *If G is any Hilbert $\mathcal{B}$-module, $\pi : \mathcal{A} \to \mathcal{L}(G)$ is a $*$-homomorphism, $w \in \mathcal{L}(F, G)$ such that $\rho(a) = w^*\pi(a)w$* for all $a \in \mathcal{A}$, *and furthermore the $\mathcal{B}$-linear span of $\{\pi(a)wf : a \in \mathcal{A}, f \in F\}$ is dense in G, then there exists a unitary $u \in \mathcal{L}(F_\rho, G)$ such that $\pi(a) = u\pi_\rho(a)u^*$ and $w = uv_\rho$.*

Proof:
The proof uses very similar ideas as the proof of Theorem 2.2.9. Without loss of generality, we can assume that $\mathcal{A}$ is unital, since if necessary we can replace $\mathcal{A}$ by the unital C^*-algebra $\mathcal{M}(\mathcal{A})$, and by the continuity of ρ with respect to the appropriate strict topology, can lift ρ on $\mathcal{M}(\mathcal{A})$, which is the closure of $\mathcal{A}$ with respect to the strict topology. Consider the set S of formal $\mathbb{C}$-linear combinations of the pairs of the form (a, f) with $a \in \mathcal{A}, f \in F$. Define a right $\mathcal{B}$-module structure on S by setting $(a, f)b := (a, fb)$ and by extending linearly. Furthermore, equip S with a bilinear $\mathcal{B}$-valued form $\langle \cdot, \cdot \rangle_\rho$ given by

$$\langle (a, f), (a', f') \rangle_\rho := \langle f, \rho(a^*a')f' \rangle,$$

$a, a' \in \mathcal{A}$, $f, f' \in F$, and extending linearly. From the assumption that ρ is completely positive it is possible to verify that $\langle \cdot, \cdot \rangle_\rho$ is indeed nonnegative definite. So, by the usual procedure of taking the quotient by the submodule $\{X \in S : \langle X, X \rangle_\rho = 0\}$ and completing the quotient, we obtain a Hilbert $\mathcal{B}$-module F_ρ. We shall denote the equivalence class in F_ρ corresponding to the element $(a, f) \in S$ by $[a, f]$. Now, define $\pi_\rho : \mathcal{A} \to \mathcal{L}(F_\rho)$ by $\pi_\rho(a)([a', f]) := [aa', f]$. It is simple to verify that $\pi_\rho(ab) = \pi_\rho(a)\pi_\rho(b)$ for all $a, b \in \mathcal{A}$. Furthermore, $\langle \pi_\rho(a)[a_1, f_1], [a_2, f_2] \rangle_\rho = \langle f_1, \rho((aa_1)^*a_2)f_2 \rangle = \langle f_1, \rho(a_1^* a^* a_2) f_2 \rangle = \langle [a_1, f_1], \pi_\rho(a^*)[a_2, f_2] \rangle_\rho$. This proves that $\pi_\rho(a)^* = \pi_\rho(a^*)$. Finally, we define $v_\rho : F \to F_\rho$ by $v_\rho(f) = [1, f]$. We can verify that $\rho(a) = v_\rho^* \pi_\rho(a) v_\rho$ by a direct and straightforward calculation. This proves (i). The proof of (ii) is also routine and hence omitted. □

The triple $(F_\rho, \pi_\rho, v_\rho)$ is called the KSGNS triple associated with ρ. In case $F = \mathcal{B} = \mathbb{C}$, we recover the GNS construction, whereas the Stinespring's theorem is obtained by putting $\mathcal{B} = \mathbb{C}$.

4.2 Hilbert von Neumann modules

If $\mathcal{A}$ is a concrete C^*-algebra in $\mathcal{B}(h)$ for some Hilbert space h, then for any Hilbert space $\mathcal{H}$, the pre-Hilbert module $\mathcal{A} \otimes_{\text{alg}} \mathcal{H}$ may be viewed as a subset of $\mathcal{B}(h, h \otimes \mathcal{H})$ and $\mathcal{A} \otimes_{C^*} \mathcal{H}$ is the closure of this subset under the operator-norm inherited from $\mathcal{B}(h, h \otimes \mathcal{H})$. Instead, we may inherit one of the locally convex topologies from $\mathcal{B}(h, h \otimes \mathcal{H})$, e.g. the topology of strong operator convergence, and require the closure of $\mathcal{A} \otimes_{\text{alg}} \mathcal{H}$ under this topology. This will lead to another topological module, in general bigger than $\mathcal{A} \otimes_{C^*} \mathcal{H}$. To be precise, let us consider the topology of strong operator convergence, that is, the one given by the seminorms $\{X \mapsto \|Xu\|\ , u \in h\}$. We denote the closure by $\mathcal{A}'' \otimes_s \mathcal{H}$ or simply by $\mathcal{A}'' \otimes \mathcal{H}$ when there is no possibility of confusion. $\mathcal{A}'' \otimes \mathcal{H}$ has a natural $\mathcal{A}''$ module action from both sides and has a natural $\mathcal{A}''$-valued inner product. In view of this, we may assume that $\mathcal{A}$ itself is a unital von Neumann algebra in $\mathcal{B}(h)$. Thus we define a *Hilbert (von Neumann) $\mathcal{A}$-module* to be a right Hilbert $\mathcal{A}$-module E, such that E is equipped with the weakest locally convex topology in which the map $E \ni \xi \mapsto \langle \xi, \xi \rangle^{\frac{1}{2}} \in \mathcal{A}$ is continuous (with respect to the strong operator topology on $\mathcal{A}$), and assume furthermore that E is complete in this topology. It should be noted that for any Hilbert space $\mathcal{H}$, $\mathcal{B}(h, \mathcal{H})$ has a canonical Hilbert von Neumann $\mathcal{B}(h)$-module structure, with the inner product $\langle R, S \rangle := R^*S$ for $R, S \in \mathcal{B}(h, \mathcal{H})$.

Remark 4.2.1 *Consider the Theorem 4.1.6 and assume that $\mathcal{A}$ is a von Neumann algebra and E is a Hilbert von Neumann $\mathcal{A}$-module. Then it is not difficult to see from the proof of the Theorem 4.1.6 that the image of E under Γ will be closed under the weak operator topology (and hence strong operator topology as well), inherited from $\mathcal{B}(h, \mathcal{H})$.*

We are now going to show that Hilbert von Neumann modules are very well-behaved in contrast to their C^*-counterparts, namely they possess the crucial property that every closed submodule is complemented, and hence most of the results valid for a Hilbert space can be translated with little or no modification in the framework of Hilbert von Neumann modules, which is not true for a general Hilbert C^*-module. We begin with proving the existence of a sort of orthonormal basis in any Hilbert von Neumann module (see [121] for more details).

Theorem 4.2.2 *Given a von Neumann algebra $\mathcal{A} \subseteq \mathcal{B}(h)$ and a Hilbert von Neumann $\mathcal{A}$-module E, looked upon as a submodule of $\mathcal{B}(h, \mathcal{H})$ for some Hilbert space $\mathcal{H}$ as in the Theorem 4.1.6, there exists a family $(e_\beta)_{\beta \in I}$ of partial isometries in E, (to be called a complete quasi orthonormal system for E), indexed by some set I, with the following properties:*

(i) $\langle e_\beta, e_\alpha \rangle = e_\beta^* e_\beta \delta_{\beta\alpha}$; and
(ii) for any $e \in E$, the net $e_J := \sum_{\beta \in J} e_\beta \langle e_\beta, e \rangle$, indexed by the finite subsets $J \subseteq I$, partially ordered by inclusion, converges in the topology of E to e.

Proof:
Without loss of generality (by the Theorem 4.1.6 and Remark 4.2.1) we can assume that $\{eu, e \in E, u \in h\}$ is total in $\mathcal{H}$, and the topology of E is inherited from the strong-operator topology of $\mathcal{B}(h, \mathcal{H})$. Now, let us call (following the terminology of [121]) a collection $(e_\beta)_{\beta \in T}$ for some indexed set T to be a *quasi-orthonormal system* if (e_β) satisfy the property (*i*) in the statement of the present theorem, but not necessarily (*ii*). By the standard argument involving Zorn's Lemma it can be proved that there is a maximal element (where we have chosen the natural partial ordering by inclusion for the class of all quasi-orthonormal systems of E), and let us denote this maximal quasi-orthonormal system by $(e_\beta)_{\beta \in I}$. We claim that this system will satisfy the property (*ii*) in the statement of the theorem, and thereby completing the proof. Note that if we take $p_\alpha = e_\alpha e_\alpha^*$, then $p_\alpha e_\alpha = e_\alpha$ and $p_\alpha p_\beta = 0$ for different α, β. Therefore, for any finite subset J of I the operator $P_J = \sum_{\beta \in J} e_\beta e_\beta^* = \sum_J p_\beta$ is a finite rank projection in $\mathcal{L}(E)$, and since the p_β elements are mutually orthogonal, it is clear that P_J is a nondecreasing net of projections, hence weakly Cauchy. From this, it can be seen that for each fixed $x \in E$, $(P_J x)_J$ converges in the topology of E. If possible, suppose that there is an element $e \in E$ for which $\lim_J P_J e$ is not equal to e. Let $y = (e - \lim_J(P_J e))$. Clearly, $\langle y, e_\alpha \rangle = e^* e_\alpha - \sum_\beta e^* e_\beta e_\beta^* e_\alpha = e^* e_\alpha - e^* e_\alpha = 0$, for all $\alpha \in I$. Since y is an element of $\mathcal{B}(h, \mathcal{H})$, we can write the polar decomposition of y as $y = v|y|$, where v is a partial isometry in $\mathcal{B}(h, \mathcal{H})$ and $|y| = \langle y, y \rangle^{\frac{1}{2}} \in \mathcal{A}$. Choose $v_n = yn(1 + n|y|)^{-1} \in E, n = 1, 2, \ldots$, we see that $v_n \to v$ in the strong operator topology, so that $v \in E$ (since E is closed under the strong operator topology). Furthermore, since $v_n^* e_\alpha = n(1 + n|y|)^{-1} y^* e_\alpha = 0$ for all α, we have that $\langle v, e_\alpha \rangle = v^* e_\alpha = 0$. Thus, we can form a quasi orthonormal system in E by adding v to $(e_\beta)_I$ which will be strictly larger than $(e_\beta)_I$ and thus contradicts the assumed maximality, unless $v = 0$. So, $v = 0$, that is, $y = 0$, proving our claim. □

Using Zorn's Lemma, we can deduce similarly the following, the proof of which is omitted.

Lemma 4.2.3 *Given any quasi-orthonormal system $(e_\alpha)_{\alpha \in I_1}$ for some Hilbert von Neumann $\mathcal{A}$-module E (as defined in the proof of the previous theorem), we can always find a complete quasi-orthonormal system $(e_\beta)_{\beta \in I}$ for E with $I_1 \subseteq I$.*

This lemma enables us to prove the following theorem.

Theorem 4.2.4 *Let $\mathcal{A} \subseteq \mathcal{B}(h)$ be a von Neumann algebra and E be a Hilbert von Neumann $\mathcal{A}$-module. Then, for any closed (with respect to the locally* *convex topology of E) submodule F of E, there is a direct-sum decomposition of E as $E = F \oplus F^{\perp}$, where $F^{\perp} := \{e \in E : \langle e, f \rangle = 0$ for all $f \in F\}$.*

Proof:
By Theorem 4.2.2 and Lemma 4.2.3, we choose a complete quasi-orthonormal system $(e_\alpha)_{\alpha \in I_1}$ for F and enlarge it to a complete quasi-orthonormal system $(e_\beta)_{\beta \in I}$ for E. Let $I_2 := I - I_1$, and it is clear that for $\beta \in I_2$, $e_\beta \in F^{\perp}$. Now, given any $e \in E$, we can write it as $e = e_1 + e_2$, where $e_1 = \lim_{J_1 \subseteq I_1} \sum_{\beta \in J_1} e_\beta \langle e_\beta, e \rangle$, and $e_2 = \lim_{J_2 \subseteq I_2} \sum_{\beta \in J_2} e_\beta \langle e_\beta, e \rangle$, with J_1, J_2 finite subsets of I_1 and I_2 respectively. Clearly, $e_1 \in F$, and $e_2 \in F^{\perp}$, thus completing the proof. □

The existence of complete quasi-orthonormal systems leads to the following result about Hilbert von Neumann modules, which is analogous to the Riesz' representation theorem in Hilbert spaces.

Proposition 4.2.5 *Let E be a Hilbert von Neumann $\mathcal{A}$-module, where $\mathcal{A} \subseteq \mathcal{B}(h)$ is a von Neumann algebra. Suppose that $\Phi : E \to \mathcal{A}$ is an $\mathcal{A}$-linear continuous map. Then there exists an element $e_\Phi \in E$ such that $\Phi(e) = \langle e_\Phi, e \rangle$ for all $e \in E$. This means that a Hilbert von Neumann module is self-dual.*

We omit the proof since we shall have no occasion to use this result and the reader is referred to [121] for a proof.

Next we shall obtain a characterization of Hilbert von Neumann $\mathcal{A}$-modules by showing that they always arise canonically from some normal $*$-representa tion of the commutant von Neumann algebra $\mathcal{A}'$ of $\mathcal{A}$. To this end, we first introduce the notion of the *intertwiner module* as follows.

Theorem 4.2.6 *Let $\mathcal{B} \subseteq \mathcal{B}(h)$ be a von Neumann algebra and $\pi : \mathcal{B} \to \mathcal{B}(\mathcal{H})$ be some normal $*$-representation of $\mathcal{B}$. Let $\mathcal{I} \equiv \mathcal{I}(\mathcal{B}, \pi) := \{R \in \mathcal{B}(h, \mathcal{H}) : \pi(b)R = Rb,$ for all $b \in \mathcal{B}\}$. Then $\mathcal{I}$ can be given a Hilbert $\mathcal{B}'$-module structure (where $\mathcal{B}'$ is the commutant of $\mathcal{B}$ in $\mathcal{B}(h)$) by inheriting the right action of $\mathcal{B}(h)$ on $\mathcal{B}(h, \mathcal{H})$ and the canonical $\mathcal{B}(h)$ valued inner product of $\mathcal{B}(h, \mathcal{H})$. $\mathcal{I}$ is also complete in the strong operator topology inherited from $\mathcal{B}(h, \mathcal{H})$.*

Proof:
For $a \in \mathcal{B}', b \in \mathcal{B}$ and $R \in \mathcal{I}$, we have $\pi(b)Ra = Rba = (Ra)b$, since a and b commute with each other. This proves $Ra \in \mathcal{I}$. Furthermore, for $R, S \in \mathcal{I}$ and $b \in \mathcal{B}$, we have that $bR^*S = (Rb^*)^*S = (\pi(b)^*R)^*S = R^*\pi(b)S = R^*Sb$, that is, $R^*S \in \mathcal{B}'$. Finally, the completeness mentioned in the statement of the present theorem follows by standard arguments. □

The following theorem shows how Hilbert von Neumann $\mathcal{A}$ modules are in one-to-one correspondence with the normal $*$-representations (modulo unitary isomorphism) of $\mathcal{A}'$.

Theorem 4.2.7 *Let $\mathcal{A}$ be a von Neumann algebra in $\mathcal{B}(h)$ and let E be a Hilbert von Neumann $\mathcal{A}$-module $\subseteq \mathcal{B}(h, \mathcal{H})$, where $\mathcal{H}$ is a Hilbert space. Assume furthermore that $\{eu, e \in E, u \in h\}$ is a total subset of $\mathcal{H}$. Then we can find a unique normal $*$-representation $\rho : \mathcal{A}' \to \mathcal{B}(\mathcal{H})$ such that $E = \mathcal{I}(\mathcal{A}', \rho)$.*

Proof:
The uniqueness of ρ is clear, since the action $\rho(a)$ is uniquely determined on the total set of vectors consisting of $\{eu, e \in E, u \in h\}$. To prove the existence, we begin with the obvious definition $\rho(a)(eu) = e(au), e \in E, u \in h$. However, we need to prove that $\rho(a)$ is well-defined and extends to a unique bounded linear operator on $\mathcal{H}$. Fix $e_1, \ldots, e_n \in E$ and $u_1, \ldots, u_n \in h$, n positive integer. We have

$$\begin{aligned}&\left\| \rho(a) \left(\sum_i e_i u_i \right) \right\|^2 \\ &= \sum_{i,j} \langle u_i, a^* \langle e_i, e_j \rangle a u_j \rangle \\ &= \sum_{i,j} \langle u_i, a^* a \langle e_i, e_j \rangle u_j \rangle,\end{aligned}$$

since a and $\langle e_i, e_j \rangle$ commute. Now, the $n \times n$ $\mathcal{A}$-valued matrix $(\langle e_i, e_j \rangle)_{i,j=1,\ldots,n}$ is a nonnegative element in $M_n(\mathcal{A}) = M_n \otimes \mathcal{A} \subseteq M_n \otimes \mathcal{B}(h)$, and its each entry (with respect to an orthonormal basis of $\mathbb{C}^n$) commutes with each entry of the nonnegative element $diag(\|a\|^2 I - a^* a)$ of $M_n \otimes \mathcal{B}(h)$. So, the product of them, that is, $(((\|a\|^2 I - a^* a)\langle e_i, e_j \rangle))_{ij}$ is nonnegative. Using this fact, we see that $\|\rho(a)(\sum_i e_i u_i)\|^2 \leq \|a\|^2 \| \sum_i e_i u_i \|^2$ from which it follows that $\rho(a)$ is well-defined and extends to a bounded linear operator on $\mathcal{H}$ with operator-norm less than or equal to that of a. Furthermore, it is clear from the definition that $a \mapsto \rho(a)$ is a $*$-representation of $\mathcal{A}'$. Its normality can also be verified by routine arguments. Indeed, if a_α is an nondecreasing net of nonnegative elements in $\mathcal{A}'$, with $a_\alpha \uparrow a$, then from the definition of ρ it is obvious that $\langle \xi, \rho(a_\alpha)\xi \rangle \uparrow \langle \xi, \rho(a)\xi \rangle$ for ξ in the algebraic linear span of $\{eu, e \in E, u \in h\}$. However, since this linear span is dense in $\mathcal{H}$ and also since $\|\rho(a_\alpha)\| \leq \|\rho(a)\|$ for all α, it follows that the same conclusion holds for any $\xi \in \mathcal{H}$, hence the normality of ρ is verified.

Finally, it is clear that $E \subseteq \mathcal{I}(\mathcal{A}', \rho)$. By Theorem 4.2.4, $\mathcal{I}(\mathcal{A}', \rho) = E \oplus E^\perp$, where $E^\perp$ is taken in $\mathcal{I}(\mathcal{A}', \rho)$. It is enough to show $E^\perp = \{0\}$, but it is clear that for any element $e' \in E^\perp$ and $u \in h$, we have $\langle e'u, ev \rangle = \langle u, \langle e', e \rangle v \rangle = 0$ for any $e \in E, v \in h$, and by totality of the set $\{eu, e \in E, u \in h\}$, $\langle e'u, \eta \rangle = 0$ for all $\eta \in \mathcal{H}$, so $e'u = 0$ for all $u \in h$,, that is, $e' = 0$. This completes the proof. □

This result, combined with the result about the structure of normal $*$-representations of von Neumann algebras mentioned in Chapter 2 (Proposition 2.1.7), we obtain an analogue of the Kasparov's stabilization theorem for any Hilbert von Neumann module. However, before that, we need to note a few simple but useful facts about the Hilbert von Neumann module of the form $\mathcal{A} \otimes \mathcal{H}$, where $\mathcal{H}$ is a Hilbert space, possibly nonseparable. For this, let us first introduce some notations, which will be very useful in subsequent chapters also. Let $\mathcal{H}_1, \mathcal{H}_2$ be two Hilbert spaces and A be a (possibly unbounded) linear operator from $\mathcal{H}_1$ to $\mathcal{H}_1 \otimes \mathcal{H}_2$ with domain $\mathcal{D}$. For each $f \in \mathcal{H}_2$, we define a linear operator $\langle f, A \rangle$ with domain $\mathcal{D}$ and taking value in $\mathcal{H}_1$ such that

$$\langle \langle f, A \rangle u, v \rangle = \langle Au, v \otimes f \rangle, \tag{4.1}$$

for $u \in \mathcal{D},\ v \in \mathcal{H}_1$. This definition makes sense because we have, $|\langle Au, v \otimes f \rangle| \leq \|Au\| \, \|f\| \, \|v\|$, and thus $\mathcal{H}_1 \ni v \to \langle Au, v \otimes f \rangle$ is a bounded linear functional. Moreover, $\|\langle f, A \rangle u\| \leq \|Au\| \, \|f\|$, for all $u \in \mathcal{D},\ f \in \mathcal{H}_2$. Similarly, for each fixed $u \in \mathcal{D}, v \in \mathcal{H}_1,\ \mathcal{H}_2 \ni f \to \langle Au, v \otimes f \rangle$ is bounded linear functional, and hence there exists a unique element of $\mathcal{H}_2$, to be denoted by $A_{v,u}$, satisfying

$$\langle A_{v,u}, f \rangle = \langle Au, v \otimes f \rangle = \langle \langle f, A \rangle u, v \rangle. \tag{4.2}$$

We shall denote by $\langle A, f \rangle$ the adjoint of $\langle f, A \rangle$, whenever it exists. Clearly, if A is bounded, then so is $\langle f, A \rangle$ and $\|\langle f, A \rangle\| \leq \|A\| \, \|f\|$. Similarly, for any $T \in \mathcal{B}(\mathcal{H}_1 \otimes \mathcal{H}_2)$ and $f \in \mathcal{H}_2$, one can define $T_f \in \mathcal{B}(\mathcal{H}_1, \mathcal{H}_1 \otimes \mathcal{H}_2)$ by setting $T_f u = T(u \otimes f)$. With the above notations at our disposal, let us give a brief sketch of some properties of $\mathcal{A} \otimes \mathcal{H}$.

Lemma 4.2.8 *Any element X of $\mathcal{A} \otimes \mathcal{H}$ can be written as, $X = \sum_{\alpha \in J} x_\alpha \otimes \gamma_\alpha$, where $\{\gamma_\alpha\}_{\alpha \in J}$ is an orthonormal basis of $\mathcal{H}$ and $x_\alpha \in \mathcal{A}$. The above sum over a possibly uncountable index set J makes sense in the usual way: it is strongly convergent and* for all $u \in h$, *there exists an at most countable subset J_u of J such that $Xu = \sum_{\alpha \in J_u} (x_\alpha u) \otimes \gamma_\alpha$. Moreover, once $\{\gamma_\alpha\}$ is fixed, x_α's are uniquely determined by X.*

Proof:
Set $x_\alpha = \langle \gamma_\alpha, X \rangle$. Clearly, if $X \in \mathcal{A} \otimes_{\text{alg}} \mathcal{H}$, $x_\alpha \in \mathcal{A}$ for all α. Since any element of $\mathcal{A} \otimes \mathcal{H}$ is a strong limit of elements from $\mathcal{A} \otimes_{\text{alg}} \mathcal{H}$; and since $\mathcal{A}$ is strongly closed, it follows that $x_\alpha \in \mathcal{A}$ for an arbitrary $X \in \mathcal{A} \otimes \mathcal{H}$. Now, for a fixed $u \in h$, let J_u be the (at most countable) set of indices such that for all $\alpha \in J_u$, there exists $v_\alpha \in h$ with $\langle Xu, v_\alpha \otimes \gamma_\alpha \rangle \neq 0$. Then for any

$v \in h$ and $\gamma \in \mathcal{H}$, we have with $c_\alpha^\gamma = \langle \gamma_\alpha, \gamma \rangle$,

$$\begin{aligned}\langle Xu, v \otimes \gamma \rangle &= \sum_{\alpha \in J_u} c_\alpha^\gamma \langle Xu, v \otimes \gamma_\alpha \rangle = \sum_{\alpha \in J_u} c_\alpha^\gamma \langle \langle \gamma_\alpha, X \rangle u, v \rangle \\ &= \sum_{\alpha \in J_u} \langle x_\alpha u, v \rangle \langle \gamma_\alpha, \gamma \rangle = \left\langle \sum_{\alpha \in J_u} (x_\alpha \otimes \gamma_\alpha) u, \, v \otimes \gamma \right\rangle;\end{aligned}$$

that is, $X = \sum_{\alpha \in J} x_\alpha \otimes \gamma_\alpha$ in the sense described in the statement of the lemma. Given $\{\gamma_\alpha\}$, the choice of x_α values is unique, because for any fixed α_0, $\langle \gamma_{\alpha_0}, X \rangle = x_{\alpha_0}$, which follows from the previous computation if we take γ to be γ_{α_0}. □

Corollary 4.2.9 *Let $X, Y \in \mathcal{A} \otimes \mathcal{H}$ be given by $X = \sum_{\alpha \in J} x_\alpha \otimes \gamma_\alpha$ and $Y = \sum_{\alpha \in J} y_\alpha \otimes \gamma_\alpha$ as in the lemma above. For any finite subset I of J, if we denote by X_I and Y_I the elements $\sum_{\alpha \in I} x_\alpha \otimes \gamma_\alpha$ and $\sum_{\alpha \in I} y_\alpha \otimes \gamma_\alpha$ respectively, then* $\lim_I \langle X_I, Y_I \rangle = \langle X, Y \rangle$ *where the limit is taken over the directed family of finite subsets of J with usual partial ordering by inclusion.*

Proof:
The proof is an adaptation of Lemma 27.7 in [100], hence the details are omitted.

□

We give below a convenient necessary and sufficient criterion for verifying whether an element of $\mathcal{B}(h, h \otimes \mathcal{H})$ belongs to $\mathcal{A} \otimes \mathcal{H}$.

Lemma 4.2.10 *Let $X \in \mathcal{B}(h, h \otimes \mathcal{H})$. Then X belongs to $\mathcal{A} \otimes \mathcal{H}$ if and only if $\langle \gamma, X \rangle \in \mathcal{A}$ for all γ in some dense subset $\mathcal{E}$ of $\mathcal{H}$.*

Proof:
That $X \in \mathcal{A} \otimes \mathcal{H}$ implies $\langle \gamma, X \rangle \in \mathcal{A}$ for all $\gamma \in \mathcal{H}$ has already been observed in the proof of the previous lemma. For the converse, first we claim that $\langle \gamma, X \rangle \in \mathcal{A}$ for all γ in $\mathcal{E}$ (where $\mathcal{E}$ is dense in $\mathcal{H}$) will imply $\langle \gamma, X \rangle \in \mathcal{A}$ for all $\gamma \in \mathcal{H}$. Indeed, for any $\gamma \in \mathcal{H}$ there exists a net $\gamma_\alpha \in \mathcal{E}$ such that $\gamma_\alpha \to \gamma$, and hence $\| \langle \gamma, X \rangle - \langle \gamma_\alpha, X \rangle \| \leq \| \gamma_\alpha - \gamma \| \| X \| \to 0$. Now let us fix an orthonormal basis $\{\gamma_\alpha\}_{\alpha \in J}$ of $\mathcal{H}$ and write $X = \sum_{\alpha \in J} \langle \gamma_\alpha, X \rangle \otimes \gamma_\alpha$ by Lemma 4.2.8. Clearly, the net $X_{\mathcal{I}}$ indexed by finite subsets $\mathcal{I}$ of J (partially ordered by inclusion) converges strongly to X. Since $X_{\mathcal{I}} \in \mathcal{A} \otimes_{\mathrm{alg}} \mathcal{H}$ for any such finite subset $\mathcal{I}$ (as $\langle \gamma_\alpha, X \rangle \in \mathcal{A}$ for all α), the proof follows by noting that $\mathcal{A}$ is strongly closed. □

In case $\mathcal{H} = \Gamma(k)$, the Fock space over some Hilbert space k (see Section 2.4), we call the module $\mathcal{A} \otimes \Gamma(k)$ the right Fock $\mathcal{A}$-module over $\Gamma(k)$, for short the *Fock module*, and denote it by $\mathcal{A} \otimes \Gamma$.

The next theorem spells out the generic structure of a Hilbert von Neumann $\mathcal{A}$-module.

Theorem 4.2.11 *Given any Hilbert von Neumann $\mathcal{A}$-module E, there is a Hilbert space (possibly nonseparable) $\mathcal{H}_0$ such that E is isometrically isomorphic with some closed (Hilbert von Neumann)-submodule of the Hilbert von Neumann $\mathcal{A}$-module $\mathcal{A} \otimes \mathcal{H}_0$.*

Proof:
By Theorem 4.2.7, we may assume that $E = \mathcal{I}(\mathcal{A}', \rho)$ for some normal $*$-representation of $\rho : \mathcal{A}' \to \mathcal{B}(\mathcal{H})$ for some Hilbert space $\mathcal{H}$, such that $\rho(1) = 1_{\mathcal{H}}$. Now, by appealing to the Proposition 2.1.7 of Chapter 2, we obtain some Hilbert space $\mathcal{H}_0$ and a Hilbert space isometry $\Sigma_0 : \mathcal{H} \to h \otimes \mathcal{H}_0$ (where h is such that $\mathcal{A} \subseteq \mathcal{B}(h)$ and $\mathcal{A}'$ is the commutant of $\mathcal{A}$ in $\mathcal{B}(h)$) such that $\rho(a) = \Sigma_0^*(a \otimes 1)\Sigma_0$ for all $a \in \mathcal{A}'$ and $P = \Sigma_0 \Sigma_0^*$ is a projection in $\mathcal{A} \otimes \mathcal{B}(\mathcal{H}_0)$. We shall now explicitly calculate the space $\mathcal{I}(\mathcal{A}', \rho)$. Let $R \in \mathcal{B}(h, \mathcal{H})$ be such that $Ra = \rho(a)R$ for all $a \in \mathcal{A}'$. Applying Σ_0 on both sides of this equality, we get that $\Sigma_0 Ra = P(a \otimes 1)\Sigma_0 R$, and since P commutes with $(\mathcal{A}' \otimes 1)$ and $P\Sigma_0 = \Sigma_0$, it follows that $\Sigma_0 Ra = (a \otimes 1)\Sigma_0 R$, for all $a \in \mathcal{A}'$, that is, $\Sigma_0 R \in \mathcal{A} \otimes \mathcal{H}_0$. In fact, every step in the above calculation is clearly reversible, and thus we get that

$$\mathcal{I}(\mathcal{A}', \rho) = \{R \in \mathcal{B}(h, \mathcal{H}) : \Sigma_0 R \in \mathcal{A} \otimes \mathcal{H}_0\}.$$

We now define $\Sigma : \mathcal{I}(\mathcal{A}', \rho) \to \mathcal{A} \otimes \mathcal{H}_0$ by $\Sigma(R) := \Sigma_0 R$, and clearly this is an $\mathcal{A}$-linear isometry, with the adjoint given by $\mathcal{A} \otimes \mathcal{H}_0 \ni S \mapsto \Sigma_0^* S \in \mathcal{I}(\mathcal{A}', \rho)$. We note that $\Sigma_0^* S$ is in $\mathcal{I}(\mathcal{A}', \rho)$ because $\Sigma_0 \Sigma_0^* S = PS$, which is in $\mathcal{A} \otimes \mathcal{H}_0$ as $P \in \mathcal{A} \otimes \mathcal{B}(\mathcal{H}_0)$. Thus, $E = \mathcal{I}(\mathcal{A}', \rho)$ is isometrically isomorphic with the closed submodule $P(\mathcal{A} \otimes \mathcal{H}_0)$. □

For various applications of Hilbert modules in quantum probability and related fields, we refer the reader to [7] and [120].

4.3 Group actions on Hilbert modules

4.3.1 The case of Hilbert C^*-modules

Let us first discuss the case of Hilbert C^*-modules. Let G be a locally compact group, $\mathcal{A}$ be a C^*-algebra, and assume that there is a strongly continuous representation $\alpha : G \to \mathrm{Aut}(\mathcal{A})$ as discussed in Chapter 2. Following the terminology of [90], we introduce the concept of a Hilbert C^* $G - \mathcal{A}$-module as follows.

Definition 4.3.1 A *Hilbert C^* $G - \mathcal{A}$ module* (or *$G - \mathcal{A}$ module* for short) is a pair (E, β) where E is a Hilbert C^* $\mathcal{A}$-module and β is a map from G into the set of $\mathbb{C}$-linear (caution : **not** $\mathcal{A}$-linear !) maps from E to E, such that $\beta_g, g \in G$ satisfies the following:

(i) $\beta_{gh} = \beta_g \circ \beta_h$ for $g, h \in G$, $\beta_e = \mathrm{Id}$, where e is the identity element of G;
(ii) $\beta_g(\xi a) = \beta_g(\xi)\alpha_g(a)$ for $\xi \in E, a \in \mathcal{A}$;
(iii) $g \mapsto \beta_g(\xi)$ is continuous for each fixed $\xi \in E$;
(iv) $\langle \beta_g(\xi), \beta_g(\eta) \rangle = \alpha_g(\langle \xi, \eta \rangle)$ for all $\xi, \eta \in E$, where $\langle \cdot, \cdot \rangle$ denotes the $\mathcal{A}$-valued inner product of E.

When β is understood from the context, we may refer to E as a $G - \mathcal{A}$ module, without explicitly mentioning the pair (E, β). Given two $G - \mathcal{A}$ modules (E_1, β) and (E_2, γ), there is a natural G-action π induced on $\mathcal{L}(E_1, E_2)$, given by $\pi_g(T)(\xi) := \gamma_g(T(\beta_{g^{-1}}(\xi)))$ for $g \in G, \xi \in E_1, T \in \mathcal{L}(E_1, E_2)$. $T \in \mathcal{L}(E_1, E_2)$ is said to be *G-equivariant* if $\pi_g(T) = T$ for all $g \in G$. It is clear that for each fixed $T \in \mathcal{L}(E_1, E_2)$ and $\xi \in E_1$, $g \mapsto \pi_g(T)\xi$ is continuous. We say that T is *G-continuous* if $g \mapsto \pi_g(T)$ is continuous with respect to the norm topology on $\mathcal{L}(E_1, E_2)$. We say that E_1 and E_2 are *isomorphic as $G - \mathcal{A}$-modules*, or that they are *equivariantly isomorphic* if there is a G-equivariant unitary map $T \in \mathcal{L}(E_1, E_2)$.

We shall now prove an analogue of Kasparov's theorem in the framework of $G - \mathcal{A}$ modules. Before that, it is necessary to introduce some notation. Following [90], for any Hilbert $\mathcal{A}$-module F, let us denote by F^∞ the Hilbert $\mathcal{A}$-module obtained by taking a direct sum of countably infinitely many copies of F. More precisely, F^∞ as a set is defined to consist of all sequences $(f_1, f_2, \ldots)$, $f_i \in F$, such that the series $\sum_n \langle f_n, f_n \rangle$ converges in the norm of $\mathcal{A}$. The right $\mathcal{A}$-module structure on F^∞ is the obvious one: $(f_1, f_2, \ldots)a := (f_1 a, f_2 a, \ldots)$, and the $\mathcal{A}$-valued inner product is given be the sum of componentwise inner product. We note that for a separable infinite dimensional Hilbert space $\mathcal{H}$, we have $\mathcal{H}_\mathcal{A} = \mathcal{A}^\infty$, viewing $\mathcal{A}$ as an $\mathcal{A}$-module in the obvious way. Furthermore, in case F is a $G - \mathcal{A}$ module with the corresponding G-action β, then so is F^∞ by taking the G-action β_g^∞ as $\beta_g^\infty(f_1, f_2, \ldots) := (\beta_g(f_1), \beta_g(f_2), \ldots)$.

A word of caution about this notation: $\mathcal{A}^\infty$ should not be confused with the notation $\mathcal{A}_\infty$ which will be introduced and used in Chapters 8 and 9 to denote 'smooth' subalgebras.

For a $G - \mathcal{A}$ module (E, β), we define a right $\mathcal{A}$-module structure on $C_c(G, E)$ (the set of continuous compactly supported functions from G to E) by setting $(fa)(t) := f(t)a, t \in G, a \in \mathcal{A}$. We also define an $\mathcal{A}$-valued inner product by

$$\langle f_1, f_2 \rangle := \int_G \langle f_1(t), f_2(t) \rangle dt,$$

where dt denotes integration with respect to some fixed left Haar measure on G. We denote the completion of the pre-Hilbert $\mathcal{A}$-module $C_c(G, E)$ by $L^2(G, E)$, which is a Hilbert C^* $\mathcal{A}$-module. Furthermore, we make it into a

$G - \mathcal{A}$module by defining a G-action on $C_c(G, E)$ by $\tilde{\beta}_g(f)(t) = \beta_g(f(g^{-1}t))$, $g, t \in G$, and extending it by continuity on $L^2(G, E)$. In fact, it is clear that $L^2(G, E)$ is the completion of the algebraic tensor product of $L^2(G)$ and E, with the G-action being the tensor product of the left regular representation on $L^2(G)$ and the given action β on E. It may also be noted that since any Hilbert $\mathcal{A}$ module can be viewed as a $G - \mathcal{A}$ module by taking the trivial G-action, we can actually perform the above construction for any $\mathcal{A}$-module E to obtain $L^2(G, E)$, which will be a $G - \mathcal{A}$ module.

Lemma 4.3.2 *Let (E_1, β) and (E_2, γ) be two $G - \mathcal{A}$ modules such that E_1, E_2 are isomorphic as Hilbert $\mathcal{A}$-modules, that is, there is an $\mathcal{A}$-linear unitary (not necessarily equivariant) U from E_1 onto E_2. Then $L^2(G, E_1)$ and $L^2(G, E_2)$ are isomorphic as $G - \mathcal{A}$-modules.*

Proof:
Define $V : C_c(G, E_1) \to C_c(G, E_2)$ by

$$(Vf)(t) = \gamma_t(U(\beta_{t^{-1}}(f(t)))).$$

It is not difficult to observe that V is $\mathcal{A}$-linear, since

$$\begin{aligned} (V(fa))(t) &= \gamma_t(U(\beta_{t^{-1}}(f(t)a))) \\ &= \gamma_t(U(\beta_{t^{-1}}(f(t))\alpha_{t^{-1}}(a))) = \gamma_t(U(\beta_{t^{-1}}(f(t)))\alpha_{t^{-1}}(a)) \\ &= \gamma_t(U(\beta_{t^{-1}}(f(t))))\alpha_t(\alpha_{t^{-1}}(a)) = (Vf)(t)a. \end{aligned}$$

Similarly, by using the property (iv) in the definition of a $G - \mathcal{A}$ module, one can verify that V is an isometry. That V is onto and G-equivariant follow also by straightforward arguments. Thus, V extends to an equivariant unitary from $L^2(G, E_1)$ onto $L^2(G, E_2)$. □

Using this lemma, we state and prove the following equivariant analogue of Kasparov's theorem. However, we need the compactness of G for technical purposes.

Theorem 4.3.3 *Let (E, β) be a Hilbert C^* $G - \mathcal{A}$ module and assume that E is countably generated as a Hilbert $\mathcal{A}$-module, that is, there is a countable set $S = \{e_1, e_2, \cdots\}$ of elements of E such that the right $\mathcal{A}$-linear span of S is dense in E. Assume furthermore that G is compact. Then, there is a G-equivariant unitary from $E \oplus L^2(G, \mathcal{A}^\infty)$ onto $L^2(G, \mathcal{A}^\infty)$.*

Proof:
Let us first show that $E \oplus L^2(G, E)^\infty \cong L^2(G, E)^\infty$ as $G - \mathcal{A}$ modules, by explicitly constructing an equivariant unitary map. Define $V : E \to L^2(G, E)$

given by $(V\xi)(t) := \xi$, for all $t \in G$. Clearly V is a G-equivariant isometry. Now, define $U : E \oplus L^2(G, E)^\infty \to L^2(G, E)^\infty$ by

$$U(\xi \oplus (f_1, f_2, \ldots)) := (V\xi + (1 - VV^*)f_1, VV^* f_1 + (1 - VV^*)f_2, \ldots),$$

where $\xi \in E$ and $f_1, f_2, \ldots \in L^2(G, E)$. It is clear that U is G-equivariant as V is so. Furthermore, by direct computation it is easy to see that the adjoint U^* of U is given by,

$$U^*(f_1, f_2, \ldots) = (V^* f_1 \oplus (VV^* f_2 + (1 - VV^*)f_1, VV^* f_3 + (1 - VV^*)f_3, \ldots)).$$

Now, by using the fact that $V^*V = \mathrm{Id}$, it can be verified that U and U^* are inverses of each other, that is, U is a unitary, which proves the claim that $E \oplus L^2(G, E)^\infty$ and $L^2(G, E)^\infty$ are isomorphic as $G - \mathcal{A}$-modules.

Let us now note that for any $G - \mathcal{A}$ module F, $L^2(G, F)^\infty \cong (L^2(G) \otimes F) \oplus (L^2(G) \otimes F) \oplus \cdots \cong L^2(G) \otimes (F \oplus F \oplus \cdots) \cong L^2(G, F^\infty)$, as $G - \mathcal{A}$-modules. In particular, $L^2(G, \mathcal{A})^\infty \cong L^2(G, \mathcal{A}^\infty)$. Also, it is a straightforward observation that $\mathcal{A}^\infty$ and $(\mathcal{A}^\infty)^\infty$ are isomorphic as Hilbert C^* $\mathcal{A}$-modules, hence by the Lemma 4.3.2, we have that $L^2(G, \mathcal{A}^\infty)^\infty \cong L^2(G, (\mathcal{A}^\infty)^\infty)) \cong L^2(G, \mathcal{A}^\infty) \cong L^2(G, \mathcal{A})^\infty$ as $G - \mathcal{A}$ modules. But on the other hand, by the Kasparov's theorem (Theorem 4.1.10) proved for C^*-modules without any group action, we know that $\mathcal{A}^\infty$ and $E \oplus \mathcal{A}^\infty$ are isomorphic as Hilbert $\mathcal{A}$-modules, hence by Lemma 4.3.2, we have the isomorphisms of $G - \mathcal{A}$-modules: $L^2(G, \mathcal{A}^\infty) \cong L^2(G, E \oplus \mathcal{A}^\infty)$. Thus, we have, $L^2(G, \mathcal{A})^\infty \cong L^2(G, \mathcal{A}^\infty)^\infty \cong L^2(G, E \oplus \mathcal{A}^\infty)^\infty \cong L^2(G, E)^\infty \oplus L^2(G, \mathcal{A}^\infty)^\infty \cong L^2(G, E)^\infty \oplus L^2(G, \mathcal{A})^\infty$. Finally, using the earlier proven fact that $E \oplus L^2(G, E)^\infty$ is isomorphic with $L^2(G, E)^\infty$, we have the following isomorphisms of $G - \mathcal{A}$ modules:

$$\begin{aligned} E \oplus L^2(G, \mathcal{A})^\infty &\cong E \oplus L^2(G, E)^\infty \oplus L^2(G, \mathcal{A})^\infty \\ &\cong L^2(G, E)^\infty \oplus L^2(G, \mathcal{A})^\infty \cong L^2(G, \mathcal{A})^\infty \\ &\cong L^2(G, \mathcal{A}^\infty). \end{aligned}$$

This completes the proof. □

Remark 4.3.4 *We observe that $L^2(G, \mathcal{H}_\mathcal{A}) \cong \mathcal{A} \otimes_{C^*} (\mathcal{H} \otimes L^2(G))$ as $G - \mathcal{A}$ module with the G-action on $\mathcal{A} \otimes_{C^*} (\mathcal{H} \otimes L^2(G))$ given by $\alpha_g \otimes 1_\mathcal{H} \otimes L_g$, where L_g is the left regular representation. Thus, the above theorem in particular implies that E is isometrically isomorphic with a closed complemented submodule of the Hilbert $\mathcal{A}$-module $\mathcal{A} \otimes_{C^*} (\mathcal{H} \otimes L^2(G))$ (where $\mathcal{H}$ is separable), and the G-action on E is inherited from the action $\alpha_g \otimes 1 \otimes L_g$ on $\mathcal{A} \otimes_{C^*} (\mathcal{H} \otimes L^2(G))$.*

4.3.2 The case of Hilbert von Neumann modules

Let us assume now that $\mathcal{A} \subseteq \mathcal{B}(h)$ is a von Neumann algebra, where h is a Hilbert space and G is a locally compact group with a strongly continuous unitary representation $g \mapsto u_g \in \mathcal{B}(h)$. Let $\alpha_g(a) := u_g a u_g^*$, for all $g \in G$, $a \in \mathcal{B}(h)$, and assume that $\alpha_g(\mathcal{A}) \subseteq \mathcal{A}$, for all $g \in G$. We define a *Hilbert von Neumann* $G - \mathcal{A}$ *module* (E, β) in a similar way as we have done for the C^*-module case, that is, in the Definition 4.3.1, but with the requirement that E is now a Hilbert von Neumann $\mathcal{A}$-module equipped with the natural locally convex topology described earlier, and that we replace the norm-continuity in (iii) of the above definition by a weaker continuity: namely, the continuity of $g \mapsto \beta_g(\xi)$ (for fixed $\xi \in E$) with respect to the locally convex topology of E. We shall now prove an equivariant version of Theorem 4.2.11, and we shall not need compactness of the group G in this case.

Theorem 4.3.5 *Let (E, β) be a Hilbert von Neumann $G - \mathcal{A}$ module, where $\mathcal{A}$ and G are as above. Then we can find a Hilbert space k_0 and a $\mathcal{A}$-linear isometry $\Gamma : E \to \mathcal{A} \otimes \mathcal{H}_0$, where $\mathcal{H}_0 := L^2(G) \otimes k_0$, such that $\Gamma \beta_g \Gamma^* = (\alpha_g \otimes L_g \otimes I_{k_0})P$, for all $g \in G$, where P denotes the projection $\Gamma\Gamma^*$ and L_g is the left regular representation of G in $L^2(G)$.*

Proof:
The proof of this theorem is adapted from [30]. We shall prove the theorem in two steps. First, we shall prove an equivariant version of Theorem 4.2.7, using the same notation. Assume, as in Theorem 4.2.7, that E is a submodule of $\mathcal{B}(h, \mathcal{H})$ for some Hilbert space $\mathcal{H}$ and $\{eu, e \in E, u \in h\}$ is total in $\mathcal{H}$. Define $\rho : \mathcal{A}' \to \mathcal{B}(\mathcal{H})$ as in 4.2.7, so that $E = \mathcal{I}(\mathcal{A}', \rho)$. Note that since $\alpha_g(a) \in \mathcal{A}$ for all $a \in \mathcal{A}$ and $g \in G$, we indeed have that $\alpha_g(\mathcal{A}') \subseteq \mathcal{A}'$. To see this, fix $b \in \mathcal{A}'$ and $g \in G$. Since $u_g a u_g^*$ belongs to $\mathcal{A}$ for every $a \in \mathcal{A}$, we have that $u_g a u_g^* b = b u_g a u_g^*$. Multiplying both sides by u_g^* on the left and by u_g on the right, and since u_g is unitary, we get $a(u_g^* b u_g) = (u_g^* b u_g)a$ for all $a \in \mathcal{A}$. That is, $u_g^* b u_g$ belongs to $\mathcal{A}'$. Since this is valid for all $g \in G$, by replacing g by g^{-1}, we conclude that $\alpha_g(b) := u_g b u_g^* \in \mathcal{A}'$.

Now, our aim is to define a unitary representation of G in $\mathcal{H}$ such that ρ is covariant with respect to that representation, where G-action on $\mathcal{A}'$ is taken to be α_g. Define an operator V_g on $\mathcal{H}$ for each $g \in G$ by setting

$$V_g(\xi v) := \beta_g(\xi) u_g v, \xi \in E, v \in h;$$

and extending it linearly on the span of $\{\xi v, \xi \in E, v \in h\}$. We have to check that this is well-defined and admits a unitary extension to the whole of $\mathcal{H}$. Fix $\xi_1, \ldots, \xi_m, \eta_1, \ldots, \eta_n \in E$, $v_1, \ldots, v_m, v_1', \ldots, v_n' \in h$, m, n are integers.

We have

$$\begin{aligned}
&\left\langle \sum_{i=1}^{m} \xi_i v_i, V_g \left(\sum_{j=1}^{n} \eta_j v'_j \right) \right\rangle \\
&= \sum_{i,j} \langle v_i, \langle \xi_i, \beta_g(\eta_j) \rangle u_g(v'_j) \rangle \\
&= \sum_{i,j} \langle v_i, \langle \beta_g(\beta_{g^{-1}}(\xi_i)), \beta_g(\eta_j) \rangle u_g(v'_j) \rangle \\
&= \sum_{i,j} \langle v_i, \alpha_g(\langle \beta_{g^{-1}}(\xi_i), (\eta_j) \rangle) u_g(v'_j) \rangle \\
&= \sum_{i,j} \langle v_i, u_g \langle \beta_{g^{-1}}(\xi_i), (\eta_j) \rangle v'_j \rangle \\
&= \left\langle \left(\sum_{i=1}^{m} \beta_{g^{-1}}(\xi_i) u_g^* v_i \right), \left(\sum_{j=1}^{n} \eta_j v'_j \right) \right\rangle.
\end{aligned}$$

Thus, if $X \equiv \sum_j \eta_j v'_j = 0$, then $\langle Y, V_g(X) \rangle = 0$ for all Y belonging to the span of the total set $S \equiv \{\xi v, \xi \in E, v \in h\}$, hence $V_g(X) = 0$. Proceeding in a very similar way we can show that $\langle V_g(Y), V_g(X) \rangle = \langle Y, X \rangle$ for X, Y as above. Furthermore, since β_g and u_g are bijections on E and h, respectively, it is clear that V_g maps the total set S onto itself. So, V_g has a unique unitary extension on $\mathcal{H}$, to be denoted again by the same notation. Furthermore, $V_g V_h = V_{gh}$ and $V_e = \mathrm{Id}$ (where e is the identity of G) on S, and hence on the whole of $\mathcal{H}$. To prove the strong continuity of $g \mapsto V_g$, it is enough to prove that $g \mapsto V_g X$ is continuous for any X of the form $\xi v, \xi \in E, v \in h$. But

$$\|V_g(\xi v) - \xi v\|^2 = 2\langle \xi v, \xi v \rangle - \langle V_g(\xi v), \xi v \rangle - \langle \xi v, V_g(\xi v) \rangle.$$

Now, we have

$$\langle V_g(\xi v), \xi v \rangle - \langle \xi v, \xi v \rangle = \langle \beta_g(\xi)(u_g v - v), \xi v \rangle + \langle (\beta_g(\xi) - \xi) v, \xi v \rangle.$$

By assumption $\lim_{g \to e}(\beta_g(\xi) - \xi) = 0$ with respect to the weak operator topology, so $\langle (\beta_g(\xi) - \xi) v, \xi v \rangle \to 0$ as $g \to e$. Furthermore,

$$\begin{aligned}
&|\langle \beta_g(\xi)(u_g v - v), \xi v \rangle| \\
&\leq \|\xi v\| \|\beta_g(\xi)\| \|u_g v - v\| \\
&= \|\xi v\| \|\langle \beta_g(\xi), \beta_g(\xi) \rangle\|^{\frac{1}{2}} \|u_g v - v\|,
\end{aligned}$$

which goes to 0 as $g \to e$, since u_g is strongly continuous and

$$\|\langle \beta_g(\xi), \beta_g(\xi) \rangle\| = \|\alpha_g(\langle \xi, \xi \rangle)\| = \|u_g \langle \xi, \xi \rangle u_g^*\| \leq \|\xi\|.$$

It is to be noted that ρ is also covariant with respect to V_g, since for $b \in \mathcal{A}'$ and $\xi \in E, v \in h$, we have

$$\begin{aligned} V_g \rho(b) V_{g^{-1}}(\xi v) &= V_g \rho(b)(\beta_{g^{-1}}(\xi) u_g^* v) \\ = V_g(\beta_{g^{-1}}(\xi) b u_g^* v) &= \beta_g(\beta_{g^{-1}}(\xi)) u_g b u_g^* v \\ = \xi \alpha_g(b) v &= \rho(\alpha_g(b))(\xi v). \end{aligned}$$

In the next step, we shall appeal to the theory of crossed product von Neumann algebras as discussed in Chapter 2. Recall that by Proposition 2.1.19, $\mathcal{A}'_G \equiv \mathcal{A}' >\!\!\lhd_\alpha G$ is the weak closure of the set $\{(u_g \otimes L_g), (b \otimes 1), g \in G, b \in \mathcal{A}'\} \subseteq \mathcal{B}(h) \otimes \mathcal{B}(L^2(G))$, and using the general theory described in Subsection 2.1.3, there is a normal $*$-representation $\tilde{\rho} : \mathcal{A}'_G \to \mathcal{B}(\mathcal{H})$ satisfying $\tilde{\rho}(b \otimes 1) = \rho(b)$ and $\tilde{\rho}(u_g \otimes L_g) = V_g$. By Proposition 2.1.7, we can find some Hilbert space k_0 and an isometry (as $\tilde{\rho}$ is clearly unital) $\Gamma : \mathcal{H} \to h \otimes L^2(G) \otimes k_0 \equiv h \otimes \mathcal{H}_0$ such that $\Gamma^*(X \otimes 1_{k_0})\Gamma = \tilde{\rho}(X)$, for all $X \in \mathcal{A}'_G$. We claim that this Γ satisfies the properties mentioned in the statement of the theorem, thereby completing the proof. For this, note that by Proposition 2.1.7, $P := \Gamma\Gamma^*$ commutes with $X \otimes 1_{k_0}$ for all $X \in \mathcal{A}'_G$, so in particular with $(u_g \otimes L_g \otimes 1)$, $g \in G$. So we have

$$\Gamma V_g \Gamma^* = P(u_g \otimes L_g \otimes 1)P = (u_g \otimes L_g \otimes 1)\Gamma\Gamma^*,$$

which implies that $\Gamma V_g = (u_g \otimes L_g \otimes 1)\Gamma$. Thus,

$$\Gamma \beta_g(\xi) = \Gamma V_g(\xi) u_g^* = (u_g \otimes L_g \otimes 1)\Gamma \xi u_g^* = (\alpha_g \otimes L_g \otimes 1)(\Gamma(\xi)),$$

hence $\Gamma \beta_g \Gamma^* = (\alpha_g \otimes L_g \otimes 1)\Gamma\Gamma^* = (\alpha_g \otimes L_g \otimes 1)P$. Furthermore, the fact that $\Gamma(E) \subseteq \mathcal{A} \otimes \mathcal{H}_0$ can be verified as in the proof of the Theorem 4.2.11. □

5

Quantum stochastic calculus with bounded coefficients

5.1 Basic processes

In this chapter, we shall first discuss the construction of the four basic processes (creation, annihilation, number and time) for a fairly general class of coefficients (not necessarily bounded), and the construction and properties of left stochastic integrals with respect to such processes. However, we shall confine ourselves to the class of bounded coefficients for defining right integrals, and also for studying quantum stochastic differential equations in this chapter, leaving the discussion on such equations with more general (unbounded) coefficients for Chapter 7 of the book.

Let us recall the notations introduced in Section 4.2 of Chapter 4, particularly formulae (4.1), (4.2) and related notation. Let $\mathcal{H}_1$ and $\mathcal{H}_2$ be two Hilbert spaces (possibly non-separable). Let $\mathcal{E}(\mathcal{V})$ denote the complex linear span of exponential vectors $e(g)$ for g belonging to some subspace (not necessarily closed) $\mathcal{V}$ of $\mathcal{H}_2$.

Now, we define a map $S : \Gamma^f(\mathcal{H}_2) \to \Gamma(\mathcal{H}_2)$ by setting,

$$S(g_1 \otimes g_2 \otimes \cdots \otimes g_n) = \frac{1}{(n-1)!} \sum_{\sigma \in S_n} g_{\sigma(1)} \otimes \cdots \otimes g_{\sigma(n)}, \tag{5.1}$$

and linearly extending it to $\mathcal{H}_2^{\otimes^n}$, where S_n is the group of permutations of n objects. Clearly, $\|S|_{\mathcal{H}_2^{\otimes n}}\| \leq n$. We denote by $\tilde{S}$ the operator $1_{\mathcal{H}_1} \otimes S$.

Let us denote by $\mathrm{Lin}(\mathcal{V}_1, \mathcal{V}_2)$ the space of linear (without any continuity or boundedness assumption) maps from a vector space $\mathcal{V}_1$ to another vector space $\mathcal{V}_2$, where $\mathcal{V}_1, \mathcal{V}_2$ need not be Hilbert spaces. We also adopt the convention of denoting algebraic and topological tensor product of vector spaces by the same symbol $\otimes$ whenever there is no chance of confusion. In particular, we shall use the symbol $\mathcal{V}_1 \otimes \mathcal{V}_2$ to mean algebraic tensor product whenever at least one of the spaces involved is a nonclosed subspace of a Hilbert space. However, when both of them are closed subspaces, we shall notationally distinguish between the algebraic and topological tensor products, and use the symbol $\otimes_{\rm alg}$ for the algebraic tensor product. It should also be noted that we shall often omit the tensor product symbol $\otimes$ between two or more vectors when there is no confusion.

Let A be a linear map from $\mathcal{H}_1$ to $\mathcal{H}_1 \otimes \mathcal{H}_2$ with domain $\mathcal{D}$, that is, $A \in \mathrm{Lin}(\mathcal{D}, \mathcal{H}_1 \otimes \mathcal{H}_2)$. Let us now define the creation operator $a^\dagger(A)$ abstractly which will act on the linear span of vectors of the form $vg^{\otimes^n}$ and $ve(g)$ (where $g^{\otimes^n}$ denotes $\underbrace{g \otimes \cdots \otimes g}_{n \text{ times}}$), $n \geq 0$, with $v \in \mathcal{D},\ g \in \mathcal{H}_2$. We define,

$$a^\dagger(A)(vg^{\otimes^n}) = \frac{1}{\sqrt{n+1}} \tilde{S}((Av) \otimes g^{\otimes^n}). \tag{5.2}$$

We observe that $\sum_{n \geq 0} \frac{1}{n!} \| a^\dagger(A)(vg^{\otimes^n}) \|^2 < \infty$, which allows us to define $a^\dagger(A)(ve(g))$ as the direct sum $\oplus_{n \geq 0} \frac{1}{(n!)^{\frac{1}{2}}} a^\dagger(A)(vg^{\otimes^n})$. We have the following simple but useful observation.

Lemma 5.1.1 *For $v \in \mathcal{D},\ u \in \mathcal{H}_1$ and $g, w \in \mathcal{H}_2$,*

$$\langle a^\dagger(A)(ve(g)),\ ue(w) \rangle = \langle A_{u,v},\, w \rangle \langle e(g),\ e(w) \rangle = \frac{d}{d\varepsilon} \langle e(g + \varepsilon A_{u,v}), e(w) \rangle |_{\varepsilon = 0}. \tag{5.3}$$

Proof:
First observe that

$$\langle a^\dagger(A)(vg^{\otimes^n}), ue(w) \rangle = \left\langle \frac{1}{\sqrt{n+1}} \tilde{S}((Av) \otimes g^{\otimes^n}),\ \frac{1}{\sqrt{(n+1)!}} u \otimes w^{\otimes^{n+1}} \right\rangle.$$

It is clear that the adjoint S^* of the operator S is given by $S^*(f^{\otimes^n}) = nf^{\otimes^n}$. Thus, we have that

$$\begin{aligned} \langle \tilde{S}((Av) \otimes g^{\otimes^n}), u \otimes w^{\otimes^{n+1}} \rangle &= \langle ((Av) \otimes g^{\otimes^n}), u \otimes S^*(w^{\otimes^{n+1}}) \rangle \\ &= (n+1) \langle Av, u \otimes w \rangle \langle g, w \rangle^n \\ &= (n+1)\sqrt{n!} \langle A_{u,v}, w \rangle \langle g^{\otimes^n}, e(w) \rangle. \end{aligned}$$

Hence

$$\begin{aligned} \langle a^\dagger(A)(vg^{\otimes^n}), ue(w) \rangle &= \frac{(n+1)\sqrt{n!}}{\sqrt{(n+1)!}\sqrt{n+1}} \langle A_{u,v}, w \rangle \langle g^{\otimes^n}, e(w) \rangle \\ &= \langle A_{u,v}, w \rangle \langle g^{\otimes^n}, e(w) \rangle. \end{aligned}$$

From this the result follows. □

We now define the annihilation operator $a(B)$ where B is a densely defined linear map from $\mathcal{H}_1$ to $\mathcal{H}_1 \otimes \mathcal{H}_2$, with the domain $\mathrm{Dom}(B)$, say, that is, $B \in \mathrm{Lin}(\mathrm{Dom}(B), \mathcal{H}_1 \otimes \mathcal{H}_2)$. The domain of $a(B)$ is taken to be the complex linear span of vectors of the form $ue(\xi)$ and $u\xi^{\otimes^n}$ $(n = 1, 2, \cdots)$, such that $u \in \mathrm{Dom}(\langle B, \xi\rangle)$ $(\langle B, \xi\rangle := \langle \xi, B\rangle^*)$; and

$$a(B)ue(\xi) := (\langle B, \xi\rangle u)e(\xi),$$

$$a(B)u\xi^{\otimes^n} := \frac{1}{\sqrt{n!}} \frac{d^n}{ds^n}|_{s=0} a(B)(ue(s\xi)) = \sqrt{n}(\langle B, \xi\rangle u)\xi^{\otimes^{n-1}}.$$

Furthermore, we define the number operator $\Lambda(T)$ for $T \in \mathrm{Lin}(\mathrm{Dom}(T), \mathcal{H}_1 \otimes \mathcal{H}_2)$, where $\mathrm{Dom}(T) \subseteq \mathcal{H}_1 \otimes \mathcal{H}_2$. We take $\mathrm{Dom}(\Lambda(T))$ to be the complex linear span of vectors of the form $ue(\xi)$ with (u, ξ) satisfying $u \otimes \xi \in \mathrm{Dom}(T)$, that is, $u \in \mathrm{Dom}(T_\xi)$, and set

$$\Lambda(T)ue(\xi) := a^\dagger(T_\xi)(ue(\xi)).$$

We also define $\Lambda(T)(u\xi^{\otimes^n}) := \frac{1}{\sqrt{n!}} \frac{d^n}{ds^n}|_{s=0} \Lambda(T)(ue(s\xi))$, which clearly exists.

Remark 5.1.2 *If there are subspaces (not necessarily closed) $\mathcal{D}_1, \mathcal{D}_2$ of $\mathcal{H}_1$, $\mathcal{H}_2$, respectively, such that $\mathcal{D}_1 \subseteq \mathrm{Dom}(B) \cap_{f \in \mathcal{D}_2} \mathrm{Dom}(\langle B, f\rangle)$, then both $a^\dagger(B)$ and $a(B)$ have $\mathcal{D}_1 \otimes \mathcal{E}(\mathcal{D}_2)$ in their domains and one can also verify that $a^\dagger(A)$ is the adjoint of $a(A)$ on $\mathcal{D}_1 \otimes \mathcal{E}(\mathcal{D}_2)$.*

Next, to define the basic processes, we need some more notations. Let h and k_0 be Hilbert spaces, $k = L^2(\mathbb{R}_+, k_0)$, $k_t = L^2([0, t]) \otimes k_0$, $k^t = L^2((t, \infty)) \otimes k_0$, $\Gamma_t = \Gamma(k_t)$, $\Gamma^t = \Gamma(k^t)$, $\Gamma = \Gamma(k)$. We assume that $\mathcal{D}_0, \mathcal{V}_0$ are dense subspaces of h and k_0 respectively, and $R, R' \in \mathrm{Lin}(\mathcal{D}_0, h \otimes k_0)$, $T \in \mathrm{Lin}(\mathcal{D}_0 \otimes \mathcal{V}_0, h \otimes k_0)$ are closable operators. Assume furthermore that $\mathcal{D}_0 \subseteq \cap_{\xi \in \mathcal{V}_0} \mathrm{Dom}(\langle R', \xi\rangle)$. We define $R_t^\Delta : \mathcal{D}_0 \otimes \Gamma_t \to h \otimes \Gamma_t \otimes k^t$ for $t \geq 0$ and a bounded interval Δ in (t, ∞) by

$$R_t^\Delta(u\psi) = P((1_h \otimes \chi_\Delta)(Ru) \otimes \psi),$$

where $\chi_\Delta : k_0 \to k^t$ is the operator which takes α to $\chi_\Delta(\cdot)\alpha$ for $\alpha \in k_0$, and P is the canonical unitary isomorphism from $h \otimes k \otimes \Gamma$ to $h \otimes \Gamma \otimes k$. We denote by L^p_{loc} the set of $f \in k$ such that $\int_0^t \|f(s)\|^p ds < \infty$ for all $t \geq 0$, and by $\mathcal{C}$ the set of bounded continuous functions in k. For a measurable k_0-valued function f in $k = L^2(\mathbb{R}_+, k_0)$, we shall denote by f_t and f^t respectively the functions $f\chi_{[0,t]} \in k_t$ and $f\chi_{(t,\infty)} \in k^t$.

We now define the creation field $a_R^\dagger(\Delta)$ on either of the domains consisting of finite linear combinations of vectors of the form $u_t \otimes f^{t\otimes^n}$ or of $u_t \otimes e(f^t)$ for $u_t \in \mathcal{D}_0 \otimes \Gamma_t,\ f^t \in \Gamma^t,\ n \geq 0$, as

$$a_R^\dagger(\Delta) = a^\dagger(R_t^\Delta), \tag{5.4}$$

where $a^\dagger(R_t^\Delta)$ carries the meaning discussed before Lemma 5.1.1, with $\mathcal{H}_1 = h \otimes \Gamma_t, \mathcal{H}_2 = k^t$ and $\mathcal{D} = \mathcal{D}_0 \otimes \Gamma_t$.

Let $T_{f^t}^\Delta$ be the densely defined linear map from $h \otimes \Gamma_t$ to $h \otimes \Gamma_t \otimes k^t$, with the domain consisting of linear span of vectors of the form $u\psi$, with $\psi \in L^2((t,\infty), k_0)$ satisfying $\psi(s) \in \mathcal{V}_0$ for almost all s; and $T_{f^t}^\Delta$ is given by

$$T_{f^t}^\Delta(u\alpha_t) = P(1 \otimes \hat{\chi}_\Delta)(\hat{T}(uf^t) \otimes \alpha_t). \tag{5.5}$$

Here, $\hat{T}$ is the densely defined map on $(h \otimes L^2((t,\infty), k_0)) \equiv L^2((t,\infty), h \otimes k_0)$ with the domain consisting of L^2-functions $\eta : (t,\infty) \to h \otimes k_0$ with $\eta(s) \in \mathcal{D}_0 \otimes \mathcal{V}_0$ for almost all s, and given by, $\hat{T}(\eta)(s) = T(\eta(s)),\ s > t$. $\hat{\chi}_\Delta$ is the multiplication by $\chi_\Delta(\cdot)$ on $L^2((t,\infty), k_0)$. Clearly, in case T is bounded, $\|\hat{T}\| \leq \|T\|$, which makes $T_{f^t}^\Delta$ also bounded.

With the above notation, we define the annihilation and number fields $a_{R'}(\Delta)$ and $\Lambda_T(\Delta)$ as follows

$$a_{R'}(\Delta)(u_t e(f^t)) = ((\int_\Delta \langle R', f(s)\rangle ds)u_t)e(f^t), \tag{5.6}$$

and

$$\Lambda_T(\Delta)(u_t e(f^t)) = a^\dagger(T_{f^t}^\Delta)(u_t e(f^t)); \tag{5.7}$$

for f such that $f(s) \in \mathcal{V}_0$ for almost all $s > t$. We note here that objects similar to $a_R(\cdot), a_{R'}^\dagger(\cdot)$ and $\Lambda_T(\cdot)$ were used in [71], however in a coordinatized form.

For Hilbert spaces $\mathcal{H}_1, \mathcal{H}_2, \mathcal{H}_3$ and a closable densely defined operator $B \in \mathrm{Lin}(\mathcal{H}_1, \mathcal{H}_2)$, we denote by $B \otimes I_{\mathcal{H}_3}$ (or simply $B \otimes I$ if $\mathcal{H}_3$ is understood from the context) the *closure* of the densely defined closable operator $B \otimes_{\mathrm{alg}} I : \mathrm{Dom}(B) \otimes_{\mathrm{alg}} \mathcal{H}_3 \to \mathcal{H}_2 \otimes_{\mathrm{alg}} \mathcal{H}_3$. Note that $\mathrm{Dom}(B) \otimes_{\mathrm{alg}} \mathcal{H}_3$ is a core for $B \otimes I$.

Definition 5.1.3 Let $\mathcal{D}_1$ be a dense subspace of h and $\mathcal{V}_0$ be a dense subspace of k_0. Let $\mathcal{V}$ denote the space of $\mathcal{V}_0$-valued simple, right continuous functions on $[0,\infty)$. That is,

$$\mathcal{V} = \left\{ f = \sum_{i=0}^{k} \chi_{[t_i, t_{i+1})}\xi_i : 0 = t_0 \leq t_1 \leq \cdots < t_{k+1} = \infty;\ k \text{ positive integer},\ \xi_i \in \mathcal{V}_0 \right\}.$$

We say that a family $(H_t)_{t\geq 0}$ of linear maps on $h \otimes \Gamma$ is an *adapted* $(\mathcal{D}_1, \mathcal{V}_0)$-*admissible process* if

(i) the domain of H_t contains vectors of the form $v f_t^{\otimes^n} \psi^t, n = 1, 2, \ldots$ and $ve(f_t)\psi^t$, where $v \in \mathcal{D}_1$, $f_t \in k_t$, such that f_t is a simple $\mathcal{V}_0$-valued right continuous function, and $\psi^t \in \Gamma^t$;

(ii) there are operators $\hat{H}_t \in \text{Lin}(\mathcal{D}_1 \otimes \mathcal{E}(k_t), h \otimes \Gamma_t)$ such that $H_t = \hat{H}_t \otimes I|_{\Gamma^t}$;

(iii) given $f \in \mathcal{V}$ there exist Hilbert space h_1, and closable maps $r_t \equiv r_t^f \in \text{Lin}(\mathcal{D}_1, h_1)$ depending only on t and f such that

$$\sup_{0 \leq s \leq t} \|H_s(ue(f))\| \leq \|r_t u\|, \quad \text{for all } t \geq 0,\ u \in \mathcal{D}_1,\ f \in \mathcal{V}. \tag{5.8}$$

Clearly, the definition of $H_t ue(f)$ can be extended to u belonging to the domain of the closure of r_t (again denoted by r_t, if no confusion arises), and we shall denote this extension of H_t by the same symbol again. If $\mathcal{D}_1 = h$, it follows from the closed graph theorem that r_t is bounded for each t, and replacing r_t^f by $c(t, f)I = \|r_t\|I$, we can re-write (5.8) as

$$\sup_{0 \leq s \leq t} \|H_s(ue(f))\| \leq c(t, f)\|u\|, \quad \text{for all } u \in h,\ f \in \mathcal{V}. \tag{5.9}$$

If, moreover, $\mathcal{V}_0 = k_0$ and the inequality (5.8) holds for all $f \in k$, we call (H_t) an adapted admissible (or simply admissible if adaptedness is understood) process.

We shall often denote an operator B and its trivial extension $B \bigotimes I$ to some bigger space by the same notation, unless there is any confusion in doing so. We also denote the unitary isomorphism from $h \otimes k_0 \otimes \Gamma(k)$ onto $h \otimes \Gamma(k) \otimes k_0$ and that from $h \otimes k \otimes \Gamma(k)$ onto $h \otimes \Gamma(k) \otimes k$ by the same letter P. Clearly, for an adapted $(\mathcal{D}_1, \mathcal{V}_0)$-admissible process (H_t), $H_t P$ acts on any vector of the form $w \otimes e(g)$ where $w \in \mathcal{D}_1 \otimes_{\text{alg}} k_0$, $g \in \mathcal{V}$ and we have the estimate

$$\sup_{0 \leq s \leq t} \|H_s P(we(g))\| \leq \|(r_t^g \otimes I_{k_0})w\|.$$

This allows one to extend $H_t P$ on the whole of the domain containing vectors of the form $\bar{w}e(g)$, $\bar{w} \in \text{Dom}((r_t \otimes I_{k_0})) \subseteq h \otimes k_0$, $g \in \mathcal{V}$. We denote this extension again by $H_t P$. When P is taken to be the isomorphism from $h \otimes k \otimes \Gamma(k)$ onto $h \otimes \Gamma(k) \otimes k$, we define $H_t P$ in an exactly parallel manner.

Next we prove a few preliminary results which will be needed for establishing the quantum Itô formula in the next subsection. In what follows, (H_t) and (H'_t) are assumed to be adapted $(\mathcal{D}_1, \mathcal{V}_0)$-admissible processes for some $\mathcal{D}_1, \mathcal{V}_0$ with closable operators $r_t^f : \mathcal{D}_1 \to h_1$, $r_t'^f : \mathcal{D}_1 \to h'_1$, so that we have

$$\sup_{0 \leq s \leq t} \|H_s(ue(f))\| \leq \|r_t^f u\|, \quad \text{for all } u \in \mathcal{D}_1,\ f \in \mathcal{V}; \tag{5.10}$$

and

$$\sup_{0 \leq s \leq t} \|H'_s(ue(f))\| \leq \|r_t'^f u\|, \quad \text{for all } u \in \mathcal{D}_1,\ f \in \mathcal{V}. \tag{5.11}$$

Lemma 5.1.4 *Let $\Delta, \Delta' \subseteq (t, \infty)$ be intervals of finite length, $R, S \in \mathrm{Lin}(\mathcal{D}_0, h \otimes k_0)$, $u, v \in \mathcal{D}_0$; $g, f \in \mathcal{V}$. Assume furthermore that*

$$R(\mathcal{D}_0) \subseteq \cap_{t \geq 0} \mathrm{Dom}(r_t \otimes I), \text{ and } S(\mathcal{D}_0) \subseteq \cap_{t \geq 0} \mathrm{Dom}(r'_t \otimes I).$$

Then we have

$$\begin{aligned}
&\langle H_t a_R^\dagger(\Delta)(ve(g)),\ H'_t a_S^\dagger(\Delta')(ue(f))\rangle \\
&= e^{\langle g^t, f^t\rangle}\{\langle H_t R_t^\Delta(ve(g_t)),\ H'_t S_t^{\Delta'}(ue(f_t))\rangle + \langle\langle f^t, H_t R_t^\Delta\rangle ve(g_t), \\
&\quad \langle g^t, H'_t S_t^{\Delta'}\rangle ue(f_t)\rangle\} \\
&= \int_{\Delta\cap\Delta'} \langle (H_t P R)(ve(g)),\ (H'_t P S)(ue(f))\rangle ds + \\
&\int_\Delta \int_{\Delta'} \langle\langle f(s), H_t P R\rangle(ve(g)), \langle g(s'),\ H'_t P S\rangle(ue(f))\rangle ds\, ds'. \qquad (5.12)
\end{aligned}$$

Proof:
We fix f, g and denote r_t^f, $r_t'^g$ etc. simply by r_t and r'_t respectively. For the present proof, we make the convention of writing $\frac{df(\varepsilon)}{d\varepsilon}|_{\varepsilon=\varepsilon_0}$ for the following limit (whenever it exists):

$$\lim_{n\to\infty} n\left(f\left(\varepsilon_0 + \frac{1}{n}\right) - f(\varepsilon_0)\right)$$

Let R_Δ denote $(1 \otimes \chi_\Delta)R \in \mathrm{Lin}(\mathcal{D}_0, h \otimes k)$ for $R \in \mathrm{Lin}(\mathcal{D}_0, h \otimes k_0)$. Let us now choose and fix orthonormal bases $\{e_\nu\}_{\nu\in J}$ and $\{k_\alpha\}_{\alpha\in I}$ of $h \otimes \Gamma_t$ and Γ^t respectively ($t \geq 0$). We also choose subsets J_0 and I_0, which are at most countable, of J and I, respectively, as follows. Let J_0 be such that

$$\langle H_t P R_\Delta(ve(g)),\ e_\nu \otimes k_\alpha\rangle = 0 = \langle e_\nu \otimes k_\alpha,\ H'_t P S_{\Delta'}(ue(f))\rangle$$

for all $\alpha \in I$, whenever $\nu \notin J_0$. Fixing this J_0, we choose I_0 to be the union of $I_{\nu,n}$, $\nu \in J_0$, $n = 1, 2, \cdots, \infty$, such that

$$\langle e(g^t + \frac{1}{n}(H_t P R_\Delta)_{e_\nu, ve(g_t)}),\ k_\alpha\rangle = 0 = \langle k_\alpha,\ e(f^t + \frac{1}{n}(H'_t P S_{\Delta'})_{e_\nu, ue(f_t)})\rangle$$

for all $\alpha \notin I_{\nu,n}$ when $n < \infty$, and

$$\langle e(g^t),\ k_\alpha\rangle = 0 = \langle k_\alpha, e(f^t)\rangle \ \text{ for } \alpha \notin I_{\nu,\infty}.$$

We now have

$$\langle H_t a_R^\dagger(\Delta)(ve(g)),\ H'_t a_S^\dagger(\Delta')(ue(f))\rangle$$

$$= \sum_{\substack{\nu \in J_0 \\ \alpha \in I_0}} \langle H_t a_R^\dagger(\Delta)(ve(g)),\ e_\nu \otimes k_\alpha \rangle \langle e_\nu \otimes k_\alpha,\ H_t' a_S^\dagger(\Delta')(ue(f)) \rangle$$

$$= \sum_{\substack{\nu \in J_0 \\ \alpha \in I_0}} \Big(\frac{d}{d\varepsilon} \langle e(g^t + \varepsilon (H_t P R_\Delta)_{e_\nu, ve(g_t)}), k_\alpha \rangle |_{\varepsilon=0} \Big)$$

$$\times \Big(\frac{d}{d\eta} \langle k_\alpha, e(f^t + \eta (H_t' P S_{\Delta'})_{e_\nu, ue(f_t)}) \rangle |_{\eta=0} \Big)$$

$$= \sum_{\nu \in J_0} \frac{\partial^2}{\partial\varepsilon \partial\eta} \Big(\sum_{\alpha \in I_0} \langle e(g^t + \varepsilon (H_t P R_\Delta)_{e_\nu, ve(g_t)} \Big),\ k_\alpha \rangle$$

$$\langle k_\alpha, e(f^t + \eta (H_t' P S_{\Delta'})_{e_\nu, ue(f_t)}) \rangle |_{\varepsilon=0=\eta})$$

$$= \sum_{\nu \in J_0} \frac{\partial^2}{\partial\varepsilon \partial\eta} (\langle e(g^t + \varepsilon (H_t P R_\Delta)_{e_\nu, ve(g_t)}),$$

$$e(f^t + \eta (H_t' P S_{\Delta'})_{(e_\nu, ue(f_t)})) \rangle |_{\varepsilon=0=\eta})$$

$$= \sum_{\nu \in J_0} e^{\langle g^t, f^t \rangle} (\langle (H_t P R_\Delta)_{e_\nu, ve(g_t)}, (H_t' P S_{\Delta'})_{e_\nu, ue(f_t)} \rangle$$

$$+ \langle (H_t P R_\Delta)_{e_\nu, ve(g_t)}, f^t \rangle \langle g^t, (H_t' P S_{\Delta'})_{e_\nu, ue(f_t)} \rangle)$$

Before proceeding further, let us justify the intermediate step in the above calculations, which involves an interchange of summation and limit, by appealing to the Dominated Convergence theorem. Indeed, for any fixed $\alpha \in I_0$, ψ, ψ' $\in k^t$, if we write $k_\alpha^{(n)}$ for the projection of k_α on $k^{t^{\otimes n}}$ $(n \geq 0)$, then $\langle e(g^t + \varepsilon\psi), k_\alpha \rangle$ can be expressed as $\sum_{i \geq 0} c_i^{(\alpha)} \varepsilon^i$, where $c_i^{(\alpha)} = \sum_{n \geq i} \frac{1}{\sqrt{n!}} \binom{n}{i} \langle g^{t \otimes^{(n-i)}}$ $\otimes\ \psi^{\otimes^{(i)}}, k_\alpha^{(n)} \rangle$, where $g^{t \otimes^{(n-i)}} \equiv \underbrace{g^t \otimes \cdots \otimes g^t}_{(n-i)\text{-times}}$, and $\psi^{\otimes^{(i)}} \equiv \underbrace{\psi \otimes \cdots \otimes \psi}_{i\text{-times}}$. It can be verified that the above is an absolutely summable power series in ε, converging uniformly for $\varepsilon \in [0, M]$, say, for any fixed $M > 0$. Similar analysis can be done for $\langle k_\alpha, e(f^t + \eta\psi') \rangle$. By the Mean Value theorem and some simple estimation, and since $\{k_\alpha^{(n)}\}_{\alpha \in I_0}$ are mutually orthogonal for any fixed n, with $\|k_\alpha^{(n)}\| \leq 1$ for all α, we have that for ε, η, ε', η' in $[0, M]$,

$$\frac{1}{|(\varepsilon - \varepsilon')(\eta - \eta')|} \sum_{\alpha \in I_0} |(\langle e(g^t + \varepsilon\psi), k_\alpha \rangle - \langle e(g^t + \varepsilon'\psi), k_\alpha \rangle)$$

$$\times\ (\langle k_\alpha, e(f^t + \eta\psi') \rangle - \langle k_\alpha, e(f^t + \eta'\psi') \rangle)|$$

$$\leq \sum_{\substack{n \geq 0, m \geq 0, \\ 0 \leq i \leq n, 0 \leq j \leq m}} \frac{i.j.M^{i+j-2}}{\sqrt{n!m!}} \binom{n}{i} \binom{m}{j} \sum_{\alpha \in I_0} |\langle g^{t \otimes^{(n-i)}} \otimes \psi^{\otimes^{(i)}}, k_\alpha^{(n)} \rangle$$

$$\langle k_\alpha^{(n)}, f^{t \otimes^{(m-j)}} \otimes \psi'^{\otimes^{(j)}} \rangle|$$

$$\leq \sum_{n,m,i,j} \frac{ijM^{i+j-2}}{\sqrt{n!m!}} \binom{n}{i}\binom{m}{j} \|g^{t\otimes^{(n-i)}} \otimes \psi^{\otimes^{(i)}}\| \, \|f^{t\otimes^{(m-j)}} \otimes \psi'^{\otimes^{(j)}}\|$$

$$\leq \sum_{\substack{n\geq 0\\ m\geq 0}} \frac{mn\|\psi\| \, \|\psi'\| \, (\|g^t\| + M\|\psi\|)^{n-1}(\|f^t\| + M\|\psi'\|)^{m-1}}{\sqrt{m!n!}} < \infty.$$

This allows us to apply the Dominated Convergence theorem.

Let us now choose a countable subset I_0' of I so that $0 = \langle (H_t P R_\Delta)_{e_\nu, ve(g_t)}, k_\alpha \rangle = \langle k_\alpha, (H_t' P S_{\Delta'})_{e_\nu, ue(f_t)} \rangle$ for α not in I_0', and for all $\nu \in J_0$.
Clearly, we have

$$\sum\nolimits_{\nu\in J_0} \langle (H_t P R_\Delta)_{e_\nu, ve(g_t)}, (H_t' P S_{\Delta'})_{e_\nu, ue(f_t)} \rangle$$

$$= \sum_{\nu\in J_0, \alpha\in I_0'} \langle (H_t P R_\Delta)(ve(g_t)), e_\nu \otimes k_\alpha \rangle \, \langle e_\nu \otimes k_\alpha, (H_t' P S_{\Delta'})(ue(f_t)) \rangle$$

$$= \langle (H_t P R_\Delta)(ve(g_t)), (H_t' P S_{\Delta'})(ue(f_t)) \rangle.$$

Since $\mathcal{D}_1 \otimes_{\rm alg} k_0$ is a core for $r_t \otimes I$ and $r_t' \otimes I$, we can choose sequences $\omega^{(n)}$, $\omega'^{(n)}$ of vectors which can be written as finite sums of the form, $\omega^{(n)} = \sum v_i^{(n)} \otimes \beta_i^{(n)}$, $\omega'^{(n)} = \sum_i u_i^{(n)} \otimes \alpha_i^{(n)}$, where $u_i^{(n)}, v_i^{(n)} \in \mathcal{D}_1$, $\beta_i^{(n)}, \alpha_i^{(n)} \in k_0$, and as $n \to \infty$, we have the following convergence:

$$\omega^{(n)} \to Rv, \; \omega'^{(n)} \to Su,$$

$$(r_t \otimes I)\omega^{(n)} \to (r_t \otimes I)Rv, \;\; (r_t' \otimes I)\omega'^{(n)} \to (r_t' \otimes I)Su.$$

Then we have

$$\|H_t P(1 \otimes \chi_\Delta) \, (\omega^{(n)} \otimes e(g_t)) - H_t P R_\Delta(ve(g_t))\|$$

$$\leq \|(r_t \otimes I)(\omega^{(n)} - (Rv))\| \, |\Delta| \; \to 0 \text{ as } n \to \infty,$$

where $|\Delta|$ denotes the Lebesgue measure of Δ. Similarly,
$\|H_t' P(1 \otimes \chi_{\Delta'}) \, (\omega'^{(n)} \otimes e(f_t)) - (H_t' P S_{\Delta'})(ue(f_t))\| \to 0$ as $n \to \infty$. Hence we obtain

$$\langle (H_t P R_\Delta)(ve(g_t)), (H_t' P S_{\Delta'}(ue(f_t))) \rangle$$

$$= \lim_{n\to\infty} \langle H_t P(1 \otimes \chi_\Delta)(\omega^{(n)} e(g_t)), H_t' P(1 \otimes \chi_{\Delta'})(\omega'^{(n)} e(f_t)) \rangle$$

$$= \lim_{n\to\infty} \int \Big\langle H_t\Big(\sum_i v_i^{(n)} \otimes e(g_t) \otimes \beta_i^{(n)}\Big), H_t'\Big(\sum_i u_i^{(n)} \otimes e(f_t) \otimes \alpha_i^{(n)}\Big)\Big\rangle \chi_{\Delta\cap\Delta'}(s)ds$$

$$= \lim_{n\to\infty} |\Delta \cap \Delta'| \langle (H_t P)(\omega^{(n)} e(g_t)), (H_t' P)(\omega'^{(n)} e(f_t)) \rangle$$

$$= |\Delta \cap \Delta'| \langle (H_t P R)(ve(g_t)), (H_t' P S)(ue(f_t)) \rangle$$

$$= \int_{\Delta\cap\Delta'} \langle H_t P R(ve(g_t)), H_t' P S(ue(f_t)) \rangle ds.$$

Moreover,

$$\sum_{\nu\in J_0} \langle (H_t P R_\Delta)_{e_\nu, ve(g_t)}, f^t\rangle \langle g^t, (H_t' P S_{\Delta'})_{e_\nu, ue(f_t)}\rangle$$

$$= \sum_{\nu\in J_0} \langle\langle f^t, H_t P R_\Delta\rangle (ve(g_t)), e_\nu\rangle \langle e_\nu, \langle g^t, H_t' P S_{\Delta'}\rangle (ue(f_t))\rangle$$

$$= \langle\langle f^t, H_t P R_\Delta\rangle (ve(g_t)), \langle g^t, (H_t' P S_{\Delta'})\rangle (ue(f_t))\rangle,$$

where the last step follows by Parseval's identity, noting the fact that for $\nu \notin J_0$, $\langle\langle f^t, H_t P R_\Delta\rangle(ve(g_t)), e_\nu\rangle = 0$ because for such ν, $\langle (H_t P R_\Delta)(ve(g_t)), e_\nu \otimes k_\alpha\rangle = 0$ for all $\alpha \in I$; and similarly $\langle e_\nu, \langle g^t, (H_t' P S_{\Delta'})\rangle(ue(f_t))\rangle = 0$ for all $\nu \notin J_0$.

We complete the proof by observing that

$$\langle f^t, H_t R_t^\Delta\rangle = \int_\Delta \langle f(s), H_t P R\rangle ds, \text{ and}$$

$$\langle g^t, H_t' S_t^{\Delta'}\rangle = \int_{\Delta'} \langle g(s'), H_t' P S\rangle ds'.$$

To see this, it is enough to note that for $\omega \in h$, $\xi_t \in k_t$, we have,

$\langle\langle f^t, H_t R_t^\Delta\rangle(ve(g_t)), \omega e(\xi_t)\rangle$
$= \int_\Delta \langle (H_t P R)(ve(g_t)), \omega e(\xi_t) \otimes f^t(s)\rangle ds$, which can be justified by considering $\omega^{(n)}$ as before and applying the Dominated Convergence theorem.

Since $\Delta \subseteq (t, \infty)$ and hence $f^t(s) = f(s)$ for $s \in \Delta$, the above expression can now be written as

$$\int_\Delta \langle (H_t P R)(ve(g_t)), we(h_t) f(s)\rangle ds$$

$$= \int_\Delta \langle\langle f(s), H_t P R\rangle (ve(g_t)), we(h_t)\rangle ds.$$

This completes the proof. □

Remark 5.1.5 *If H_t, H_t', R and S are bounded, then (5.12) of Lemma 5.1.4 holds without the restriction on the vectors u, v and on the functions f, g. Moreover, in this case one can replace h and k by $h \otimes \Gamma_t$ and k^t, respectively.*

Lemma 5.1.6 *Let T, T' be closable linear maps from $\mathcal{D}_0 \otimes \mathcal{V}_0$ to $h \otimes k_0$ with* $\mathrm{Ran}(T) \subseteq \cap_{t \geq 0}\mathrm{Dom}(r_t \otimes I)$, $\mathrm{Ran}(T') \subseteq \cap_{t \geq 0}\mathrm{Dom}(r'_t \otimes I)$, *$u, v \in \mathcal{D}_0$, and $g, f \in \mathcal{V}$, that is, $f(s), g(s) \in \mathcal{V}_0$ for all s. Then we have*

$$\langle (H_t T^{\Delta}_{g^t})(ve(g)), (H'_t T^{\Delta'}_{f^t})(ue(f)) \rangle = \int\limits_{\Delta \cap \Delta'} \langle H_t P T P^*(ve(g)g(s)), H'_t P T' P^*(ue(f)f(s)) \rangle ds,$$

and

$$\langle g^t, H'_t T^{\Delta'}_{f^t} \rangle = \int\limits_{\Delta'} \langle g(s), H'_t T'_{f(s)} \rangle ds.$$

Proof:

We choose sequences $w^{(n)}, w'^{(n)}$ in $h \otimes L^2((t, \infty), k_0)$, which can be written as finite sums, $w^{(n)} = \sum_i v_i^{(n)} \otimes g_i^{(n)}$ and $w'^{(n)} = \sum_j u_j^{(n)} \otimes f_j^{(n)}$; $v_i^{(n)}, u_j^{(n)} \in \mathcal{D}_1$, $g_i^{(n)}, f_j^{(n)} \in L^2((t, \infty), k_0)$ with $g_i^{(n)}(s), f_j^{(n)}(s) \in \mathcal{V}_0$; and $w^{(n)} \to \hat{T}(vg^t)$, $w'^{(n)} \to \hat{T}'(uf^t)$, $(r_t \otimes I_{k^t})w^{(n)} \to (r_t \otimes I)\hat{T}(vg^t)$, $(r'_t \otimes I)w'^{(n)} \to (r'_t \otimes I)\hat{T}'(vg^t)$ as $n \to \infty$. Clearly we have

$$\begin{aligned}
&\langle H_t T^{\Delta}_{g^t}(ve(g)), H'_t T'^{\Delta'}_{f^t}(ue(f)) \rangle \\
&= \lim_{n \to \infty} \int \Big\langle \sum_i H_t(v_i e(g)) \chi_{\Delta}(s) g_i(s), \sum_j H'_t(u_j e(f)) \chi_{\Delta'}(s) f_j(s) \Big\rangle ds \\
&= \lim_{n \to \infty} \int\limits_{\Delta \cap \Delta'} \langle H_t P(w^{(n)}(s)e(g)), H'_t P(w'^{(n)}(s)e(f)) \rangle ds \\
&= \int\limits_{\Delta \cap \Delta'} \langle H_t P(\hat{T}(vg^t)(s)e(g)), H'_t P(\hat{T}'(uf^t)(s)e(f)) \rangle ds.
\end{aligned}$$

The last step follows because $\int\limits_{\Delta \cap \Delta'} \| H_t P(w^{(n)}(s)e(g)) - H_t P(\hat{T}(vg^t)(s)e(g)) \|^2 ds$ is majorised by $\|(r_t^g \otimes I)(w^{(n)} - \hat{T}(vg^t))\|^2$ which goes to 0, and similar statement holds for H'_t and $w'^{(n)}$. Since for $s \in \Delta \cap \Delta' \subseteq (t, \infty)$, $H_t P(\hat{T}(vg^t)(s)e(g)) = H_t P(T(vg(s))e(g)) = H_t P T P^*(ve(g)g(s))$, and similarly $H'_t P(\hat{T}'(uf^t)(s)e(f)) = H'_t P T P^*(ue(f)f(s))$, the proof of the first part of the lemma is complete. The proof of the other part of the lemma is similar. □

Lemma 5.1.7 *In the notation of Lemma 5.1.4, we have*

$$\langle \eta, H_t P R \rangle ve(g) = H_t((\langle \eta, R \rangle v)e(g)),$$

where $v \in \mathcal{D}_0$, $\eta \in k_0$, $g \in \mathcal{V}$.

Proof:
Let us recall that we can replace r_t in (5.8) by its closure, to be denoted by the same notation, and thus replace $\mathcal{D}_1$ by the domain of the closure of r_t. In other words, we can assume without loss of generality that r_t is closed. It follows from the estimate (5.8) that for every fixed $g, f \in \mathcal{V}, t \geq 0$, there is a unique linear map M_t with the domain $\mathcal{D}_1$ satisfying $\langle H_t(ve(g)), ue(f)\rangle = \langle M_t v, u\rangle$ for $v \in \mathcal{D}_1, u \in h$. Let $\tilde{M}_t = M_t \otimes 1_{k_0}$.

For $\eta \in k_0$ and any Hilbert space $\mathcal{H}$, we introduce bounded linear map $C_\eta : \mathcal{H} \otimes k_0 \to \mathcal{H}$ given by $C_\eta(u \otimes \xi) := u\langle \eta, \xi\rangle$. Thus, $\langle \eta, R\rangle v = C_\eta(Rv)$. We claim that for $\xi \in \mathrm{Dom}(r_t \otimes I)$, $C_\eta(\xi) \in \mathrm{Dom}(r_t)$ and

$$C_\eta(r_t \otimes I)\xi = r_t C_\eta \xi.$$

Clearly, the above equality holds for $\xi \in \mathcal{D}_1 \otimes_{\mathrm{alg}} k_0$. Using the fact that $\mathcal{D}_1 \otimes_{\mathrm{alg}} k_0$ is a core for $r_t \otimes I$, given $\xi \in \mathrm{Dom}(r_t \otimes I)$, we choose sequence ξ_n from $\mathcal{D}_1 \otimes_{\mathrm{alg}} k_0$ such that $\xi_n \to \xi$, $(r_t \otimes I)\xi_n \to (r_t \otimes I)\xi$. Hence we have $C_\eta \xi_n \to C_\eta \xi$, $r_t C_\eta \xi_n = C_\eta(r_t \otimes I)\xi_n \to C_\eta(r_t \otimes I)\xi$. Since r_t is closed, the claim is proved. By a similar argument using the estimate (5.8), we conclude that $C_\eta \xi \in \mathrm{Dom}(M_t)$ with

$$C_\eta \tilde{M}_t \xi = M_t C_\eta \xi.$$

Furthermore,

$$\langle H_t P(we(g)),\ P(w'e(f))\rangle = \langle \tilde{M}_t w, w'\rangle$$

for all $w \in \mathrm{Dom}(r_t \otimes I)$, $w' \in h \otimes k_0$. This again follows by first observing it for $w, w' \in \mathcal{D}_1 \otimes_{\mathrm{alg}} k_0$ and then using the estimate (5.8), together with the core property of $\mathcal{D}_1 \otimes_{\mathrm{alg}} k_0$ mentioned earlier.

Thus, we have for $v \in \mathcal{D}_0$, $u \in h$

$$\langle\langle \eta, H_t P R\rangle ve(g),\ ue(f)\rangle = \langle H_t P((Rv)e(g)),\ ue(f)\eta\rangle$$

$$= \langle H_t P((Rv)e(g)),\ P(u\eta e(f))\rangle = \langle \tilde{M}_t(Rv),\ (u\eta)\rangle$$

$$= \langle M_t\langle \eta, R\rangle v,\ u\rangle = \langle H_t((\langle \eta, R\rangle v)e(g)), ue(f)\rangle.$$

This completes the proof, since the vectors of the form $ue(f)$ are total in $h \otimes \Gamma(k)$. □

5.2 Stochastic integrals and quantum Itô formulae

Following the ideas of [70] and [100], we call an adapted $(\mathcal{D}_1, \mathcal{V}_0)$-admissible process $(H_t)_{t\geq 0}$ satisfying $\sup_{0\leq s\leq t} \|H_s ve(g)\| \leq \|r_t^g v\|$ (for all $v \in \mathcal{D}_1,\ g \in \mathcal{V}$), *simple* if H_t is of the form,

$$H_t = \sum_{i=0}^{m} H_{t_i} \chi_{[t_i, t_{i+1})}(t)$$

where m is a positive integer and $0 \equiv t_0 < t_1 < \cdots < t_m < t_{m+1} \equiv \infty$. If M denotes one of the four basic processes $a_R, a_R^\dagger$ and Λ_T and tI, and if (H_t) is simple, then we define the left quantum stochastic integral $\int_0^t H_s M(ds)$ in the natural manner:

$$\int_0^t H_s M(ds) = \sum_{i=0}^{m} H_{t_i} M([t_i, t_{i+1}) \cap [0, t]).$$

In the case when R, S, T are bounded, we also define the right quantum stochastic integral

$$\int_0^t M(ds) H_s = \sum_{i=0}^{m} M([t_i, t_{i+1}) \cap [0, t]) H_{t_i}.$$

Definition 5.2.1 We call an adapted $(\mathcal{D}_1, \mathcal{V}_0)$-admissible process H_t $(\mathcal{D}_1, \mathcal{V}_0)$-*regular* if $t \mapsto H_t(ue(f))$ is continuous for all fixed $u \in \mathcal{D}_1$ and $f \in \mathcal{V}$. It is called *regular* if for all $u \in h$ and $f \in k$ satisfying $\int_0^t \|f(s)\|^4 ds < \infty$ for all t, we have $ue(f) \in \text{Dom}(H_t)$ for all $t \geq 0$, $t \mapsto H_t ue(f)$ is continuous and there is some constant $c(t, f)$ such that $\sup_{s \leq t} \|H_s ue(f)\| \leq c(t, f)\|u\|$.

Lemma 5.2.2 *If H_t is $(\mathcal{D}_1, \mathcal{V}_0)$-regular, then so is the extension $H_t P$ in the sense that $t \mapsto H_t P\xi \otimes e(f)$ is continuous for any $\xi \in \cap_{t \geq 0} \text{Dom}(r_t \otimes I) \subseteq h \otimes k_0$, $f \in \mathcal{V}$.*

Proof:
The continuity of $t \mapsto H_t P(\xi e(f))$ is obvious for ξ in the algebraic tensor product $\mathcal{D}_1 \otimes_{\text{alg}} k_0$. Consider now a general $\xi \in \cap_{s \geq 0} \text{Dom}(r_s \otimes I)$ and a sequence $t_n \to t \in [0, \infty)$. Let t_0 be any number such that $t_n, t \leq t_0$ for all $n \geq 1$. Since $\mathcal{D}_1 \otimes_{\text{alg}} k_0$ is a core for $r_{t_0} \otimes I$, for a given $\epsilon > 0$, we can choose ξ_0 from $h \otimes_{\text{alg}} k_0$ such that $\|\xi - \xi_0\| < \epsilon$, $\|(r_{t_0} \otimes I)(\xi - \xi_0)\| < \epsilon$; and then an integer N depending on ξ_0 such that $\|H_{t_n} P(\xi_0 e(f)) - H_t P(\xi_0 e(f))\| < \epsilon$ for all $n \geq N$. Using the triangle inequality and the fact that $\|H_s P((\xi - \xi_0)e(f))\| \leq \epsilon$ for all $s \leq t_0$, we get the estimate $\|(H_{t_n} P - H_t P)(\xi e(f))\| \leq 3\epsilon$ for all $n \geq N$, which completes the proof. □

The next theorem gives the quantum Itô formulae for simple integrands.

Theorem 5.2.3 *Let $\mathcal{D}_0, \mathcal{D}_1$ be dense subspaces of h and $\mathcal{V}_0$ dense subspace of k_0, $R, S, R', S' \in \mathrm{Lin}(\mathcal{D}_0, h \otimes k_0)$, $T, T' \in \mathrm{Lin}(\mathcal{D}_0 \otimes \mathcal{V}_0, h \otimes k_0)$, $f \in \mathcal{V}$ and $r_t^f \in \mathrm{Lin}(\mathcal{D}_1, h')$ $(t \geq 0)$ be closable operators, where $h' = h_t^f$ is some Hilbert space (depending on f, t). Assume furthermore the following*

(a) $\mathcal{D}_0 \subseteq \cap_{\xi \in \mathcal{V}_0} \mathrm{Dom}(\langle R, \xi \rangle) \cap \mathrm{Dom}(\langle R', \xi \rangle)$.
(b) E, F, G, H *and* E', F', G', H' *are adapted* $(\mathcal{D}_1, \mathcal{V}_0)$*-admissible simple processes satisfying* $\sup_{0 \leq s \leq t} \{\|E_s ue(f)\| + \|F_s ue(f)\| + \|G_s ue(f)\| + \|H_s ue(f)\| + \|E'_s ue(f)\| + \|F'_s ue(f)\| + \|G'_s ue(f)\| + \|H'_s ue(f)\|\} \leq \|r_t^f u\|$ for all $u \in \mathcal{D}_1$, $f \in \mathcal{V}$.
(c) The ranges of R, R', S, S', T, T' *are contained in the domain of* $\widetilde{r_t} \equiv r_t \otimes I_{k_0}$ *for all* $t \geq 0$.

Let

$$X_t = \int_0^t \left(E_s \Lambda_T(ds) + F_s a_R(ds) + G_s a_S^\dagger(ds) + H_s ds \right),$$
$$X'_t = \int_0^t \left(E'_s \Lambda_{T'}(ds) + F'_s a_{R'}(ds) + G'_s a_{S'}^\dagger(ds) + H' ds \right).$$

Then for $u, v \in \mathcal{D}_0$; $f, g \in \mathcal{V}$, *we have the following.*

(i) (First fundamental formula)

$$\langle X_t ve(g), ue(f) \rangle = \int_0^t ds \langle \{\langle f(s), E_s P T_{g(s)} \rangle + F_s \langle R, g(s) \rangle + G_s \langle f(s), S \rangle + H_s\}(ve(g)), ue(f) \rangle \tag{5.13}$$

(ii) (Second fundamental formula or quantum Itô formula)
For this part suppose that $f, g : [0, \infty) \to \mathcal{V}_0$ *are simple right continuous functions. Then*

$$\begin{aligned}
&\langle X_t ve(g), X'_t ue(f) \rangle \\
&= \int_0^t ds \left[\langle X_s ve(g), \{\langle g(s), E'_s P T'_{f(s)} \rangle + F'_s \langle R', f(s) \rangle + G'_s \langle g(s), S' \rangle + H'_s\}(ue(f)) \rangle \right] \\
&+ \int_0^t ds \left[\langle \{\langle f(s), E_s P T_{g(s)} \rangle + F_s \langle R, g(s) \rangle + G_s \langle f(s), S \rangle + H_s\}(ve(g)), X'_s ue(f) \rangle \right] \\
&+ \int_0^t ds \Big[\langle E_s P T_{g(s)}(ve(g)), E'_s P T'_{f(s)}(ue(f)) \rangle + \langle E_s P T_{g(s)}(ve(g)), G'_s P S'(ue(f)) \rangle \\
&+ \langle G_s P S(ve(g)), E'_s P T'_{f(s)}(ue(f)) \rangle + \langle G_s P S(ve(g)), G'_s P S'(ue(f)) \rangle \Big].
\end{aligned} \tag{5.14}$$

(iii) In case the coefficients R, S, T, R', S', T' *are bounded operators and the processes* $E, F, G, H, E', F', G', H'$ *are admissible, (i), (ii) hold for all* $u, v \in h$, *and* $f, g \in k$.

Proof:
The proof is very similar in spirit to the proof in [70], [100]. First, a comment with regard to the notation used above is in order. For example, for almost

all $s \in \mathbb{R}_+$, the expression $E_s PT_{g(s)}(ve(g))$ is to be understood as $(E_s \otimes I_{k_0})P(T_{g(s)}v\otimes e(g)) = (E_s\otimes I_{k_0})P(T(v\otimes g(s))\otimes e(g)) \in h\otimes\Gamma\otimes k_0$. Thus the operator $E_s PT_{g(s)}$ maps $\mathcal{D}_0\otimes\Gamma$ into $h\otimes\Gamma\otimes k_0$ and therefore $\langle f(s), E_s PT_{g(s)}\rangle$ maps $\mathcal{D}_0 \otimes \Gamma$ into $h \otimes \Gamma$.

We denote the essential supremum of f and g by $\|f\|_\infty$ and $\|g\|_\infty$, respectively. We fix $t \geq 0$, and without loss of generality, take $\mathcal{P} \equiv \{t_i\}_{i=0,1,\ldots,m+1}$ to be a partition of $[0, t]$ such that $0 = t_0 < t_1 < \cdots < t_{m+1} = t$; and $L_s = \sum_{i=o}^m L_{t_i}\chi_{[t_i,t_{i+1})}(s)$, $L'_s = \sum_{i=0}^m L'_{t_i}\chi_{[t_i,t_{i+1})}(s)$, where L is one of the four coefficient processes E, F, G, H and L' is one of the processes E', F', G', H'. Observe that the definition of stochastic integrals for simple adapted processes does not depend on the choice of the partition as long as L_s and L'_s take constant values in any subinterval of the partition; and this allows us to refine $\mathcal{P}$ arbitrarily.

By definition of the basic processes as given in the previous section, we have that

$$
\begin{aligned}
&\langle X_t ve(g), ue(f)\rangle \\
&= \sum_{i=0}^m \langle \{E_{t_i}\Lambda_T([t_i,t_{i+1})) + F_{t_i}a_R([t_i,t_{i+1})) + G_{t_i}a_S^\dagger([t_i,t_{i+1})) \\
&\quad + H_{t_i}(t_{i+1}-t_i)\}, ue(f)\rangle \\
&= \sum_{i=0}^m \langle \{E_{t_i} a^\dagger_{T_{g^{t_i}}^{[t_i,t_{i+1})}} + G_{t_i} a^\dagger_{S_{t_i}^{[t_i,t_{i+1})}}\}(ve(g)), ue(f)\rangle + \\
&\quad \int_{[t_i,t_{i+1})} \langle \{F_{t_i}\langle R, g(s)\rangle + H_{t_i}\}(ve(g)), ue(f)\rangle ds.
\end{aligned}
$$

Now, note that by Lemma 5.1.1,

$$
\begin{aligned}
&\langle G_{t_i} a^\dagger_{S_{t_i}^{[t_i,t_{i+1})}}(ve(g)), ue(f)\rangle \\
&= \langle G_{t_i}\langle f, S_{t_i}^{[t_i,t_{i+1})}\rangle(ve(g)), ue(f)\rangle \\
&= \int_{[t_i,t_{i+1})} \langle G_{t_i}\langle f(s), S\rangle ve(g), ue(f)\rangle ds.
\end{aligned}
$$

Similarly,

$$
\begin{aligned}
&\langle E_{t_i} a^\dagger_{T_{g^{t_i}}^{[t_i,t_{i+1})}}(ve(g)), ue(f)\rangle \\
&= \int_{[t_i,t_{i+1})} \langle E_{t_i}\langle f(s), T_{g^{t_i}(s)}\rangle(ve(g)), ue(f)\rangle ds
\end{aligned}
$$

$$= \int\limits_{[t_i,t_{i+1})} \langle f(s), E_s P T_{g(s)} \rangle (ve(g)), ue(f) \rangle.$$

From this (i) follows.

We shall now briefly sketch the proof of (ii). It is clear that the left-hand side of (5.14) can be written as $S_1^{\mathcal{P}} + S_2^{\mathcal{P}} + S_3^{\mathcal{P}}$, where

$$\begin{aligned} S_1^{\mathcal{P}} &= \sum_{i=0}^{m} \langle X_{t_i}(ve(g)), \{E'_{t_i} \Lambda_{T'}([t_i, t_{i+1})) + F'_{t_i} a_{R'}([t_i, t_{i+1})) + G'_{t_i} a^{\dagger}_{S'}([t_i, t_{i+1})) \\ &\quad + H'_{t_i}(t_{i+1} - t_i)\}(ue(f)) \rangle, \end{aligned}$$

$$\begin{aligned} S_2^{\mathcal{P}} &= \sum_{i=0}^{m} \langle \{E_{t_i} \Lambda_T([t_i, t_{i+1})) + F_{t_i} a_R([t_i, t_{i+1})) + G_{t_i} a^{\dagger}_S([t_i, t_{i+1})) \\ &\quad + H_{t_i}(t_{i+1} - t_i)\}(ve(g)), X'_{t_i}(ue(f)) \rangle, \end{aligned}$$

$$\begin{aligned} S_3^{\mathcal{P}} &= \sum_{i=0}^{m} \langle \{E_{t_i} \Lambda_T([t_i, t_{i+1})) + F_{t_i} a_R([t_i, t_{i+1})) + G_{t_i} a^{\dagger}_S([t_i, t_{i+1})) \\ &\quad + H_{t_i}(t_{i+1} - t_i)\}(ve(g)), \{E'_{t_i} \Lambda_{T'}([t_i, t_{i+1})) + F'_{t_i} a_{R'}([t_i, t_{i+1})) \\ &\quad + G'_{t_i} a^{\dagger}_{S'}([t_i, t_{i+1})) + H'_{t_i}(t_{i+1} - t_i)\}(ue(f)) \rangle. \end{aligned}$$

Similarly, denote by S_1, S_2, S_3 respectively the first, second and third integral in the right-hand side of (5.14). We want to show $S_i^{\mathcal{P}} \to S_i$ as $\mathcal{P}$ is made of arbitrarily small norm, for $i = 1, 2, 3$. Let $\|\mathcal{P}\|$ denote the norm of the partition $\mathcal{P}$. We have

$$\begin{aligned} &|S_1^{\mathcal{P}} - S_1| \\ &\leq \sum_{i=0}^{m} \left| \int_{t_i}^{t_{i+1}} \langle (X_{t_i} - X_s)(ve(g)), B_s(ue(f)) \rangle ds \right| \\ &\quad (\text{where } \mathrm{B_s} = \langle \mathrm{g(s)}, \mathrm{E'_s P T'_{f(s)}} \rangle + \mathrm{F'_s} \langle \mathrm{R'}, \mathrm{f(s)} \rangle + \mathrm{G'_s} \langle \mathrm{g(s)}, \mathrm{S'} \rangle + \mathrm{H'_s} \\ &\leq \sum_{i=0}^{m} \int_{t_i}^{t_{i+1}} \|(X_{t_i} - X_s)(ve(g))\| \{ \|g\|_\infty \|\widetilde{r_t^f} T'(uf(s))\| + \|r_t^f \langle R', f(s) \rangle u\| \\ &\quad + \|g\|_\infty \|\widetilde{r_t^f} S' u\| + \|r_t^f u\| \} ds. \end{aligned} \tag{5.15}$$

Recall at this point that f, g have been assumed to be $\mathcal{V}_0$-valued simple right continuous functions, of the form $g = \sum_k g_k \chi_{[s_k, s_{k+1})}$, $f = \sum_l f_l \chi_{[s'_l, s'_{l+1})}$, where p, q are positive integers, $\{0 = s_1, \ldots, s_{p+1} = \infty\}$ and $\{0 = s'_1, \ldots, s'_{q+1} = \infty\}$ are two finite partitions of $[0, \infty)$ and f_l, g_k are elements of $\mathcal{V}_0$. Thus, taking

$$M'(u, f) := \mathrm{Max}\{\|\widetilde{r_t^f} T'(uf_l))\|, \|r_t^f \langle R', f_l \rangle u\| : l = 1, \ldots, q\},$$

we can estimate the integrand in the expression (5.15) by the s-independent quantity $\{(\|g\|_\infty + 1)M'(u, f) + \|g\|_\infty \|\widetilde{r_t^f} S'u\| + \|r_t^f u\|\}$. On the other hand, for $s \in [t_i, t_{i+1})$,

$$\begin{aligned}
&\|(X_{t_i} - X_s)(ve(g))\| \\
&\leq \|E_{t_i} \Lambda_T([t_i, s))(ve(g))\| + \|F_{t_i} a_R([t_i, s))(ve(g))\| \\
&\quad + \|G_{t_i} a_S^\dagger([t_i, s))(ve(g))\| + \|H_{t_i}(s - t_i)(ve(g))\|.
\end{aligned}$$

We have by Lemma 5.1.4,

$$\begin{aligned}
&\|G_{t_i} a_S^\dagger([t_i, s))(ve(g))\|^2 \\
&= (s - t_i)\|G_{t_i} PS(ve(g))\|^2 \\
&\quad + \int_{[t_i,s)\times[t_i,s)} \langle\langle g(\tau), G_{t_i} PS\rangle(ve(g)), \langle g(\tau'), G_{t_i} PS\rangle(ve(g))\rangle d\tau d\tau' \\
&\leq (t_{i+1} - t_i)\|\widetilde{r_t^g} Sv\|^2\{1 + (t_{i+1} - t_i)\|g\|_\infty^2\}.
\end{aligned}$$

Similarly,

$$\begin{aligned}
&\|E_{t_i} \Lambda_T([t_i, s))(ve(g))\|^2 \\
&\leq (t_{i+1} - t_i)M^2(v, g)\{1 + (t_{i+1} - t_i)\|g\|_\infty^2\},
\end{aligned}$$

and

$$\|F_{t_i} a_R([t_i, s))(ve(g))\|^2 \leq (t_{i+1} - t_i)^2 M^2(v, g),$$

where $M(v, g) := \text{Max}\{\|\widetilde{r_t^g} T(vg_k))\|, \|r_t^g \langle R, g_k\rangle v\| : k = 1, \ldots, p\}$.

These estimates allow us to majorize the last expression in (5.15) by $Ct\|\mathcal{P}\|^{\frac{1}{2}}$ when $\|\mathcal{P}\|$ $(:= \text{Max}\{(t_{i+1} - t_i)\}) \leq 1$, where

$$\begin{aligned}
C := &\{(\|g\|_\infty + 1)M'(u, f) + \|g\|_\infty \|\widetilde{r_t^f} S'u\| + \|r_t^f u\|\}\times \\
&\{(\|\widetilde{r_t^g} Sv\| + M(v, g))(1 + \|g\|_\infty^2)^{\frac{1}{2}} + M(v, g) + \|r_t^g v\|\}.
\end{aligned}$$

This shows that $S_1^{\mathcal{P}} \to S_1$ as $\|\mathcal{P}\| \to 0$. Similar analysis can be carried out for $S_2^{\mathcal{P}}$ and $S_3^{\mathcal{P}}$. For example, let us consider one typical term in the expansion of $S_3^{\mathcal{P}}$, say $\sum_i \langle G_{t_i} a_S^\dagger([t_i, t_{i+1}))(ve(g)), G'_{t_i} a_{S'}^\dagger([t_i, t_{i+1}))(ue(f))\rangle$. By Lemma 5.1.4, this is equal to

$$\begin{aligned}
&\sum_i \langle G_{t_i} a_S^\dagger([t_i, t_{i+1}))(ve(g)), G'_{t_i} a_{S'}^\dagger([t_i, t_{i+1}))(ue(f))\rangle \\
&= \int_0^t \langle G_s PS(ve(g)), G'_s PS'(ue(f))\rangle ds \\
&\quad + \sum_{i=0}^m \int_{[t_i,t_{i+1})\times[t_i,t_{i+1})} \langle\langle f(s), G_{t_i} PS\rangle(ve(g)), \langle g(s'), G'_{t_i} PS'\rangle(ue(f))\rangle ds ds',
\end{aligned}$$

the second term of which can be majorized by some constant (independent of $\mathcal{P}$) times $\|\mathcal{P}\|$, and hence goes to 0 as $\|\mathcal{P}\| \to 0$. A similar analysis can be carried out for showing that $S_2^{\mathcal{P}}$ converges to S_2. This completes the proof of (*ii*).

The proof of (*iii*) can be given by a modification of the estimates involved at each step in the proof of (*i*) and (*ii*). Let us show, for example, the convergence $S_1^{\mathcal{P}} \to S_1$ as $\|\mathcal{P}\| \to 0$. Assume that R, S, T, R', S', T' are bounded and assume without loss of generality that $r_t^f = c(t, f)I$, $r_t^g = c(t, g)I$ for all t. We have

$$\begin{aligned}
&|S_1^{\mathcal{P}} - S_1| \\
&\leq \sum_{i=0}^{m} \left| \int_{t_i}^{t_{i+1}} \langle (X_{t_i} - X_s)(ve(g)), B_s(ue(f))\rangle ds \right| \\
&\leq \sum_{i=0}^{m} \int_{t_i}^{t_{i+1}} c(t, f)\|u\| \|(X_{t_i} - X_s)ve(g)\| (\|g(s)\| \|f(s)\| \|T'\| \\
&+ \|f(s)\| \|R'\| + \|g(s)\| \|S'\| + 1) ds. \qquad (5.16)
\end{aligned}$$

For $s \in [t_i, t_{i+1})$, we can estimate $\|(X_{t_i} - X_s)ve(g)\|$ as follows. As before,

$$\begin{aligned}
&\|G_{t_i} a_S^{\dagger}([t_i, s))(ve(g))\|^2 \\
&= c^2(t, g)\|v\|^2 \|S\|^2 \left\{ (t_{i+1} - t_i) + \left(\int_{t_i}^{t_{i+1}} \|g(\tau)\| d\tau \right)^2 \right\} \\
&\leq c^2(t, g)\|v\|^2 \|S\|^2 \{(t_{i+1} - t_i) + (t_{i+1} - t_i)\|g\|^2_{[t_i, t_{i+1})}\},
\end{aligned}$$

where $\|g\|_{[t_i,t_{i+1})} := \left(\int_{t_i}^{t_{i+1}} \|g(\tau)\|^2 d\tau \right)^{\frac{1}{2}}$. Similarly,

$$\|E_{t_i} \Lambda_T([t_i, s))(ve(g))\|^2 \leq c^2(t, g)\|v\|^2 \|T\|^2 \|g\|^2_{[t_i,t_{i+1})} (1 + \|g\|^2_{[t_i,t_{i+1})});$$

$$\|F_{t_i} a_R([t_i, s))(ve(g))\|^2 \leq (t_{i+1} - t_i)^2 c^2(t, g)\|R\|^2 \|v\|^2 \|g\|^2_{[t_i,t_{i+1})}.$$

Thus,

$$\|(X_{t_i} - X_s)ve(g)\| \leq c(t, g)\|v\| M_i(\mathcal{P}, g),$$

where

$$\begin{aligned}
M_i(\mathcal{P}, g) :&= \|S\|(t_{i+1} - t_i)^{\frac{1}{2}} (1 + \|g\|^2_{[t_i,t_{i+1})})^{\frac{1}{2}} \\
&+ \|T\| \|g\|_{[t_i,t_{i+1})} (1 + \|g\|^2_{[t_i,t_{i+1})})^{\frac{1}{2}} \\
&+ (t_{i+1} - t_i)(\|R\| \|g\|_{[t_i,t_{i+1})} + 1).
\end{aligned}$$

Since $s \mapsto \int_0^s \|g(\tau)\|^2 d\tau$ is uniformly continuous on the compact interval $[0, t]$, for a given $\epsilon > 0$, we can choose $\delta > 0$ with the property that Max_i

$M_i(\mathcal{P}, g) \leq \epsilon$ whenever $\mathcal{P}$ is a partition with $\|\mathcal{P}\| \leq \delta$. For such a $\mathcal{P}$, we have from (5.16)

$$\begin{aligned}
&|S_1^{\mathcal{P}} - S_1| \\
&\leq \epsilon c(t, f)c(t, g)\|u\|\|v\| \int_0^t (\|g(s)\|\|f(s)\|\|T'\| \\
&+ \|f(s)\|\|R'\| + \|g(s)\|\|S'\| + 1)ds,
\end{aligned}$$

which completes the proof of (*iii*) since f and g are in $L^1([0, t], k_0)$. □

Next we give the well-known result due to Gronwall.

Lemma 5.2.4 (Gronwall's Lemma) *Let F, G, α be nonnegative continuous functions on $\mathbb{R}_+$ and F, G be monotone nondecreasing. Suppose $F(0) = 0$ and*

$$\alpha(t) \leq G(t) + \int_0^t \alpha(s)dF(s) \text{ for all } t \geq 0.$$

Then we have,

$$\alpha(t) \leq G(t)\exp(F(t)) \text{ for all } t \geq 0.$$

Proof:
We have that

$$\begin{aligned}
\alpha(t) &\leq G(t) + \int_0^t \alpha(s)dF(s) \\
&\leq G(t) + \int_0^t G(s)dF(s) + \int_0^t \int_0^s \alpha(s_1)dF(s_1)dF(s) \\
&\leq G(t)(1 + F(t)) + \int_0^t \int_0^s \alpha(s_1)dF(s_1)dF(s),
\end{aligned}$$

since G is nondecreasing and $F(0) = 0$. This leads us to make the following claim

$$\begin{aligned}
\alpha(t) \leq\ & G(t)\left\{1 + F(t) + \frac{F(t)^2}{2} + \cdots + \frac{(F(t))^n}{n!}\right\} \\
&+ \int_0^t \int_0^{s_1} \cdots \int_0^{s_n} \alpha(s_{n+1})dF(s_{n+1}) \cdots dF(s_1). \qquad (5.17)
\end{aligned}$$

It is verified by induction. We have already shown it for $n = 1$, and the inductive step from n to $n + 1$ follows by using

$$\alpha(s_{n+1}) \leq G(s_{n+1}) + \int_0^{s_{n+1}} \alpha(s_{n+2})dF(s_{n+2}) \leq G(t) + \int_0^{s_{n+1}} \alpha(s_{n+1})dF(s_{n+2}),$$

as G is nondecreasing, and noting that

$$\int_0^t \int_0^{s_1} \cdots \int_0^{s_n} dF(s_{n+1}) \cdots dF(s_1) = \frac{F(t)^{n+1}}{(n + 1)!}.$$

Now, we let n go to ∞, so that

$$\int_0^t \int_0^{s_1} \cdots \int_0^{s_n} \alpha(s_{n+1}) dF(s_{n+1}) \cdots dF(s_1) \leq \alpha(t) \frac{(F(t))^{n+1}}{(n+1)!} \to 0,$$

and thus the right-hand side of the inequality (5.17) tends to $G(t)\exp(F(t))$, which completes the proof of the lemma. □

For a simple integrand H_t, one can derive the following estimate by using Gronwall's Lemma as in [100].

Lemma 5.2.5 *Let υ, g, X_t be as in the Theorem 5.2.3 (ii). Then one has*

$$\begin{aligned} &||X_t \upsilon e(g)||^2 \\ &\leq e^t \int_0^t ds \big[|| \{E_s P T_{g(s)} + G_s P S\}(\upsilon e(g))||^2 + \\ &||\{\langle g(s), E_s P T_{g(s)}\rangle + F_s \langle R, g(s)\rangle + \langle g(s), G_s P S\rangle + H_s\}(\upsilon e(g))||^2 \big]. \end{aligned} \tag{5.18}$$

Proof:
Write $A(s) = \{\langle g(s), E_s P T_{g(s)}\rangle + F_s \langle R, g(s)\rangle + \langle g(s), G_s P S\rangle + H_s\}(\upsilon e(g))$, $B(s) = (E_s P T_{g(s)} + G_s P S)(\upsilon e(g))$. Then, by the second fundamental formula (5.14) we have,

$$\begin{aligned} &\|X_t \upsilon e(g)\|^2 \\ &\leq 2\mathrm{Re}\left(\int_0^t \langle X_s(\upsilon e(g)), A(s)\rangle ds \right) + \int_0^t \|B(s)\|^2 ds \\ &\leq \int_0^t \|X_s(\upsilon e(g))\|^2 ds + \int_0^t (\|A(s)\|^2 + \|B(s)\|^2) ds. \end{aligned}$$

The last step follows by the inequality $2\mathrm{Re}(\langle x, y\rangle) \leq \|x\|^2 + \|y\|^2$. Now, by Lemma 5.2.4 we complete the proof, by taking $\alpha(t) = \|X_t(\upsilon e(g))\|^2$, $F(t) = t$ and $G(t) = \int_0^t (\|A(s)\|^2 + \|B(s)\|^2) ds$. □

The extension of the definition of X_t to the case when (E, F, G, H) are $(\mathcal{D}_1, \mathcal{V}_0)$-regular can now be treated similarly, and we have the following result.

Theorem 5.2.6 *Let $\mathcal{D}_1, \mathcal{D}_0, \mathcal{V}_0, R, S, T$ be as in the Theorem 5.2.3, but assume (E, F, G, H) to be $(\mathcal{D}_1, \mathcal{V}_0)$-regular instead of simple. Then the integral $X_t = \int_0^t \{E_s \Lambda_T(ds) + F_s a_R(ds) + G_s a_S^\dagger(ds) + H_s ds\}$ can be defined as a linear map on the domain spanned by the vectors of the form $\upsilon e(g)$, where υ, g satisfy the conditions in (b) of the Theorem 5.2.3. The first and second fundamental formulae as well as the estimate (5.18) remain valid and $t \mapsto X_t \upsilon e(g)$ is continuous for such υ and g. Furthermore, if $\widetilde{r^g}_t T$, $\widetilde{r^g}_t S$ and $r_t^g R^*$*

are assumed to be closable (where we use the notation $\widetilde{A}$ for the ampliation $A \otimes I$ of an operator A), then (X_t) is an adapted $(\mathcal{D}_0, \mathcal{V}_0)$-regular process.

If R, S, T are bounded, and (E, F, G, H) are adapted regular processes, the above integral X_t can be defined on the space spanned by vectors of the form $ve(g)$ where $v \in h, g \in L^4_{\mathrm{loc}}$ and where $L^4_{\mathrm{loc}} := \{g \in k : \int_0^t \|g(s)\|^4 ds < \infty$ for all $t\}$. *Furthermore, (X_t) is an adapted regular process in this case.*

Proof:
Fix $v \in \mathcal{D}_1$, $g \in \mathcal{V}$ as in the statement of the theorem. For $n = 1, 2, \ldots$ define $L_t^{(n)} = L_{\frac{j}{n}}$ if $\frac{j}{n} \leq t < \frac{j+1}{n}$ for $j = 0, 1, 2, \ldots$, where L denotes one of the four processes E, F, G, H. It follows from the assumed regularity of E, F, G, H that $L_t^{(n)}(ve(g))$ converges to $L_t(ve(g))$ uniformly over bounded subintervals of $\mathbb{R}_+$. By Lemma 5.2.2, same statement as above remains true if we replace $L_t^{(n)}$ and L_t by $L_t^{(n)} P$, $L_t P$ respectively, and v by $\xi \in h \otimes k_0$. Clearly, each $L^{(n)}$ is simple adapted process and thus we can form $X_t^{(n)} = \int_0^t \{E_s^{(n)} \Lambda_T(ds) + F_s^{(n)} a_R(ds) + G_s^{(n)} a_S^\dagger(ds) + H_s^{(n)} ds\}$. By the estimate obtained in Lemma 5.2.5, we have

$$
\begin{aligned}
&||(X_t^{(m)} - X_t^{(n)})ve(g)||^2 \\
&\leq e^t \int_0^t ds\big[||\{(E_s^{(m)} - E_s^{(n)})PT_{g(s)} + (G_s^{(m)} - G_s^{(n)})PS\}(ve(g))||^2 \\
&+ ||\{\langle g(s), (E_s^{(m)} - E_s^{(n)})PT_{g(s)}\rangle + (F_s^{(m)} - F_s^{(n)})\langle R, g(s)\rangle \\
&+ \langle g(s), (G_s^{(m)} - G_s^{(n)})PS\rangle + (H_s^{(m)} - H_s^{(n)})\}(ve(g))||^2\big]. \qquad (5.19)
\end{aligned}
$$

The integrand on the right-hand side of the above clearly goes to 0 as m, $n \to \infty$, and we can estimate the integrand by $C(t, g, v)$ given by

$$
\begin{aligned}
C(t, g, v) := \{&4(M^2(v, g) + \|\widetilde{r_t^g} Sv\|^2) + 8(\|g\|_\infty^2 M^2(v, g) + M^2(v, g) \\
&+\|g\|_\infty^2 \|\widetilde{r_t^g} Sv\|^2 + \|r_t^g v\|^2)\},
\end{aligned}
$$

which can be seen to be independent of the integration variable s. We have used here the notation in the proof of the Theorem 5.2.3 and the fact that

$$
\|(L_s^{(m)} - L_s^{(n)})P(\xi e(g))\| \leq \|\widetilde{r_t^g} \xi\|
$$

for all $s \leq t$ and $\xi \in \mathrm{Dom}(\widetilde{r_t^g})$. From this it follows, by the Dominated convergence theorem, that $\{X_t^{(m)}(ve(g))\}$ is a Cauchy sequence and hence its limit defines $X_t(ve(g))$. It can be seen that the two fundamental formulae in Theorem 5.2.3 and the estimate in the Lemma 5.2.5 are valid for X_t as well. To show the continuity of $t \mapsto X_t(ve(g))$, it suffices to observe that for $0 \leq t_1 \leq t_2 \leq t$, we have the following, which can be obtained from Lemma 5.2.5 replacing the processes E, F, G, H by $\chi_{(t_1,t_2)}E,\ \chi_{(t_1,t_2)}F,\ \chi_{(t_1,t_2)}G,\ \chi_{(t_1,t_2)}H$, respectively:

$$\|(X_{t_1} - X_{t_2})(ve(g))\|^2 \leq \exp(t) \int_{t_1}^{t_2} C(t, g, v)ds = \exp(t)C(t, v, g)(t_2 - t_1).$$

To prove the $(\mathcal{D}_0, \mathcal{V}_0)$-regularity of X_t when $\widetilde{r^g}_t T$, $\widetilde{r^g}_t S$ and $r_t^g R^*$ are assumed to be closable, it suffices to obtain closable operators β_t^g defined on $\mathcal{D}_0$ such that $\sup_{s \leq t} \|X_s ve(g)\| \leq \|\beta_t^g v\|$. To this end, we take $\beta_t^g : \mathcal{D}_0 \to L^2([0, t], h'')$, with

$$h'' = (h_t^g \otimes k_0) \oplus (h_t^g \otimes k_0) \oplus h_t^g \oplus h_t^g,$$

such that β_t^g is given by

$$\begin{aligned}(\beta_t^g v)(s) := \ & e^{\frac{t}{2}}\{\sqrt{2}(1 + 2\|g(s)\|^2)^{\frac{1}{2}} \widetilde{r_t^g}(T(vg(s))) \oplus \\ & \sqrt{2}(1 + 2\|g(s)\|^2)^{\frac{1}{2}} \widetilde{r_t^g}(Sv) \ \oplus 2r_t^g(\langle R, g(s)\rangle v) \ \oplus 2r_t^g v\}.\end{aligned}$$

Since g is simple, it can be seen that $\|\beta_t^g v\|^2 = \int_0^t \|(\beta_t^g v)(s)\|^2 ds < \infty$ for $v \in \mathcal{D}_0$ and that $\sup_{0 \leq s \leq t} \|X_s ve(g)\| \leq \|\beta_t^g v\|$, which follows from (5.18). Also, β_t^g is closable since $\widetilde{r_t^g} T$, $\widetilde{r_t^g} S$, $r_t^g R^*$ and r_t^g are so by assumption.

In case R, S, T are assumed to be bounded and $\mathcal{D}_1 = h$, without loss of generality we can also assume that $r_t^g = c(t, g)I$ for all t and g. Then for $v \in h$, g satisfying $\int_0^t \|g(s)\|^4 ds < \infty$, we can estimate the integrand on the right-hand side of (5.19) by the following quantity:

$$\begin{aligned}\Psi(s) := \ & c^2(t, g)\|v\|^2\{4(\|g(s)\|^2\|T\|^2 + \|S\|^2) + 8(\|g(s)\|^4\|T\|^2 \\ & + \|g(s)\|^2(\|R\|^2 + \|S\|^2) + 1\},\end{aligned}$$

for some constant $c(t, g)$. Since $\int_0^t \Psi(s)ds < \infty$ by our assumption on g, it follows that $X_t ve(g)$ can be defined and $t \mapsto X_t ve(g)$ is continuous for all v and g. Moreover, we have from Lemma 5.2.5

$$\begin{aligned}& \|X_t(ve(g))\|^2 \\ & \leq \|v\|^2 c^2(t, g)e^t \int_0^t \{2(\|T\|^2\|g(s)\|^2 + \|S\|^2) + 4(\|g(s)\|^4\|T\|^2 + \\ & \quad \|g(s)\|^2\|R\|^2 + \|g(s)\|^2\|S\|^2 + 1)\}ds.\end{aligned}$$

The integral on the right-hand side is finite by the assumption on g, and thus we get a finite positive constant $K(t, g)$ such that $\|X_t ve(g)\| \leq \|v\|K(t, g)$. This completes the proof. □

Corollary 5.2.7 *(i) Along with the assumptions in the statement of Theorem 5.2.6, assume furthermore that R, S, T are functions of t such that for ξ*

in $\mathcal{V}_0$ *and* u *in* $\mathcal{D}_0 \subseteq \mathrm{Dom}(\langle \underset{\sim}{R(t)}, \xi\rangle) \cap \mathrm{Dom}(\widetilde{S(t)}) \cap \mathrm{Dom}(T(t)_\xi)$, *each of the maps* $t \mapsto r_t^g \langle R(t), \xi\rangle u, r_t^g S(t)u, r_t^g T(t)(u\xi)$ *is continuous. Then the integral*

$$X(t) = \int_0^t (E_s \Lambda_T(ds) + F_s a_R(ds) + G_s a_S^\dagger(ds) + H_s ds)$$

is defined on linear span of vectors of the form $ve(g)$ *with* $v \in \mathcal{D}_0$ *and* $g \in \mathcal{V}$. *Moreover,* $t \mapsto X_t ve(g)$ *is continuous for such* v *and* g, *and the adapted process* (X_t) *satisfies the estimate (5.18) with the constant coefficients* T, R, S *replaced by* $T(s), R(s)$ *and* $S(s)$, *respectively.*

(ii) *In part (i), if we replace the domain* $\mathcal{D}_0$ *of* $T(t)_\xi, \langle R(t), \xi\rangle, S(t)$ $(\xi \in \mathcal{V}_0)$ *by the* t*-dependent domain* $\mathcal{D}_0 \otimes_{\text{alg}} \mathcal{E}(\mathcal{V}_t)$, *where* $\mathcal{V}_t = \mathcal{V} \cap k_t$, *and if we assume that* $t \mapsto r_t^g \langle R(t), \xi\rangle ve(g_t), r_t^g S(t)ve(g_t), r_t^g T(t)(ve(g_t)\xi)$ *are continuous for* $g \in \mathcal{V}$, $v \in \mathcal{D}_0$, $\xi \in \mathcal{V}_0$, *then the conclusions in (i) remain valid.*

(iii) *Assume that the time-dependent coefficients* $L(t)$ $(= R(t), S(t), T(t))$ *in (i) are bounded operators,* $t \mapsto \|L(t)\|$ *is uniformly bounded over compact intervals and* E, F, G, H *are adapted regular processes. Then* $X_t ve(g)$ *can be defined for all* $v \in h$ *and* g *in* L^4_{loc}, *and* (X_t) *is also an adapted regular process.*

(iv) *The conclusions in (iii) remain valid under the hypotheses of (ii) if the following additional assumptions are made:*

(a) E, F, G, H *are adapted regular;*

(b) R, S, T *are also regular in the sense that for* $s \geq 0$, $v \in h$ *and* $g \in L^4_{\text{loc}}$, *we have the following:*

$$ve(g_s) \in \mathrm{Dom}(S(s)) \cap_{\xi \in k_0} \mathrm{Dom}(\langle R(s), \xi\rangle) \cap \mathrm{Dom}(T(s)_\xi),$$
$$\sup_{s \leq t} \|S(s)ve(g_s)\| \leq c'(t, g)\|v\|,$$

$$\sup_{s \leq t} \mathrm{Max}\{\|T(s)(ve(g_s)\xi)\|, \|\langle R(s), \xi\rangle ve(g_s)\|\}$$
$$\leq c'(t, g)\|v\|\|\xi\| \text{ for all } \xi \in k_0,$$

where $c'(t, g)$ *is a constant;*

(c) $s \mapsto S(s)ve(g_s), T(s)_\xi(ve(g_s)), \langle R(s), \xi\rangle ve(g_s)$ *are continuous for* v, g *as in (b) and all* $\xi \in k_0$.

Proof:
(*i*) It may be noted that if we replace constant coefficients in the Theorem 5.2.6 by simple coefficients then the integral is just a finite linear combination of

integrals with constant coefficients. Thus, it is possible to deduce the existence of integral and the fundamental formulae, hence also the estimate (5.18) for simple coefficients. To consider more general time-dependent coefficients as in the statement of the present corollary, recall that $g \in \mathcal{V}$ is right continuous simple function, thus we can choose sequences $T^{(n)}, R^{(n)}, S^{(n)}$ of simple coefficients such that $\widetilde{r_t^g T^{(n)}}(s)(vg(s))$, $r_t^g \langle R^{(n)}(s), g(s)\rangle v$ and $\widetilde{r_t^g S^{(n)}}(s)v$ converge to $\widetilde{r_t^g T}(s)(vg(s))$, $r_t^g \langle R(s), g(s)\rangle v$ and $\widetilde{r_t^g S}(s)v$ respectively for every $v \in \mathcal{D}_0$ and $g \in \mathcal{V}$. With these choices, we can define the integral $X^{(n)}(t)$ on $v \otimes e(g)$ in a natural way using Theorem 5.2.6. The hypotheses of continuity of the coefficients will allow one to pass to the limit in the integral as well by using the estimate (5.18). For example, $\sup_{s \leq t} || \int_0^t E_s(\Lambda_{T^{(n)}}(ds) - \Lambda_{T^{(m)}}(ds))ve(g)||^2 \leq \int_0^t (1 + ||g(s)||^2)||)\|\widetilde{r_t^g}(T^{(n)}(s) - T^{(m)}(s))(vg(s))\|^2 ds \to 0$ as $m, n \to \infty$, by the Dominated Convergence theorem, since $\int_0^t (1 + \|g(s)\|^2)\|\widetilde{r_t^g T}(s)(vg(s))\|^2 ds < \infty$ and since $s \mapsto \widetilde{r_t^g T}(s)(v\xi)$ is continuous for $\xi \in \mathcal{V}_0$ and g a simple $\mathcal{V}_0$-valued function.

(*ii*) This part follows from (*i*) with some natural modifications. For instance, in the estimate above we shall have instead $|| \int_0^t E_s[\Lambda_{T^{(n)}}(ds) - \Lambda_{T^{(m)}}(ds)]ve(g)||^2 \leq \int_0^t ds(1 + ||g(s)||^2)||\widetilde{r_t^g}(T^{(n)}(s) - T^{(m)}(s))(v \otimes e(g) \otimes g(s))||^2$. Note that for $\xi \in \mathcal{V}_0$, the map $s \mapsto \widetilde{r_t^g T}(s)(ve(g)\xi) = P(T(s)(ve(g_s)\xi) \otimes e(g^s))$ (where P denotes the identification between $h \otimes \Gamma_s \otimes k_0 \otimes \Gamma^s$ and $h \otimes \Gamma_s \otimes \Gamma^s \otimes k_0 = h \otimes \Gamma \otimes k_0$) is continuous since both the maps $s \mapsto \widetilde{r_t^g T}(s)(ve(g_s))$ and $s \mapsto e(g^s)$ are. Thus, for a simple function g, so that $g(s)$ varies over a finite set of vectors in $\mathcal{V}_0$ for $s \in [0, t]$, it is clear that $\int_0^t ds(1 + ||g(s)||^2)||\widetilde{r_t^g}(T^{(n)}(s) - T^{(m)}(s))(v \otimes e(g) \otimes g(s))||^2 \to 0$ as $m, n \to \infty$, by the Dominated Convergence theorem as in the proof of (*i*).

The proofs of (*iii*) and (*iv*) can be given by an applications of these estimates and arguments used in the proof of the last part of the Theorem 5.2.6. □

Theorem 5.2.8 Right integrals

Assume that $\mathcal{D}_0$, $\mathcal{V}_0$ are dense subspaces of h and k_0 respectively, $A \in \mathrm{Lin}(\mathcal{D}_0, h)$, $R, S \in \mathrm{Lin}(\mathcal{D}_0, h \otimes k_0)$, $T \in \mathrm{Lin}(\mathcal{D}_0 \otimes \mathcal{V}_0, h \otimes k_0)$, and (Z_s) is an adapted process such that for $\xi \in \mathcal{V}_0$, we have $\mathcal{D}_0 \otimes_{\mathrm{alg}} \mathcal{E}(\mathcal{V}) \subseteq \cap_{s \geq 0} \mathrm{Dom}(AZ_s) \cap \mathrm{Dom}(\langle R, \xi\rangle Z_s) \cap \mathrm{Dom}(SZ_s) \cap \mathrm{Dom}(T_\xi Z_s)$. Note that as before, we have denoted an operator B and $B \otimes I$ by the same symbol, where B stands for one of the maps $A, \langle R, \xi\rangle, S, T_\xi$. Furthermore, suppose that $s \mapsto AZ_s ve(g)$, $\langle R, \xi\rangle Z_s ve(g)$, $SZ_s ve(g)$, $TZ_s ve(g)$ are continuous for $v \in \mathcal{D}_0$, $g \in \mathcal{V}$. Then we can define the right integral of the form $\int_0^t M(ds)Z_s$ on the linear span of vectors of the form $ve(g)$, with $v \in \mathcal{D}_0$, $g \in \mathcal{V}$, where $M(ds) = a_R(ds) + a_S^\dagger(ds) + \Lambda_T(ds) + Ads$; and can also obtain formulae similar to those in the Theorem 5.2.3 and Theorem 5.2.6.

Proof:
Indeed, we can apply Corollary 5.2.7 (ii) with $E = F = G = I$, and with the choices $R(t) = RZ_t$, $S(t) = SZ_t$, and $T(t) = TZ_t$ to define $\int_0^t (a_R(ds) + a_S^\dagger(ds) + \Lambda_T(ds)Z_s)ve(g)$ for v, g as in the statement of this theorem. $(\int_0^t AZ_s ds)ve(g) := \int_0^t (AZ_s ve(g))ds$ is defined as $s \mapsto AZ_s ve(g)$ is continuous by assumption. The analogues of first and second fundamental formulae and the estimate as in the Theorems 5.2.3 and 5.2.6 follow similarly.

□

Remark 5.2.9 *It should be noted that for more general unbounded R, S, T, we may have domain problems to define the right integrals : the domains of R, S, T may not be left invariant under the action of Z. This problem can be solved by the 'time reversal principle', which we shall present in Chapter 7, where we deal with Q.S.D.E. with unbounded coefficients.*

From now on, for the rest of the present chapter (in fact for Chapter 6 as well), we shall assume that the coefficients R, S, T are bounded operators.

Remark 5.2.10 *When R, S, T, A in the Theorem 5.2.8 are bounded and Z is a regular adapted process, the right integral is defined and satisfies the fundamental formulae and the estimates in Theorems 5.2.3 and 5.2.6.*

Remark 5.2.11 *(i) The Itô formulae derived in Theorem 5.2.6 can be put in a convenient symbolic form. Let $\tilde{\pi}_0(x)$ denote $x \otimes 1_{\Gamma(k)}$ and $\pi_0(x)$ denote $x \otimes 1_{k_0}$. Then the Itô formulae are:*
$a_R(dt)\tilde{\pi}_0(x)a_S^\dagger(dt) = R^*\pi_0(x)Sdt$, $\Lambda_T(dt)\tilde{\pi}_0(x)\Lambda_{T'}(dt) = \Lambda_{T\pi_0(x)T'}(dt)$, $\Lambda_T(dt)\tilde{\pi}_0(x)a_S^\dagger(dt) = a_{T\pi_0(x)S}^\dagger(dt)$, $a_S(dt)\tilde{\pi}_0(x)\Lambda_T(dt) = a_{T^*\pi_0(x^*)S}(dt)$, *and the products of all other types are* 0.

(ii) The coordinate-free approach of quantum stochastic calculus developed here includes the coordinatized version as presented in [100]. Let us consider for example, for $f \in k_0$, the operator R_f defined by $R_f u = u \otimes f$, for $u \in h$. We observe that the creation and annihilation operators $a^\dagger(R_f)$, $a(R_f)$ coincide with the creation and annihilation operators $a^\dagger(f)$ and $a(f)$ (respectively) defined in [100] associated with f. Indeed, it is straightforward to see that $(R_f)_{u,v} = \langle u, v\rangle f$ for $u, v \in h$. Thus, for $g, l \in k$, $\langle a^\dagger(R_f)ve(g), ue(l)\rangle = \frac{d}{d\varepsilon}(\langle e(g + \varepsilon\langle u, v\rangle f), e(l)\rangle)|_{\varepsilon=0} = e^{\langle g,l\rangle}\langle v, u\rangle\langle f, l\rangle = \langle v, u\rangle \frac{d}{d\varepsilon}\langle e(g + \varepsilon f), e(l)\rangle|_{\varepsilon=0} = \langle v, u\rangle\langle a^\dagger(f)e(g), e(l)\rangle$. It is also clear that $\langle R_f, g\rangle = \langle f, g\rangle$ and hence $a(R_f)(ve(g)) = \langle f, g\rangle ve(g) = v(a(f)e(g))$. Finally, the number operator $\Lambda(T)$ in the sense of [100] for $T \in \mathcal{B}(k_0)$ can be identified with $\Lambda_{1_h \otimes T}$. We can also recover the coordinate-version of the Itô formulae as in [100]. Let us record the Itô formulae in this form too for future use. Let us assume that

k_0 is separable and introduce the basic processes in the notation of [100]:

$$\Lambda_0^0(t) := tI,$$

$$\Lambda_0^i(t) := a_{R_{e_i}}([0,t]),$$

$$\Lambda_i^0(t) := a_{R_{e_i}}^\dagger([0,t]),$$

$$\Lambda_j^i(t) := \Lambda_{|e_j><e_i|}([0.t]),$$

for $i, j = 1, 2, \ldots$, where $\{e_i, i = 1, 2, \ldots\}$ is an orthonormal basis of k_0. Then the Itô formula (formally) becomes

$$\Lambda_\beta^\alpha(dt)\Lambda_\nu^\mu(dt) = \hat{\delta}_\nu^\alpha \Lambda_\beta^\mu(dt),$$

for $\alpha, \beta = 0, 1, 2, \ldots$, and where $\hat{\delta}_\beta^\alpha = 0$ if at least one of the two indices α, β is 0, or if the two indices are unequal, otherwise $\hat{\delta}_\beta^\alpha = 1$.

5.3 Hudson–Parthasarathy (H–P) type equations

We consider the quantum stochastic differential equations (Q.S.D.E.) of the form,

$$dX_t = X_t(a_R(dt) + a_S^\dagger(dt) + \Lambda_T(dt) + Adt), \tag{5.20}$$

$$dY_t = (a_R(dt) + a_S^\dagger(dt) + \Lambda_T(dt) + Adt)Y_t, \tag{5.21}$$

with prescribed initial values $\tilde{X}_0 \otimes 1$ and $\tilde{Y}_0 \otimes 1$ respectively, with $\tilde{X}_0, \tilde{Y}_0 \in \mathcal{B}(h)$ where $R, S \in \mathcal{B}(h, h \otimes k_0)$, $T \in \mathcal{B}(h \otimes k_0)$, $A \in \mathcal{B}(h)$.

We note that the first of the above two equations has to be interpreted as the strong integral equation

$$X_t(ve(g)) = X_0 ve(g) + \left(\int_0^t X_s(a_R(ds) + a_S^\dagger(ds) + \Lambda_T(ds) + Ads) \right) ve(g)$$

for all $v \in h$, $g \in L^2(\mathbb{R}_+, k_0)$ with $\int_0^t \|g(s)\|^4 ds < \infty$ for all t. The second equation is to be interpreted in a similar way.

Let us at this point introduce some useful compact notation for the quadruple (R, S, T, A). Let $\hat{k_0} := \mathbb{C} \oplus k_0$ and $Z \in \mathcal{B}(h \otimes \hat{k_0})$ be defined by

$$Z = \begin{pmatrix} A & R^* \\ S & T \end{pmatrix}, \tag{5.22}$$

with respect to the decomposition $h \otimes \hat{k_0} = h \oplus h \otimes k_0$. We shall call Z *the matrix of coefficients* or *coefficient matrix*, and the equation (5.20) ((5.21), respectively) will be referred to as the left (respectively, right) Q.S.D.E. with the coefficient matrix Z, and the words 'left' or 'right' may be dropped if understood from the context. We shall denote by $\hat{Q}$ the projection $\begin{pmatrix} 0 & 0 \\ 0 & I_{h \otimes k_0} \end{pmatrix}$. For $\xi \in k_0$, we shall denote by $\hat{\xi}$ the vector $1 \oplus \xi \in \hat{k_0}$.

Let us now re-write the first and second fundamental formulae (5.13) and (5.14) for the choices $E_t = F_t = G_t = X_t$, $H_t = X_t A$, in terms of Z and $\hat{Q}$:

$$\langle X_t v e(g), u e(f) \rangle = \langle X_0 v e(g), u e(f) \rangle + \int_0^t \langle X_s \langle \hat{f}(s), Z_{\hat{g}(s)} \rangle v e(g), u e(f) \rangle ds, \tag{5.23}$$

$$\begin{aligned} &\langle X_t v e(g), X_t u e(f) \rangle \\ &= \langle X_0 v e(g), X_0 u e(f) \rangle + \int_0^t \{ \langle X_s \langle \hat{f}(s), Z_{\hat{g}(s)} \rangle v e(g), X_s u e(f) \rangle \\ &\quad + \langle X_s v e(g), X_s \langle \hat{g}(s), Z_{\hat{f}(s)} \rangle u e(f) \rangle \\ &\quad + \langle X_s \hat{Q} Z_{\hat{g}(s)} v e(g), X_s \hat{Q} Z_{\hat{f}(s)} u e(f) \rangle \} ds. \end{aligned} \tag{5.24}$$

These formulae can be verified by expanding the two sides of the equation in terms of the constituents of Z and comparing with the relevant expressions in equations (5.13) and (5.14).

Similarly, one gets for the solution Y_t of the right Q.S.D.E. (5.21) the following:

$$\langle Y_t v e(g), u e(f) \rangle = \langle Y_0 v e(g), u e(f) \rangle + \int_0^t \langle \langle \hat{f}(s), Z_{\hat{g}(s)} \rangle Y_s v e(g), u e(f) \rangle ds, \tag{5.25}$$

$$\begin{aligned} &\langle Y_t v e(g), Y_t u e(f) \rangle \\ &= \langle Y_0 v e(g), Y_0 u e(f) \rangle + \int_0^t \{ \langle \langle \hat{f}(s), Z_{\hat{g}(s)} \rangle Y_s v e(g), Y_s u e(f) \rangle \\ &+ \langle Y_s v e(g), \langle \hat{g}(s), Z_{\hat{f}(s)} \rangle Y_s u e(f) \rangle + \langle \hat{Q} Z_{\hat{g}(s)} Y_s v e(g), \hat{Q} Z_{\hat{f}(s)} Y_s u e(f) \rangle \} ds. \end{aligned} \tag{5.26}$$

Before we proceed to study existence and other properties of solutions of the above equations, let us describe the Q.S.D.E. in the 'coordinatized' formalism (see Remark 5.2.11 (*ii*)), in case when k_0 is separable. This is just to keep open the possibility of switching from one formalism to another. Indeed, even though the coordinate free language is in general more transparent and neat, sometimes it may be better or more natural to use the coordinates.

Let $\{e_i\}_{i=1,2,\ldots}$ be an orthonormal basis of k_0 and let Z^α_β, $\alpha, \beta = 0, 1, 2, \cdots$ be defined as bounded operators on h as follows:

$$Z^0_0 = A, \; Z^0_i = \langle e_i, R \rangle^*, \; Z^i_0 = \langle e_i, S \rangle, \; Z^i_j = \langle e_i, T_{e_j} \rangle,$$

for $i, j = 1, 2, \ldots,$ where it may be recalled that $T_{e_j} \in \mathcal{B}(h, h \otimes k_0)$ is given by $T_{e_j}(u) = T(u \otimes e_j)$. In this language, an alternative notation for the Q.S.D.E. (5.20) and (5.21) will be respectively $dX_t = X_t Z^{\alpha}_{\beta} \Lambda^{\beta}_{\alpha}(dt)$ and $dY_t = Z^{\alpha}_{\beta} Y_t \Lambda^{\beta}_{\alpha}(dt)$, where we adopt the convention that repeated indices mean summation over them.

Theorem 5.3.1 *The Q.S.D.E.'s (5.20) and (5.21) admit unique solutions as adapted regular processes. Furthermore, X_t^* and Y_t^* are adapted regular processes.*

Proof:
The standard proofs of existence and uniqueness of solutions along the lines of that given in [100] (Section 26 for the left equation and Section 27 for the right equation) work here too. We set up the iteration by taking $X_t^{(0)} = I$ and $X_t^{(n+1)} = \int_0^t X_s^{(n)}(a_R(ds) + a_S^{\dagger}(ds) + \Lambda_T(ds) + A ds)$ for $n = 0, 1, \ldots$. Fix $t_0 \geq 0$, $g \in L^{\infty}(\mathbb{R}_+, k_0)$ satisfying $\int_0^t \|g(s)\|^4 ds < \infty$ for all t. By the inequality $\|p + q + r + s\|^2 \leq 4(\|p\|^2 + \|q\|^2 + \|r\|^2 + \|s\|^2)$, the estimate (5.18) and Corollary 5.2.7 we have that for $t \leq t_0$,

$$
\begin{aligned}
||X_t^{(1)}(ve(g))||^2 &\leq e^t \int_0^t ds\big[||(T_{g(s)} + S)(ve(g))||^2 \\
&+ ||(\langle g(s), T_{g(s)}\rangle + \langle R, g(s)\rangle + \langle g(s), S\rangle + A)(ve(g))||^2\big] \\
&\leq 4e^{t_0} C(t) \|v\|^2 \|e(g)\|^2,
\end{aligned}
$$

where

$$
C(t) = \int_0^t \{\|T\|^2 \|g(s)\|^2 (1 + \|g(s)\|^2) + \|S\|^2 (1 + \|g(s)\|^2) + \|R\|^2 \|g(s)\|^2 + \|A\|^2\} ds.
$$

Now assume the induction hypothesis that

$$
\|X_t^{(n)}(ve(g))\|^2 \leq (4e^t)^n \frac{C(t)^n}{n!} \|v\|^2 \|e(g)\|^2.
$$

Observe that for $\xi \in h \otimes k_0$, $\|X_t^{(n)} P(\xi e(g))\|^2 \leq (4e^{t_0})^n \frac{C(t)^n}{n!} \|\xi\|^2 \|e(g)\|^2$, and hence we have

$$
\begin{aligned}
&||X_t^{(n+1)} ve(g)||^2 \\
&\leq e^t \int_0^t ds\Big[||\{X_s^{(n)} P T_{g(s)} + X_s^{(n)} P S\}(ve(g))||^2 + \\
&||\{\langle g(s), X_s^{(n)} P T_{g(s)}\rangle + X_s^{(n)} \langle R, g(s)\rangle + \langle g(s), X_s^{(n)} P S\rangle + X_s^{(n)} A\}(ve(g))||^2\Big] \\
&\leq (4e^{t_0})^{n+1} \|v\|^2 \|e(g)\|^2 \int_0^t \{\|T\|^2 \|g(s)\|^2 (1 + \|g(s)\|^2) + \|S\|^2 (1 + \|g(s)\|^2) \\
&+ \|R\|^2 \|g(s)\|^2 + \|A\|^2\} \frac{C(s)^n}{n!} ds
\end{aligned}
$$

$$= (4e^{t_0})^{n+1}\|v\|^2\|e(g)\|^2 \int_0^t \frac{C(s)^n}{n!} dC(s)$$

$$= (4e^{t_0})^{n+1}\|v\|^2\|e(g)\|^2 \frac{C(t)^{n+1}}{(n+1)!}.$$

This proves the induction hypothesis for all n. Clearly, the process X_t given by $X_t(ve(g)) := \sum_{n=0}^{\infty} X_t^{(n)}(ve(g))$ is well defined, since it follows from the above estimates that the sum converges absolutely and uniformly over compact subsets of $[0, \infty)$, which in particular implies the continuity of $t \mapsto X_t(ve(g))$. Moreover, we have,

$$\|X_t ve(g)\| \leq \|v\|\|e(g)\| \sum_{n \geq 0} \frac{(4e^{t_0} C(t_0))^{\frac{n}{2}}}{\sqrt{n!}}.$$

From this, it is clearly seen that (X_t) is an adapted regular process. That it is a solution of (5.20) is also clear. This proves the existence of the left equation. To show the uniqueness, suppose that X'_t is another solution of (5.20). Since $X_t - X'_t = \int_0^t (X_s - X'_s)(a_R(ds) + a_S^\dagger(ds) + \Lambda_T(ds) + A ds)$, an argument similar to the above shows that $\|(X_t - X'_t)(ve(g))\|^2 \leq \frac{4e^{t_0}C(t_0)^n}{n!}\|v\|^2\|e(g)\|^2$ for all n and hence $X_t ve(g) = X'_t ve(g)$.

The proof for the right equation (5.21) is similar. For the iteration process in the case of the right equation to make sense, one has to take into account the Remark 5.2.10 while interpreting the right integrals involved.

Finally, let us prove the regularity of (X_t^*) and (Y_t^*). It is enough to sketch the arguments for (X_t^*), as (Y_t^*) can be treated similarly. Consider the iterates $X_t^{(n)}$. We claim that

$$\|X_t^{(n)*} ve(g)\|^2 \leq \frac{4e^{t_0} C'(t_0)^n}{n!} \|ve(g)\|^2,$$

where $C'(t)$ is given by an expression similar to $C(t)$, with R, S, T, A replaced by S, R, T^*, A^* respectively. This claim can be verified by induction as before, using that $X_t^{(n)*}$ satisfies a right Q.S.D.E. of the form

$$X_t^{(n+1)*} = \int_0^t (a_R^\dagger(ds) + a_S(ds) + \Lambda_{T^*}(ds) + A^* ds) X_s^{(n)*}.$$

From this, the regularity of X_t^* follows. □

Theorem 5.3.2 *Let X_t satisfy the Q.S.D.E. (5.20) with the initial condition $X_0 = I$ and with the coefficient matrix Z as in the beginning of the present section. Then we have the following.*

(i) X_t is a contraction for each $t \geq 0$ if and only if the following operator inequality holds:

$$Z + Z^* + Z\hat{Q}Z^* \leq 0. \tag{5.27}$$

(ii) X_t is a co-isometry for each t if and only if the above inequality is replaced by an equality, that is, $Z + Z^* + Z\hat{Q}Z^* = 0$.

Proof:
It is a straightforward observation that $W_t = X_t^*$ satisfies the right Q.S.D.E. of the form (5.21) with the coefficient matrix Z^*. For $u, v \in h$ and $f, g \in \mathcal{C}$, by a simple calculation using the formula (5.26) we get the following:

$$\langle W_t ve(g), W_t ue(f)\rangle - \langle ve(g), ue(f)\rangle = \int_0^t \langle W_s ve(g), \langle \hat{g}(s), (Z+Z^*+Z\hat{Q}Z^*)_{\hat{f}(s)}\rangle W_s ue(f)\rangle. \tag{5.28}$$

Let us denote the operator $Z+Z^*+Z\hat{Q}Z^*$ by B. If $B \leq 0$, we have for vectors $u_1, u_2, \ldots, u_n \in h$ and $f_1, f_2, \ldots, f_n \in \mathcal{C}$,

$$\begin{aligned}
&\left\|\sum_{i=1}^n W_t(u_i e(f_i))\right\|^2 - \left\|\sum_{i=1}^n u_i e(f_i)\right\|^2 \\
&= \sum_{i,j} \int_0^t \langle W_s u_i e(f_i), \langle \hat{f}_i(s), B_{\hat{f}_j(s)}\rangle W_s u_j e(f_j)\rangle ds \\
&= \int_0^t \sum_{i,j} -\langle (-B)^{\frac{1}{2}}_{\hat{f}_i(s)}(W_s u_i e(f_i)), (-B)^{\frac{1}{2}}_{\hat{f}_j(s)}(W_s u_j e(f_j))\rangle ds \\
&= -\int_0^t \left\|\sum_i (-B)^{\frac{1}{2}}_{\hat{f}_i(s)}(W_s u_i e(f_i))\right\|^2 ds \leq 0;
\end{aligned}$$

where we have used the fact that $B \leq 0$ to take the square root of $-B$. This clearly implies that W_t (and hence X_t) is a contraction for each t.

Conversely, if X_t is a contraction, we have $\langle W_t ve(g), W_t ve(g)\rangle - \langle ve(g), ve(g)\rangle \leq 0$ for all v, g. Thus, for $t > 0$, we have

$$\frac{1}{t}\int_0^t \langle W_s ve(g), \langle \hat{g}(s), B_{\hat{g}(s)}\rangle W_s ve(g)\rangle \leq 0. \tag{5.29}$$

For $v' \in h$ and $\xi' = (\lambda \oplus \xi) \in \hat{k_0}$, where $\lambda \in \mathbb{C}, \lambda \neq 0$ and $\xi \in k_0$, choose a bounded continuous function $g : \mathbb{R}_+ \to k_0$ with $g(0) = \lambda^{-1}\xi$, and also $v = \lambda v'$. It follows by taking $t \to 0+$ in (5.29) that $\langle v'\xi', B(v'\xi')\rangle \leq 0$. By taking limit $\lambda \to 0$, we get the above inequality also for $\xi' = (0 \oplus \xi)$. Similarly, it can be shown that for $v_1', \cdots, v_n' \in h$ and $\xi_1', \cdots, \xi_n' \in k_0$ $\sum_{ij}\langle v_i'\xi_i', B(v_j'\xi_j')\rangle \leq 0$. This proves that $B \leq 0$.

Moreover, it is also immediate from the above calculation, by replacing the inequalities involved by equalities, that $\langle W_t ve(g), W_t ue(f)\rangle = \langle ve(g), ue(f)\rangle$ for all u, v, f, g; if and only if $B = 0$. This completes the proof. □

The condition $Z + Z^* + Z\hat{Q}Z^* = 0$ is clearly seen to be equivalent to the following set of conditions in terms of R, S, T, A:

$$A + A^* + R^*R = 0, \tag{5.30}$$

$$R + S + TR = 0, \tag{5.31}$$

$$T + T^* + TT^* = 0. \tag{5.32}$$

We can write (5.32) as $(T + I)(T^* + I) = I$, which is equivalent to saying that $T = \Sigma - I$ where $\Sigma := T + I$ is a co-isometry. Furthermore, the relation (5.31) can now be written as $S = -\Sigma R$, and (5.30) leads to the form $A = iH - \frac{1}{2}R^*R$ for some self-adjoint H. Motivated by this, we now consider the following special Q.S.D.E.:

$$dU_t = U_t(a_R(dt) - a^{\dagger}_{\Sigma R}(dt) + \Lambda_{\Sigma - 1}(dt) + (iH - \frac{1}{2}R^*R)dt), \; U_0 = I,$$

where Σ is a unitary in $\mathcal{B}(h \otimes k_0)$, $R \in \mathcal{B}(h, h \otimes k_0)$ and H is a self-adjoint element of $\mathcal{B}(h)$. Then we have:

Theorem 5.3.3 (see [91], [94] also)
The unique solution U_t of the above Q.S.D.E. is unitary for all $t \geq 0$.

Proof:
It follows from Lemma 5.3.2 that U_t is a co-isometry for all t, so it remains to show that it is also an isometry for all t. To this end, we note the following facts:

(a) For fixed $g, f \in L^2(\mathbb{R}_+, k_0) \cap L^\infty(\mathbb{R}_+, k_0)$ and $t \geq 0$, there exists a unique operator $M_t^{f,g} \in \mathcal{B}(h)$ such that

$$\langle v, M_t^{f,g} u \rangle = \langle U_t(ve(g)), \; U_t(ue(f)) \rangle.$$

(b) Setting $\widetilde{M}_t^{f,g} = M_t^{f,g} \otimes 1_{k_0}$, we have that for all $w, w' \in h \otimes k_0$,

$$\langle U_t P(we(g)), \; U_t P(w'e(f)) \rangle = \langle w, \widetilde{M}_t^{f,g} w' \rangle.$$

It is a simple computation using the quantum Itô formulae ((5.14) in Theorem 5.2.6) to verify that

$$\begin{aligned}
\langle v, M_t^{f,g} u \rangle - \langle ve(g), ue(f) \rangle = \int_0^t ds[\langle v, M_s^{f,g} \langle R, f(s) \rangle u \rangle + \langle v, \langle g(s), R \rangle M_s^{f,g} u \rangle \\
-\langle v, M_s^{f,g} \langle g(s), \Sigma R \rangle u \rangle + \langle vg(s), \widetilde{M}_s^{f,g}((\Sigma - 1)(uf(s))) \rangle + \langle v, M_s^{f,g}(iH - \frac{1}{2}R^*R)u \rangle \\
-\langle v, \langle \Sigma R, f(s) \rangle M_s^{f,g} u \rangle + \langle vg(s), (\Sigma^* - 1)\widetilde{M}_s^{f,g}(uf(s)) \rangle + \langle v, (-iH - \frac{1}{2}R^*R) M_s^{f,g} u \rangle \\
+\langle \Sigma Rv, \widetilde{M}_s^{f,g}(\Sigma Ru) \rangle - \langle \Sigma Rv, \widetilde{M}_s^{f,g}(\Sigma - 1)(uf(s)) \rangle - \langle vg(s), (\Sigma^* - 1)\widetilde{M}_s^{f,g}(\Sigma Ru) \rangle \\
+\langle vg(s), (\Sigma^* - 1)\widetilde{M}_s^{f,g}(\Sigma - 1)(uf(s)) \rangle].
\end{aligned}$$

Let us consider maps $Y_i, i = 1, \ldots, 5$ from $[0, \infty) \times \mathcal{B}(h)$ to $\mathcal{B}(h)$ given by:

$$Y_1(s, X) = X\langle R, f(s)\rangle + \langle g(s), R\rangle X - X\langle g(s), \Sigma R\rangle - \frac{1}{2}(XR^*R + R^*RX)+$$
$$i[X, H] - \langle \Sigma R, f(s)\rangle X, \qquad Y_2(s, X) = R^*\Sigma^*\tilde{X}\Sigma R,$$
$$Y_3(s, X) = \langle g(s), \{(\Sigma^* - 1)\tilde{X} + \tilde{X}(\Sigma - 1) + (\Sigma^* - 1)\tilde{X}(\Sigma - 1)\}f(s)\rangle,$$
$$Y_4(s, X) = -\langle(\Sigma^* - I)\tilde{X}^*\Sigma R, f(s)\rangle, \quad Y_5(s, X) = -\langle(\Sigma^* - I)\tilde{X}\Sigma R, g(s)\rangle^*,$$

where $\tilde{X} = (X \otimes 1_{k_0})$. Then it follows that,

$$\langle v, M_t^{f,g}u\rangle - \langle ve(g), ue(f)\rangle = \int_s^t \langle v, \sum_{i=1}^5 Y_i(s, M_s^{f,g})u\rangle ds,$$

that is,

$$\frac{dM_t^{f,g}}{dt} = \sum_{i=1}^5 Y_i(t, M_t^{f,g}).$$

We note that $M_0^{f,g} =< e(g), e(f) > I$ is a solution of the above equation since $\sum_{i=1}^5 Y_i(t, I) = 0$, Σ being unitary. Moreover, $Y_i(t, \cdot)$ is linear and bounded for each i. Thus by the uniqueness of a solution of the Banach space-valued initial value problem, we conclude that $M_t^{f,g} = M_0^{f,g}$ for all t, or equivalently that U_t is an isometry. □

5.4 Map-valued, Evans–Hudson (E–H)-type quantum stochastic calculus

In the previous subsections, we have considered Q.S.D.E.'s on the Hilbert space $h \otimes \Gamma$. Now we shall study an associated class of Q.S.D.E.'s, but on the Fock module $\mathcal{A} \otimes \Gamma$. This is closely related to the Evans–Hudson (E–H) type of Q.S.D.E.'s ([44, 100]).

First we have to formulate the notion of map-valued quantum stochastic processes and the associated integrals. We shall give the definition of integrator processes with coefficients belonging to a class of maps more general than bounded. After that, we study a special class of map-valued Q.S.D.E., but this time with bounded coefficients only. The existence and properties of solutions of similar map-valued Q.S.D.E. with unbounded coefficients will be taken up in later chapters.

5.4.1 The formalism of map-valued quantum stochastic integration

Let $\mathcal{A} \subseteq \mathcal{B}(h)$ be a von Neumann algebra, k_0 be a Hilbert space, and let us consider a quadruplet of linear (but not necessarily bounded) maps $(\mathcal{L}, \delta, \delta', \sigma)$

defined on an ultra-weak dense $*$-subalgebra $\mathcal{A}_0$ of $\mathcal{A}$, such that for $x \in \mathcal{A}_0$, we have $\delta(x), \delta'(x) \in \mathcal{A} \otimes k_0$, $\mathcal{L}(x) \in \mathcal{A}$ and $\sigma(x) \in \mathcal{A} \otimes \mathcal{B}(k_0)$. As before, let us denote by k the Hilbert space $L^2(\mathbb{R}_+, k_0)$, by Γ the symmetric Fock space on k and by L^p_{loc} (where $p \geq 1$) the set of $f \in k$ such that $\int_0^t \|f(s)\|^p ds < \infty$ for all $t \geq 0$.

We now introduce the basic map-valued processes. Fix $t \geq 0$, a bounded interval $\Delta \subseteq (t, \infty)$, elements $x_1, x_2, \ldots, x_n \in \mathcal{A}_0$ and vectors $f_1, f_2, \ldots, f_n \in k$; $u \in h$. We define the following:

$$\left(a_\delta(\Delta)\left(\sum_{i=1}^n x_i \otimes e(f_i)\right)\right)u = \sum_{i=1}^n a_{\delta(x_i^*)}(\Delta)(ue(f_i)),$$

$$\left(a^\dagger_{\delta'}(\Delta)\left(\sum_{i=1}^n x_i \otimes e(f_i)\right)\right)u = \sum_{i=1}^n a^\dagger_{\delta'(x_i)}(\Delta)(ue(f_i)),$$

$$\left(\Lambda_\sigma(\Delta)\left(\sum_{i=1}^n x_i \otimes e(f_i)\right)\right)u = \sum_{i=1}^n \Lambda_{\sigma(x_i)}(\Delta)(ue(f_i)),$$

$$\left(\mathcal{I}_\mathcal{L}(\Delta)\left(\sum_{i=1}^n x_i \otimes e(f_i)\right)\right)u = \sum_{i=1}^n |\Delta|(\mathcal{L}(x_i)u) \otimes e(f_i)),$$

where $|\Delta|$ denotes the length of Δ.

Lemma 5.4.1 *The above processes are well defined on $\mathcal{A}_0 \otimes_{\text{alg}} \mathcal{E}(k)$ and they take values in $\mathcal{A} \otimes \Gamma(k)$.*

Proof:
First note that $e(f_1), \ldots, e(f_n)$ are linearly independent whenever $f_1, \ldots, f_n$ are distinct, from which it is clear that $\sum_{i=1}^n x_i \otimes e(f_i) = 0$ implies $x_i = 0$ for all i, whenever f_i's are distinct. This will establish that the processes are well defined. The second part of the lemma will follow from Lemma 4.2.10 with the choice of the dense set $\mathcal{E}$ to be $\mathcal{E}(k)$ and $\mathcal{H} = \Gamma(k)$ and by noting the fact that $\mathcal{L}, \delta, \delta'$ and σ have appropriate ranges. For example,

$$\langle e(g), a^\dagger_{\delta'}(\Delta)(x \otimes e(f))\rangle = \langle e(g), e(f)\rangle \int_\Delta \langle g(s), \delta'(x)\rangle ds,$$

which belongs to $\mathcal{A}$, so the range of $a^\dagger_{\delta'}(\Delta)$ is contained in $\mathcal{A} \otimes \Gamma(k)$. Similarly, one verifies that

$$\langle e(g), \Lambda_\sigma(\Delta)(x \otimes e(f))\rangle = \langle e(g), e(f)\rangle \int_\Delta \langle g(s), \sigma(x)_{f(s)}\rangle ds,$$

which belongs to $\mathcal{A}$ since $\sigma(x) \in \mathcal{A} \otimes \mathcal{B}(k_0)$. □

Definition 5.4.2 A family of (possibly unbounded) linear maps $(Y(t))_{t \geq 0}$: $\mathrm{Dom}(Y(t)) \subseteq \mathcal{A} \otimes \Gamma$ to $\mathcal{A} \otimes \Gamma$ is said to be a *map-valued process*. It is said to be:

(i) *adapted*, if there is a family of (possibly unbounded, linear) maps $Y_0(t)$ from $\mathcal{A} \otimes \Gamma_t$ to $\mathcal{A} \otimes \Gamma_t$, such that $\mathrm{Dom}(Y_0(t)) \otimes_{\mathrm{alg}} \Gamma^t \subseteq \mathrm{Dom}(Y(t))$, with $Y(t) = Y_0(t) \otimes_{\mathrm{alg}} I_{\Gamma^t}$ on $\mathrm{Dom}(Y_0(t)) \otimes_{\mathrm{alg}} \Gamma^t$ (where $\Gamma^t = \Gamma(L^2((t, \infty), k_0)))$, for all $t \geq 0$;
(ii) *regular*, if $Y(t)$ is defined on $\mathcal{A} \otimes_{\mathrm{alg}} \mathcal{E}(k)$, $t \mapsto Y(t)(x \otimes e(f))u$ is continuous for every fixed $x \in \mathcal{A}$, $u \in h$, $f \in L^4_{\mathrm{loc}}$ and furthermore, there exist positive constants C_t, Hilbert space $\mathcal{H}''$, and bounded operators $r_t \in \mathcal{B}(h, h \otimes \mathcal{H}'')$ such that

$$||Y(t)(x \otimes e(f))u|| \leq C_{t_0} \|e(f)\| ||(x \otimes 1_{\mathcal{H}''}) r_{t_0} u|| \text{ for all } t \leq t_0, \tag{5.33}$$

for all $x \in \mathcal{A}$, u, f as before.

Let us now assume that $\mathcal{A}_0 = \mathcal{A}$ and the maps $(\mathcal{L}, \delta, \delta', \sigma)$ are bounded. We want to define the integral of the form $\int_0^t Y(s) \circ (a_\delta + a^\dagger_{\delta'} + \Lambda_\sigma + \mathcal{I}_\mathcal{L})(ds)$ where $Y(t)$ is an adapted regular map-valued process. We need the following lemma for this.

Lemma 5.4.3 *(The Lifting lemma) Let $\mathcal{H}$ be a Hilbert space and $\mathcal{V}$ be a vector space. Let $\beta : \mathcal{A} \otimes_{\mathrm{alg}} \mathcal{V} \to \mathcal{A} \otimes \mathcal{H}$ be a linear map satisfying the estimate*

$$||\beta(x \otimes \eta)u|| \leq c_\eta ||(x \otimes 1_{\mathcal{H}''}) r u|| \tag{5.34}$$

for some Hilbert space $\mathcal{H}''$ and $r \in \mathcal{B}(h, h \otimes \mathcal{H}'')$ (both independent of η) and for $u \in h$, $\eta \in \mathcal{V}$ and some positive constant c_η depending on η. Then, for any Hilbert space $\mathcal{H}'$, we can define a map $\tilde{\beta} : (\mathcal{A} \otimes \mathcal{H}') \otimes_{\mathrm{alg}} \mathcal{V} \to \mathcal{A} \otimes (\mathcal{H} \otimes \mathcal{H}')$ by $\tilde{\beta}(x \otimes f \otimes \eta) = \beta(x \otimes \eta) \otimes f$ for $x \in \mathcal{A}$, $\eta \in \mathcal{V}$, $f \in \mathcal{H}'$. Moreover, $\tilde{\beta}$ satisfies the estimate

$$||\tilde{\beta}(X \otimes \eta)u|| \leq c_\eta ||(X \otimes 1_{\mathcal{H}''}) r u||, \tag{5.35}$$

where $X \in \mathcal{A} \otimes \mathcal{H}'$.

Proof:
Let $X \in \mathcal{A} \otimes \mathcal{H}'$ be given by the strongly convergent sum $X = \sum x_\alpha \otimes e_\alpha$, where $x_\alpha \in \mathcal{A}$ and $\{e_\alpha\}$ is an orthonormal basis of $\mathcal{H}'$. We can verify that

$$\begin{aligned} ||\tilde{\beta}(\textstyle\sum x_\alpha \otimes e_\alpha \otimes \eta)u||^2 &= \textstyle\sum ||\beta(x_\alpha \otimes \eta)u||^2 \\ \leq c_\eta^2 \sum_\alpha ||(x_\alpha \otimes 1_{\mathcal{H}''}) r u||^2 &= c_\eta^2 ||(X \otimes 1_{\mathcal{H}''}) r u||^2 \end{aligned}$$

and thus $\tilde{\beta}$ is well defined and satisfies the required estimate. □

Assume now that $(Y(t))$ is a map-valued adapted regular process, as in the Definition 5.4.2. For every $s \geq 0$, we have by the Lifting lemma 5.4.3 an extension $\widetilde{Y(s)} : \mathcal{A} \otimes k_0 \otimes \mathcal{E}(k_s \cap L^4_{\mathrm{loc}}) \to \mathcal{A} \otimes \Gamma_s \otimes k_0$ of the map $(Y(s) \otimes_{\mathrm{alg}} \mathrm{id})P$, where P interchanges the second and third tensor components.

Lemma 5.4.4 *For fixed $x \in \mathcal{A}$ and $f \in L^4_{\mathrm{loc}}$, the families of operators $S(s) \equiv S_x(s) : h \otimes_{\mathrm{alg}} \mathcal{E}(L^4_{\mathrm{loc}} \cap k_s) \to h \otimes \Gamma_s \otimes k_0$ and $T(s) \equiv T_x(s) : h \otimes_{\mathrm{alg}} \mathcal{E}(L^4_{\mathrm{loc}} \cap k_s) \otimes k_0 \to h \otimes \Gamma_s \otimes k_0$ defined below satisfy the hypotheses of Corollary 5.2.7(iv), with the choices $E = G = I$, $F = H = R(s) = 0$:*

$$S_x(s)(ue(f_s)) = \widetilde{Y(s)}(\delta'(x) \otimes e(f_s))u, \tag{5.36}$$

$$T_x(s)(ue(f_s) \otimes \xi) = \widetilde{Y(s)}(\sigma(x)_\xi \otimes e(f_s))u. \tag{5.37}$$

Moreover, the map $s \mapsto Y(s)((\mathcal{L}(x) + \langle \delta(x^*), f(s) \rangle) \otimes e(f))u$ is strongly integrable over $[0, t]$ for every $t \geq 0$, and for fixed $x \in \mathcal{A}$, $u \in h$ and $f \in L^4_{\mathrm{loc}}$.

Proof:
The first part follows from the hypotheses on $Y(s)$, the Lifting lemma and the fact that $s \mapsto e(f_s)$ is strongly continuous for $f \in L^4_{\mathrm{loc}}$. Let us verify the conditions of Corollary 5.2.7(iv) for S_x only, as the proof for T_x will be very similar. By the Lifting lemma, we have

$$\|S_x(s)(ue(f_s))\| \leq C_t \|(\delta'(x) \otimes 1_{\mathcal{H}''})r_t u\| \|e(f_s)\|$$

for all $s \leq t$. Thus, $\|S_x(s)(ue(f_s))\| \leq C_t \|e(f)\| \|r_t\| \|\delta'(x)\| \|u\|$ for $s \leq t$. Furthermore, it is clear that $s \mapsto \widetilde{Y(s)}(X \otimes e(f_s))u$ is continuous for $X \in \mathcal{A} \otimes_{\mathrm{alg}} k_0$. In fact, for $t_0 > 0$ and $\epsilon > 0$, using the density of $\mathcal{A} \otimes_{\mathrm{alg}} k_0$ in the Hilbert von Neumann module $\mathcal{A} \otimes k_0$, we can choose $X \in \mathcal{A} \otimes_{\mathrm{alg}} k_0$ such that $\|((X - \delta'(x)) \otimes 1)r_{t_0} u\| \leq \epsilon$. Thus, we have for $s, t \leq t_0$

$$\begin{aligned}
&\|S_x(s)(ue(f_s)) - S_x(t)(ue(f_t))\| \\
&\leq \|\widetilde{Y(s)}(X \otimes e(f_s))u - \widetilde{Y(t)}(X \otimes e(f_t))u\| + \|\widetilde{Y(s)}((\delta'(x) - X) \otimes e(f_s))u\| \\
&\quad + \|\widetilde{Y(t)}((\delta'(x) - X) \otimes e(f_t))u\| \\
&\leq \|\widetilde{Y(s)}(X \otimes e(f_s))u - \widetilde{Y(t)}(X \otimes e(f_t))u\| + 2C_{t_0} \|e(f_{t_0})\| \epsilon,
\end{aligned}$$

which proves the continuity of $s \mapsto S_x(s)(ue(f_s))$. Furthermore, we note that the strong measurability of the map $s \mapsto Y(s)((\mathcal{L}(x) + \langle \delta(x^*), f(s) \rangle) \otimes e(f))u$ follows from the strong continuity of $s \mapsto Y(s)(y \otimes e(f))u$ for fixed $y \in \mathcal{A}$, $u \in h$ and $f \in L^4_{\mathrm{loc}}$. It also follows from the estimate (5.33) and the fact that $f \in L^4_{\mathrm{loc}} \subseteq L^1_{\mathrm{loc}}$ that

$$\int_0^t \|Y(s)((\mathcal{L}(x) + \langle \delta(x^*), f(s) \rangle) \otimes e(f))u\| ds < \infty.$$

□

In view of this lemma, by map-valued process we shall mean adapted regular map-valued process in the rest of the present chapter. This lemma also allows us to define the integral $\int_0^t Y(s) \circ (a_\delta + a^\dagger_{\delta'} + \Lambda_\sigma(ds) + \mathcal{I}_\mathcal{L})(ds)(x \otimes e(f))$ for $x \in \mathcal{A}$ and $f \in L^4_{\text{loc}}$.

Definition 5.4.5 For fixed $x \in \mathcal{A}$, $u \in h$ and $f \in L^4_{\text{loc}}$, we define the integral $\int_0^t Y(s) \circ (a_\delta + \mathcal{I}_\mathcal{L})(ds)(x \otimes e(f))u$ by setting it to be equal to

$$\int_0^t Y(s)((\mathcal{L}(x) + \langle \delta(x^*), f(s) \rangle) \otimes e(f))u \, ds \tag{5.38}$$

This integral exists and is finite since $s \mapsto Y(s)((\mathcal{L}(x) + \langle \delta(x^*), f(s) \rangle) \otimes e(f))u$ is strongly integrable over $[0, t]$. We define the integral involving the other two processes, that is, $\int_0^t Y(s) \circ (\Lambda_\sigma + a^\dagger_{\delta'})(ds)(x \otimes e(f))u$ by setting it to be equal to

$$\left(\int_0^t \Lambda_{T_x}(ds) + a^\dagger_{S_x}(ds) \right) ue(f), \tag{5.39}$$

which is well-defined by Corollary 5.2.7*(iv)*.

5.4.2 Solution of a class of map-valued quantum stochastic differential equations (Q.S.D.E.) with bounded coefficients

For this part of the theory, we assume that we are given the *structure maps*, that is, the triple of linear maps $(\mathcal{L}, \delta, \sigma)$ where $\mathcal{L} \in \mathcal{B}(\mathcal{A})$, $\delta \in \mathcal{B}(\mathcal{A}, \mathcal{A} \otimes k_0)$ and $\sigma \in \mathcal{B}(\mathcal{A}, \mathcal{A} \otimes \mathcal{B}(k_0))$ satisfying:

(S1) $\sigma(x) = \pi(x) - x \otimes I_{k_0} \equiv \Sigma^*(x \otimes I_{k_0})\Sigma - x \otimes I_{k_0}$, where Σ is a partial isometry in $h \otimes k_0$ such that π is a $*$-representation on $\mathcal{A}$.

(S2) $\delta(x) = Rx - \pi(x)R$, where $R \in \mathcal{B}(h, h \otimes k_0)$ so that δ is a π-derivation, that is, $\delta(xy) = \delta(x)y + \pi(x)\delta(y)$.

(S3) $\mathcal{L}(x) = R^*\pi(x)R + lx + xl^*$, where $l \in \mathcal{A}$ with the condition $\mathcal{L}(1) = 0$ so that $\mathcal{L}$ satisfies the second order cocycle relation with δ as coboundary, that is

$$\mathcal{L}(x^*y) - x^*\mathcal{L}(y) - \mathcal{L}(x)^*y = \delta(x)^*\delta(y) \text{ for all } x, y \in \mathcal{A}.$$

Given the generator $\mathcal{L}$ of a Q.D.S., it will be established in the next chapter that one can choose k_0, R and Σ such that the hypotheses **(S1)**–**(S3)** are satisfied.

To describe E–H flow in this language, it is convenient to introduce a map Θ (to be referred to as the *structure matrix*) encompassing the triple $(\mathcal{L}, \delta, \sigma)$ as follows:

$$\Theta(x) = \begin{pmatrix} \mathcal{L}(x) & \delta^\dagger(x) \\ \delta(x) & \sigma(x) \end{pmatrix}, \tag{5.40}$$

where $x \in \mathcal{A}$, $\delta^\dagger(x) = \delta(x^*)^* : h \otimes k_0 \to h$, so that $\Theta(x)$ can be looked upon as a bounded linear map from $h \otimes \hat{k_0} \equiv h \otimes (\mathbb{C} \oplus k_0)$ into itself. It is also clear from **(S1)**–**(S3)** that Θ maps $\mathcal{A}$ into $\mathcal{A} \otimes \mathcal{B}(\hat{k_0})$. The next lemma sums up important properties of Θ.

Lemma 5.4.6 *Let Θ be as above. Then one has the following.*

(i)

$$\Theta(x) = \Psi(x) + K(x \otimes 1_{\hat{k_0}}) + (x \otimes 1_{\hat{k_0}})K^*, \tag{5.41}$$

where $\Psi : \mathcal{A} \to \mathcal{A} \otimes \mathcal{B}(\hat{k_0})$ is a completely positive map and $K \in \mathcal{B}(h \otimes \hat{k_0})$.

(ii) Θ is conditionally completely positive and satisfies the structure relation:

$$\Theta(xy) = \Theta(x)(y \otimes 1_{\hat{k_0}}) + (x \otimes 1_{\hat{k_0}})\Theta(y) + \Theta(x)\hat{Q}\Theta(y), \tag{5.42}$$

where $\hat{Q} = \begin{pmatrix} 0 & 0 \\ 0 & 1_{h \otimes k_0} \end{pmatrix}$.

(iii) There exists $D \in \mathcal{B}(h \otimes \hat{k_0}, h \otimes k_0')$ (where $k_0' = \hat{k_0} \otimes \mathbb{C}^3$) such that

$$||\Theta(x)\zeta|| \leq ||(x \otimes 1_{k_0'})D\zeta|| \tag{5.43}$$

for all $\zeta \in h \otimes \hat{k_0}$.

Proof:
Define the following maps with respect to the direct sum decomposition $h \otimes \hat{k_0} = h \oplus (h \otimes k_0)$:

$$\tilde{R} = \begin{pmatrix} 0 & 0 \\ R & -I \end{pmatrix}, \tilde{\pi}(x) = \begin{pmatrix} x & 0 \\ 0 & \pi(x) \end{pmatrix},$$

$$K = \begin{pmatrix} l & 0 \\ R & -\frac{1}{2}1_{h \otimes k_0} \end{pmatrix}, \tilde{\Sigma} = \begin{pmatrix} 1_h & 0 \\ 0 & \Sigma \end{pmatrix}.$$

Then it is straightforward to see that (i) is verified with $\Psi(x) = \tilde{R}^*\tilde{\Sigma}^*(x \otimes 1_{\hat{k_0}})\tilde{\Sigma}\tilde{R} = \tilde{R}^*\tilde{\pi}(x)\tilde{R}$. Clearly, Ψ is completely positive. That Θ is conditionally completely positive and satisfies the structure relation (5.42) is also a consequence of (i) and **(S1)**–**(S3)**. The estimate in (iii) follows from the structure of Ψ given above with the choice of D from $h \otimes \hat{k_0}$ to $h \otimes \hat{k_0} \otimes \mathbb{C}^3 \equiv (h \otimes \hat{k_0}) \oplus (h \otimes \hat{k_0}) \oplus (h \otimes \hat{k_0})$ given by

$$D\xi = 2\left(||\tilde{\Sigma}\tilde{R}||\ \tilde{\Sigma}\tilde{R}\xi \oplus ||K||\xi \oplus K^*\xi\right), \quad \xi \in h \otimes \hat{k_0}. \qquad \square$$

Theorem 5.4.7 *The integral $Z(t) \equiv \int_0^t Y(s) \circ (a_\delta^\dagger + a_\delta + \Lambda_\sigma + \mathcal{I}_\mathcal{L})(ds)$, where $Y(s)$ is an adapted regular map-valued process satisfying (5.33), is well defined* *on $\mathcal{A} \otimes_{\text{alg}} \mathcal{E}(L^4_{\text{loc}})$ and is an adapted regular map-valued process. Moreover, the integral satisfies the following estimate for all $x \in \mathcal{A}$, $u \in h$ and $f \in L^4_{\text{loc}}$:*

$$\begin{aligned} &\|\{Z(t)(x \otimes e(f)\}u\|^2 \\ &\leq 2e^t \int_0^t \exp(\|f^s\|^2)\{\|\hat{Y}(s)(\Theta(x)_{\hat{f}(s)} \otimes e(f_s))u\|^2 + \\ &\quad \|\langle f(s), \hat{Y}(s)(\Theta(x)_{\hat{f}(s)} \otimes e(f_s))\rangle u\|^2\}ds, \end{aligned} \tag{5.44}$$

where Θ was as defined earlier, $\hat{Y}(s) = Y(s) \oplus \widetilde{Y(s)} : \mathcal{A} \otimes \hat{k_0} \otimes_{\text{alg}} \mathcal{E}(L^4_{\text{loc}} \cap k_s) \to \mathcal{A} \otimes \Gamma_s \otimes \hat{k_0}$, $\hat{f}(s) = 1 \oplus f(s)$ and $f(s)$ is identified with $0 \oplus f(s)$ in $\hat{k_0}$.

Proof:
We have already seen that the integral is well defined. To obtain the estimate, fix $x \in \mathcal{A}$, $u \in h$ and $f \in L^4_{\text{loc}}$. By the inequality $\|a+b\|^2 \leq 2(\|a\|^2 + \|b\|^2)$, we have

$$\begin{aligned} &\|(Z(t)(x \otimes e(f))\,u\|^2 \\ &\leq 2\left\|\left(\int_0^t Y(s) \circ (a_\delta + \mathcal{I}_{\mathcal{L}})(ds)(x \otimes e(f))u\right)\right\|^2 \\ &+ 2\left\|\left(\int_0^t Y(s) \circ (a_\delta^\dagger + \Lambda_\sigma)(ds)(x \otimes e(f))u\right)\right\|^2. \end{aligned}$$

To complete the proof, we need to estimate the two terms on the right-hand side of the above inequality separately. For this, observe that

$$\begin{aligned} &\left\|\left(\int_0^t Y(s) \circ (a_\delta + \mathcal{I}_{\mathcal{L}})(ds)(x \otimes e(f))u\right)\right\|^2 \\ &= \left\|\left(\int_0^t Y(s)((\mathcal{L}(x) + \langle\delta(x^*), f(s)\rangle) \otimes e(f))u\, ds\right)\right\|^2 \\ &\leq \left(\int_0^t \|Y(s)((\mathcal{L}(x) + \langle\delta(x^*), f(s)\rangle) \otimes e(f))u\|ds\right)^2 \\ &\leq t\int_0^t \|Y(s)((\mathcal{L}(x) + \langle\delta(x^*), f(s)\rangle) \otimes e(f))u\|^2 ds. \\ &\leq e^t \int_0^t \|Y(s)((\mathcal{L}(x) + \langle\delta(x^*), f(s)\rangle) \otimes e(f))u\|^2 ds. \end{aligned}$$

Furthermore, by (5.18) and Corollary 5.2.7 with $E = G = I$, $F = H = 0$, we get

$$\begin{aligned} &\left\|\left(\int_0^t Y(s) \circ (a_\delta^\dagger + \Lambda_\sigma)(ds)(x \otimes e(f))u\right)\right\|^2 \\ &= \left\|\left(\int_0^t (a_{S_x}^\dagger + \Lambda_{T_x})(ds)(ue(f))\right)\right\|^2 \end{aligned}$$

$$\leq e^t \int_0^t [\|\widetilde{Y(s)}((\sigma_{f(s)}(x) + \delta(x)) \otimes e(f))u\|^2$$
$$+ \|Y(s)((\langle f(s), \sigma_{f(s)}(x)\rangle + \langle f(s), \delta(x)\rangle) \otimes e(f))u\|^2]ds.$$

Finally, to prove the continuity of the map $t \mapsto Z(t)(x \otimes e(f))u$ we remark that it is possible to define an integral of the form $\int_{t_1}^{t_2} Y(s) \circ (a_\delta^\dagger + a_\delta + \Lambda_\sigma + \mathcal{I}_\mathcal{L})(ds)$ by adapting the arguments for defining $Z(t)$, and one can similarly prove an analogue of the inequality (5.44) with the range of integration on the right-hand side being $[t_1, t_2]$. Thus, by the inequality (5.33) satisfied by $Y(t)$ and the Lifting lemma, we have for $0 \leq t_1 \leq t_2$ and $x \in \mathcal{A}$, $u \in h$, $f \in L^4_{\text{loc}}$ the following:

$$\|(Z(t_1) - Z(t_2))(x \otimes e(f))u\|^2$$
$$\leq 2e^{t_2} C_{t_2}^2 \|r_{t_2}\|^2 \|\Theta(x)\|^2 \|e(f)\|^2 \|u\|^2 \int_{t_1}^{t_2} \|\hat{f}(s)\|^2 (1 + \|f(s)\|^2) ds,$$

which proves the continuity of the map $t \mapsto Z(t)(x \otimes e(f))u$. □

Next, we want to consider the solution of an equation of the E–H type which in our notation can be written as

$$J_t = \text{Id}_{\mathcal{A}\otimes\Gamma} + \int_0^t J_s \circ (a_\delta^\dagger + a_\delta + \Lambda_\sigma + \mathcal{I}_\mathcal{L})(ds), \ \ 0 \leq t \leq t_0. \tag{5.45}$$

Remark 5.4.8 *The map-valued Q.S.D.E. (5.45) can also be described in the coordinatized formalism. Let k_0 be separable and let $\{e_i, i \in \mathbb{N}\}$ be an orthonormal basis of k_0, and $\{e_0 := 1, e_i : i \in \mathbb{N}\}$ be the corresponding basis for $\hat{k_0}$. Let $\Theta(x) \in \mathcal{A} \otimes \mathcal{B}(\hat{k_0})$ be written as $\Theta(x) = \left(\left(\theta_\beta^\alpha(x)\right)\right)_{\alpha,\beta\in\{0\}\cup\mathbb{N}}$, with respect to the basis $\{e_\alpha : \alpha = \{0\} \cup \mathbb{N}\}$, and where $\theta_\beta^\alpha(x) \in \mathcal{A}$. Then the E-H type Q.S.D.E. (5.45) is equivalent to the following Q.S.D.E.:*

$$dj_t(x) = j_t(\theta_\beta^\alpha(x))d\Lambda_\alpha^\beta, \ \ j_0(x) = x \otimes I;$$

$j_t(x)(ue(f)) := J_t(x \otimes e(f))u$ gives the relation between j_t and J_t.

Now we are ready to prove the main result of this section.

Theorem 5.4.9 *(see also [44])*

(i) There exists a unique solution J_t of equation (5.45), which is an adapted regular process mapping $\mathcal{A} \otimes \mathcal{E}(L^4_{\text{loc}})$ into $\mathcal{A} \otimes \Gamma$. Furthermore, one has an estimate

$$\sup_{0\leq t\leq t_0} ||J_t(x \otimes e(f))u|| \leq C'(f)||(x \otimes 1_{\Gamma^f(k')})E_{t_0}^f u||, \tag{5.46}$$

where $f \in L^4_{\text{loc}}$, $k' = L^2([0, t_0], k_0')$, $E_t^f \in \mathcal{B}(h, h \otimes \Gamma^f(k'))$, $C'(f)$ is a constant and $\Gamma^f(k')$ is the full Fock space over k'.

(ii) *Setting $j_t(x)(ue(g)) = J_t(x \otimes e(g))u$, we have*

(a) $\langle j_t(x)ue(g), j_t(y)ve(f)\rangle = \langle ue(g), j_t(x^*y)ve(f)\rangle$ for all $g, f \in L^4_{\mathrm{loc}}$, *and*

(b) j_t *extends uniquely to a normal $*$-homomorphism from $\mathcal{A}$ into $\mathcal{A} \otimes \mathcal{B}(\Gamma)$.*

(iii) *If $\mathcal{A}$ is commutative, then the algebra generated by $\{j_t(x) : x \in \mathcal{A}, 0 \leq t \leq t_0\}$ is commutative.*

(iv) $j_t(1) = 1$ for all $t \in [0, t_0]$ *if and only if* $\Sigma^*\Sigma = 1_{h \otimes k_0}$.

Proof:
(*i*) We write for $\Delta \subseteq [0, \infty)$, $M(\Delta) \equiv a_\delta(\Delta) + a^\dagger_\delta(\Delta) + \Lambda_\sigma(\Delta) + \mathcal{I}_\mathcal{L}(\Delta)$, and set up an iteration by

$$J_t^{(n+1)}(x \otimes e(f)) = \int_0^t J_s^{(n)} \circ M(ds)(x \otimes e(f)), \;\; J_t^{(0)}(x \otimes e(f)) = x \otimes e(f),$$

with $x \in \mathcal{A}$ and $f \in L^4_{\mathrm{loc}}$ fixed. Since $J_t^{(1)} = M([0, t])$, $J_t^{(1)}$ it satisfies the following estimate (by the definition of $M(\Delta)$, estimate (5.18) and Lemma 5.4.6(iii)):

$$\begin{aligned} &\|J_t^{(1)}(x \otimes e(f))u\|^2 \\ &\leq 2e^{t_0}\|e(f)\|^2 \int_0^t ds \|\Theta(x)(u \otimes \hat{f}(s))\|^2 \|\hat{f}(s)\|^2 \\ &\leq 2\|e(f)\|^2 e^{t_0} \int_0^t ds \|\hat{f}(s)\|^2 \|(x \otimes 1_{k_0'})D(u \otimes \hat{f}(s))\|^2. \end{aligned}$$

For the given f, define $E_t^{(1)} : h \to h \otimes k'$ by $(E_t^{(1)}u)(s) = D(u \otimes \hat{f}(s)\|\hat{f}_t(s)\|)$, where $\hat{f}_t(s) = \chi_{[0,t]}(s)\hat{f}(s)$. Then the above estimate reduces to

$$\|J_t^{(1)}(x \otimes e(f))u\|^2 \leq 2\|e(f)\|^2 e^{t_0} \|(x \otimes 1_{k'})E_t^{(1)}u\|^2. \tag{5.47}$$

It is also possible to see from the definition of $E_t^{(1)}$ that

$$\|(x \otimes 1_{k'})E_t^{(1)}u\|^2 \leq \|(x \otimes 1_{k'})E_{t_0}^{(1)}u\|^2$$

whenever $t_0 \geq t$. This shows that $(J_t^{(1)})$ is indeed adapted regular process, so that $\int_0^t J_s^{(1)} \circ M(ds)$ is well-defined.

Now, an application of the Lifting lemma leads to

$$\|\hat{J}_t^{(1)}(X \otimes e(f))u\|^2 \leq 2\|e(f)\|^2 e^{t_0} \|(X \otimes 1_{k'})E_t^{(1)}u\|^2,$$

for $X \in \mathcal{A} \otimes \hat{k_0}$, where as before, $\hat{J}_t^{(1)} = J_t^{(1)} \oplus \widetilde{J_t^{(1)}}$. As an induction hypothesis, assume that $J_t^{(n)}$ is an adapted regular process satisfying the estimate

$$\|J_t^{(n)}(x \otimes e(f))u\|^2 \leq C^n \|e(f)\|^2 \|(x \otimes 1_{k'^{\otimes n}})E_t^{(n)}u\|^2,$$

where $C = 2e^{t_0}$ and the map $E_t^{(n)} : h \to h \otimes k'^{\otimes^n}$ is defined by,

$$(E_t^{(n)} u)(s_1, s_2, \cdots, s_n) = (D \otimes 1_{k'^{\otimes^{n-1}}}) P_n \{(E_{s_1}^{(n-1)} u)(s_2, \cdots, s_n) \otimes \hat{f}(s_1) \| \hat{f}_t(s_1) \|\}.$$

Here, $P_n : h \otimes k_0'^{\otimes^{(n-1)}} \otimes \hat{k_0} \to h \otimes \hat{k_0} \otimes k_0'^{\otimes^{(n-1)}}$ is the (unitary) operator which interchanges the second and third tensor components and $E_t^{(0)} := 1_h$. Then it follows from the Theorem 5.4.7 that $J_t^{(n+1)}$ is an adapted regular process and moreover, by (5.44) we have the following inequality

$$\begin{aligned} &\| J_t^{(n+1)}(x \otimes e(f))u \|^2 \\ &\leq 2e^t \int_0^t \exp(\| f^s \|^2)(1 + \| f(s) \|^2) \| \hat{J}_s^{(n)}(\Theta_{\hat{f}(s)}(x) \otimes e(f_s))u \|^2 ds. \end{aligned}$$

By the induction hypothesis and the Lifting lemma, it follows that

$$\begin{aligned} &\int_0^t \| e(f^s) \|^2 (1 + \| \hat{f}(s) \|^2) \| \hat{J}_s^{(n)}(\Theta_{\hat{f}(s)}(x) \otimes e(f_s))u \|^2 ds \\ &\leq C^n \| e(f) \|^2 \int_0^t \| \hat{f}(s) \|^2 \| (\Theta_{\hat{f}(s)}(x) \otimes 1_{k'^{\otimes^n}}) E_s^{(n)} u \|^2 ds. \end{aligned}$$

Applying (5.43) to the right-hand side of the above inequality, we get the required estimate for $J_t^{(n+1)}$, thereby proving the induction hypothesis for $n+1$.

Thus, if we put $J_t = \sum_{n=0}^{\infty} J_t^{(n)}$, then

$$\begin{aligned} \| J_t(x \otimes e(f))u \| &\leq \sum_{n=0}^{\infty} \| J_t^{(n)}(x \otimes e(f))u \| \\ &\leq \| e(f) \| \sum_{n=0}^{\infty} C^{\frac{n}{2}} (n!)^{-\frac{1}{4}} \| (x \otimes 1_{\hat{k}^{\otimes^n}})(n!)^{\frac{1}{4}} E_t^{(n)} u \| \\ &\leq \| e(f) \| \left(\sum_{n=0}^{\infty} \frac{C^n}{\sqrt{n!}} \right)^{\frac{1}{2}} \| (x \otimes 1_{\Gamma^f(k')}) E_t^f u \|, \end{aligned} \tag{5.48}$$

where we have set $E_t^f : h \to h \otimes \Gamma^f(\hat{k})$ by

$$E_t^f u = \oplus_{n=0}^{\infty} (n!)^{\frac{1}{4}} E_t^{(n)} u.$$

We observe that

$$\begin{aligned} \| E_t^f u \|^2 &= \sum_{n=0}^{\infty} (n!)^{\frac{1}{2}} \| E_t^{(n)} u \|^2 \\ &\leq \| u \|^2 \sum_{n=0}^{\infty} (n!)^{\frac{1}{2}} \| D \|^{2n} \left\{ \int_{0 < s_n < s_{n-1} < \cdots s_1 < t} ds_n \cdots ds_1 \| \hat{f}(s_n) \|^4 \cdots \| \hat{f}(s_1) \|^4 \right\} \\ &= \| u \|^2 \sum_{n=0}^{\infty} (n!)^{-\frac{1}{2}} \| D \|^{2n} \mu_f(t)^n, \end{aligned}$$

where $\mu_f(t) = \int_0^t \|\hat{f}(s)\|^4 ds$. The estimate (5.48) proves the existence of the solution of equation (5.45), as well as its continuity relative to the strong operator topology in $\mathcal{B}(h)$.

For showing the uniqueness of the solution of (5.45), suppose that J'_t is another adapted regular map-valued process satisfying (5.45). Then $\Xi_t := J_t - J'_t$ is also an adapted regular process and

$$\Xi_t = \int_0^t \Xi_s \circ M(ds).$$

By the definition of regularity, for fixed t_0 and $f \in L^4_{\text{loc}}$, we can choose a Hilbert space $\mathcal{H}_1$, a positive constant c and bounded operator $r_1 : h \to h \otimes \mathcal{H}_1$ (depending on t_0 and f) such that

$$\|\Xi_t(x \otimes e(f))u\| \leq c\|e(f)\| \|(x \otimes 1_{\mathcal{H}_1})r_1 u\|$$

for all $t \leq t_0$, $u \in h$ and $x \in \mathcal{A}$. Now, by repeated application of (5.44) of Theorem 5.4.7 along with the Lifting lemma, for any positive integer n we get the following:

$$\begin{aligned}
&\|\Xi_t(x \otimes e(f))u\|^2 \\
&\leq 2e^t \int_0^t (1 + \|f(s)\|^2) \|\hat{\Xi}_s(\Theta_{\hat{f}(s)}(x) \otimes e(f))u\|^2 ds \\
&\leq (2e^t)^2 \int_0^t \|\hat{f}(s)\|^2 \int_0^s \|\hat{f}(w)\|^2 \|\hat{\hat{\Xi}}_w((\Theta_{\hat{f}(w)} \otimes \mathrm{Id}_{\hat{k_0}}) \circ \Theta_{\hat{f}(s)}(x) \otimes e(f))u\|^2 dw ds \\
&\leq (c\|e(f)\| \, \|r_1\| \, \|u\| \, \|x\|)^2 \frac{(2e^t \|\Theta\|^2 \mu_f(t))^n}{n!},
\end{aligned}$$

where we have denoted by $\hat{\hat{\Xi}}_w$ the map $(\Xi_w \otimes \mathrm{Id}_{\hat{k_0}^{\otimes 2}}) \circ P$, $P : \mathcal{A} \otimes \hat{k_0}^{\otimes 2} \otimes \Gamma \to \mathcal{A} \otimes \Gamma \otimes \hat{k_0}^{\otimes 2}$ being the unitary which interchanges the second and third tensor components. Since the left-hand side of the above inequality is independent of n, while the right-hand side goes to 0 as $n \to \infty$, we conclude that $\|\Xi_t(x \otimes e(f)u\| = 0$, which completes the proof of the uniqueness.

(*ii*) First we prove the following identity:

$$\langle J_t(x \otimes e(f))u, J_t(y \otimes e(g))v\rangle = \langle ue(f), J_t(x^* y \otimes e(g))v\rangle. \tag{5.49}$$

For this it is convenient to lift the maps J_t to the module $\mathcal{A} \otimes \Gamma^f(\hat{k_0}) \otimes_{\text{alg}} \mathcal{E}(L^4_{\text{loc}})$, that is, replace $\mathcal{A}$ by $\mathcal{A} \otimes \Gamma^f(\hat{k_0})$. We define $\hat{J}_t : \mathcal{A} \otimes \Gamma^f(\hat{k_0}) \otimes_{\text{alg}} \mathcal{E}(L^4_{\text{loc}}) \to \mathcal{A} \otimes \Gamma \otimes \Gamma^f(\hat{k_0})$ by $\hat{J}_t = (J_t \otimes \mathrm{Id})P$, where P interchanges the second and third tensor components. Recalling from the discussions following Theorem 4.2.7 and Section 5.1 that $\Theta(x)_\zeta \in \mathcal{B}(h, h \otimes \hat{k_0})$ for $x \in \mathcal{A}, \zeta \in \hat{k_0}$, we can define $\Theta_\zeta : \mathcal{A} \to \mathcal{A} \otimes \hat{k_0}$ by $\Theta_\zeta(x) = \Theta(x)_\zeta$, and extend as above to $\widehat{\Theta_\zeta} : \mathcal{A} \otimes \Gamma^f(\hat{k_0}) \to \mathcal{A} \otimes \Gamma^f(\hat{k_0})$ by setting $\widehat{\Theta_\zeta}|_{\mathcal{A} \otimes \hat{k_0}^{\otimes n}} = \Theta_\zeta \otimes \mathrm{Id}_{\hat{k_0}^{\otimes n}}$. By the Lifting lemma, both $\hat{J}_t$ and $\widehat{\Theta_\zeta}$ are well defined and enjoy the estimates as in (i) of the present Theorem 5.4.9 and Lemma 5.4.6 (iii), respectively.

Next, note that for fixed $f, g \in L^4_{\text{loc}}$ and $x, y \in \mathcal{A}$, one can get the following by using the equation (5.45) for J_t and quantum Itô formula (5.14) and the structure relation in Lemma 5.4.6 (ii):

$$\begin{aligned} &\langle J_t(x \otimes e(f))u, J_t(y \otimes e(g))v\rangle \\ &= \langle xu\otimes e(f), yv\otimes e(g)\rangle + \int_0^t ds\{\hat{J}_s(\Theta_{\hat{f}(s)}(x)\otimes e(f))u, \hat{J}_s(y\otimes \hat{g}(s)\otimes e(g))v\rangle \\ &\quad + \langle \hat{J}_s(x \otimes \hat{f}(s) \otimes e(f))u, \hat{J}_s(\Theta_{\hat{g}(s)}(y) \otimes e(g))v\rangle \\ &\quad + \langle \hat{J}_s(\Theta_{f(s)}(x) \otimes e(f))u, \hat{J}_s(\Theta_{g(s)}(y) \otimes e(g))v\rangle\}, \end{aligned} \tag{5.50}$$

where $f(s)$ and $g(s)$ in k_0 are identified with $0 \oplus f(s)$ and $0 \oplus g(s)$ in $\hat{k_0}$, respectively. We claim that the identity above remains valid even when we replace x, y by $X, Y \in \mathcal{A} \otimes \Gamma^f(\hat{k_0})$ and $\Theta_\zeta(x), \Theta_\zeta(y)$ by $\widehat{\Theta}_\zeta(X), \widehat{\Theta}_\zeta(Y)$, respectively, where ζ is one of the vectors $\hat{f}(s), \hat{g(s)}, f(s)$ and $g(s)$. To see this, it suffices to observe that in the resulting identity, both left- and right-hand sides vanish if $X \in \mathcal{A} \otimes \hat{k_0}^{\otimes n}$ and $Y \in \mathcal{A} \otimes \hat{k_0}^{\otimes m}$ with $m \neq n$, and then use the definition of $\hat{J}_t$ and $\widehat{\Theta}_\zeta$ to prove the identity for $X, Y \in \mathcal{A} \otimes_{\text{alg}} \hat{k_0}^{\otimes m}$. Finally, use Corollary 4.2.9 and strong continuity of $\hat{J}_t$ obtained from the estimate in (i) to extend the identity from $X = \sum x_\alpha \otimes e_\alpha$, $Y = \sum y_\alpha \otimes e_\alpha$ (finite sums) to arbitrary X and Y. Thus one has upon setting

$$\Phi_t(X, Y) \equiv \langle \hat{J}_t(X \otimes e(f))u, \hat{J}_t(Y \otimes e(g))v\rangle - \langle ue(f), \hat{J}_t(\langle X, Y\rangle \otimes e(g))v\rangle$$

the equation

$$\Phi_t(X, Y) = \int_0^t ds\{\Phi_s(\widehat{\Theta}_{\hat{f}(s)}(X), \mathcal{J}_{\hat{g}(s)}(Y)) + \Phi_s(\mathcal{J}_{\hat{f}(s)}(X), \widehat{\Theta}_{\hat{g}(s)}(Y)) + \Phi_s(\widehat{\Theta}_{f(s)}(X), \widehat{\Theta}_{g(s)}(Y))\}, \tag{5.51}$$

where $\langle X, Y\rangle$ is the module inner product in $\mathcal{A} \otimes \Gamma^f(\hat{k_0})$ and we have set for $\zeta, \eta_1, \cdots, \eta_n \in \hat{k_0}$ the map $\mathcal{J}_\zeta(x \otimes \eta_1 \cdots \otimes \eta_n) = x \otimes \eta_1 \cdots \otimes \eta_n \otimes \zeta$, and extend it naturally as a map from $\mathcal{A} \otimes \Gamma^f(\hat{k_0})$ to itself. It is clear that the estimates in Lemma 5.4.6 (iii) and Theorem 5.4.9 (i) extend to

$$\|\widehat{\Theta}_\zeta(X)u\| \leq \|(X \otimes 1_{k_0'})D(u \otimes \zeta)\|$$

and

$$\sup_{0\leq t\leq t_0} \|\hat{J}_t(X \otimes e(f))u\| \leq C'(f)\|(X \otimes 1_{\Gamma^f(k')})E^f_{t_0}u\|.$$

From the above estimates and definition of Φ_t, it is clear that

$$|\Phi_t(X, Y)| \leq \|u\|\|v\|\|X\|\|Y\|\|E^g_{t_0}\|C'(g)\{\|E^f_{t_0}\|C'(f) + \|e(f)\|\}.$$

This implies, by iterating the expression (5.51) a sufficient number of times, that $\Phi_t(X, Y) = 0$ which leads to $\Phi_t(x, y) = 0$ for all $x, y \in \mathcal{A}$. Since

$$\left\langle ve(g), j_t(x)\left(\sum_{i=1}^{n} u_i e(f_i)\right)\right\rangle = \langle J_t(x^* \otimes e(g))v, \sum u_i e(f_i)\rangle$$

by the above identity, it follows that $j_t(x)$ is well defined on $h \otimes_{\mathrm{alg}} \mathcal{E}(L^4_{\mathrm{loc}})$, and thus *(ii)*(*a*) is proved.
For the proof of *(ii)*(*b*), we extend $j_t(x)$ linearly to the dense set (say $\mathcal{D}$) of vectors of the form $\sum_i u_i e(f_i)$ (finite sum) with $u_i \in h$ and $f_i \in L^4_{\mathrm{loc}}$, and by the part *(ii)*(*a*) of this theorem,

$$\left\langle j_t(x)\left(\sum_i u_i e(f_i)\right), j_t(x)\left(\sum_j u_j e(f_j)\right)\right\rangle = \sum_{i,j} \langle u_i e(f_i), j_t(x^*x) u_j e(f_j)\rangle,$$

thereby showing the positivity of the map j_t. For $\psi \in \mathcal{D}$, it follows by *(ii)*(*a*) that

$$\| j_t(1)\psi \|^2 = \langle \psi, j_t(1)\psi\rangle \leq \|\psi\| \| j_t(1)\psi\|,$$

or equivalently

$$\| j_t(1)\psi\|(\|\psi\| - \| j_t(1)\psi\|) \geq 0,$$

which implies that $\| j_t(1)\psi\| \leq \|\psi\|$ for all ψ belonging to the dense domain ψ. So $j_t(1)$ extends to a bounded operator with $\| j_t(1)\| \leq 1$. Now by the positivity of the map j_t and the fact that $\|x\|^2 1 - x^*x \geq 0$ for every $x \in \mathcal{A}$, we get that

$$\begin{aligned} \| j_t(x)\psi\|^2 &= \langle \psi, j_t(x^*x)\psi\rangle \\ &\leq \|x\|^2 \langle \psi, j_t(1)\psi\rangle \leq \|x\|^2 \|\psi\| \| j_t(1)\psi\| \leq \|x\|^2 \|\psi\|^2, \end{aligned}$$

proving that $j_t(x)$ is bounded on the dense domain $\mathcal{D}$ and therefore admits a bounded extension on whole of $h \otimes \Gamma$. Denoting the extension by the same notation, we get from the relation *(ii)*(*a*) the homomorphism property of the map j_t and also the bound $\| j_t(x)\| \leq \|x\|$ for all $x \in \mathcal{A}$. Similarly, one can prove that j_t is a $*$-preserving map. To prove that j_t is normal, let us first observe that by (5.46), the map $\mathcal{A} \ni x \mapsto j_t(x)ue(f) \in h \otimes \Gamma$ (where $u \in h$ and $f \in L^4_{\mathrm{loc}}$ are fixed) is continuous with respect to the ultra-strong topology on $\mathcal{A}$, and hence the continuity of $x \mapsto j_t(x)\psi$ for a fixed $\psi \in \mathcal{D}$ also follows. Since j_t is a contractive map and $\mathcal{D}$ is dense in $h \otimes \Gamma$, we conclude that $x \mapsto j_t(x)\vartheta$ is continuous (with respect to the ultra-strong topology on $\mathcal{A}$) for an arbitrary vector $\vartheta \in h \otimes \Gamma$, which is equivalent to the normality of j_t. We refer the reader to [100] for the proof of *(iii)*.

For (*iv*), we note that $j_t(1) = 1$ for all t if and only if $dJ_t(1 \otimes e(f))u = 0$ for all $u \in h$, $f \in L^4_{\mathrm{loc}}$, and from equation (5.45) and (**S1**) it is clear that this can happen if and only if

$$0 = \int_0^t ds \langle ue(f), J_s \circ (\Lambda_{\Sigma^*\Sigma - I}(ds)(ve(f))\rangle$$
$$= \int_0^t ds \langle ue(f), J_s(\langle f(s), (\Sigma^*\Sigma - I)_{f(s)}\rangle \otimes e(f))v\rangle$$

for all t, since $\pi(1)\delta = \delta$. But this is possible if and only if

$$\langle \zeta, (\Sigma^*\Sigma - I)_\zeta\rangle = 0 \text{ for all } \zeta \in k_0,$$

which is equivalent to $\Sigma^*\Sigma = I$. □

6

Dilation of quantum dynamical semigroups with bounded generator

We begin with a definition of dilation of a quantum dynamical semigroup (Q.D.S.) (not necessarily with bounded generator) before we proceed to construct such a dilation for Q.D.S. with a bounded generator. Most of the material presented in this chapter can be found in [30], [60] and [62].

Definition 6.0.1 A *Hudson–Parthasarathy (H–P) dilation* of a Q.D.S. $(T_t)_{t \geq 0}$ on a C^* or von Neumann algebra $\mathcal{A} \subseteq \mathcal{B}(h)$ is given by a family $(U_t)_{t \geq 0}$ of unitary operators acting on $h \otimes \Gamma(L^2(\mathbb{R}_+, k_0))$, where k_0 is a Hilbert space (called the noise or multiplicity space for the dilation), such that the following holds

(*i*) U_t satisfies a Q.S.D.E. of the form

$$dU_t = U_t(a_R(dt) + a_S^\dagger(dt) + \Lambda_T(dt) + Adt), \;\; U_0 = I,$$

for some (possibly unbounded) operators R, S, T, A, with the initial condition $U_0 = I$.

(*ii*) For all $u, v \in h$, $x \in \mathcal{A}$,

$$\langle ve(0), U_t(x \otimes I)U_t^* ue(0)\rangle = \langle v, T_t(x)u\rangle.$$

Definition 6.0.2 An *Evans–Hudson (E–H) dilation* of a Q.D.S. $(T_t)_{t \geq 0}$ on a C^* or von Neumann algebra $\mathcal{A} \in \mathcal{B}(h)$ is given by a family $(j_t)_{t \geq 0}$ of $*$-homomorphisms (normal in case $\mathcal{A}$ is a von Neumann algebra) from $\mathcal{A}$ into the von Neumann algebra $\mathcal{A}'' \otimes \mathcal{B}(\Gamma(L^2(\mathbb{R}_+, k_0)))$, where k_0 is a Hilbert space (called the *noise* or *multiplicity space* for the dilation), such that we have the following.

(*i*) There exists an ultra-weak dense $*$-subalgebra $\mathcal{A}_0$ of $\mathcal{A}$ such that the map-valued process $J_t : \mathcal{A} \otimes_{\text{alg}} \mathcal{E}(L^2(\mathbb{R}_+, k_0)) \to \mathcal{A}'' \otimes \Gamma(L^2(\mathbb{R}_+, k_0))$ defined by $J_t(x \otimes e(f))u := j_t(x)(ue(f))$ for $x \in \mathcal{A},\ u \in h,\ f \in L^2(\mathbb{R}_+, k_0)$, satisfies a map-valued Q.S.D.E. of the form

$$dJ_t = J_t \circ \left(a_\delta(dt) + a_\delta^\dagger(dt) + \Lambda_\sigma(dt) + \mathcal{I}_{\mathcal{L}}(dt) \right), \quad J_0 = \text{Id};$$

on $\mathcal{A}_0 \otimes \mathcal{E}(\mathcal{C})$, where $\delta : \mathcal{A}_0 \to \mathcal{A} \otimes k_0$, $\sigma : \mathcal{A}_0 \to \mathcal{A} \otimes \mathcal{B}(k_0)$ are linear maps and where $\mathcal{L}$ is the generator of (T_t), with $\mathcal{A}_0 \subseteq \text{Dom}(\mathcal{L})$.

(*ii*) For all $u, v \in h,\ x \in \mathcal{A}$,

$$\langle ve(0), j_t(x)ue(0) \rangle = \langle v, T_t(x)u \rangle.$$

Remark 6.0.3 *The map-valued Q.S.D.E. in the condition (i) of the above definition should be understood to imply that the integral* $\int_0^t J_s \circ (a_\delta + a_\delta^\dagger + \Lambda_\sigma + \mathcal{I}_{\mathcal{L}})(ds)$ *can be defined and it is equal to* $J_t - \text{Id}$. *It must be noted that when* $\mathcal{A}_0$ *is strictly smaller than* $\mathcal{A}$ *and the maps* $(\mathcal{L}, \delta, \sigma)$ *are not bounded, the definition of such a map-valued integral has not yet been given, since the theory of map-valued integrals in the previous chapter was built under the assumption of boundedness of the coefficients and regularity of the integrands. The purpose of allowing a class of maps in the above definition, more general than necessary for the present chapter, is to cover the more general framework of map-valued Q.S.D.E. to be considered in Chapter 8, where we shall obtain E–H dilation for a class of Q.D.S. with unbounded generator.*

The phrase *quantum stochastic dilation* (or simply *dilation*) will be used to mean one of above two kinds of dilation.

Let us now consider a Q.D.S. (T_t) on a unital von Neumann algebra $\mathcal{A} \subseteq \mathcal{B}(h)$ and consider the problem of its quantum stochastic dilation. We remark that without loss of generality we may assume T_t to be conservative, since the case of a nonconservative T_t can be reduced to that of a conservative one by replacing the original algebra $\mathcal{A}$ by the bigger algebra $\mathcal{A} \oplus \mathbb{C}$ and replacing T_t by an appropriate extension of given by $\tilde{T}_t(x \oplus c) \equiv (T_t(x) + c.(1 - T_t(1))) \oplus c$. It is clear that $\tilde{T}_t$ is a conservative quantum dynamical semigroup on $\mathcal{A} \oplus \mathbb{C}$ with the generator $\tilde{\mathcal{L}}$ given by $\tilde{\mathcal{L}}(x \oplus c) = (\mathcal{L}(x) - c\mathcal{L}(1)) \oplus 0$, for x in the domain $\text{Dom}(\mathcal{L}) \subseteq \mathcal{A}$ of the generator $\mathcal{L}$ of T_t, provided that 1 is also in $\text{Dom}(\mathcal{L})$.

Let us assume in this chapter that T_t is uniformly continuous, or equivalently, $\mathcal{L}$ is norm-bounded. In this chapter, at first a H–P dilation U_t is constructed in $h \otimes \Gamma \equiv h \otimes \Gamma(L^2(\mathbb{R}_+, k_0))$ for a suitable multiplicity space k_0. However, $j_t^0(x) := U_t(x \otimes I)U_t^*$ in general will not satisfy a flow equation of the E–H type. It is shown that there exists a suitable choice of a partial isometry in $h \otimes k_0$ such that the E–H type of flow equation can be implemented by a partial isometry-valued process in $h \otimes \Gamma$. It should be mentioned here that in [73] an E–H-type dilation was achieved with $\mathcal{A} = \mathcal{B}(h)$ for a countably infinite dimensional h only.

6.1 Hudson–Parthasarathy (H–P) dilation

Let $(\rho, \mathcal{K}, \alpha, H, R)$ be a quintuple associated with T_t obtained from Theorem 3.1.8 and (k_1, Σ_1) be the pair for the representation ρ as in the Proposition 2.1.7. Denote the projection $\Sigma_1 \Sigma_1^*$ by P_1. Now set $\tilde{R} = \Sigma_1 R$, $\tilde{R} \in \mathcal{B}(h, h \otimes k_1)$ so that $\tilde{R}^* = R^* \Sigma_1^*$ and we have

$$\tilde{R}^*(x \otimes 1_{k_1})\tilde{R} = R^* \Sigma_1^*(x \otimes 1_{k_1})\Sigma_1 R = R^* \rho(x) R.$$

Also,

$$\tilde{R}^* \tilde{R} = R^* \Sigma_1^* \Sigma_1 R = R^* R, \text{ as } \Sigma_1^* \Sigma_1 = 1_{\mathcal{K}}.$$

Now, we take the unitary process U_t which satisfies the following Q.S.D.E. (as in Section 3.1)

$$dU_t = U_t \left(a_{\tilde{R}}^\dagger(dt) - a_{\tilde{R}}(dt) + \left(iH - \frac{1}{2} \tilde{R}^* \tilde{R} \right) dt \right), \quad U_0 = I. \tag{6.1}$$

Let $\tilde{\Gamma}$ denote $\Gamma(L^2(\mathbb{R}_+, k_1))$. Taking $j_t^0(x) = U_t(x \otimes 1_{\tilde{\Gamma}})U_t^*$, we see that for each t, $j_t^0(\cdot)$ is a $*$-homomorphism. We now claim that $\langle ve(0), j_t^0(x)ue(0)\rangle = \langle v, T_t(x)u\rangle$. To prove this, it is enough to show that

$$\langle ve(0), \frac{d}{dt} j_t^0(x)(ue(0))\rangle = \langle v, T_t(\mathcal{L}(x))u\rangle,$$

and this follows from the quantum Itô formula for right integrals, as mentioned in Remark 5.2.10. Indeed, we have,

$$\langle U_t^*(ve(0)), (x \otimes 1_{\tilde{\Gamma}})U_t^*(ue(0))\rangle = \int_0^t ds \langle ve(0), j_s^0(\mathcal{L}(x))(ue(0))\rangle,$$

where

$$\begin{aligned} \mathcal{L}(x) &= R^* \rho(x) R - \frac{1}{2} R^* R x - \frac{1}{2} x R^* R + i[H, x] \\ &= \tilde{R}^*(x \otimes 1_{k_1})\tilde{R} - \frac{1}{2} \tilde{R}^* \tilde{R} x - \frac{1}{2} x \tilde{R}^* \tilde{R} + i[H, x]. \end{aligned}$$

Thus, if we denote by $\mathbb{E}_0$ the vacuum expectation map which takes an element G of $\mathcal{B}(h \otimes \tilde{\Gamma})$ to an element $\mathbb{E}_0 G$ in $\mathcal{B}(h)$ satisfying $\langle v, (\mathbb{E}_0 G)u\rangle = \langle ve(0), G(ue(0))\rangle$ (for $u, v \in h$), then

$$\frac{d}{dt} \mathbb{E}_0 j_t^0(x) = \mathbb{E}_0 j_t^0(\mathcal{L}(x)),$$

which implies that $\mathbb{E}_0 j_t^0(x) = T_t(x)$, since $\mathcal{L}$ is bounded.

A simple calculation using the quantum Itô formula and equation (6.1) shows that

$$dj_t^0(x) = U_t[a^\dagger_{\alpha(x)}(dt) - a_{\alpha(x)}(dt) + \mathcal{L}(x)dt]U_t^*, \tag{6.2}$$

where $\alpha(x) = \tilde{R}x - (x \otimes 1_{k_1})\tilde{R} = \Sigma_1[Rx - \rho(x)R]$, for $x \in \mathcal{A}$. In general, $\alpha(x)$ may not be in $\mathcal{A} \otimes k_1$ and therefore the equation (6.2) is not a flow equation of the E–H type. However, in case $\mathcal{A} = \mathcal{B}(h)$, it is trivially a flow equation.

6.2 Existence of structure maps and Evans–Hudson (E–H) dilation of T_t

In the context of the Theorem 3.1.8 and the Proposition 2.1.7, it should be noted that in general $\mathcal{K}$ need not be of the form $h \otimes k'$ and neither ρ nor α need be structure maps as defined in Chapter 5, that is, ρ need not be in $\mathcal{A} \otimes \mathcal{B}(k')$ nor $\alpha(x)$ be in $\mathcal{A} \otimes k'$. However, the following theorem asserts that one can 'rotate' the whole structure appropriately so that the 'rotated' ρ and α (denoted π and δ, respectively) become structure maps without changing $\mathcal{L}$ (see also [102]).

Theorem 6.2.1 *Let T_t be a conservative norm-continuous Q.D.S. with generator $\mathcal{L}$. Then there exist a Hilbert space k_0, a normal $*$-representation $\pi : \mathcal{A} \to \mathcal{A} \otimes \mathcal{B}(k_0)$ and a π-derivation δ of $\mathcal{A}$ into $\mathcal{A} \otimes k_0$ such that the hypotheses* (**S1**)*–*(**S3**) *in Chapter 5 are satisfied.*

Proof:

(*i*) Let $(\rho, \mathcal{K}, \alpha, H, R)$ be a quintuple for T_t as in Theorem 3.1.8. We define a map $\rho' : \mathcal{A}' \to \mathcal{B}(\mathcal{K})$, where $\mathcal{A}'$ denotes the commutant of $\mathcal{A}$ in $\mathcal{B}(h)$, by

$$\rho'(a)(\alpha(x)u) = \alpha(x)\, au, x \in \mathcal{A}, \quad u \in h, \quad a \in \mathcal{A}', \tag{6.3}$$

and extend it linearly to the algebraic span of $\mathcal{D} \equiv \{\alpha(x)u \ : \ u \in h, x \in \mathcal{A}\}$.

To show that it is well defined, we need to show that whenever $\sum_{i=1}^m \alpha(x_i)u_i = 0$ for $x_i \in \mathcal{A}$, $u_i \in h$, one has $\rho'(a)\left(\sum_{i=1}^m \alpha(x_i)u_i\right) = 0$. Since

$$\alpha(x_i)^*\alpha(y) = \mathcal{L}(x_i^* y) - \mathcal{L}(x_i^*)y - x_i^*\mathcal{L}(y) \in \mathcal{A}$$

for $y \in \mathcal{A}$, we have for $a \in \mathcal{A}'$,

$$\begin{aligned}&\left\langle \rho'(a)\left(\sum_{i=1}^m \alpha(x_i)u_i\right), \ \alpha(y)v\right\rangle \\ &= \sum_{i=1}^m \langle \alpha(x_i)au_i, \alpha(y)v\rangle = \sum_{i=1}^m \langle u_i, a^*\alpha(x_i)^*\alpha(y)v\rangle\end{aligned}$$

$$= \sum_{i=1}^{m} \langle u_i, \alpha(x_i)^* \alpha(y) a^* v \rangle = \left\langle \sum_{i=1}^{m} \alpha(x_i) u_i, \quad \alpha(y) a^* v \right\rangle, \tag{6.4}$$

thereby proving that ρ' is well defined. A similar computation gives

$$\left\| \rho'(a) \left(\sum_{i=1}^{m} \alpha(x_i) u_i \right) \right\|^2 = \sum_{i=1}^{m} \sum_{j=1}^{m} \langle u_i, \alpha(x_i)^* \alpha(x_j) a^* a u_j \rangle. \tag{6.5}$$

Denoting the operator $\alpha(x_i)^* \alpha(x_j)$ by A_{ij}, and noting that $A \equiv ((A_{ij}))_{i,j=1,\dots m}$ acts as a positive operator on $\underbrace{h \oplus \cdots \oplus h}_{\text{m copies}}$, which commutes with the positive operator $C \otimes I_m$, where $C = \|a\|^2 . 1 - a^* a$ and I_m denotes the identity matrix of order m, we observe that $A(C \otimes I_m)$ is also a positive operator. Thus, considering the vector $u_1 \oplus u_2 \oplus \cdots \oplus u_m \in \underbrace{h \oplus \ldots \oplus h}_{\text{m copies}}$, the right-hand side of (6.5) can be estimated as:

$$\begin{aligned} & \sum_{i=1}^{m} \sum_{j=1}^{m} \langle u_i, \alpha(x_i)^* \alpha(x_j) a^* a u_j \rangle \\ & \leq \|a\|^2 \sum_{i=1}^{m} \sum_{j=1}^{m} \langle u_i, \alpha(x_i)^* \alpha(x_j) u_j \rangle \\ & = \left\| \sum_{i=1}^{m} \alpha(x_i) u_i \right\|^2 \|a\|^2, \end{aligned}$$

proving that $\rho'(a)$ extends to a bounded operator on $\mathcal{K}$ satisfying $||\rho'(a)|| \leq ||a||$ since $\mathcal{D}$ is total in $\mathcal{K}$. It can be observed from the definition of ρ' and (6.4) that it is a unital $*$-representation of $\mathcal{A}'$ in $\mathcal{K}$. Next we show that ρ' is normal. For this, take a net $\{a_\nu\}$ such that $0 \leq a_\nu \uparrow a$ where $a_\nu, a \in \mathcal{A}'$. It is clear from the definition of ρ' that $\rho'(a_\nu)\alpha(x)u \to \rho'(a)\alpha(x)u$ for all $x \in \mathcal{A}$, $u \in h$, and thus, $\rho'(a_\nu) \stackrel{s}{\to} \rho'(a)$ on $\mathcal{K}$ by totality of $\mathcal{D}$ in $\mathcal{K}$ and since $\|\rho'(a_\nu)\| \leq \|a_\nu\| \leq ||a||$ for all ν.

(*ii*) By (*i*), $\rho' : \mathcal{A}' \to \mathcal{B}(\mathcal{K})$ is a unital normal $*$-representation. By Proposition 2.1.7, there exist a Hilbert space k_2, an isometry $\Sigma_2 : \mathcal{K} \to h \otimes k_2$ with $\mathcal{K}_2 = \operatorname{Ran} \Sigma_2 \stackrel{\sim}{=} \mathcal{K}$ such that

$$\rho'(a) = \Sigma_2^* (a \otimes 1_{k_2}) \Sigma_2, \tag{6.6}$$

and for all $a \in \mathcal{A}'$, $a \otimes 1_{k_2}$ commutes with $P_2 \equiv \Sigma_2 \Sigma_2^*$. Let us now take

$$\tilde{\delta}(x) = \Sigma_2 \alpha(x), \quad \tilde{\pi}(x) = \Sigma_2 \rho(x) \Sigma_2^*.$$

It is clear that $\tilde{\delta}$ is a $\tilde{\pi}$-derivation. Moreover,

$$\tilde{\delta}(x^*)^*\tilde{\delta}(y) = \alpha(x^*)^*\Sigma_2^*\Sigma_2\alpha(y) = \alpha(x^*)^*\alpha(y),$$

and hence

$$\mathcal{L}(xy) - x\mathcal{L}(y) - \mathcal{L}(x)y = \tilde{\delta}(x^*)^*\tilde{\delta}(y).$$

Taking $\tilde{R} = \Sigma_2 R \in \mathcal{B}(h, h \otimes k_2)$, we observe that

$$\tilde{\delta}(x) = \Sigma_2\alpha(x) = \Sigma_2(Rx - \rho(x)R) = \tilde{R}x - \tilde{\pi}(x)\tilde{R}.$$

It is also clear that

$$\begin{aligned}\mathcal{L}(x) &= \tilde{R}^*\tilde{\pi}(x)\tilde{R} - \frac{1}{2}\tilde{R}^*\tilde{R}x - \frac{1}{2}x\tilde{R}^*\tilde{R} + i[H, x] \\ &= R^*\rho(x)R - \frac{1}{2}R^*Rx - \frac{1}{2}xR^*R + i[H, x].\end{aligned}$$

To show that $\tilde{\delta}(x) \in \mathcal{A} \otimes k_2$ for all $x \in \mathcal{A}$, it is enough (by Lemma 4.2.10) to verify that for any $f \in k_2$, $\langle f, \tilde{\delta}(x)\rangle \in \mathcal{A}$, or equivalently that $\langle f, \tilde{\delta}(x)\rangle$ commutes with all $a \in \mathcal{A}'$. For $f \in k_2$, $a \in \mathcal{A}'$, $u, v \in h$, $x \in \mathcal{A}$, since P_2 and $(a \otimes 1_{k_2})$ commute,we have,

$$\begin{aligned}&\langle\langle f, \tilde{\delta}(x)\rangle au, v\rangle = \langle\tilde{\delta}(x)au, v \otimes f\rangle = \langle\Sigma_2\alpha(x)au, v \otimes f\rangle \\ &= \langle\Sigma_2\rho'(a)(\alpha(x)u), v \otimes f\rangle = \langle\Sigma_2\Sigma_2^*(a \otimes 1_{k_2})\Sigma_2\alpha(x)u, v \otimes f\rangle \\ &= \langle P_2(a \otimes 1_{k_2})\Sigma_2\alpha(x)u, v \otimes f\rangle = \langle(a \otimes 1_{k_2})\Sigma_2\Sigma_2^*\Sigma_2\alpha(x)u, v \otimes f\rangle \\ &= \langle\Sigma_2\alpha(x)u, (a^*v) \otimes f\rangle = \langle\langle f, \tilde{\delta}(x)\rangle u, a^*v\rangle = \langle a\langle f, \tilde{\delta}(x)\rangle u, v\rangle.\end{aligned}$$

Next, we want to show that $\tilde{\pi}(x) \in \mathcal{A} \otimes \mathcal{B}(k_2)$ for $x \in \mathcal{A}$; and for this it is enough to verify $\tilde{\pi}(x)(a \otimes 1_{k_2}) = (a \otimes 1_{k_2})\tilde{\pi}(x)$ for all $a \in \mathcal{A}'$. Since $\Sigma_2^*(1 - P_2) = 0$ and P_2 commutes with $(a \otimes 1_{k_2})$, it is clear that

$$\begin{aligned}\tilde{\pi}(x)(a \otimes 1_{k_2})(1 - P_2) &= \Sigma_2\rho(x)\Sigma_2^*(1 - P_2)(a \otimes 1_{k_2}) \\ &= 0 = (a \otimes 1_{k_2})\tilde{\pi}(x)(1 - P_2).\end{aligned}$$

Thus, it suffices to verify that

$$\tilde{\pi}(x)(a \otimes 1_{k_2})P_2 = (a \otimes 1_{k_2})\tilde{\pi}(x)P_2,$$

or equivalently (since $\Sigma_2\mathcal{D}$ is total in $\mathcal{K}_2$) that

$$\tilde{\pi}(x)(a \otimes 1_{k_2})\Sigma_2\alpha(y)u = (a \otimes 1_{k_2})\tilde{\pi}(x)\Sigma_2\alpha(y)u,$$

for all $y \in \mathcal{A}$, $u \in h$. For this, observe that since $a \in \mathcal{A}'$,

$$\tilde{\pi}(x)(a \otimes 1_{k_2})\Sigma_2 \alpha(y)u = \Sigma_2 \rho(x)\Sigma_2^*(a \otimes 1_{k_2})\Sigma_2 \alpha(y)u = \Sigma_2 \rho(x)\rho'(a)\alpha(y)u$$
$$= \Sigma_2 \rho(x)\alpha(y)au = \Sigma_2 \alpha(xy)au - \Sigma_2 \alpha(x)yau = \Sigma_2 \rho'(a)(\alpha(xy) - \alpha(x)y)u$$
$$= \Sigma_2 \rho'(a)\rho(x)\alpha(y)u = \Sigma_2 \Sigma_2^*(a \otimes 1_{k_2})\Sigma_2 \rho(x)\Sigma_2^*(\Sigma_2 \alpha(y)u)$$
$$= P_2(a \otimes 1_{k_2})\tilde{\pi}(x)(\Sigma_2 \alpha(y)u) = (a \otimes 1_{k_2})\tilde{\pi}(x)(\Sigma_2 \alpha(y)u).$$

(iii) Recall that by the Proposition 2.1.7, we can write ρ as $\rho(x) = \Sigma_1^*(x \otimes 1_{k_1})\Sigma_1$ as in the previous section. It follows that $\tilde{\pi}(x) = \Sigma_2 \Sigma_1^*(x \otimes 1_{k_1})\Sigma_1 \Sigma_2^* \equiv \tilde{\Sigma}^*(x \otimes 1_{k_1})\tilde{\Sigma}$ on $h \otimes k_2$ so that $\tilde{\Sigma}$ is a partial isometry with initial set $P_2(h \otimes k_2)$ and final set $P_1(h \otimes k_1)$, where $P_j (j = 1, 2) = \Sigma_j \Sigma_j^*$. Now set $k_0 = k_1 \oplus k_2$ and $\Sigma = 0 \oplus \tilde{\Sigma} : h \otimes k_0 \to h \otimes k_0$ with initial set $(0 \oplus P_2)(h \otimes k_0)$ and final set $(P_1 \oplus 0)(h \otimes k_0)$ and $\pi(x) = 0 \oplus \tilde{\pi}(x)$, $\delta(x)u = \tilde{\delta}(x)u$ for $x \in \mathcal{A}$, $u \in h$. It is clear that $\delta(x) \in \mathcal{A} \otimes k_0$, $\pi(x) \in \mathcal{A} \otimes \mathcal{B}(k_0)$ and (**S1**)–(**S3**) are satisfied. □

Remark 6.2.2 *Although ρ was assumed to be unital, π chosen by us is not unital. However, in some cases it may be possible to choose Σ, k_0 in such a manner that π is unital.*

We summarize the main result of this section in form of the following theorem.

Theorem 6.2.3 *Let $(T_t)_{t \geq 0}$ be a conservative norm-continuous Q.D.S. with $\mathcal{L}$ as its generator. Then there is a flow $J_t : \mathcal{A} \otimes \mathcal{E}(\mathcal{C}) \to \mathcal{A} \otimes \Gamma$ satisfying an E–H type Q.S.D.E. (5.45) with structure maps $(\mathcal{L}, \delta, \sigma)$ satisfying (**S1**)–(**S3**) of subsection 5.4.2, where $\Gamma \equiv \Gamma(L^2(\mathbb{R}_+, k_0))$ and $\mathcal{C}$ consists of bounded continuous functions in $L^2(\mathbb{R}_+, k_0)$, such that $j_t(x)$ defined in Theorem 5.4.9 is a (not necessarily unital) $*$-homomorphism of $\mathcal{A}$ into $\mathcal{A} \otimes \mathcal{B}(\Gamma)$ and $\mathbb{E}_0 j_t(x) = T_t(x)$* for all $x \in \mathcal{A}$.

Proof:
The proof is immediate by (i) observing the existence of structure maps δ and σ satisfying (**S1**), (**S2**) from Theorem 6.2.1, (ii) observing that $\mathcal{L}$ satisfies (**S3**), and finally (iii) constructing the solution J_t of equation (5.45) with structure maps $(\mathcal{L}, \delta, \sigma)$ as in Theorem 5.4.9. That $\mathbb{E}_0 j_t(x) = T_t(x)$ for all $x \in \mathcal{A}$ follows from the Q.S.D.E. (5.45). □

Remark 6.2.4 *It may be observed from the proof of Theorem 6.2.1 that the Hilbert space k_0 in the statement of Theorem 6.2.1 can be chosen to be separable if the initial Hilbert space h is separable. In such a case, if we choose an orthonormal basis $\{e_j\}$ in k_0, then the estimate for δ in (**S2**) is precisely the* *coordinate-free form of the condition*

$$\sum_{i=0}^{\infty} ||\theta_0^i(x)u||^2 \leq \sum_{i \in \mathcal{I}_0} ||x D_0^i u||^2,$$

with $\sum_{i \in \mathcal{I}_0} ||D_0^i u||^2 \leq \alpha_0 ||u||^2$ *as in [94], [91], where* $\mathcal{I}_0$ *is a countable index set. The similar conditions on* $\theta_j^i (j \neq 0)$ *as in [94] are trivially satisfied by* $((\theta_j^k))_{k,j=1}^{\infty} \equiv \sigma$ *and for* $\theta_0^0 \equiv \mathcal{L}$ *as can be seen from* (**S1**) *and* (**S3**)*. It may also be noted that* j_t *satisfies the E–H equation* $dj_t(x) = \sum_{i,j} j_t(\theta_j^i(x)) d\Lambda_i^j(t)$, with $j_0 = \mathrm{Id}$, *in the coordinatized form with the appropriate choices of* θ_j^i*'s in terms of* $\mathcal{L}$, δ *and* σ *as above. The flow equation (5.45) is in fact a coordinate-free modification of the coordinatized E–H equation given above.*

6.3 A duality property

In the previous section, starting with a $*$-homomorphism ρ of $\mathcal{A}$ and ρ-derivation α, we constructed a $*$-homomorphism ρ' of $\mathcal{A}'$ which satisfies $\rho'(a)\alpha(x) = \alpha(x)a$ for all $a \in \mathcal{A}', x \in \mathcal{A}$. Let us now observe that $\rho(x)$ and $\rho'(a)$ commute for each $x \in \mathcal{A}$ and $a \in \mathcal{A}'$. Due to the totality of vectors of the form $\alpha(y)u, \ y \in \mathcal{A}, u \in h$ in $\mathcal{K}$, it is enough to verify that $\rho'(a)\rho(x)\alpha(y) = \rho(x)\rho'(a)\alpha(y)$. But we have

$$\begin{aligned} \rho'(a)\rho(x)\alpha(y) &= \rho'(a)\alpha(xy) - \rho'(a)\alpha(x)y = \alpha(xy)a - \alpha(x)ay \\ &= (\alpha(xy) - \alpha(x)y)a \ = \ \rho(x)\alpha(y)a \ = \ \rho(x)\rho'(a)\alpha(y). \end{aligned}$$

Denote by $\mathcal{E}_\rho$ and $\mathcal{E}_{\rho'}$ respectively the spaces of intertwiners of ρ and ρ', that is,

$$\mathcal{E}_\rho \equiv \{L \in \mathcal{B}(h, \mathcal{K}) \ : \ Lx = \rho(x)L \ \text{ for all } x \in \mathcal{A}\},$$

$$\mathcal{E}_{\rho'} \equiv \{S \in \mathcal{B}(h, \mathcal{K}) \ : \ Sa = \rho'(a)S \ \text{ for all } a \in \mathcal{A}'\}.$$

Clearly, $\mathcal{E}_\rho$ is a Hilbert von Neumann $\mathcal{A}'$-module and $\mathcal{E}_{\rho'}$ is a Hilbert von Neumann $\mathcal{A}$-module. The right module actions are given by $(L, a) \mapsto La$ and $(S, x) \mapsto Sx$ for $a \in \mathcal{A}'$ and $x \in \mathcal{A}$, respectively. Furthermore, it may be observed that inner products of $\mathcal{E}_\rho$ and $\mathcal{E}_{\rho'}$, inherited from that of $\mathcal{B}(h, \mathcal{K})$, take values in $\mathcal{A}'$ and $\mathcal{A}$ respectively. To see this, note that for all $x \in \mathcal{A}$ and $L, M \in \mathcal{E}_\rho$, we have $\langle L, M\rangle x = L^*Mx = L^*\rho(x)M = xL^*M = x\langle L, M\rangle$; and similarly for all $a \in \mathcal{A}'$, $S, T \in \mathcal{E}_{\rho'}$, $\langle S, T\rangle a = a\langle S, T\rangle$.

Clearly, $\alpha(x) \in \mathcal{E}_{\rho'}$ for all $x \in \mathcal{A}$, and hence there is an implementer R of α which belongs to $\mathcal{E}_{\rho'}$, since such an R can be chosen from the ultra-weak closure of $\{\alpha(x)y \ : \ x, y \in \mathcal{A}\}$. Now, choose and fix any L from $\mathcal{E}_\rho$ and a self-adjoint element $H' \in \mathcal{A}'$. Consider the ρ'-derivation given by $\beta_L(a) \equiv La - \rho'(a)L$ and the CCP map given by $\mathcal{L}'_{(L,H')}(a) = L^*\rho'(a)L - \frac{1}{2}L^*La - \frac{1}{2}aL^*L + i[H', a]$. Since $\rho'(a)$ and $\rho(x)$ commute for all $a \in \mathcal{A}'$ and $x \in \mathcal{A}$, it is clear that $\beta_L(a) \in \mathcal{E}_\rho$ for all $a \in \mathcal{A}'$. Furthermore, $L^*\rho'(a)Lx = L^*\rho'(a)\rho(x)L = L^*\rho(x)\rho'(a)L = xL^*\rho'(a)L$ for $x \in \mathcal{A}, a \in \mathcal{A}'$, which shows that $L^*\rho'(a)L \in \mathcal{A}'$, and hence the range of the map $\mathcal{L}'_{(L,H')}$ is in $\mathcal{A}'$. Thus, given the semigroup $T_t \equiv e^{t\mathcal{L}}$ on $\mathcal{A}$, we are able to construct a family of semigroups $T_t'^{(L,H')} \equiv e^{t\mathcal{L}'_{(L,H')}}$ on $\mathcal{A}'$, indexed by L, H', such that each member of this family is 'conjugate' or 'dual' to T_t in some suitable sense. To make this precise, let us make the following definition.

Definition 6.3.1 A pair of uniformly continuous Q.D.S. (S_t, S_t') on $\mathcal{A}$ and $\mathcal{A}'$, respectively, are said to be *conjugate* to each other if there exist a Hilbert space $\mathcal{K}$, $*$-homomorphisms η of $\mathcal{A}$ and η' of $\mathcal{A}'$ into $\mathcal{B}(\mathcal{K})$, an η-derivation β of $\mathcal{A}$ and η'- derivation β' of $\mathcal{A}'$ into $\mathcal{B}(h, \mathcal{K})$ and self-adjoint elements $K \in \mathcal{A}$, $K' \in \mathcal{A}'$ such that the following holds.

(*i*) $\eta(x)$ and $\eta'(a)$ commute for each $x \in \mathcal{A}$ and $a \in \mathcal{A}'$.

(*ii*) There exist $W, W' \in \mathcal{B}(h, \mathcal{K})$ such that for all $x \in \mathcal{A}$ and $a \in \mathcal{A}'$, we have

$$\beta(x) = Wx - \eta(x)W, \quad \beta'(a) = W'a - \eta'(a)W',$$

$$Wa = \eta'(a)W, \text{ and } W'x = \eta(x)W'.$$

(*iii*) The generators $\mathcal{L}^S$ and $\mathcal{L}^{S'}$ of S_t and S_t' (respectively) have the forms

$$\mathcal{L}^S(x) = W^*\eta(x)W - \frac{1}{2}W^*Wx - \frac{1}{2}xW^*W + i[K, x],$$

and

$$\mathcal{L}^{S'}(a) = W'^*\eta'(a)W' - \frac{1}{2}W'^*W'a - \frac{1}{2}aW'^*W' + i[K', a].$$

It is clear that T_t and $T_t'^{(L,H')}$ are conjugate to each other for any L, H' according to the above definition. Note that in this definition, we do not require that either of the sets $\{\beta(x)u \ : x \in \mathcal{A}, u \in h\}$ and $\{\beta'(a)u \ : a \in \mathcal{A}', u \in h\}$ is total in $\mathcal{K}$. In fact, in the present context, there need not be any $L \in \mathcal{E}_\rho$ such that $\{\beta_L(a)u \ : a \in \mathcal{A}', u \in h\}$ is total in $\mathcal{K}$. For example, if $\mathcal{A} = \mathcal{B}(h)$, then $\mathcal{A}'$ is isomorphic to $\mathbb{C}$, and hence for any L, β_L will be identically zero. However, if $\mathcal{A}'$ is not too small compared to $\mathcal{A}$, one may expect that for some L, the above-mentioned sets may be total.

6.4 Appearance of Poisson terms in the dilation

Given the semigroup T_t, we have obtained a quantum stochastic dilation j_t which satisfies E–H-type Q.S.D.E. involving the deterministic (time) integrator $\mathcal{I}_{\mathcal{L}}(dt)$ and nondeterministic integrators $a_\delta(dt)$, $a_\delta^\dagger(dt)$ and $\Lambda_\sigma(dt)$. We shall now investigate the necessity of the Λ-coefficient, which we call conservation or Poisson term. We say that the semigroup T_t is *Poisson-free* if there exists an E–H dilation for T_t which has no Poisson term. It is clear that T_t with the generator $\mathcal{L}$ is Poisson-free if and only if it is possible to obtain a triplet of structure maps $(\mathcal{L}, \delta, \sigma)$ with σ being identically zero. It should be noted here that there may exist some other E–H dilations for a Poisson-free semigroup T_t which involves a nonzero Poisson term. We first state and prove a criterion for Poisson-free nature of a semigroup.

Theorem 6.4.1 *Let T_t be a uniformly continuous conservative Q.D.S. on a unital von Neumann algebra $\mathcal{A}$ with the generator $\mathcal{L}$. Denote by $\mathcal{Z}$ the center of $\mathcal{A}$. Let $D : \mathcal{A}\times\mathcal{A}\times\mathcal{A} \to \mathcal{A}$ be the trilinear map introduced in the proof of the Theorem 3.1.8 of Chapter 3, that is, $D(a, b, c) = \mathcal{L}(abc) - \mathcal{L}(ab)c - a\mathcal{L}(bc) + a\mathcal{L}(b)c$. Then, the following condition is necessary for T_t to be Poisson-free. For all $x, y \in \mathcal{A}$ and $z \in \mathcal{Z}$,*

$$D(x^*, z, y) = D(x^*, 1, y)z. \tag{6.7}$$

Furthermore, when either $\mathcal{A}$ or $\mathcal{A}'$ is commutative, then the above condition is also sufficient for T_t to be Poisson-free.

Proof:
To prove necessity, assume that T_t is Poisson-free. Hence there will exist a Hilbert space k_0 and a $(x \otimes 1_{k_0})$-derivation $\delta \in \mathcal{B}(\mathcal{A}, \mathcal{A} \otimes k_0)$ satisfying the cocycle identity $\mathcal{L}(x^*y) = \mathcal{L}(x^*)y + x^*\mathcal{L}(y) + \delta(x)^*\delta(y)$ for all $x, y \in \mathcal{A}$. From this it can be verified that $D(a, b, c) = \delta(a^*)^*(b \otimes 1_{k_0})\delta(c)$. But for $x \in \mathcal{A}$, $\delta(x) \in \mathcal{A}\otimes k_0$, which implies that $\delta(x)z = (z\otimes 1_{k_0})\delta(x)$ for all $z \in \mathcal{Z}$. Hence we have

$$D(x^*, z, y) = \delta(x)^*\delta(y)z = D(x^*, 1, y)z.$$

For the converse (sufficiency) part, recall the notations used in the proof of the Theorem 6.2.1. Clearly, $D(a, b, c) = \alpha(a^*)^*\rho(b)\alpha(c)$ for all $a, b, c \in \mathcal{A}$. For $z \in \mathcal{Z}$ and $x \in \mathcal{A}$, we have

$$\begin{aligned}
&(\alpha(x)z - \rho(z)\alpha(x))^*(\alpha(x)z - \rho(z)\alpha(x)) \\
&= z^*D(x^*, 1, x)z - D(x^*, z^*, x)z - z^*D(x^*, z, x) + D(x^*, z^*z, x) \\
&= 0,
\end{aligned}$$

using the condition (6.7) and the fact that z and $D(x^*, 1, x)$ commute. This shows that $\rho(z)\alpha(x) = \alpha(x)z$. However, since $\rho'(z)\alpha(x) = \alpha(x)z$, we obtain that $\rho(z)$ and $\rho'(z)$ agree on the total set of vectors of the form $\alpha(x)u, x \in \mathcal{A}, u \in h$ and hence $\rho(z) = \rho'(z)$ for all $z \in \mathcal{Z}$. As in the proof of the Theorem 6.2.1, write $\rho'(a) = \Sigma_2{}^*(a \otimes 1_{k_2})\Sigma_2$, for some Hilbert space k_2 and an isometry $\Sigma_2 : \mathcal{K} \to h \otimes k_2$ such that $P_2 = \Sigma_2\Sigma_2{}^*$ belongs to $\mathcal{A} \otimes \mathcal{B}(k_2)$. It has been shown that $\tilde{\delta}(\cdot) \equiv \Sigma_2\alpha(\cdot) \in \mathcal{A} \otimes k_2$, $\tilde{\pi}(\cdot) \equiv \Sigma_2\rho(\cdot)\Sigma_2{}^* \in \mathcal{A} \otimes \mathcal{B}(k_2)$. However, in the present situation, $\rho(z) = \rho'(z) = \Sigma_2{}^*(z \otimes 1_{k_2})\Sigma_2$ for all $z \in \mathcal{Z}$, and hence we have $\tilde{\pi}(z) = (z \otimes 1_{k_2})P_2$. Assume now that $\mathcal{A}$ is commutative, that is, $\mathcal{A} = \mathcal{Z}$. Since $P_2\tilde{\delta} = \tilde{\delta}$ and $\tilde{\delta}$ is a $\tilde{\pi}$-derivation, it is clear that $\tilde{\delta}$ is $(z \otimes 1_{k_2})$-derivation, which proves the existence of a Poisson-free E–H dilation.

In case when $\mathcal{A}' = \mathcal{Z}$, we choose an isometry $\Sigma' : \mathcal{K} \to h \otimes k'$ for some Hilbert space k', such that $\Sigma'\Sigma'^* \in \mathcal{Z} \otimes \mathcal{B}(k')$, and $\rho(x) = \Sigma'^*(x \otimes 1_{k'})\Sigma'$. Then $\rho'(z) = \Sigma'^*(z \otimes 1_{k'})\Sigma'$ for $z \in \mathcal{Z} = \mathcal{A}'$, and it follows that $\Sigma'\alpha(\cdot)$ is a $(x \otimes 1_{k'})$-derivation belonging to $\mathcal{A} \otimes \mathcal{B}(k_2)$. This completes the proof in case when $\mathcal{A}'$ is commutative. □

Let us now investigate the two extreme cases of von Neumann algebras, namely $\mathcal{B}(h)$ and commutative von Neumann algebras.

Corollary 6.4.2 *Any uniformly continuous conservative Q.D.S. on $\mathcal{B}(h)$ is Poisson-free. On the other hand, a uniformly continuous conservative Q.D.S. T_t on a commutative von Neumann algebra is Poisson-free if and only if T_t is trivial, that is, $T_t(x) = x$ for all $x \in \mathcal{A}$.*

Proof:
First consider the case $\mathcal{A} = \mathcal{B}(h)$. Since $\mathcal{A}' = \mathcal{Z}$ is isomorphic to $\mathbb{C}$, and $D(x, ., y)$ is clearly $\mathbb{C}$-linear, the condition (6.7) follows trivially.

Now, consider the case when $\mathcal{A}$ is commutative, and hence $\mathcal{Z} = \mathcal{A}$. Assume that (6.7) holds. We claim that this condition forces $\mathcal{L}$ to be a derivation. Take $z = y$ to be a projection in $\mathcal{A}$. By expanding both sides of the condition (6.7), we have:

$$\begin{aligned} \mathcal{L}(x^*y) - \mathcal{L}(x^*y)y - x^*\mathcal{L}(y) + x^*\mathcal{L}(y)y &= (\mathcal{L}(x^*y) - x^*\mathcal{L}(y) - \mathcal{L}(x^*)y)y, \\ \text{or,}\quad [\mathcal{L}(x^*y) - x^*\mathcal{L}(y)](1-y) &= [\mathcal{L}(x^*y) - x^*\mathcal{L}(y) - \mathcal{L}(x^*)y]y, \\ \text{or,}\quad [\mathcal{L}(x^*y) - x^*\mathcal{L}(y) - \mathcal{L}(x^*)y](1-y) &= [\mathcal{L}(x^*y) - x^*\mathcal{L}(y) - \mathcal{L}(x^*)y]y, \end{aligned}$$

where the last step follows because $y(1-y) = 0$. But the above implies

$$[\mathcal{L}(x^*y) - \mathcal{L}(x^*)y - x^*\mathcal{L}(y)](2y-1) = 0,$$

which shows (since $(2y-1)^2 = 1$) $\mathcal{L}(x^*y) = \mathcal{L}(x^*)y + x^*\mathcal{L}(y)$ for all $x \in \mathcal{A}$ and all projections $y \in \mathcal{A}$. However, the fact that any element of $\mathcal{A}$ can be approximated in the strong topology by a norm-bounded sequence of elements from the linear span of the projections in $\mathcal{A}$ and that $\mathcal{L}$ is continuous in strong topology on any norm-bounded set imply that $\mathcal{L}(x^*y) = \mathcal{L}(x^*)y + x^*\mathcal{L}(y)$ for all $x, y \in \mathcal{A}$, that is, $\mathcal{L}$ is a derivation of $\mathcal{A}$ into itself. But for von Neumann algebras, all derivations are inner (see [40]), that is, there is a self-adjoint $H \in \mathcal{A}$ such that $\mathcal{L}(x) = [H, x]$ for all $x \in \mathcal{A}$, which shows that $\mathcal{L}(x) = 0$ since $\mathcal{A}$ is commutative. Therefore, $T_t(x) = x$ for all $x \in \mathcal{A}$. □

However, the statement of the above corollary for commutative von Neumann algebras does not extend to the case when T_t is not uniformly continuous, that is, when its generator is unbounded. The simplest example is provided by the heat semigroup on $\mathcal{A} \equiv L^\infty(\mathbb{R})$. Let $w(t)$, $t \geq 0$ denote the one-dimensional standard Brownian motion defined on the Wiener space $(\Omega, \mathcal{F}, I\!P)$, where $I\!P$ denotes the one-dimensional Wiener measure. It is well-known (see [100]) that $L^2(I\!P)$ is naturally isomorphic with $\Gamma(L^2(\mathbb{R}_+))$. Denoting $L^2(\mathbb{R})$ by h, we define a time-indexed family j_t of $*$-homomorphisms from $\mathcal{A}$ to $\mathcal{A} \otimes \mathcal{B}(\Gamma(L^2(\mathbb{R}_+)) \cong L^\infty(\mathbb{R}) \otimes \mathcal{B}(L^2(I\!P))$ by setting $j_t(\phi)$ to be the multiplication by $\phi(\cdot + w(t))$ in $L^2(\mathbb{R}) \otimes L^2(I\!P)$. Denote furthermore by $\mathcal{A}^{(2)}$ the set of functions ϕ in $L^\infty(\mathbb{R})$ which have second-order continuous derivatives such that ϕ' and ϕ'' belong to $L^\infty(\mathbb{R})$. It is then clear that for $\phi \in \mathcal{A}^{(2)}$, $j_t(\phi)$ satisfies the stochastic differential equation

$$dj_t(\phi) = j_t(\phi')dw(t) + \frac{1}{2} j_t(\phi'')dt; \quad j_0(\phi) = \phi.$$

Translating this into the Fock space language and noting that $dw(t)$ corresponds to $a(dt) + a^\dagger(dt)$ (see [100]), we obtain the following E–H flow equation:

$$dJ_t = J_t \circ (a_\delta(dt) + a_\delta^\dagger(dt) + \mathcal{I}_\mathcal{L}(dt)), \quad J_0 = \mathrm{Id},$$

where J_t is defined by $J_t(\phi \otimes e(f))u = j_t(\phi)(ue(f))$ for $u \in h$, $f \in L^2(\mathbb{R}_+)$, and $\delta(\phi) = \phi'$, $\mathcal{L}(\phi) = \frac{1}{2}\phi''$ for $\phi \in \mathcal{A}^{(2)}$. This is clearly a Poisson-free E–H dilation of the semigroup generated by $\mathcal{L}$. We shall now show that it is possible to construct a sequence $T_t^{(n)}$ of uniformly continuous Q.D.S. on $\mathcal{A}$ which approximates T_t in a suitable sense and $T_t^{(n)}$ admits an E–H dilation $j_t^{(n)}$ such that the Poisson or Λ-term in the flow equation for $j_t^{(n)}$ tends to zero as $n \to \infty$ in an appropriate sense.

Let $\mathcal{L}_n : \mathcal{A} \to \mathcal{A}$ be defined by

$$\mathcal{L}_n(\phi)(x) = \frac{n}{2}(\phi(x + \frac{1}{\sqrt{n}}) + \phi(x - \frac{1}{\sqrt{n}}) - 2\phi(x)).$$

Clearly, for $\phi \in \mathcal{A}^{(2)}$, $\mathcal{L}_n(\phi)$ Converges to $\mathcal{L}(\phi)$ pointwise (since ϕ'' is continuous) and hence by an application of the Dominated Convergence theorem it also converges strongly in h. To obtain the structure maps for constructing an E–H dilation of $T_t^{(n)} \equiv e^{t\mathcal{L}_n}$, let us consider the unitary operator $\mathcal{T}_n$ in $\mathcal{B}(h)$ given by $\mathcal{T}_n f(x) = f(x + \frac{1}{\sqrt{n}})$. Let $k_0 = \mathbb{C}^2$ and $R_n \in \mathcal{B}(h, h \otimes k_0)$ be the operator

$$R_n f = \sqrt{n/2}((\mathcal{T}_n - I)f \oplus (\mathcal{T}_n^* - I)f).$$

Define a $*$-homomorphism π_n of $\mathcal{A}$ into $\mathcal{A} \otimes \mathcal{B}(k_0)$ by setting

$$\pi_n(\phi) = \begin{pmatrix} \mathcal{T}_n \phi \mathcal{T}_n^* & 0 \\ 0 & \mathcal{T}_n^* \phi \mathcal{T}_n \end{pmatrix}.$$

Since it can be observed that $\mathcal{T}_n \phi \mathcal{T}_n^*$ is multiplication by $\phi(\cdot + \frac{1}{\sqrt{n}})$ and $\mathcal{T}_n^* \phi \mathcal{T}_n$ is multiplication by $\phi(\cdot - \frac{1}{\sqrt{n}})$, it follows that $\pi_n \in \mathcal{B}(\mathcal{A}, \mathcal{A} \otimes \mathcal{B}(k_0))$. Now, define a π_n-derivation δ_n by setting $\delta_n(\phi) = R_n \phi - \pi_n(\phi) R_n$. Clearly,

$$\delta_n(\phi)(x) = \sqrt{n/2} \left\{ \phi\left(x + \frac{1}{\sqrt{n}}\right) - \phi(x), \phi\left(x - \frac{1}{\sqrt{n}}\right) - \phi(x) \right\},$$

which in particular shows that $\delta_n(\phi) \in \mathcal{A} \otimes k_0$. Then we verify that

$$\mathcal{L}_n(\phi) = R_n^* \pi_n(\phi) R_n - \frac{1}{2} R_n^* R_n \phi - \frac{1}{2} \phi R_n^* R_n.$$

Thus, having got the structure maps $(\mathcal{L}_n, \delta_n, \pi_n - \mathrm{Id})$, we can construct an E–H flow $j_t^{(n)}$ for $T_t^{(n)}$. The Λ-coefficient in this flow equation is $\pi_n - \mathrm{Id}$. It can be shown that for any continuous function ϕ in $\mathcal{A}$, $\pi_n(\phi) - \phi$ converges to 0 strongly in h. If ϕ is everywhere continuously differentiable with its derivative in $\mathcal{A}$, then this convergence will take place in the norm of $\mathcal{A}$. In fact, if we denote by $\mathcal{A}^\infty$ the set of all smooth functions having the derivatives of all order in $\mathcal{A}$, then for any fixed $\phi \in \mathcal{A}^\infty$, $\mathcal{L}_n(\phi)$, $\delta_n(\phi)$ and $\pi_n(\phi)$ converge respectively to $\mathcal{L}(\phi)$, $\tilde{\delta}(\phi)$ and $\tilde{\pi}(\phi)$ in norm, where $\tilde{\delta}(\phi) = \frac{1}{\sqrt{2}}(\phi', -\phi')$ and $\tilde{\pi}(\phi) = \begin{pmatrix} \phi & 0 \\ 0 & \phi \end{pmatrix}$. For ϕ in $\mathcal{A}^{(2)}$, the above convergence take place strongly in h.

Let $(w_1(t), w_2(t))$, $t \geq 0$ be a two-dimensional standard Brownian motion and $I\!P_2$ be the measure induced by this process. Then $L^2(I\!P_2) \cong \Gamma(L^2(\mathbb{R}_+ \otimes \mathbb{C}^2)) = \Gamma(L^2(\mathbb{R}_+, k_0))$. Define $\widetilde{w(t)} = \frac{1}{\sqrt{2}}(w_1(t) - w_2(t))$. Clearly, $\widetilde{w(t)}$ is a one-dimensional standard Brownian motion. If we define a map $\tilde{j}_t : \mathcal{A} \to \mathcal{A} \otimes \mathcal{B}(\Gamma(L^2(\mathbb{R}_+, k_0))) \cong \mathcal{A} \otimes \mathcal{B}(L^2(I\!P_2))$ by $\tilde{j}_t(\phi) =$ multiplication by $\phi(\cdot + \widetilde{w(t)})$, then this time-indexed family of $*$-homomorphisms satisfies the following flow equation:

$$d\tilde{j}_t(\phi) = \frac{1}{\sqrt{2}} \tilde{j}_t(\phi')(dw_1(t) - dw_2(t)) + \frac{1}{2} \tilde{j}_t(\phi'')dt, \quad \tilde{j}_0(\phi) = \phi;$$

where $\phi \in \mathcal{A}^{(2)}$. In the Fock-space language, this becomes the following:

$$d\tilde{J}_t = \tilde{J}_t \circ (a_{\tilde{\delta}}(dt) + a^{\dagger}_{\tilde{\delta}}(dt) + \mathcal{I}_{\mathcal{L}}(dt)), \quad \tilde{J}_0 = \mathrm{Id},$$

where $\tilde{J}_t(\phi \otimes e(f))u := \tilde{j}_t(\phi)(ue(f))$ for $u \in h$, $f \in L^2(\mathbb{R}_+, k_0)$. Thus we see that although the approximating sequence $j_t^{(n)}$ satisfies an E–H equation with Poisson terms, in the limit the contributions of Poisson terms disappear, giving a Poisson-free flow equation for $\tilde{j}_t$.

6.5 Implementation of E–H flow

Recall the notations of Theorems 6.2.1 and 6.2.3. We have for $x \in \mathcal{A}$, $\pi(x) = \Sigma^*(x \otimes 1_{k_0})\Sigma \in \mathcal{A} \otimes \mathcal{B}(k_0)$, $\delta(x) = Rx - \pi(x)R \in \mathcal{A} \otimes k_0$ for a suitable Hilbert space k_0, where $R \in \mathcal{B}(h, h \otimes k_0)$ and Σ is a partial isometry in $h \otimes k_0$. Now let us consider the H-P type Q.S.D.E.

$$dV_t = V_t(a^{\dagger}_R(dt) + \Lambda_{\Sigma^* - I}(dt) - a_{\Sigma R}(dt) + (iH - \frac{1}{2}R^*R)dt), \; V_0 = I. \quad (6.8)$$

Then by Theorem 5.3.2, there is a contraction-valued unique solution V_t as a regular process on $h \otimes \Gamma$. The following theorem shows that every E–H type flow J_t satisfying equation (5.45) is actually implemented by a process V_t satisfying equation (6.8).

Theorem 6.5.1 *The flow J_t satisfying the equation (5.45) is implemented by a partial isometry valued process V_t satisfying (6.8), that is,*

$$J_t(x \otimes e(f))u = V_t(x \otimes 1_{\Gamma})V_t^*(ue(f)).$$

Furthermore, the projection-valued processes $P_t \equiv V_t V_t^*$ and $Q_t \equiv V_t^* V_t$ belong to $\mathcal{A} \otimes \mathcal{B}(\Gamma)$ and $\mathcal{A}' \otimes \mathcal{B}(\Gamma)$, respectively.

The next lemma is preparatory to the proof of the theorem.

Lemma 6.5.2 *If $\mathcal{B}$ is a von Neumann algebra in $\mathcal{B}(\mathcal{H})$ for some Hilbert space $\mathcal{H}$ and p is a projection in $\mathcal{B}(\mathcal{H})$ such that $\mathcal{B} \ni x \mapsto pxp$ is a $*$-homomorphism of $\mathcal{A}$, then $p \in \mathcal{B}'$.*

Proof:
Let q be any projection in $\mathcal{B}$. We have by the hypothesis that,

$$pqp = pq^n p = (pqp)^n \quad \text{for all } n \geq 1.$$

But

$$(pqp)^n = \underbrace{pqp \cdot pqp \dots}_{\text{n times}} = (pq)^n p \xrightarrow{s} (p \wedge q)p = p \wedge q,$$

by von Neumann's Alternating projection theorem (see [128]), where $p \wedge q$ denotes the projection onto $\mathrm{Ran}(p) \cap \mathrm{Ran}(q)$. Thus we have, $(qp)^* qp = pqp = p \wedge q$, which implies that qp is a partial isometry with the initial set $\mathrm{Ran}(p \wedge q)$ and hence $qp.p \wedge q = qp$. But $qp.p \wedge q = p \wedge q$, and thus $qp = p \wedge q = (p \wedge q)^* = pq$. This completes the proof because $\mathcal{B}$ is generated by its projections. □

Proof of the theorem:
Setting $J'_t(x \otimes e(f))u = V_t(x \otimes 1_\Gamma)V_t^*(ue(f))$ for $u \in h, f \in \mathcal{C}$, and using equation (6.8) it can be verified that $J'_0 = \mathrm{Id}$ and J'_t satisfies the same flow equation (5.45) as does J_t with the triple $(\mathcal{L}, \delta, \sigma)$ given earlier. By the uniqueness of the solution of the initial value problem (5.45) we conclude that $J_t = J'_t$. Now, as in the Theorem 5.4.9, if we set $j_t(x)ue(f) = J_t(x \otimes e(f))u$, it follows that $j_t(x) = V_t(x \otimes 1_\Gamma)V_t^*$ and that $j_t(\cdot)$ is a $*$-homomorphism of $\mathcal{A}$. Therefore, $V_t(xy \otimes 1_\Gamma)V_t^* = V_t(x \otimes 1_\Gamma)Q_t(y \otimes 1_\Gamma)V_t^*$ for $x, y \in \mathcal{A}$. In particular, $P_t = j_t(1) = V_t V_t^*$ is a projection, that is, V_t is a partial isometry-valued regular process. It also follows from the same identity that $Q_t(xy \otimes 1_\Gamma)Q_t = Q_t(x \otimes 1_\Gamma)Q_t(y \otimes 1_\Gamma)Q_t$, that is, $x \otimes 1_\Gamma \mapsto Q_t(x \otimes 1_\Gamma)Q_t$ is a $*$-homomorphism of $\mathcal{A} \otimes 1_\Gamma$. Therefore by the Lemma 6.5.2, $Q_t \in (\mathcal{A} \otimes 1_\Gamma)' = \mathcal{A}' \otimes \mathcal{B}(\Gamma)$. □

6.6 Dilation on a C^*-algebra

Assume that $\mathcal{A}$ is a separable unital C^*-algebra in $\mathcal{B}(h)$ and T_t is a uniformly continuous quantum dynamical semigroup acting on $\mathcal{A}$ with the bounded generator $\mathcal{L}$. Recall the discussions of Hilbert C^*-modules and related topics in Chapter 4. Let k_0 be a separable Hilbert space, and fix an orthonormal basis $\{e_1, e_2, \dots\}$ of k_0. We denote the Hilbert modules $\mathcal{A} \otimes_{C^*} k_0$ and $\mathcal{A} \otimes_{C^*} \Gamma$ by F and W respectively, where $\Gamma = \Gamma(L^2(\mathbb{R}_+, k_0))$. The C^* algebras $\mathcal{A} \otimes \mathcal{K}(k_0)$ and $\mathcal{A} \otimes \mathcal{K}(\Gamma)$ will be denoted by $\mathcal{F}$ and $\mathcal{W}$ respectively. Note that the space of adjointable linear maps are given by, $\mathcal{L}(F) = \mathcal{M}(\mathcal{F})$ and $\mathcal{L}(W) = \mathcal{M}(\mathcal{W})$, where $\mathcal{M}(\mathcal{B})$ denotes the multiplier algebra of a C^* algebra $\mathcal{B}$ as defined in Chapter 2. If $\eta \in \mathcal{A} \otimes_{\mathrm{alg}} \mathcal{K}(k_0)$, say of the form $\eta = \sum_{i,j=1}^n x_{ij} \otimes |e_i\rangle\langle e_j|$, we note that $\eta_f \in \mathcal{A} \otimes_{\mathrm{alg}} k_0$ is given by

$$\eta_f = \sum_{ij=1}^{n} \langle e_j, f\rangle x_{ij} \otimes |e_i\rangle.$$

We observe that for all $u \in h$, $\|\eta_f u\|^2 = \sum_i \|x_i^f u\|^2$, where $x_i^f = \sum_j \langle e_j, f\rangle x_{ij} \in \mathcal{A}$. On the other hand, taking $a = \frac{1\otimes|f\rangle\langle f|}{\|f\|^2}$ if f is nonzero and $a = 0$ when $f = 0$, we see that $\|\eta a(u \otimes f)\|^2 = \|\frac{1}{\|f\|^2}\sum_{i,j}\langle e_j, f\rangle x_{ij} u \otimes |e_i\rangle\langle f, f\rangle\|^2 = \sum_i \|x_i^f u\|^2$. That is, $\|\eta_f u\| = \|(\eta a)(u\otimes f)\| \leq \|\eta a\|\|u\|\|f\|$, and hence $\|\eta_f\| \leq \|\eta a\|\|f\|$. This shows that given an element $\eta \in \mathcal{L}(F) = \mathcal{M}(\mathcal{F})$ and $f \in k_0$, if we choose a net $\eta^{(\alpha)}$ from $\mathcal{A} \otimes_{\text{alg}} \mathcal{K}(k_0)$ converging in the strict topology of the multiplier to η, then $\eta_f^{(\alpha)}$ will be a Cauchy net in norm topology. This shows that η_f belongs to F for all $f \in k_0$ and $\eta \in \mathcal{L}(F)$. Furthermore, for a bounded linear map $\sigma : \mathcal{A} \to \mathcal{L}(F)$ and $f \in k_0$ we set the map $\sigma_f : \mathcal{A} \to F$ by $\sigma_f(x) = \sigma(x)_f$.

The next theorem proves the existence of a canonical E–H dilation.

Theorem 6.6.1 *Given a uniformly continuous, conservative quantum dynamical semigroup T_t on a separable unital C^*-algebra $\mathcal{A}$ with generator $\mathcal{L}$, there exists a separable Hilbert space k_0 and a $*$-homomorphism $\pi : \mathcal{A} \to \mathcal{L}(F) = \mathcal{M}(\mathcal{F})$, a π-derivation $\delta : \mathcal{A} \to F$ such that $\mathcal{L}(x^*y) - \mathcal{L}(x^*)y - x^*\mathcal{L}(y) = \delta(x)^*\delta(y)$.*

Furthermore, we can extend the maps $\mathcal{L}, \delta, \pi$ to the universal enveloping von Neumann algebra $\tilde{\mathcal{A}}$ of $\mathcal{A}$ such that the E-H type Q.S.D.E.

$$dJ_t = J_t \circ (a_\delta(dt) + a_\delta^\dagger(dt) + \Lambda_{\pi-\text{Id}}(dt) + \mathcal{I}_\mathcal{L}(dt))$$

with the initial condition $J_0 \equiv \text{Id}$ admits a unique solution as a map from $\tilde{\mathcal{A}} \otimes \Gamma$ to itself. The restriction of J_t on W takes value in W and there is a $*$-homomorphism $j_t : \mathcal{A} \to \mathcal{L}(W) = \mathcal{M}(\mathcal{W})$ satisfying $j_t(x)(ue(f)) = J_t(x \otimes e(f))u$ for all $x \in \mathcal{A}$, $u \in \tilde{h}$, $f \in L^2(\mathbb{R}_+, k_0)$, where $\tilde{h}$ denotes the universal enveloping GNS space of $\mathcal{A}$.

Proof:
Let us embed $\mathcal{A}$ in its universal enveloping GNS space $\tilde{h}$, where the weak closure of the image of $\mathcal{A}$ in $\tilde{h}$ is the universal enveloping von Neumann algebra $\tilde{\mathcal{A}}$ of $\mathcal{A}$. We identify $\mathcal{A}$ with its embedding inside $\mathcal{B}(\tilde{h})$. By the Theorem 3.1.8 and the Remark 3.1.10, we obtain a Hilbert space $\mathcal{K}$, a $*$-homomorphism $\rho : \mathcal{A} \to \mathcal{B}(\mathcal{K})$, a ρ-derivation $\alpha : \mathcal{A} \to \mathcal{B}(\tilde{h}, \mathcal{K})$ such that $\mathcal{L}(x^*y) - \mathcal{L}(x^*)y - x^*\mathcal{L}(y) = \alpha(x)^*\alpha(y)$. Consider the Hilbert $\mathcal{A}$-module E defined as the closure of the algebraic linear span of elements of the form $\alpha(x)y$, where $x, y \in \mathcal{A}$, with respect to the operator norm of $\mathcal{B}(\tilde{h}, \mathcal{K})$. $\mathcal{A}$ acts on E by the usual right multiplication, and the inner product of E is inherited from that of $\mathcal{B}(\tilde{h}, \mathcal{K})$, namely, $\langle L, M\rangle = L^*M$ for $L, M \in E$. We note that $\langle \alpha(x)y, \alpha(a)b\rangle = y^*\alpha(x)^*\alpha(a)b \in \mathcal{A}$ for $x, y, a, b \in \mathcal{A}$, and thus E is indeed a Hilbert $\mathcal{A}$-module. We identify ρ with a left action $\hat{\rho}$ given by, $\hat{\rho}(x)(\alpha(y)) = \alpha(xy) - \alpha(x)y$ and extending it $\mathcal{A}$-linearly. Furthermore, since $\mathcal{A}$ is separable, E is countably generated as

a Hilbert $\mathcal{A}$-module. To see this, one can choose any countable total family $\{x_i\}$ of $\mathcal{A}$ and note that E is the closed $\mathcal{A}$-linear span of $\{\alpha(x_i)\}$. By the Theorem 4.1.10, we obtain a separable Hilbert space k_0, an isometric $\mathcal{A}$-linear map $t : E \to F$ which embeds E as a complemented closed submodule of F. Clearly, by the Theorem 4.1.11, $t\hat{\rho}(x)t^* \in \mathcal{L}(F)$. We set $\delta(x) = t(\alpha(x))$ and $\pi(x) = t\hat{\rho}(x)t^*$ to complete the first part of the proof.

For the second part, first note that π being a $*$-homomorphism it admits an extension as a $*$-homomorphism of $\tilde{\mathcal{A}}$ into $(\mathcal{L}(F))'' = \tilde{\mathcal{A}} \otimes \mathcal{B}(k_0)$. Also note that by the Theorem 3.1.8 and the Remark 3.1.10, we obtain R in the ultra-weak closure of $\{\alpha(x)y : x, y \in \mathcal{A}\}$ such that $\mathcal{L}(x) = R^*\hat{\rho}(x)R - \frac{1}{2}R^*Rx - \frac{1}{2}xR^*R + i[H, x]$ for some self-adjoint element $H \in \tilde{\mathcal{A}}$ and $\alpha(x) = Rx - \hat{\rho}(x)R$ for all $x \in \mathcal{A}$. Setting $\tilde{R} = tR$ we observe that *(i)* $\tilde{R}$ is in the ultra-weak closure of $\{\delta(x)y : x, y \in \mathcal{A}\}$, *(ii)* $\delta(x) = \tilde{R}x - \pi(x)\tilde{R}$, and *(iii)* $\mathcal{L}(x) = \tilde{R}^*\pi(x)\tilde{R} - \frac{1}{2}\tilde{R}^*\tilde{R}x - \frac{1}{2}x\tilde{R}^*\tilde{R} + i[H, x]$. Next we extend $\mathcal{L}$ and δ to $\tilde{\mathcal{A}}$, retaining the same notation for both. Since $\tilde{R}$ is in particular in the ultra-weak closure of F, which is same as $\tilde{\mathcal{A}} \otimes k_0$, we see that the extended maps $\mathcal{L}$ and δ map $\tilde{\mathcal{A}}$ into itself. Now, by the results of Section 5.4, we obtain J_t and j_t as mentioned in the statement of the present theorem. It remains to show that $J_t(x \otimes e(f)) \in W$ for $x \in \mathcal{A}$, and $j_t(x) \in \mathcal{L}(W)$.

To this end, first note the following. If $\beta \in \mathcal{B}(\tilde{\mathcal{A}}, \tilde{\mathcal{A}} \otimes k_0)$ such that $\beta(x) \in F$ for all $x \in \mathcal{A}$, then we claim that $a_\beta^\dagger(\Delta)(x \otimes e(f))$ belongs to W for any bounded subinterval Δ of $\mathbb{R}_+$ and $x \in \mathcal{A}$, $f \in L^2(\mathbb{R}_+, k_0)$. This follows from the definition of $a_\beta^\dagger(\Delta)(x \otimes f^{\otimes^n})$ which belongs to W. This enables us to prove that $a_\delta^\dagger(\cdot)$ and $\Lambda_{\pi-\mathrm{Id}}(\cdot)$ maps $x \otimes e(f)$ into W for $x \in \mathcal{A}$. Similarly, one can show that $a_\delta(\cdot)$ and $\mathcal{I}_\mathcal{L}(\cdot)$ also have the same property. Now, recall the iterates $J_t^{(n)}$ constructed in the proof of Theorem 5.4.9 of Chapter 5 and by the estimates made in that proof, it is clear that

$$\|(J_t(x \otimes e(f)) - \sum_{m \le n} J_t^{(m)}(x \otimes e(f)))u\| \le \|u\|\|x\|\|e(f)\|\|E_t^f\| \left(\sum_{m=n+1}^{\infty} C^{\frac{m}{2}} (m!)^{-\frac{1}{2}} \right)^{\frac{1}{2}},$$

for some constant C, and thus,

$$\|J_t(x \otimes e(f)) - \sum_{m \le n} J_t^{(m)}(x \otimes e(f))\| \to 0.$$

But by iterative construction, each $J_t^{(m)}(x \otimes e(f))$ belongs to W.

We now prove that $j_t(x) \in \mathcal{L}(W) = \mathcal{M}(\mathcal{W})$ for $x \in \mathcal{A}$. By definition of the multiplier algebra, we have to show that for all $A \in \mathcal{W}$, $j_t(x)A$ and $Aj_t(x)$ belong to $\mathcal{W}$. For this it is enough to verify that $j_t(x)A \in \mathcal{W}$ for all $x \in \mathcal{A}$ and all $A \in \mathcal{W}$, since $\mathcal{W}$ is $*$-closed and $j_t(x^*) = j_t(x)^*$. Since $\mathcal{W}$ is the operator-norm closure of the finite linear combinations of elements of the form $y \otimes |e(f)\rangle\langle e(g)|$ for $y \in \mathcal{A}$, $f, g \in L^2(\mathbb{R}_+, k_0)$, it suffices to show that for fixed $x \in \mathcal{A}$ and $t \geq 0$, $j_t(x)(y \otimes |e(f)\rangle\langle e(g)|)$ is in $\mathcal{W}$ for $y \in \mathcal{A}$, $f, g \in L^2(\mathbb{R}_+, k_0)$. Since $J_t(x \otimes e(f)) \in W = \mathcal{A} \otimes_{C^*} \Gamma$, we can choose a sequence L_n of the form $\sum_{i=1}^{k_n} z_i^{(n)} \otimes \rho_i^{(n)}$ for $z_i^{(n)} \in \mathcal{A}$, $\rho_i^{(n)} \in \Gamma$ such that L_n converges in the norm of W to $J_t(x \otimes e(f))$. Now, observe that for $u \in \tilde{h}$ and $\eta \in \Gamma$,

$$\begin{aligned} j_t(x)(y \otimes |e(f)\rangle\langle e(g)|)(u \otimes \eta) &= \langle e(g), \eta\rangle J_t(x \otimes e(f))yu \\ &= \lim_{n\to\infty} \langle e(g), \eta\rangle L_n yu. \end{aligned}$$

Choose an orthonormal basis $\{\gamma_l\}$ of Γ and take a vector $\zeta \equiv \sum_l \zeta_l \otimes \gamma_l$ of $\tilde{h} \otimes \Gamma$, where $\zeta_i \in \tilde{h}$. It can be observed that

$$\begin{aligned} &\|\{j_t(x)(y \otimes |e(f)\rangle\langle e(g)|) - \sum_{i=1}^{k_n} (z_i^{(n)} y \otimes |\rho_i^{(n)}\rangle\langle e(g)|)\}\zeta\| \\ &\leq \sum_l |\langle e(g), \gamma_l\rangle| \|(J_t(x \otimes e(f))y - L_n y)\zeta_l\| \\ &\leq \|J_t(x \otimes e(f)) - L_n\| \|y\| \left(\sum_l |\langle e(g), \gamma_l\rangle|^2\right)^{\frac{1}{2}} \left(\sum_l \|\zeta_l\|^2\right)^{\frac{1}{2}} \\ &= \|J_t(x \otimes e(f)) - L_n\| \|y\| \|e(g)\| \|\zeta\|, \end{aligned}$$

and hence $j_t(x)(y \otimes |e(f)\rangle\langle e(g)|)$ is norm-limit of $\sum_{i=1}^{k_n} z_i^{(n)} y \otimes |\rho_i^{(n)}\rangle\langle e(g)|$, which belongs to $\mathcal{A} \otimes_{\text{alg}} \mathcal{K}(\Gamma)$. This completes the proof. □

6.7 Covariant dilation theory

Let us consider an additional structure in our set-up, namely that coming from the action of a locally compact group G, and let $\alpha : G \to \text{Aut}(\mathcal{A})$ be its action on $\mathcal{A}$. Furthermore, let T_t be a covariant quantum dynamical semigroup, that is,

$$T_\circ \alpha_g = \alpha_g \circ T_t \text{ for all } t \geq 0,\ g \in G,$$

or equivalently,

$$\mathcal{L} \circ \alpha_g = \alpha_g \circ \mathcal{L} \text{ for all } g \in G. \tag{6.9}$$

It is then natural to ask whether a covariant dilation exists, that is, if the dilation is given by $j_t : \mathcal{A} \to \mathcal{B}$, the question is whether we can obtain an action β of G on $\mathcal{B}$ such that $j_t(\alpha_g(x)) = \beta_g(j_t(x))$ for all $g \in G$, $x \in \mathcal{A}$ and $t \geq 0$. In the present section, we investigate this problem.

In this section, we fix a C^* or von Neumann algebra $\mathcal{A} \subseteq \mathcal{B}(h)$ (where h is some Hilbert space), a locally compact group G and an action $\alpha : G \to \text{Aut}(\mathcal{A})$

which is strongly continuous in the sense that $g \mapsto \alpha_g(x)$ is continuous in the topology of $\mathcal{A}$, which is the norm topology in case $\mathcal{A}$ is considered as a C^*-algebra, and the ultra-weak topology if $\mathcal{A}$ is a von Neumann algebra.

6.7.1 Covariant E–H theory for a C^*-algebra

In this subsection we assume that $\mathcal{A}$ is a separable unital C^*-algebra, h is the universal enveloping GNS space as in the previous section, T_t is uniformly continuous with the generator $\mathcal{L}$ (assume without loss of generality that $\mathcal{L}(1) = 0$), which is covariant under α in the sense that $\mathcal{L} \circ \alpha_g = \alpha_g \circ \mathcal{L}$. For simplicity we shall assume that α is implemented by a unitary representation v_g in h and that G is compact. We now prove the following main result of this subsection.

Theorem 6.7.1 *Let $\mathcal{A}$ be a separable unital C^*-algebra and let (T_t) be a uniformly continuous Q.D.S. on $\mathcal{A}$, covariant with respect to the action of a compact group G on $\mathcal{A}$. Then there exists an E–H dilation $j_t : \mathcal{A} \to \mathcal{A}'' \otimes \mathcal{B}(\Gamma(L^2(\mathbb{R}_+, k_0)))$ (where k_0 is a separable Hilbert space), as defined in Definition 6.0.2, such that j_t is covariant with respect to an appropriate G-action on the range of j_t.*

Proof:
Our starting point is similar to that of the Theorem 6.6.1. As before, we get a Hilbert space $\mathcal{K}$, a $*$-representation $\pi_0 : \mathcal{A} \to \mathcal{B}(\mathcal{K})$ and a π_0-derivation $\delta_0 : \mathcal{A} \to \mathcal{B}(h, \mathcal{K})$ satisfying the relations :

$$\begin{aligned} \delta_0(x^*)^*\delta_0(y) &= \mathcal{L}(xy) - \mathcal{L}(x)y - x\mathcal{L}(y), \\ \pi_0(x)\delta_0(y) &= \delta_0(xy) - \delta_0(x)\pi_0(y). \end{aligned} \tag{6.10}$$

Furthermore, the vectors of the form $\delta_0(x)u$ where $x \in \mathcal{A}$ and $u \in h$ are total in $\mathcal{K}$. As in the Theorem 6.6.1, we consider the Hilbert $\mathcal{A}$-module E obtained by completing the pre-Hilbert module E_0 spanned by $\{\delta_0(x)y,\ x, y \in \mathcal{A}\}$ in $\mathcal{B}(h, \mathcal{K})$ in the operator norm inherited from $\mathcal{B}(h, \mathcal{K})$. We also make E into an $\mathcal{A} - \mathcal{A}$ bimodule by setting the left action as the left multiplication by $\pi_0(x)$ for $x \in \mathcal{A}$.

Now we define an action of G on E_0 by :

$$\gamma_g\left(\sum_{i=1}^{n} \delta_0(x_i)y_i\right) \equiv \sum_{i=1}^{n} \delta_0(\alpha_g(x_i))\alpha_g(y_i);$$

where $n \geq 1,\ x_i, y_i \in \mathcal{A}$ for all i and $g \in G$. For $\xi, \eta \in E_0$, it is possible to verify by using the relations (6.10) and the covariance property (6.9) of $\mathcal{L}$ that

$$\langle \gamma_g(\xi), \gamma_g(\eta) \rangle = \alpha_g(\langle \xi, \eta \rangle),$$

which implies that $\|\gamma_g(\xi)\| = \|\xi\|$, and thus γ_g is well defined and extends to an isometric complex-linear map on E satisfying $\gamma_g(\xi b) = \gamma_g(\xi)\alpha_g(b)$ for all $\xi \in E$, $b \in \mathcal{A}$. It also has a countable set of generators $\{\delta_0(x_i)\}_{i=1}^{\infty}$, where $\{x_i\}_{i=1}^{\infty}$ is any countable dense subset of the separable C^*-algebra $\mathcal{A}$. Furthermore,

$$\begin{aligned}
&\gamma_g(\pi_0(x)\delta_0(y)) \\
&= \gamma_g(\delta_0(xy)) - \gamma_g(\delta_0(x)y) \\
&= \delta_0(\alpha_g(xy)) - \delta_0(\alpha_g(x))\alpha_g(y) \\
&= \pi_0(\alpha_g(x))\delta_0(\alpha_g(y)) \\
&= \pi_0(\alpha_g(x))\gamma_g(\delta_0(y)),
\end{aligned} \tag{6.11}$$

which implies that $\gamma_g\pi_0(x) = \pi_0(\alpha_g(x))\gamma_g$. Thus, it is clear that (E, γ) is a Hilbert C^* $G - \mathcal{A}$ module, satisfying the conditions for applying the covariant Kasparov's theorem (Theorem 4.3.3). By this theorem and the Remark 4.3.4, we get a separable Hilbert space k_0 (which can be taken to be $\mathcal{H} \otimes L^2(G)$ in the notation of the Remark 4.3.4), a representation w_g of G in k_0 and an $\mathcal{A}$-linear isometry $\Sigma : E \to \mathcal{A} \otimes k_0$ such that $\Sigma \circ \gamma_g = (\alpha_g \otimes w_g) \circ \Sigma$. It follows that $\hat{\pi}(x) \equiv \Sigma\pi_0(x)\Sigma^* \in \mathcal{L}(\mathcal{A} \otimes k_0)$ and $\hat{\delta}(x) \equiv \Sigma\delta(x) \in \mathcal{A} \otimes k_0$ for all $x \in \mathcal{A}$. We define the structure matrix (as in (5.40)) $\Theta : \mathcal{A} \to \mathcal{B}(h \otimes \hat{k_0})$ by

$$\Theta(x) = \begin{pmatrix} \mathcal{L}(x) & \hat{\delta}^\dagger(x) \\ \hat{\delta}(x) & \hat{\pi}(x) - x \otimes 1_{k_0} \end{pmatrix},$$

where $\hat{\delta}^\dagger(x) = (\hat{\delta}(x^*))^*$. Let $\tilde{w_g}$ be the representation of G in $\hat{k_0}$ defined by $\tilde{w_g} = 1 \oplus w_g$, and let $V_g = v_g \otimes \tilde{w_g}$ in $h \otimes \hat{k_0}$. It can be verified that

$$V_g\Theta(x)V_g^* = \Theta(\alpha_g(x)). \tag{6.12}$$

As in the proof of the Theorem 6.6.1, we can show that there exists an E–H dilation $j_t : \mathcal{A} \to \mathcal{B}(h \otimes \Gamma(L^2(\mathbb{R}_+, k_0)))$ with the structure matrix Θ. Let us define a unitary representation $\mathcal{U}_g$ on $h \otimes \Gamma(L^2(\mathbb{R}_+, k_0))$ by setting $\mathcal{U}_g = v_g \otimes \Gamma(I_{L^2(\mathbb{R}_+)} \otimes w_g)$, where $\Gamma(U)$ denotes the second quantization of a unitary operator U on $L^2(\mathbb{R}_+, k_0)$. We claim that $\mathcal{U}_g j_t(x)\mathcal{U}_g^* = j_t(\alpha_g(x))$. For $\phi, \psi \in k_0$, we define a bounded linear map $\Theta^{\phi,\psi} : \mathcal{A} \to \mathcal{A}$ by setting

$$\langle v, \Theta^{\phi,\psi}(x)u\rangle = \langle v \otimes (1 \oplus \phi), \Theta(x)(u \otimes (1 \oplus \psi))\rangle$$

for all $x \in \mathcal{A}$, $u, v \in h$. By (6.12), we have the covariance property:

$$\Theta^{\phi,\psi}(\alpha_g(x)) = \alpha_g(\Theta^{w_g^*\phi, w_g^*\psi}(x)) \text{ for all } g \in G. \tag{6.13}$$

Now, to verify our claim, it is enough to show that

$$\langle ue(f), j_t(\alpha_g(x))u'e(f')\rangle = \langle ue(f), \mathcal{U}_g j_t(x)\mathcal{U}_g^* u'e(f')\rangle$$

for all $u, u' \in h;\ f, f' \in L^2(\mathbb{R}_+, k_0)$. But we have by (6.13) that

$$\begin{aligned}
&\frac{d}{dt}\langle ue(f), j_t(\alpha_g(x))u'e(f')\rangle \\
&= \langle ue(f), j_t(\Theta^{f(t),f'(t)}(\alpha_g(x)))u'e(f')\rangle \\
&= \langle ue(f), j_t(\alpha_g(\Theta^{w_g^* f(t), w_g^* f'(t)}(x)))u'e(f')\rangle. \qquad (6.14)
\end{aligned}$$

On the other hand, if we set $\eta_t^g(x) = \mathcal{U}_g j_t(x)\mathcal{U}_g^*$, then as in (6.14) we have

$$\begin{aligned}
&\frac{d}{dt}\langle ue(f), \eta_t^g(x)u'e(f')\rangle \\
&= \frac{d}{dt}\langle v_g^* ue((I\otimes w_g^*)f), j_t(x)v_g^* u'e((I\otimes w_g^*)f')\rangle \\
&= \langle v_g^* ue((I\otimes w_g^*)f), j_t(\Theta^{w_g^* f(t), w_g^* f'(t)}(x))v_g^* u'e((I\otimes w_g^*)f')\rangle \\
&= \langle ue(f), \eta_t^g(\Theta^{w_g^* f(t), w_g^* f'(t)}(x))u'e(f')\rangle.
\end{aligned}$$

Let us fix $f, f' \in L^\infty(\mathbb{R}_+, k_0) \cap L^2(\mathbb{R}_+, k_0)$ and set $\Psi_t : \mathcal{A} \to \mathcal{A}$ by

$$\langle u, \Psi_t(x)u'\rangle = \langle ue(f), (j_t(\alpha_g(x)) - \eta_t^g(x))u'e(f')\rangle.$$

Clearly we have

$$\frac{d}{dt}\Psi_t = \Psi_t \circ \Theta^{w_g^* f(t), w_g^* f'(t)}, \qquad (6.15)$$

with the initial condition $\Psi_0(x) = 0$. Since $\|\Psi_t\| \le 2$, and since for any $g \in G$ the map $\Theta^{w_g^* f(\cdot), w_g^* f'(\cdot)} : \mathcal{A} \to \mathcal{A}$ is bounded uniformly in t, we conclude by iterating the relation

$$\Psi_t(x) = \int_o^t \Psi_s(\Theta^{w_g^* f(s), w_g^* f'(s)}(x))ds$$

that $\Psi_t \equiv 0$ for all t and this completes the proof of the theorem. □

6.7.2 The von Neumann algebra case

Here we assume that $\mathcal{A} \subseteq \mathcal{B}(h)$ is a von Neumann algebra, $\alpha_g(x) = v_g x v_g^*$ is the action of a locally compact (not necessarily compact) group G on $\mathcal{A}$ for some unitary representation v_g of G in h, and $T_t = e^{t\mathcal{L}}$ is a covariant, normal Q.D.S. on $\mathcal{A}$, which is uniformly continuous in t. Our aim is to show the existence of a covariant E–H dilation for (T_t). For this, our approach will be very similar to the C^*-algebra case, only difference being that we are now going to use the von Neumann algebra version of the equivariant Kasparov theorem (Theorem 4.3.5).

Theorem 6.7.2 *The Q.D.S. (T_t) on the von Neumann algebra $\mathcal{A}$ equipped with the action α admits a covariant E–H dilation.*

Proof:
The proof is very similar to the proof of the previous theorem and therefore only the main points are briefly sketched. As before, we apply the result of Christensen–Evans to get π_0 and δ_0 and then consider the Hilbert von Neumann bimodule E by taking the closure of the algebraic linear span of the set $\{\delta_0(x)y, x, y \in \mathcal{A}\}$ with respect to the strong operator topology. Clearly there is a natural G-action on this Hilbert von Neumann bimodule so that it becomes a Hilbert von Neumann $G-\mathcal{A}$ module satisfying the conditions of the Theorem 4.3.5. Then by applying Theorem 4.3.5, we obtain a multiplicity space k_0 and the required covariant structure maps as in the case of the C^*-algebra setting. □

Let us now assume that G is amenable and note that we can write the generator $\mathcal{L}$ of the given Q.D.S. as

$$\mathcal{L}(x) = R^*\pi(x)R - \frac{1}{2}R^*Rx - \frac{1}{2}xR^*R + i[H, x].$$

By the Christensen–Evans theorem (Theorem 3.1.8) R belongs to the ultra-weak closure of span$\{\delta(x)y, x, y \in \mathcal{A}\}$ and hence $R \in E$, and H is a self-adjoint element of $\mathcal{A}$. We consider the E-valued map on G given by $g \mapsto R(g) \equiv \gamma_g(R)$. Since G is amenable, set $\hat{R} = \eta(R(\cdot))$, where η denotes an invariant mean on E-valued functions on G. Thus by (6.11) we note that

$$\begin{aligned}\pi(x)\gamma_g(R) &= \gamma_g\pi(\alpha_{g^{-1}}(x))R \\ &= \gamma_g\left(-\delta(\alpha_{g^{-1}}(x)) + R\alpha_{g^{-1}}(x)\right) = -\delta(x) + \gamma_g(R)x;\end{aligned}$$

and by taking the mean η on both sides, we have that $\hat{R}x - \pi(x)\hat{R} = \delta(x)$.

Recalling from Theorem 6.7.1 the isometric embedding Σ of E into $\mathcal{A} \otimes k_0 \equiv \mathcal{A} \otimes k_1 \otimes L^2(G)$ and assuming for simplicity that k_1 is separable, we can write $\Sigma\hat{R} = \sum_i S_i \otimes e_i$, where $\{e_i\}_i$ is an orthonormal basis of $L^2(G)$ and each $S_i \in \mathcal{A} \otimes k_1$. Then a simple calculation shows that

$$(\alpha_g \otimes I_{k_1})(S_i) = \sum_j \langle e_j, L_g e_i\rangle S_j.$$

Furthermore, if G is compact (for example, the group of rotations), then $L^2(G)$ can be decomposed as $\oplus_{\mu \in \hat{G}} h^{(\mu)} \otimes \mathbb{C}^{n_\mu}$ with $L_g = \oplus_\mu L_g^{(\mu)} \otimes I_{\mathbb{C}^{n_\mu}}$, where each $L_g^{(\mu)}$ is an irreducible representation of the group G in $h^{(\mu)}$ with multiplicity n_μ and dim $h^{(\mu)} < \infty$. In such a case, we have the following:

$$\Sigma\hat{R} = \sum_{\mu \in \hat{G}} \sum_{i=1}^{\dim h^{(\mu)}} S_i^{(\mu)} \otimes e_i^{(\mu)}, \text{ and } (\alpha_g \otimes I_{k_1})(S_i^{(\mu)}) = \sum_j \langle e_j^{(\mu)}, L_g^{(\mu)} e_i^{(\mu)}\rangle S_j^{(\mu)}.$$

7
Quantum stochastic calculus with unbounded coefficients

In this chapter, we shall use the notations of Chapter 5 and shall adapt the methods developed there to solve Q.S.D.E. with unbounded operator coefficients. For technical reasons, we work with separable Hilbert spaces (that is, both the initial and the noise spaces are separable). Most of the materials of this chapter are taken from [92], [94] and [95].

7.1 Notation and preliminary results

Let h and k_0 be separable Hilbert spaces and let $\mathcal{D}_0, \mathcal{V}_0$ be dense subspaces of h and k_0 respectively. Let us recall the framework for definition of basic processes and left quantum stochastic integrals with respect to them as introduced in Chapter 5. Choose and fix an orthonormal basis $\{e_1, e_2, \cdots\} \subseteq \mathcal{V}_0 \subseteq k_0$, and for $t \geq 0$, define the *time reversal operator* R_t on $k = L^2(\mathbb{R}_+, k_0)$ by

$$(R_t f)(s) := f(t-s)\chi_{[0,t]}(s) + f(s)\chi_{(t,\infty)}(s).$$

Also set $\mathcal{U}_t := \Gamma(R_t)$, where we remind the reader that $\Gamma(B)$ denotes the second quantization of a bounded operator B. In this case, note that both R_t and $\mathcal{U}_t$ are reflections (self-adjoint unitary). Recall also that we denote by Γ, Γ_t and Γ^t the Hilbert spaces $\Gamma(L^2(\mathbb{R}_+, k_0))$, $\Gamma(L^2([0,t], k_0))$ and $\Gamma(L^2((t,\infty), k_0))$, respectively, and define the 'time-shift' operator $\theta_t : L^2(\mathbb{R}_+) \to L^2(\mathbb{R}_+)$ $(t \geq 0)$ as follows:

$$(\theta_t f)(x) := f(x-t)\chi_{[t,\infty)}(x), \quad f \in L^2(\mathbb{R}_+).$$

Then $\Gamma(\theta_t)$ maps Γ onto Γ^t isometrically for every $t \geq 0$, and if we denote an operator B on Γ and its trivial ampliation $I_h \otimes B$ on $h \otimes \Gamma$ by the same notation, then it is clear that for $X \in \mathcal{B}(h \otimes \Gamma_s)$ and $t \geq 0$, the operator

$\Gamma(\theta_t)(X \otimes I_{\Gamma^s})\Gamma(\theta_t^*)$ is of the form $P_{12}(|\Omega_t >< \Omega_t| \otimes \hat{X} \otimes I_{\Gamma^{t+s}})P_{12}$ for some $\hat{X} \in \mathcal{B}(h \otimes \Gamma(L^2([t, t+s], k_0)))$, where $\Omega_t = e(0_{[0,t]})$ and P_{12} denotes the canonical isomorphism from $h \otimes \Gamma_t \otimes \Gamma^t$ onto $\Gamma_t \otimes h \otimes \Gamma^t$ which permutes the first two tensor components. We denote by $[\Gamma(\theta_t)(X \otimes I)\Gamma(\theta_t^*)]$ (or simply $[\Gamma(\theta_t)X\Gamma(\theta_t^*)]$) the operator $P_{12}(I_{\Gamma_t} \otimes \hat{X} \otimes I_{\Gamma^{t+s}})P_{12}$, that is, the projection onto the vacuum vector up to time t in the earlier expression is replaced by I_{Γ_t}. Even when X is not bounded, but densely defined and closable, so that the operator $S = P_{12}(I_{\Gamma_t} \otimes \hat{X} \otimes I_{\Gamma^{t+s}})P_{12}$ is also densely defined and closable, we denote the closure of S by $[\Gamma(\theta_t)X\Gamma(\theta_t^*)]$, as was done for a bounded X. To make the notation clear, let us observe that for $X = B \otimes |e(\alpha) >< e(\beta)|$, where $B \in \mathcal{B}(h)$, $\alpha, \beta \in L^2([0, s], k_0)$, and for $u, v \in h$, $f, g \in L^1(\mathbb{R}_+, k_0) \cap L^2(\mathbb{R}_+, k_0)$, we have

$$\begin{aligned}
&\langle ve(g), [\,\Gamma(\theta_t)X\Gamma(\theta_t^*)]ue(f)\rangle \\
&= \langle v, Bu\rangle \exp\left\{\int_0^t \langle g(x), f(x)\rangle dx + \int_t^{t+s} \langle g(x), \alpha(x-t)\rangle dx\right. \\
&\quad \left. + \int_t^{t+s} \langle \beta(x-t), f(x)\rangle dx + \int_{t+s}^\infty \langle g(x), f(x)\rangle dx\right\}.
\end{aligned}$$

Lemma 7.1.1 *For* $s, t \geq 0$, *and* $X \in \mathcal{B}(h \otimes \Gamma_s)$, *we have*

$$(i)\quad \mathcal{U}_{t+s}[\,\Gamma(\theta_t)(X \otimes I_{\Gamma^s})\Gamma(\theta_t^*)]\mathcal{U}_{t+s} = \mathcal{U}_s(X \otimes I_{\Gamma^s})\mathcal{U}_s,$$

$$(ii)\quad \mathcal{U}_{t+s}(X \otimes I_{\Gamma^t})\mathcal{U}_{t+s} = [\,\Gamma(\theta_s)\mathcal{U}_t(X \otimes I_{\Gamma^t})\mathcal{U}_t\Gamma(\theta_s^*)].$$

Proof:
Let us only prove the first of the two equalities, as the proof of the other can be obtained by similar arguments. Furthermore, it is enough to verify for X of the form $B \otimes |e(\alpha) >< e(\beta)|$, where $B \in \mathcal{B}(h)$, $\alpha, \beta \in L^2([0, s], k_0)$. Note that for $f \in L^2(\mathbb{R}_+, k_0)$, $(\theta_t^* f)(x) = f(x+t)$ for almost all $x \in [0, \infty)$, and it can be verified that

$$(\theta_t^* R_{t+s} f)(x) = f(s-x)\chi_{[0,s]}(x) + f(x+t)\chi_{(s,\infty)}(x),$$

that is, $\theta_t^* R_{t+s} f = (R_s f)\chi_{[0,s]} + (\theta_t^* f)\chi_{(s,\infty)}$. Using this for $f, g \in L^2(\mathbb{R}_+, k_0)$ and $u, v \in h$, we have that

$$\begin{aligned}
&\langle ve(g), \mathcal{U}_{t+s}[\,\Gamma(\theta_t)(X \otimes I_{\Gamma^s})\Gamma(\theta_t^*)]\mathcal{U}_{t+s}(ue(f))\rangle \\
&= \langle ve(R_{t+s}g), [\,\Gamma(\theta_t)(X \otimes I_{\Gamma^s})\Gamma(\theta_t^*)](ue(R_{t+s}f))\rangle \\
&= \langle v, Bu\rangle \exp\left\{\int_s^\infty \langle g(y), f(y)\rangle dy + \int_0^s \langle g(s-x), \alpha(x)\rangle dx\right.
\end{aligned}$$

$$+ \int_0^s \langle \beta(x), f(s-x)\rangle dx \Big\}$$
$$= \langle v, Bu\rangle \langle e(R_s g), (|e(\alpha)><e(\beta)| \otimes I_{\Gamma^s})(e(R_s f))\rangle,$$

from which (i) follows. □

Definition 7.1.2

(i) A bounded adapted process V_t (see Chapter 5 for the definition of such a process) defined on the whole of $h \otimes \Gamma(L^2(\mathbb{R}_+, k_0))$ admits a bounded adjoint and we define the *dual process* $\tilde{V}_t$ by

$$\tilde{V}_t := \mathcal{U}_t V_t^* \mathcal{U}_t^{-1} (= \mathcal{U}_t V_t^* \mathcal{U}_t).$$

(ii) We say that an adapted (not necessarily bounded) operator-process $(V_t)_{t\geq 0}$ is a *cocycle* if

$$V_{t+s} = V_s[\,\Gamma(\theta_s) V_t \Gamma(\theta_s^*)] \text{ for all } t, s \geq 0. \tag{7.1}$$

Lemma 7.1.3 *Let $(V_t)_{t\geq 0}$ be a bounded operator-valued, adapted regular process satisfying a Q.S.D.E. of the form*

$$dV_t = V_t(a_R(dt) + a_S^\dagger(dt) + \Lambda_T(dt) + A dt), \, V_0 = I,$$

where (R, S, T, A) are bounded. Then

(i) V_t is a cocycle;

(ii) the dual process $\tilde{V}_t$ will satisfy a Q.S.D.E. of the similar form with R, S, T, A replaced by S, R, T^*, A^* respectively.

Proof:
(i) Fix s and define $X_t := V_{t+s} - V_s\,[\,\Gamma(\theta_s)V_t\Gamma(\theta_s^*)]$ for $t \geq 0$. It can be seen that $X_0 = 0$ and for $u, v \in h$ and $f, g \in L^2(\mathbb{R}_+, k_0)$ the following holds:

$$\langle ve(g), X_t ue(f)\rangle = \int_0^t \langle ve(g), X_\tau\{\langle R, f(s+\tau)\rangle$$
$$+\langle g(s+\tau), S\rangle + \langle g(s+\tau), T_{f(s+\tau)}\rangle + A\} ue(f)\rangle d\tau. \tag{7.2}$$

Iterating (7.2) we can conclude that $\langle ve(g), X_t ue(f)\rangle = 0$, which proves (i).
(ii) Next we show that like V_t, $\tilde{V}_t$ also is a bounded operator-valued cocycle. Indeed, using Lemma 7.1.1 (i) and also (ii), we have that

$$\begin{aligned}\tilde{V}_{t+s} &= \mathcal{U}_{t+s} V_{t+s}^* \mathcal{U}_{t+s} \\ &= \mathcal{U}_{t+s}\,[\,\Gamma(\theta_t) V_s^* \Gamma(\theta_t^*)]\mathcal{U}_{t+s}\mathcal{U}_{t+s} V_t^* \mathcal{U}_{t+s} \\ &= \mathcal{U}_s V_s^* \mathcal{U}_s\,[\,\Gamma(\theta_s)\mathcal{U}_t V_t^* \mathcal{U}_t \Gamma(\theta_s^*)] \\ &= \tilde{V}_s\,[\,\Gamma(\theta_s)\tilde{V}_t\Gamma(\theta_s^*)].\end{aligned}$$

For $\xi, \eta \in k_0$ and $t \geq 0$, define $\tau_t^{\xi,\eta} \in \mathcal{B}(h)$ by

$$\langle v, \tau_t^{\xi,\eta} u \rangle := \langle v e(\chi_{[0,t]}\xi), \tilde{V}_t u e(\chi_{[0,t]}\eta) \rangle,$$

that is,

$$\tau_t^{\xi,\eta} = \langle e(\chi_{[0,t]}\xi), (\tilde{V}_t)_{e(\chi_{[0,t]}\eta)} \rangle.$$

From the cocycle property (7.1) of $\tilde{V}_t$, it follows that

$$\begin{aligned}
\langle v, \tau_{t+s}^{\xi,\eta} u \rangle &= \langle v e(\xi\chi_{[0,t+s]}), \tilde{V}_{t+s} u e(\eta\chi_{[0,t+s]}) \rangle \\
&= \langle v e(\xi\chi_{[0,s]}) \otimes e(\xi\chi_{(s,t+s]}), \tilde{V}_s [\Gamma(\theta_s)\tilde{V}_t \Gamma(\theta_s^*)] \\
&\qquad \times u e(\eta\chi_{[0,s]}) \otimes e(\eta\chi_{(s,t+s]}) \rangle \\
&= \langle v, \langle e(\xi\chi_{[0,s]}), (\tilde{V}_s)_{e(\eta\chi_{[0,s]})} \rangle \langle e(\xi\theta_s^* \chi_{(s,t+s]}), (\tilde{V}_t)_{e(\eta\theta_s^*\chi_{(s,t+s]})} \rangle u \rangle \\
&= \langle v, \tau_s^{\xi,\eta} \tau_t^{\xi,\eta} u \rangle, \text{ since } \theta_s^* \chi_{(s,t+s]} = \chi_{[0,t]}.
\end{aligned}$$

Thus, $\tau_t^{\xi,\eta}$ is a semigroup of bounded operators on h. Clearly, $\tau_0^{\xi,\eta} = I$. Furthermore, for $u, v \in h$, we have

$$\begin{aligned}
&\lim_{t\to 0+} \frac{\langle v, \tau_t^{\xi,\eta} u \rangle - \langle v, u \rangle}{t} \\
&= \lim_{t\to 0+} \frac{\langle v e(\chi_{[0,t]}\xi), \tilde{V}_t (u e(\chi_{[0,t]}\eta)) \rangle - \langle v e(\chi_{[0,t]}\xi), u e(\chi_{[0,t]}\eta) \rangle}{t} \\
&\quad + \lim_{t\to 0+} \frac{\langle v e(\chi_{[0,t]}\xi), u e(\chi_{[0,t]}\eta) \rangle - \langle v, u \rangle}{t} \\
&= \lim_{t\to 0+} \frac{\langle V_t (v e(R_t\chi_{[0,t]}\xi)), u e(R_t\chi_{[0,t]}\eta) \rangle - \langle v e(R_t\chi_{[0,t]}\xi), u e(R_t\chi_{[0,t]}\eta) \rangle}{t} \\
&\quad + \langle u, u \rangle \lim_{t\to 0+} \frac{\exp(t\langle \xi, \eta \rangle) - 1}{t} \\
&= \lim_{t\to 0+} \frac{\int_0^t \langle V_s \{\langle \eta, T_\xi \rangle + \langle R, \xi \rangle + \langle \eta, S \rangle + A\} (v e(\chi_{[0,t]}\xi)), u e(\chi_{[0,t]}\eta) \rangle ds}{t} \\
&\quad + \langle v, u \rangle \langle \xi, \eta \rangle \\
&= \langle \{\langle \eta, T_\xi \rangle + \langle R, \xi \rangle + \langle \eta, S \rangle + A + \langle \eta, \xi \rangle 1\} v, u \rangle \\
&= \langle \{\langle \hat{\eta}, Z_{\hat{\xi}} \rangle + \langle \eta, \xi \rangle 1\} v, u \rangle \\
&= \langle v, \{\langle \hat{\xi}, Z_{\hat{\eta}}^* \rangle + \langle \xi, \eta \rangle 1\} u \rangle,
\end{aligned}$$

where Z denotes the coefficient matrix associated with the Q.S.D.E. satisfied by V_t (see (5.22)). Thus, $\tau_t^{\xi,\eta}$ must be of the form

$$\tau_t^{\xi,\eta} = e^{t\langle \xi, \eta \rangle} e^{\langle \hat{\xi}, Z_{\hat{\eta}}^* \rangle}.$$

Since $t \mapsto V_t^*(ve(g))$ is continuous by Theorem 5.3.1, $\tilde{V}_t$ is an adapted regular process. Thus, the integral $\int_0^t \tilde{V}_s\{a_S(ds) + a_R^\dagger(ds) + \Lambda_{T^*}(ds) + A^* ds\}$ exists as a regular process and to show that $\tilde{V}_t$ satisfies a Q.S.D.E. with the coefficient matrix Z^*, it is enough to prove that for almost all t,

$$\frac{d}{dt}\langle ve(g), \tilde{V}_t ue(f)\rangle = \langle ve(g), \tilde{V}_t \langle \hat{g}(t), Z^*_{\hat{f}(t)}\rangle ue(f)\rangle$$

for all $u, v \in h$ and g, f varying over some suitable total set of vectors in k. We choose g, f to be simple functions, say of the form $g = \sum_{i=0}^p \chi_{[t_i,t_{i+1})}\xi_i$, $f = \sum_{i=0}^p \chi_{[t_i,t_{i+1})}\eta_i$, where $\xi_i, \eta_i \in k_0$ and $0 = t_0 < t_1 < \cdots < t_{p+1} < \infty$. Let $\zeta(t) := \langle ve(g), \tilde{V}_t ue(f)\rangle$. We have by (7.1) that for $t_i < t < t_{i+1}$,

$$\begin{aligned}
\zeta(t) &= \langle e(g^t), e(f^t)\rangle \langle v, \langle e(g\chi_{[0,t_i]}), (\tilde{V}_{t_i})_{e(f\chi_{[0,t_i]})}\rangle \\
&\quad \times \langle e(\xi_i \theta^*_{t_i}\chi_{(t_i,t]}), (\tilde{V}_{t-t_i})_{e(\eta_i\theta^*_{t_i}\chi_{(t_i,t]})}\rangle u\rangle \\
&= \exp\left(\int_t^{t_{p+1}} \langle g(s), f(s)\rangle ds\right) \langle v, \langle e(g\chi_{[0,t_i]}), (\tilde{V}_{t_i})_{e(f\chi_{[0,t_i]})}\rangle \circ \tau_{t-t_i}^{\xi_i,\eta_i} u\rangle \\
&= \exp\left(\int_t^{t_{p+1}} \langle g(s), f(s)\rangle ds\right) \langle v, \langle e(g\chi_{[0,t_{i-1}]}), \\
&\quad (\tilde{V}_{t_{i-1}})_{e(f\chi_{[0,t_{i-1}]})}\rangle \circ \tau_{t_i - t_{i-1}}^{\xi_{i-1},\eta_{i-1}} \circ \tau_{t-t_i}^{\xi_i,\eta_i} u\rangle \\
&\quad \ldots \\
&= e^{\int_t^{t_{p+1}} \langle g(s), f(s)\rangle ds} \langle v, \tau_{t_1}^{\xi_0,\eta_0} \circ \ldots \tau_{t_i-t_{i-1}}^{\xi_{i-1},\eta_{i-1}} \circ \tau_{t-t_i}^{\xi_i,\eta_i} u\rangle.
\end{aligned}$$

Thus,

$$\begin{aligned}
\zeta'(t) &= -\langle g(t), f(t)\rangle \zeta(t) + e^{\int_t^{t_{p+1}} \langle g(s), f(s)\rangle ds} \\
&\quad \langle v, \tau_{t_1}^{\xi_0,\eta_0} \circ \ldots \tau_{t_i-t_{i-1}}^{\xi_{i-1},\eta_{i-1}} \circ \tau_{t-t_i}^{\xi_i,\eta_i} (\langle \hat{\xi}_i, Z^*_{\hat{\eta}_i}\rangle + \langle \xi_i, \eta_i\rangle) u\rangle \\
&= -\langle \xi_i, \eta_i\rangle \zeta(t) + \langle \xi_i, \eta_i\rangle \zeta(t) + \langle ve(g), \tilde{V}_t \langle \hat{g}(t), Z^*_{\hat{f}(t)}\rangle ue(f)\rangle,
\end{aligned}$$

which completes the proof. □

7.2 Q.S.D.E. with unbounded coefficients

We shall now study a set of sufficient conditions for existence of a unitary solution of a class of Q.S.D.E. with unbounded coefficients. Let us introduce some notation at this point. Let us denote by $\mathcal{Z}_c$ the set $\{Z \in \mathcal{B}(h \otimes \hat{k_0}) : Z + Z^* + Z\hat{Q}Z^* \leq 0\}$, where $\hat{k_0}$ and $\hat{Q}$ are as in Chapter 5. Recall from Theorem 5.3.2 of Chapter 5 that the Q.S.D.E.

$$dX_t = X_t(a_{Z_{01}^*}(dt) + a_{Z_{10}}^\dagger(dt) + \Lambda_{Z_{11}}(dt) + Z_{00}dt), \;\; X_0 = I$$

admits a unique contractive solution, where we write

$$Z = \begin{pmatrix} Z_{00} & Z_{01} \\ Z_{10} & Z_{11} \end{pmatrix}, \tag{7.3}$$

with respect to the decomposition $h \otimes \hat{k_0} \equiv h \oplus (h \otimes k_0)$.

Assumptions

Fix dense subspaces $\mathcal{D}_0 \subseteq h$ and $\mathcal{V}_0 \subseteq k_0$, and for the quadruple (R, S, T, A) assume that $A \in \text{Lin}(\mathcal{D}_0, h)$, $R, S \in \text{Lin}(\mathcal{D}_0, h \otimes k_0)$, $T \in \text{Lin}(\mathcal{D}_0 \otimes \mathcal{V}_0, h \otimes k_0)$ satisfying the condition that $\mathcal{D}_0 \subseteq \cap_{\xi \in \mathcal{V}_0} \text{Dom}(\langle R, \xi \rangle)$. This implies that $(u \otimes \xi) \in \text{Dom}(R^*)$ for all $u \in \mathcal{D}_0$, $\xi \in \mathcal{V}_0$. We write the coefficient matrix $Z : \mathcal{D}_0 \otimes (\mathbb{C} \oplus \mathcal{V}_0) \to h \otimes \hat{k_0}$ by

$$Z = \begin{pmatrix} A & R^* \\ S & T \end{pmatrix}. \tag{7.4}$$

We also denote by $\mathcal{Z}$ the set of the above quadruples with the associated coefficient matrix Z such that we can find a sequence $Z^{(n)} \in \mathcal{Z}_c$, $n = 1, 2, \ldots,$ satisfying the following for all $\xi, \eta \in \mathcal{V}_0$ and $u \in \mathcal{D}_0$:

$$\lim_{n \to \infty} \langle \hat{\xi}, Z^{(n)}_{\hat{\eta}} \rangle u = \langle \hat{\xi}, Z_{\hat{\eta}} \rangle u, \tag{7.5}$$

$$\sup_{n \geq 1} \| Z^{(n)}_{\hat{\eta}} u \| < \infty. \tag{7.6}$$

Theorem 7.2.1 *Let (R, S, T, A) be a quadruple in $\mathcal{Z}$. Then there exists a solution V_t of the Q.S.D.E.*

$$dV_t = V_t(a_R(dt) + a_S^\dagger(dt) + \Lambda_T(dt) + Adt), \quad V_0 = I; \tag{7.7}$$

such that V_t is contractive for all t and $t \mapsto V_t(ve(g))$ is continuous for any fixed $v \in h$ and $g \in k$.

Proof:

Let $Z^{(n)} = \begin{pmatrix} A^{(n)} & R^{(n)*} \\ S^{(n)} & T^{(n)} \end{pmatrix}$ be the elements of $\mathcal{B}(h \otimes \hat{k_0})$ satisfying the conditions (7.5) and (7.6), and let $(V_t^{(n)})_{t \geq 0}$ be the solution of the initial value problem associated with the Q.S.D.E.

$$dV_t^{(n)} = V_t^{(n)} \left(a_{R^{(n)}}(dt) + a^\dagger_{S^{(n)}}(dt) + \Lambda_{T^{(n)}}(dt) + A^{(n)} dt \right), \quad V_0^{(n)} = I.$$

By assumption that $Z^{(n)} \in \mathcal{Z}_c$, $V_t^{(n)}$ is contractive for all t and each n. Let $v \in \mathcal{D}_0$, $g \in L^2(\mathbb{R}_+, k_0)$ be such that $g(\tau) = \sum_{k=1}^p \chi_{[t_k, t_{k+1})}(\tau) \beta_k$, for some finite partition $0 = t_1 \leq t_2 \leq \cdots < t_{p+1} = \infty$ with $\beta_k \in \mathcal{V}_0$. Let

$$M = \sup_n \text{Max}\{\|T^{(n)}(v\beta_k)\|, \|\langle R^{(n)}, \beta_k \rangle v\|, \|S^{(n)} v\|, \|A^{(n)} v\|, k = 1, \ldots, p\},$$

which is finite by (7.6). Fix $0 \leq s \leq t$. Then we have by the estimate (5.18)

$$\begin{aligned}
&\left\| \left(V_t^{(n)} - V_s^{(n)}\right) ve(g) \right\|^2 \\
&\leq e^t \int_s^t d\tau \big[\| \{V_\tau^{(n)} P T_{g(\tau)}^{(n)} + V_\tau^{(n)} P S^{(n)}\}(ve(g)) \|^2 \\
&\quad + \| \{ \langle g(\tau), V_\tau^{(n)} P T_{g(\tau)}^{(n)} \rangle + V_\tau^{(n)} \langle R^{(n)}, g(\tau) \rangle + \langle g(\tau), V_\tau^{(n)} P S^{(n)} \rangle \\
&\quad + V_\tau^{(n)} A^{(n)} \}(ve(g)) \|^2 \big] \\
&\leq e^t (t-s) M^2 \|e(g)\|^2 (2 + 8(\|g\|_\infty^2 + 1)),
\end{aligned}$$

using the contractivity of $V_\tau^{(n)}$ for each τ and n. Thus for any vector ψ in the dense subspace $\mathcal{H}_0$ of $h \otimes \Gamma$, spanned by finite linear combinations of $ve(g)$ where $v \in \mathcal{D}_0$ and g is a $\mathcal{V}_0$-valued simple function, we can get a constant $C(\psi)$ such that

$$\| \left(V_t^{(n)} - V_s^{(n)}\right) \psi \| \leq C(\psi) e^{\frac{t}{2}} (t-s)^{\frac{1}{2}}. \tag{7.8}$$

Now, using the separability of the Hilbert spaces involved (this is the first and only occasion when we use separability), we can find a countable dense set of vectors $\{\psi_i\}$ from $\mathcal{H}_0$ and observe that the function $t \mapsto \langle \psi_i, V_t^{(n)} \psi_j \rangle \equiv \rho_n^{(ij)}(t)$ is continuous for each n. Since $\{V_t^{(n)}\}$ is a family of contractions, $|\rho_n^{(ij)}(t)| \leq \|\psi_i\| \|\psi_j\|$ and since

$$|\rho_n^{(ij)}(t) - \rho_n^{(ij)}(s)| \leq \|\psi_i\| C(\psi_j) e^{\frac{t}{2}} |t-s|^{\frac{1}{2}},$$

it is clear that the sequence of functions $\{\rho_n^{(ij)}, n = 1, 2, \cdots\}$ (i, j held fixed) is equi-continuous for t varying over any compact interval. It follows by the Arzela–Ascoli theorem that there is a subsequence $n_k^{(ij)}$ such that $\rho_{n_k^{(ij)}}^{(ij)}$ is convergent uniformly over compact subsets of $[0, \infty)$ as $k \to \infty$ for fixed i and j.

We can choose a subsequence n_k, by the digitalization principle, such that $\rho_{n_k}^{(ij)}$ will converge (uniformly on compact subsets) for every i and j. By the choice of n_k, $\langle \psi_i, V_t^{(n_k)} \psi_j \rangle$ converges as $k \to \infty$, for every i, j. Since $\{V_t^{(n_k)}\}$ is a family of contractions and since $\{\psi_i\}$ is dense, it follows that $\langle \psi, V_t^{(n_k)} \psi' \rangle$ is convergent for all $\psi, \psi' \in h \otimes \Gamma$, that is, $V_t^{(n_k)}$ is weakly convergent, to some operator denoted by V_t, say. Being the weak limit of contractions, V_t itself is a contraction. Clearly, V_t is adapted since each $V_t^{(n_k)}$ is. Furthermore, for $t \geq s \geq 0$, $\psi \in \mathcal{H}_0$, we have

$$\begin{aligned}
\|(V_t - V_s)\psi\| &= \sup_{\psi' : \|\psi'\|=1} |\langle \psi', (V_t - V_s)\psi \rangle| \\
&= \sup_{\psi' : \|\psi'\|=1} \lim_{k \to \infty} |\langle \psi', (V_t^{(n_k)} - V_s^{(n_k)})\psi \rangle| \leq C(\psi) e^{\frac{t}{2}} (t-s)^{\frac{1}{2}},
\end{aligned}$$

by the estimate (7.8). It follows that $t \mapsto V_t \psi$ is continuous for $\psi \in \mathcal{H}_0$, and since V_t is contractive and $\mathcal{H}_0$ is dense, $t \mapsto V_t \psi$ is continuous for all $\psi \in h \otimes \Gamma$. Thus (V_t) is an adapted regular process, which is also bounded for each t, so it makes sense to define the integral $\int_0^t V_s(a_R(ds)+a_S^\dagger(ds)+\Lambda_T(ds)+Ads)$. The proof will be completed if we can show that

$$V_t ve(g) = ve(g) + \left(\int_0^t V_s(a_R(ds) + a_S^\dagger(ds) + \Lambda_T(ds) + Ads) \right) ve(g) \tag{7.9}$$

for $v \in \mathcal{D}_0$, and simple $\mathcal{V}_0$-valued function g, or equivalently

$$\begin{aligned} \langle V_t ve(g), ue(f) \rangle &= \langle ve(g), ue(f) \rangle \\ &+ \left\langle \left(\int_0^t V_s(a_R(ds) + a_S^\dagger(ds) + \Lambda_T(ds) + Ads) \right) ve(g), ue(f) \right\rangle \end{aligned}$$

for $u \in h$, $f \in L^2(\mathbb{R}_+, k_0)$.

To this end, observe that

$$\begin{aligned} &\langle V_t ve(g), ue(f) \rangle \\ &= \lim_{k \to \infty} \langle V_t^{(n_k)} ve(g), ue(f) \rangle \\ &= \langle ve(g), ue(f) \rangle + \lim_{k \to \infty} \left\langle \left(\int_0^t V_s(a_{R^{(n_k)}}(ds) + a^\dagger_{S^{(n_k)}}(ds) \right.\right. \\ &\qquad \left.\left. + \Lambda_{T^{(n_k)}}(ds) + A^{(n_k)}(ds) \right) ve(g), ue(f) \right\rangle \\ &= \langle ve(g), ue(f) \rangle + \lim_{k \to \infty} \int_0^t ds \langle V_s^{(n_k)} \{ \langle f(s), T^{(n_k)}_{g(s)} \rangle + \langle R^{(n_k)}, g(s) \rangle + \\ &\quad \langle f(s), S^{(n_k)} \rangle + A^{(n_k)} \} (ve(g)), ue(f) \rangle \\ &= \langle ve(g), ue(f) \rangle + \lim_{k \to \infty} \int_0^t \langle V_s^{(n_k)} \langle \hat{f}(s), Z^{(n_k)}_{\hat{g}(s)} \rangle ve(g), ue(f) \rangle ds. \end{aligned}$$

On the other hand, for fixed s, $\lim_{k \to \infty} \langle \hat{f}(s), Z^{(n_k)}_{\hat{g}(s)} \rangle v = \langle \hat{f}(s), Z_{\hat{g}(s)} \rangle v$ by (7.5), and since $V_s^{(n_k)}$ is contractive, we have that

$$\begin{aligned} &|\langle (V_s^{(n_k)} \langle \hat{f}(s), Z^{(n_k)}_{\hat{g}(s)} \rangle ve(g) - V_s \langle \hat{f}(s), Z_{\hat{g}(s)} \rangle ve(g)), ue(f) \rangle| \\ &\le \| V_s^{(n_k)} (\langle \hat{f}(s), Z^{(n_k)}_{\hat{g}(s)} \rangle v - \langle \hat{f}(s), Z_{\hat{g}(s)} \rangle v) e(g) \| \| ue(f) \| \\ &+ |\langle (V_s^{(n_k)} - V_s)(\langle \hat{f}(s), Z_{\hat{g}(s)} \rangle ve(g)), ue(f) \rangle| \\ &\le \| \langle \hat{f}(s), Z^{(n_k)}_{\hat{g}(s)} \rangle v - \langle \hat{f}(s), Z_{\hat{g}(s)} \rangle v \| \| e(g) \| \| ue(f) \| \\ &+ |\langle (V_s^{(n_k)} - V_s)(\langle \hat{f}(s), Z_{\hat{g}(s)} \rangle ve(g)), ue(f) \rangle| \\ &\to 0 \text{ as } k \to \infty. \end{aligned}$$

Moreover, since g takes only finitely many values in $\mathcal{V}_0$, it follows from (7.6) that $C := \sup_{k\geq 1}\sup_{0\leq s\leq t}\|Z^{(n_k)}_{\hat{g}(s)}v\| < \infty$, and thus we have

$$
\begin{aligned}
&|\langle V_s^{(n_k)}\langle \hat{f}(s), Z^{(n_k)}_{\hat{g}(s)}\rangle v e(g), u e(f)\rangle| \\
&\leq \|ue(f)\|\|e(g)\|\|\hat{f}(s)\| \sup_{k\geq 1}\|Z^{(n_k)}_{\hat{g}(s)}v\| \\
&\leq C\|ue(f)\|\|e(g)\|\|\hat{f}(s)\|.
\end{aligned}
$$

Since $\int_0^t \|\hat{f}(s)\|ds \leq \sqrt{t}\left(\int_0^t \|\hat{f}(s)\|^2 ds\right)^{\frac{1}{2}} \leq \sqrt{t}\|\hat{f}\| < \infty$, we conclude by the Dominated convergence theorem that

$$
\begin{aligned}
&\lim_{k\to\infty}\int_0^t \langle V_s^{(n_k)}\langle \hat{f}(s), Z^{(n_k)}_{\hat{g}(s)}\rangle v e(g), u e(f)\rangle ds \\
&= \int_0^t \langle V_s\langle \hat{f}(s), Z_{\hat{g}(s)}\rangle v e(g), u e(f)\rangle ds,
\end{aligned}
$$

which completes the proof. □

For $X \in \mathcal{B}(h\otimes\Gamma)$, $\gamma, \zeta \in \mathbb{C}\oplus\mathcal{V}_0$, we define the bilinear forms $\mathcal{L}^\gamma_\zeta(X)$ on the vector space $\mathcal{D}_0 \otimes_{\text{alg}} \Gamma$ as follows:

$$
\begin{aligned}
&\langle v\psi, \mathcal{L}^\gamma_\zeta(X) u\psi'\rangle \\
&= \langle v\psi, X\langle\gamma, Z_\zeta\rangle u\psi'\rangle + \langle\langle\gamma, Z_\zeta\rangle v\psi, Xu\psi'\rangle + \langle \hat{Q}Z_\gamma v\psi, X\hat{Q}Z_\zeta u\psi'\rangle,
\end{aligned} \tag{7.10}
$$

where $u, v \in \mathcal{D}_0$ and $\psi, \psi' \in \Gamma$ and extend linearly. Here we note that we have used the symbol X for the ampliation $(X \otimes I_{\hat{k_0}})$ as well. Clearly, we have

$$|\langle v\psi, \mathcal{L}^\gamma_\zeta(X)u\psi'\rangle| \leq C(u, v, \gamma, \zeta)\|X\|\|\psi\|\psi'\|,$$

where $C(u, v, \gamma, \zeta) := \|v\|\|\gamma\|\|Z(u\zeta)\| + \|u\|\|\gamma\|\|Z(v\zeta)\| + \|Z(v\gamma)\| \|Z(u\zeta)\|$.

We denote $\mathcal{L}^{\hat{0}}_{\hat{0}}(X)$ simply by $\mathcal{L}(X)$, where $\hat{0} = (1\oplus 0) \in \mathbb{C}\oplus\mathcal{V}_0$. Note that

$$\langle v\psi, \mathcal{L}(X)u\psi'\rangle = \langle v\psi, XAu\psi'\rangle + \langle Av\psi, Xu\psi'\rangle + \langle Rv\psi, XRu\psi'\rangle,$$

so that formally one has $\mathcal{L}(X) = XA + A^*X + R^*XR$. For $x \in \mathcal{B}(h)$, let $\mathcal{L}_0(x)$ denote the bilinear form on $\mathcal{D}_0$ given by

$$\langle v, \mathcal{L}_0(x)u\rangle := \langle v, xAu\rangle + \langle Av, xu\rangle + \langle Rv, xRu\rangle;$$

and we observe that

$$\langle v\psi, \mathcal{L}(X)u\psi'\rangle = \langle v, \mathcal{L}_0(\langle\psi, X_{\psi'}\rangle)u\rangle$$

for $X \in \mathcal{B}(h\otimes\Gamma)$, $\psi, \psi' \in \Gamma$. We also identify $\mathcal{V}_0$ naturally with $0\oplus\mathcal{V}_0$, so for $\xi, \eta' \in \mathcal{V}_0$, $\mathcal{L}^\xi_{\eta'}(X)$ will mean $\mathcal{L}^{(0\oplus\xi)}_{(0\oplus\eta')}(X)$.

Lemma 7.2.2 *Let V_t be as in the Theorem 7.2.1. Then for integers $m, n \geq 0$, $v, v \in \mathcal{D}_0$ and $\mathcal{V}_0$-valued simple functions f, g we have the following:*

$$\begin{aligned}
&\langle V_t v g^{\otimes^m}, V_t(uf^{\otimes^n})\rangle \\
&= \langle vg^{\otimes^m}, (uf^{\otimes^n})\rangle + \int_0^t \{\langle vg^{\otimes^m}, \mathcal{L}(V_s^* V_s)(uf^{\otimes^n})\rangle \\
&+ \sqrt{n}\langle vg^{\otimes^m}, \mathcal{L}_{f(s)}^{\hat{0}}(V_s^* V_s)(uf^{\otimes^{n-1}})\rangle + \sqrt{m}\langle vg^{\otimes^{m-1}}, \mathcal{L}_{\hat{0}}^{g(s)}(V_s^* V_s)(uf^{\otimes^n})\rangle \\
&+ \sqrt{m}\sqrt{n}\langle vg^{\otimes^{m-1}}, \mathcal{L}_{f(s)}^{g(s)}(V_s^* V_s)(uf^{\otimes^{n-1}})\rangle\}ds; \qquad (7.11)
\end{aligned}$$

where $f^{\otimes^{-1}} = g^{\otimes^{-1}} := 0$ and $f^{(0)} = g^{(0)} = e(0)$.

Proof:
Let $f_k, g_l; k = 1, \cdots, p; l = 1, \cdots, q$ be vectors in $\mathcal{V}_0$ such that the simple functions f, g have the form $f = \sum_{k=1}^{p} f_k \chi_{[t_k, t_{k+1})}$, $g = \sum_{l=1}^{q} g_l \chi_{[s_l, s_{l+1})}$; where $0 = t_1 < t_2 < \cdots < t_{p+1} = \infty$, $0 = s_1 < \cdots < s_{q+1} = \infty$. Let

$$\begin{aligned}
&M_1 := C(u, v, \hat{0}, \hat{0}), \qquad M_2 := \mathrm{Max}\{C(u, v, \hat{0}, f_k), k = 1, \cdots, p\}, \\
&M_3 := \mathrm{Max}\{C(u, v, g_l, \hat{0}), l = 1, \cdots, q\}, \\
&M_4 := \mathrm{Max}\{C(u, v, g_l, f_k), k = 1, \cdots, p; l = 1, \cdots, q\}.
\end{aligned}$$

For two real numbers α, β, by (5.24) we have that

$$\begin{aligned}
&\langle V_t v e(\alpha g), V_t u e(\beta f)\rangle \\
&= \langle ve(\alpha g), ue(\beta f)\rangle + \int_0^t \{\langle\langle(1 \oplus \beta f(s)), Z_{(1\oplus\alpha g(s))}\rangle ve(\alpha g), V_s^* V_s ue(\beta f)\rangle \\
&+ \langle ve(\alpha g), V_s^* V_s \langle(1 \oplus \alpha g(s)), Z_{(1\oplus\beta f(s))}\rangle ue(\beta f)\rangle \\
&+ \langle \hat{Q} Z_{(1\oplus\alpha g(s))} ve(\alpha g), V_s^* V_s \hat{Q} Z_{(1\oplus\beta f(s))} ue(\beta f)\rangle\}ds \\
&= \langle ve(\alpha g), ue(\beta f)\rangle + \int_0^t \{\langle ve(\alpha g), \mathcal{L}(V_s^* V_s) ue(\alpha f)\rangle \\
&+ \beta\langle ve(\alpha g), \mathcal{L}_{f(s)}^{\hat{0}}(V_s^* V_s) ue(\beta f)\rangle \\
&+ \alpha\langle ve(\alpha g), \mathcal{L}_{\hat{0}}^{g(s)}(V_s^* V_s) ue(\beta f)\rangle + \alpha\beta\langle ve(\alpha g), \mathcal{L}_{f(s)}^{g(s)}(V_s^* V_s) ue(\beta f)\rangle\}ds \\
&\qquad (7.12)
\end{aligned}$$

Now, since V_t is bounded for all $t \geq 0$, we can expand the left hand side of (7.12) as a power series in α, β:

$$\sum_{m,n=0}^{\infty} \frac{1}{\sqrt{m!}\sqrt{n!}} \alpha^m \beta^n \langle V_t v g^{\otimes^m}, V_t u f^{\otimes^n}\rangle.$$

Similarly, the right-hand side of (7.12) can be shown to be equal to

$$\sum_{m,n=0}^{\infty} \frac{1}{\sqrt{m!}\sqrt{n!}} \alpha^m \beta^n \Big\{ \Big\langle vg^{\otimes^m}, uf^{\otimes^n} \Big\rangle$$
$$+ \int_0^t \Big\langle vg^{\otimes^m}, \Big(\mathcal{L}(V_s^* V_s) + \beta \mathcal{L}_{f(s)}^{\hat{0}}(V_s^* V_s) + \alpha \mathcal{L}_{\hat{0}}^{g(s)}(V_s^* V_s)$$
$$+ \alpha\beta \mathcal{L}_{f(s)}^{g(s)}(V_s^* V_s) \Big) \Big(uf^{\otimes^n} \Big) \Big\rangle \Big\} ds,$$

using the fact that $\| V_s^* V_s \| \leq 1$ and the estimate

$$|\langle vg^{\otimes^m}, \Big(\mathcal{L}(V_s^* V_s) + \beta \mathcal{L}_{f(s)}^{\hat{0}}(V_s^* V_s) + \alpha \mathcal{L}_{\hat{0}}^{g(s)}(V_s^* V_s) + \alpha\beta \mathcal{L}_{f(s)}^{g(s)}(V_s^* V_s) \Big) (uf^{\otimes^n}) \rangle|$$
$$\leq \quad \|g\|^m \|f\|^n (M_1 + |\beta| M_2 + |\alpha| M_3 + |\alpha||\beta| M_4),$$

and by applying the Dominated Convergence theorem. The desired relation (7.11) now follows by comparing the coefficients of $\alpha^m \beta^n$ in both sides of (7.12). □

We next state and prove the main theorem of this section, which gives sufficient condition so that the solution V_t of the initial value problem associated with the Q.S.D.E. (7.7) in the Theorem 7.2.1 is an isometry (co-isometry or unitary respectively). For $\lambda > 0$, let us denote by β_λ the set

$$\{x \in \mathcal{B}(h) \ : \ \langle v, \mathcal{L}_0(x)u \rangle = \lambda \langle v, xu \rangle \quad \text{for all } u, v \in \mathcal{D}_0\}.$$

Theorem 7.2.3 *Let the spaces $\mathcal{D}_0$, $\mathcal{V}_0$, quadruple of maps (R, S, T, A) be as in the Theorem 7.2.1, and let V_t be the solution of (7.7). Then the following are true.*

(i) If furthermore

$$\mathcal{L}_\zeta^\gamma(I) = 0 \quad \text{for all } \gamma, \zeta \in \mathbb{C} \oplus \mathcal{V}_0, \tag{7.13}$$
$$\beta_\lambda = \{0\} \quad \text{for some } \lambda, \tag{7.14}$$

then V_t is an isometry for each t.

(ii) On the other hand, assume that there exist dense subspaces $\tilde{\mathcal{D}_0} \subseteq h$, $\tilde{\mathcal{V}_0} \subseteq k_0$ such that $\tilde{\mathcal{D}_0} \otimes \tilde{\mathcal{V}_0}$ is contained in the domain of Z^*, and the following conditions hold :

$$\lim_{n \to \infty} \langle \hat{\xi}, Z^{(n)*}_{\hat{\eta}} \rangle u = \langle \hat{\xi}, Z^*_{\hat{\eta}} \rangle u, \quad \text{for all } \xi, \eta \in \tilde{\mathcal{V}_0}, \ u \in \tilde{\mathcal{D}_0}; \tag{7.15}$$

$$\sup_{n \geq 1} \| Z^{(n)*}_{\hat{\eta}} u \| < \infty \quad \text{for all } \eta \in \tilde{\mathcal{V}_0}, u \in \tilde{\mathcal{D}_0}; \tag{7.16}$$

$$\tilde{\mathcal{L}}_\zeta^\gamma(I) = 0 \quad \text{for all } \gamma, \zeta \in \mathbb{C} \oplus \tilde{\mathcal{V}_0}; \tag{7.17}$$

$$\tilde{\beta}_\lambda = \{0\} \text{ for some } \lambda > 0; \tag{7.18}$$

where the definitions of $\tilde{\mathcal{L}}_\eta^\xi$ *and* $\tilde{\beta}_\lambda$ *are similar to the definitions of* $\mathcal{L}_\eta^\xi$ *and* β_λ, *with the replacement of* Z, $\mathcal{D}_0$ *and* $\mathcal{V}_0$ *by* Z^*, $\tilde{\mathcal{D}_0}$ *and* $\tilde{\mathcal{V}_0}$ *respectively. Then* V_t *is a co-isometry for each* t.

(iii) *If the assumptions in (i) and in (ii) are both valid, then* V_t *is unitary for all* t.

Proof:
(i) Since V_t is contractive for each t and $V_0 = I$, the operator X_t defined by $X_t = (I - V_t^* V_t)$ is nonnegative for all t and $X_0 = 0$. We need to show that $X_t = 0$ for all t, for which it suffices to show that $X_t v g^{\otimes^n} = 0$ or equivalently $\langle v g^{\otimes^n}, X_t v g^{\otimes^n}\rangle = 0$ for all $n = 0, 1, 2, \ldots$, and for all $v \in h$ and $\mathcal{V}_0$-valued simple functions g. To this end define $Y_\lambda = \int_0^\infty e^{-\lambda t} X_t dt \in \mathcal{B}(h \otimes \Gamma)$ where $\lambda > 0$ is the one for which $\beta_\lambda = \{0\}$. Let $B_\lambda^{(n)}(g) \in \mathcal{B}(h)$ be the nonnegative operator defined by

$$\langle u, B_\lambda^{(n)}(g) v\rangle := \langle u g^{\otimes^n}, Y_\lambda v g^{\otimes^n}\rangle.$$

We note that for fixed n and g, $B_\lambda^{(n)}(g) = 0$ if and only if $X_t v g^{\otimes^n} = 0$ for all $t \geq 0$, $v \in h$. We shall prove this by induction on n which will complete the proof of part (i).

Since $\mathcal{L}_\zeta^\gamma(I) = 0$ for all $\gamma, \zeta \in \mathbb{C} \oplus \mathcal{V}_0$ by (7.13), we have from the equation (7.11) the following for $u, v \in \mathcal{D}_0$:

$$\begin{aligned}
&\langle u g^{\otimes^n}, X_t(v g^{\otimes^n})\rangle \\
&= \int_0^t \{\langle u g^{\otimes^n}, \mathcal{L}(X_s)(v g^{\otimes^n})\rangle + \sqrt{n}\langle u g^{\otimes^n}, \mathcal{L}_{g(s)}^{\hat{0}}(X_s)(v g^{\otimes^{n-1}})\rangle \\
&+ \sqrt{n}\langle u g^{\otimes^{n-1}}, \mathcal{L}_{\hat{0}}^{g(s)}(X_s)(v g^{\otimes^n})\rangle + n\langle u g^{\otimes^{n-1}}, \mathcal{L}_{g(s)}^{g(s)}(X_s)(v g^{\otimes^{n-1}})\rangle\} ds.
\end{aligned} \tag{7.19}$$

Taking $n = 0$ in the above formula (and recalling that $g^{\otimes^0} := e(0)$), we get

$$\langle u e(0), X_t v e(0)\rangle = \int_0^t \langle u e(0), \mathcal{L}(X_s) v e(0)\rangle ds,$$

and an integration by parts leads to the result that

$$\langle u, B_\lambda^{(0)}(g) v\rangle = \frac{1}{\lambda}\langle u, \mathcal{L}_0(B_\lambda^{(0)}(g)) v\rangle \quad \text{for all } u, v \in \mathcal{D}_0.$$

Thus, by the hypothesis (7.14), $B_\lambda^{(0)}(g) = 0$. Now, assume that $B_\lambda^{(n-1)}(g) = 0$ for some $n \geq 1$, or equivalently, $X_t v g^{\otimes^{n-1}} = 0$ for all $v \in h$. It can be seen from the definitions of $\mathcal{L}_\zeta^\gamma$ that in this case, $\langle u\psi, \mathcal{L}_\zeta^\gamma(X_s)vg^{\otimes^{n-1}}\rangle = 0$ for all $\gamma, \zeta \in \mathbb{C} \oplus \mathcal{V}_0$, $u, v \in \mathcal{D}_0$ and $\psi \in \Gamma$. Using this fact, we get from (7.19) that

$$\langle ug^{\otimes^n}, X_t vg^{\otimes^n}\rangle = \int_0^t \langle ug^{\otimes^n}, \mathcal{L}(X_s)vg^{\otimes^n}\rangle ds,$$

hence

$$\langle u, B_\lambda^{(n)}(g)v\rangle = \frac{1}{\lambda}\langle u, \mathcal{L}_0(B_\lambda^{(n)}(g))v\rangle.$$

It follows that $B_\lambda^{(n)}(g) = 0$, which completes the proof of (i).
(ii) We have to show that V_t is a co-isometry for each t. Let $\tilde{V}_t := \mathcal{U}_t V_t^* \mathcal{U}_t$, $t \geq 0$, where $\mathcal{U}_t$ is the self-adjoint unitary operator of Section 7.1. Clearly, V_t is co-isometry if and only if $\tilde{V}_t$ is an isometry. Consider the subsequence $Z^{(n_k)}$ obtained in the proof of the Theorem 7.2.1, and recall that the subsequence of processes $V_t^{(n_k)}$, satisfying the Q.S.D.E. with coefficient matrix $Z^{(n_k)}$, has been shown to weakly converge to V_t for each t. Thus, $\widetilde{V_t^{(n_k)}} = \mathcal{U}_t V_t^{(n_k)*}\mathcal{U}_t \to \tilde{V}_t$ as $k \to \infty$ weakly. However, by Lemma 7.1.3, $\widetilde{V_t^{(n_k)}}$ satisfies the Q.S.D.E. with the coefficient matrix $Z^{(n_k)^*}$, and it follows from the proof of Theorem 7.2.1, using the hypotheses (7.15) and (7.16), that $\tilde{V}_t$ satisfies a Q.S.D.E. similar to (7.7), but with the coefficient matrix Z replaced by Z^*. Furthermore, using the hypotheses (7.17), (7.18) and part (i) of the present theorem, with $\mathcal{D}_0$, $\mathcal{V}_0$ replaced by $\tilde{\mathcal{D}_0}$, $\tilde{\mathcal{V}_0}$ respectively, we conclude that $\tilde{V}_t$ is an isometry, which completes the proof. □

7.3 Application: quantum damped harmonic oscillator

We shall briefly discuss here about a physical model where Q.S.D.E. with unbounded coefficients arise naturally. This example has been taken from [119]. Consider the well-known example of a classical damped harmonic oscillator described by the equation of motion:

$$\frac{d^2q}{dt^2} + 2\alpha\frac{dq}{dt} + \omega^2 q = 0, \tag{7.20}$$

where α, ω are positive constants satisfying $\alpha < \omega$. This is a nonconservative system, and thus cannot be described by a Hamiltonian. Nevertheless, we can introduce a pair of 'conjugate variables' (p, q) satisfying the pair of first-order differential equations

$$\frac{dq}{dt} = p - \alpha q, \quad \frac{dp}{dt} = -\delta^2 q - \alpha p, \tag{7.21}$$

where $\delta := \sqrt{\omega^2 - \alpha^2}$. It can be verified by simple computation that the second variable q of the pair (p, q) satisfying (7.21) will be a solution of (7.20).

Furthermore, we can rewrite (7.21) in a more convenient form in terms of a complex-valued function $a = a(t)$ of time t:

$$\frac{da}{dt} = -(\alpha + i\delta)a. \tag{7.22}$$

Here, (p, q) are related to a by

$$a = (2\delta)^{-\frac{1}{2}}(p - i\delta q), \;\; a^\dagger = (2\delta)^{-\frac{1}{2}}(p + i\delta q),$$

with $a^\dagger(t) := \overline{a(t)}$. Solving this equation, we get

$$p(t) = e^{-\alpha t}(p_0 \cos \delta t - \delta q_0 \sin \delta t), \quad q(t) = e^{-\alpha t}\left(q_0 \cos \delta t + \frac{p_0}{\delta} \sin \delta t\right),$$

where p_0, q_0 are the initial values. It is clear that even if (p_0, q_0) is a true conjugate pair, $(p(t), q(t))$ is not so for any positive t. This is of course expected. The quantization of the problem does not bring any change to the above situation and leads to the conclusion that there is no unitary time evolution which gives rise to the equation of motion (7.22).

Let us now change the picture and consider the situation when the damping comes from a quantum noise. To be more precise, let us replace the commutative variables (p, q) by a pair of unbounded operators, say (P, Q), such that the operator A, A^* defined by $A = (2\delta)^{-\frac{1}{2}}(P - i\delta Q)$, $A^* = (2\delta)^{-\frac{1}{2}}(P + i\delta Q)$, satisfy the canonical commutation relation (CCR)

$$[A, A^*] = I.$$

In fact, we can take $P = -i\frac{d}{dx}$ and Q to be the multiplication by x on the Hilbert space $h = L^2(\mathbb{R})$. We model the quantum damped harmonic oscillator by the equation of evolution given by the following Q.S.D.E.:

$$dU_t = U_t(a_R(dt) + a_S^\dagger(dt) - (\alpha + i\delta)A^*A dt), \quad U_0 = I; \tag{7.23}$$

where $R = \sqrt{2\alpha}A$, $S = -R$.

It should be noted here that if $\alpha = 0$, that is, if there is no damping, the above equation becomes $\frac{dU_t}{dt} = -i\omega U_t A^* A$, $U_0 = I$; so that the solution will be the evolution group for the well-known standard quantum harmonic oscillator.

Let us now prove that the equation (7.23) admits a unitary solution. To apply the criteria for existence of a unitary solution derived in the previous section, let us consider $Q_\lambda : \mathcal{B}(h) \to \mathcal{B}(h)$ for $\lambda > 0$ defined in (3.3):

$$\langle u, Q_\lambda(x)v\rangle := \int_0^\infty 2\alpha e^{-\lambda t}\langle Ae^{-tG}u, xAe^{-tG}v\rangle dt,$$

where $G = -(\alpha + i\delta)A^*A$ and $u, v \in \mathrm{Dom}(G) \subseteq \mathrm{Dom}(A)$. It is well-known that the spectrum of the operator $N := A^*A$ is $\{0, 1, 2, \ldots\}$ and the linear span $\mathcal{D}$ of its eigenspaces is dense and is contained in Dom(R). We take $\mathcal{D}_0 = \tilde{\mathcal{D}}_0 = \mathcal{D}$, $\mathcal{V}_0 = \tilde{\mathcal{V}}_0 = \mathbb{C}$. Furthermore, we choose $R_n = \sqrt{2\alpha n} A(N+n)^{-1}$, $S_n = -R_n$, $G_n = -\frac{1}{2} R_n^* R_n + i\delta n N(N+n)^{-1}$, and note that on $\mathcal{D}$ R_n, S_n and G_n converge to R, S and G respectively. In fact, with $Z^{(n)} := \begin{pmatrix} G_n & R_n^* \\ S_n & 0 \end{pmatrix}$, the conditions (7.13), (7.15), (7.16), (7.17) of the Theorem 7.2.3 are satisfied. Hence it remains to prove that $\beta_\lambda = \{0\}$, which is equivalent, by Theorem 3.2.16, to proving that $Q_\lambda^n(I) \to 0$ strongly as $n \to \infty$. To this end, we note that $Q_\lambda(I)$ in this case can be explicitly calculated as

$$Q_\lambda(I) = 2\alpha N(2\alpha N + \lambda)^{-1} = N(N + \lambda')^{-1},$$

where $\lambda' = \lambda(2\alpha)^{-1}$. By a similar calculation, we get

$$Q_\lambda^n(I) = N(N+\lambda')^{-1}(N-1)(N-1+\lambda')^{-1} \cdots (N-n+1)(N-n+1+\lambda')^{-1}.$$

Thus, if v is an eigenvector of N corresponding to the eigenvalue m (nonnegative integer), then $Q_\lambda^n(I)v = \frac{m(m-1)\cdots(m-n+1)}{(m+\lambda')\cdots(m-n+1+\lambda')} = 0$ for all $n \geq m+1$. That is, $Q_\lambda^n(I)v \to 0$ as $n \to \infty$ whenever v is any eigenvector of N, and hence for any $v \in \mathcal{D}$. Since $\mathcal{D}$ is dense in h and $\| Q_\lambda^n(I) \| \leq 1$ for all n, it follows that $Q_\lambda^n(I) \to 0$ strongly as $n \to \infty$. This completes the proof of existence and unitarity of solution U_t of (7.23).

8
Dilation of quantum dynamical semigroups with unbounded generator

In Chapter 6 we built a theory of quantum stochastic dilation 'naturally' associated with an arbitrary Q.D.S. on a von Neumann or C^*-algebra with bounded generator. There the computations involved C^* or von Neumann Hilbert modules, using the results of [24], map-valued quantum stochastic processes on modules and quantum stochastic integration with respect to them, developed in Chapter 5. It is now natural to consider the case of a Q.D.S. with unbounded generator and ask the same questions about the possibility of dilation. As one would expect, the problem is too intractable in this generality and we need to impose some further structures on it. In this chapter we shall consider a few classes of such Q.D.S. and try to construct H–P and E–H dilation for them. At first, we shall work under the framework of a Lie group action on the underlying algebra, and consider covariant Q.D.S. For H–P dilation, symmetry with respect to a trace is also assumed, whereas a general theory for E–H dilation has been built under the assumption of covariance under the action of a compact group, but without it being symmetric. Then, in the last section, we deal with a class of Q.D.S. on the U.H.F. algebra, described in Chapter 3. In this case, E–H dilation is constructed by a direct iteration using some natural estimates. However, what is common to the methods used in constructing dilation for the different kinds of Q.D.S. mentioned above is the use of a natural locally convex topology, in which the generator (unbounded in the norm topology) is continuous. In this context, the articles [6], [48], [68] and [69] are worthy of mention.

8.1 Dilation of a class of covariant Q.D.S.

Throughout this section, $\mathcal{A}$ is a separable C^*-algebra acting on a separable Hilbert space h, and G is a second countable Lie group with $\{\chi_i, i = 1, \ldots, N\}$

a basis of its Lie algebra. Moreover, $g \mapsto \alpha_g \in \mathrm{Aut}(\mathcal{A})$ is a strongly continuous representation. Let $\bar{\mathcal{A}}$ be the von Neumann algebra in $\mathcal{B}(h)$ generated by $\mathcal{A}$.

8.1.1 Notations and preliminaries

For $f \in C_c^\infty(G)$ (that is, f is smooth complex-valued function with compact support on G) and an element $a \in \mathcal{A}$, let us denote by $\alpha(f)(a)$ the norm-convergent integral $\int_G f(g)\alpha_g(a)dg$, where dg denotes the left Haar measure on G.

Let $\mathcal{A}_\infty$ denote the subset of $\mathcal{A}$ consisting of all elements a such that the map $g \mapsto \alpha_g(a)$ is infinitely differentiable with respect to the norm topology, that is, $\mathcal{A}_\infty$ is the intersection of the domains of $\partial_{i_1}\partial_{i_2}\cdots\partial_{i_k}; k \geq 1$, for all possible $i_1, i_2, \ldots \in \{1, 2, \ldots N\}$, where ∂_i denotes the closed $*$-derivation on $\mathcal{A}$ given by the generator of the one-parameter automorphism group $\alpha_{\exp(t\chi_i)}$, where exp denotes the usual exponential map for the Lie group G. The following result is essentially a consequence of the results obtained in [55] and [97].

Theorem 8.1.1 *(i) The set $\mathcal{A}_\infty$ is a dense $*$-subalgebra of $\mathcal{A}$.*

(ii) If we equip $\mathcal{A}_\infty$ with a family of norms $\|.\|_n; n = 0, 1, 2, \ldots$ given by:

$$\|a\|_n = \sum_{i_1, i_2, \cdots i_k; k \leq n} \|\partial_{i_1} \cdots \partial_{i_k}(a)\|;$$

for $n \geq 1$, and $\|a\|_0 = \|a\|$, then $\mathcal{A}_\infty$ is complete with respect to the locally convex topology induced by the above (countable) family of norms.

(iii) $\alpha_g(\mathcal{A}_\infty) \subseteq \mathcal{A}_\infty$. Furthermore, for $a \in \mathcal{A}_\infty$, $g \mapsto \alpha_g(a)$ is smooth (C^∞) in the locally convex topology mentioned above.

Proof:
(*i*) It can be seen, using the left invariance of the measure dg, that $\alpha_g(\alpha(f)(a)) = \int f(g^{-1}\mu)\alpha_\mu(a)d\mu$. If $f \in C_c^\infty(G)$, it follows from this expression that $\alpha(f)$ belongs to $\mathcal{A}_\infty$. Furthermore, for $\epsilon > 0$, choosing small enough neighborhood U of the identity of G such that $\|\alpha_g(a) - a\| \leq \epsilon$ for all $g \in U$ and $f \in C_c^\infty(G)$ with $f \geq 0$, $\int f dg = 1$ and $\mathrm{supp}(f) \subseteq U$, we get $\|\alpha(f)(a) - a\| \leq \epsilon$. This shows that $\mathcal{A}_\infty$ is dense in $\mathcal{A}$.
(*ii*) The proof of this is along standard lines, using the property that ∂_i's are closed maps on $\mathcal{A}$. Let us briefly indicate the line of reasoning. Consider a sequence $a_m \in \mathcal{A}_\infty$ so that each of the sequences a_m, $\partial_{i_1}\partial_{i_2}\cdots\partial_{i_k}(a_m)$ is Cauchy in the norm $\|\cdot\|$ of $\mathcal{A}$, for all $(i_1, \ldots, i_k)$, $k = 1, 2, \ldots$. Let a and $a^{(i_1, \ldots, i_k)}$ be the limits of a_m and $\partial_{i_1}\partial_{i_2}\cdots\partial_{i_k}(a_m)$, respectively, in the norm of

$\mathcal{A}$. Since ∂_i is a closed operator, a must belong to its domain and $\partial_i(a_m) \to \partial_i(a)$. Repeating the same argument replacing a_m and a by $\partial_i(a_m)$ and $\partial_i(a)$, respectively, we can show that for each j, $\partial_i(a)$ belongs to the domain of ∂_j and $\partial_j\partial_i(a_m) \to \partial_j\partial_i(a)$. Proceeding similarly, it follows that $a^{(i_1,\cdots,i_k)} = \partial_{i_1}\cdots\partial_{i_k}(a)$ and it belongs to the domain of ∂_i for any i. Thus $a \in \mathcal{A}_\infty$ and $a_m \to a$ in the Fréchet topology of $\mathcal{A}_\infty$.
(*iii*) First of all, by the definition of $\mathcal{A}_\infty$ and the fact that $G \times G \ni (g_1, g_2) \mapsto g_1g_2 \in G$ is C^∞ map, we observe that for $a \in \mathcal{A}_\infty$ the map $(g_1, g) \mapsto \alpha_{g_1}(\alpha_g(a)) = \alpha_{g_1g}(a)$ is C^∞ on $G \times G$, hence in particular for fixed g, $G \ni g_1 \mapsto \alpha_{g_1}(\alpha_g(a))$ is C^∞, that is, $\alpha_g(a) \in \mathcal{A}_\infty$. Similarly, for fixed $a \in \mathcal{A}_\infty$ and any positive integer k, the map $F : \mathbb{R}^k \times G \to \mathcal{A}$ given by $F(t_1, \ldots, t_k, g) = \alpha_{\exp(t_1\chi_{i_1})\cdots\exp(t_{i_k}\chi_k)g}(a)$ is C^∞. By differentiating F in its first k components at 0, we conclude that $\partial_{i_1}\cdots\partial_{i_k}(\alpha_g(a))$ is C^∞ in g, from which (iii) follows. □

Thus, $\mathcal{A}_\infty$ is a Fréchet space in the locally convex topology defined in (*ii*) of the above Theorem 8.1.1. We call $\mathcal{A}_\infty$, equipped with this topology, the *Fréchet algebra* corresponding to the action α_g. Sometimes, if we have to distinguish between various G-actions on some C^*-algebra $\mathcal{B}$, we shall use the notation $\mathcal{B}^\gamma_\infty$ to denote the Fréchet algebra $\mathcal{B}_\infty$ corresponding to the action γ_g.

Lemma 8.1.2 *For every n, the norm $\|\cdot\|_n$ satisfies*

$$\|ab\|_n \leq \|a\|_n\|b\|_n, \quad a, b \in \mathcal{A}_\infty.$$

Proof:
For a k-tuple $I = (i_1, \ldots, i_k)$ of $\{1, \ldots, N\}$, let us denote by ∂_I the operator $\partial_{i_1}\cdots\partial_{i_k}$. If I is the empty set, we shall mean by $\partial_I(x)$ just x. Let S_k denote the set of all such ordered tuples $(i_1, \ldots, i_k)$. Indeed, for $a, b \in \mathcal{A}_\infty$, we have

$$\partial_I(ab) = \sum_J \partial_J(a)\partial_{I-J}(b),$$

where J varies over all ordered subsets of I (including the empty set), and $I - J$ denotes the complement of J in I with the same order as I. Now, by definition,

$$\begin{aligned} &\|ab\|_n \\ &= \sum_{I\in S_k,\ k\leq n} \|\partial_I(ab)\| \\ &\leq \sum_{I\in S_k,\ k\leq n} \sum_{J\subseteq I} \|\partial_J(a)\|\|\partial_{I-J}(b)\| \end{aligned}$$

$$\leq \left(\sum_{I \in S_k,\ k \leq n} \|\partial_I(a)\| \right) \left(\sum_{J \in S_k,\ k \leq n} \|\partial_J(b)\| \right)$$
$$= \|a\|_n \|b\|_n.$$

□

Lemma 8.1.3 *Let* $\mathcal{A}_\infty^0 := \{a \in \mathcal{A}_\infty :$ there exist $C_a, C'_a > 0$ s.t. $\|a\|_n \leq C'_a C_a^n$ for all $n\}$. *If* G *is compact,* $\mathcal{A}_\infty^0$ *is norm-dense in* $\mathcal{A}$.

Proof:
Let us first consider the C^*-algebra $\mathcal{C} := C(G)$ with the left regular representation L_g of G, and let $\mathcal{C}_\infty$ be the corresponding Fréchet algebra, and $\mathcal{C}_\infty^0$ be defined in the same way as $\mathcal{A}_\infty^0$. We claim that $\mathcal{C}_\infty^0$ is norm-dense in $\mathcal{C}$. Let $\hat{G}$ denote the set of irreducible representations π of G and let d_π be the dimension of the representation π. Suppose that the representation $g \mapsto \pi(g)$ on $\mathbb{C}^{d_\pi}$ is given by the $d_\pi \times d_\pi$ matrix $U_g^\pi \equiv ((U_{ij}^\pi(g)))$. Let $\{\chi_1, \ldots, \chi_N\}$ be a basis of the Lie algebra of G and let $A_i(t) = \pi(\exp(t\chi_i))$ denote the one-parameter group of automorphisms corresponding to χ_i. Since d_π is finite, the generator of $A_i(t)$ must be bounded, and we denote it by X_i^π. Let $M_\pi = \|X_1^\pi\| + \cdots + \|X_N^\pi\|$. Now, by the well-known Peter–Weyl theorem (see [66]), we can get a dense subset $\{t_{ij}^\pi, ij = 1, \ldots, d_\pi;\ \pi \in \hat{G}\}$ of $\mathcal{C}$ such that

$$L_g t_{ij}^\pi = \sum_{k=1}^{d_\pi} \pi_{ik}(g) t_{kj}^\pi.$$

Thus, the d_π dimensional space $\mathcal{C}_j^\pi$ spanned by $\{t_{ij}^\pi, i = 1, \ldots, d_\pi\}$ (for fixed j) is left invariant by L_g and L_g is given by the matrix U_g^π on this space. It follows that on $\mathcal{C}_j^\pi$, the operator $\frac{d}{dt}|_{t=0} L_{\exp(t\chi_k)}$ coincides with X_k^π, and hence we have for $f \in \mathcal{C}_j^\pi$,

$$\begin{aligned}
\|f\|_n &= \sum_{i_1,\ldots,i_l\ l \leq n,\ i_p \in \{1,\ldots,N\}} \|X_{i_1}^\pi \cdots X_{i_l}^\pi f\| \\
&\leq \sum_{i_1,\ldots,i_l\ l \leq n,\ i_p \in \{1,\cdots,N\}} \|X_{i_1}^\pi\| \cdots \|X_{i_l}^\pi\| \|f\| \\
&\leq \left(\sum_{i=1}^N \|X_i^\pi\| \right)^n \|f\| \\
&= M_\pi^n \|f\|.
\end{aligned}$$

This shows that $\mathcal{C}_j^\pi \subset \mathcal{C}_\infty^0$ for every π and j, hence the density of $\mathcal{C}_\infty^0$ in $\mathcal{C}$.

Now we come to the proof of the fact that $\mathcal{A}_\infty^0$ is dense in $\mathcal{A}$, using the similar fact for $\mathcal{C}$. To this end, recall the notation $\alpha(f)(a) \equiv \int_G f(g)\alpha_g(a)dg$ for $f \in C^\infty$, $a \in \mathcal{A}$, and also the observation made in the proof of Theorem 8.1.1 that $\alpha_g(\alpha(f)(a)) = \int f(g^{-1}\nu)\alpha_\nu(a)d\nu$. Note that we work with normalized Haar measure dg, which is possible since G is compact. Using this and proceeding as in the proof of the Theorem 8.1.1, it can be deduced that $\|\alpha(f)(a)\|_n \leq \|f\|_n\|a\|$. Thus, for $f \in \mathcal{C}_\infty^0$ and any $a \in \mathcal{A}$, $\alpha(f)(a)$ belongs to $\mathcal{A}_\infty^0$. As it has been shown in the proof of Theorem 8.1.1, for a fixed $a \in \mathcal{A}$ and $\epsilon > 0$, we can find $f \in C^\infty(G)$ such that $\|\alpha(f)(a) - a\| < \epsilon$. However, we have already proved that $\mathcal{C}_\infty^0$ is norm-dense in $\mathcal{C}$, so we can choose $f_1 \in \mathcal{C}_\infty^0$ with $\|f - f_1\| < \epsilon$. So we have

$$\begin{aligned}
\|\alpha(f_1)(a) - a\| &\leq \|\alpha(f)(a) - a\| + \|\alpha(f - f_1)(a)\| \\
&\leq \epsilon + \int \|f - f_1\|\|\alpha_g(a)\|dg \\
&\leq \epsilon + \epsilon \int dg \\
&= 2\epsilon.
\end{aligned}$$

This completes the proof. □

We have a Hilbert space analogue of the above result.

Corollary 8.1.4 *Let $\mathcal{H}$ be a separable Hilbert space with a unitary representation of g. We denote the corresponding family of Hilbertian norms by $\{\|\cdot\|_{2,\,n}\}$ (to be defined in page 194) and the Fréchet space by $\mathcal{H}_\infty$. Furthermore, let $\mathcal{H}_\infty^0$ denote the following subspace of $\mathcal{H}$:*

$$\mathcal{H}_\infty^0 := \{\xi \in \mathcal{H}_\infty \ : \ \text{there exists } C_\xi, C'_\xi > 0 \text{ s.t. } \|\xi\|_{2,\,n} \leq C'_\xi C_\xi^n\}.$$

If G is compact, $\mathcal{H}_\infty^0$ is dense in $\mathcal{H}$ in the norm-topology of $\mathcal{H}$.

The proof is very similar to the proof of Lemma 8.1.3 and hence is omitted.

The following result will be useful later to extend densely defined $*$-homomorphisms.

Lemma 8.1.5 *Both $\mathcal{A}_\infty^0$ and $\mathcal{A}_\infty$ are closed under holomorphic functional calculus. In particular, if a is a positive invertible element of $\mathcal{A}_\infty$ (respectively, $\mathcal{A}_\infty^0$), then $\sqrt{a}$ again belongs to $\mathcal{A}_\infty$ (respectively $\mathcal{A}_\infty^0$).*

Proof:
Let $a \in \mathcal{A}_\infty$, and f be a holomorphic function defined on an open set containing the spectrum $\sigma(a)$ of a. We have to show that $f(a) \in \mathcal{A}_\infty$. Let us

fix some $\chi_i \in \{\chi_1, \ldots, \chi_N\}$, and denote the corresponding one-parameter group of automorphism by β_t. Note that as β_t is an automorphism, we have $\sigma(\beta_t(a)) = \sigma(a)$. Choosing a suitable contour C around $\sigma(a)$, we can write $f(a)$ as

$$f(a) = \frac{1}{2\pi i} \int_C f(z)(z-a)^{-1} dz.$$

It is also clear that $\beta_t((z-a)^{-1}) = (z - \beta_t(a))^{-1}$. Thus,

$$\begin{aligned}
&\frac{\beta_t(f(a)) - f(a)}{t} \\
&= \frac{1}{2\pi i} \int_C f(z) \frac{(z-\beta_t(a))^{-1} - (z-a)^{-1}}{t} dz \\
&= \frac{1}{2\pi i} \int_C f(z)(z-\beta_t(a))^{-1} \frac{(\beta_t(a) - a)}{t} (z-a)^{-1} dz \\
&= \frac{1}{2\pi i} \int_C g_t(z) f(z) dz \ \ \text{(say)}.
\end{aligned}$$

Moreover, we have

$$\begin{aligned}
&\frac{\|\beta_t(a) - a\|}{t} \\
&\leq \frac{1}{t} \int_0^t \|\beta_s(\partial_i(a))\| ds \\
&\leq \|\partial_i(a)\|.
\end{aligned}$$

Thus,

$$\begin{aligned}
&|f(z)| \|g_t(z)\| \\
&\leq |f(z)| \|(z-\beta_t(a))^{-1}\| \left\| \frac{\beta_t(a) - a}{t} \right\| \|(z-a)^{-1}\| \\
&\leq |f(z)| \|(z-a)^{-1}\|^2 \|\partial_i(a)\|,
\end{aligned}$$

which is integrable over C. It is also a simple observation that for all $z \in C$,

$$\lim_{t \to 0+} g_t(z) = -(z-a)^{-1} \partial_i(a) (z-a)^{-1}.$$

So by the Dominated Convergence theorem, it follows that

$$\lim_{t \to 0+} \frac{\beta_t(f(a)) - f(a)}{t} = \frac{1}{2\pi i} \int_C f(z)(z-a)^{-1} \partial_i(a)(z-a)^{-1} dz.$$

Thus, $f(a)$ is in the domain of ∂_i. By repeating similar argument, it can be shown that $f(a)$ belongs to $\mathcal{A}_\infty$. In fact, we can prove by induction on k the following expression:

$$\partial_{i_1}\cdots\partial_{i_k} f(a)$$
$$= \frac{1}{2\pi i}\int_C f(z)(z-a)^{-1}\left(\sum_{(I_1,\cdots,I_l)\in T_k}^{l} \partial_{\hat{I}_1}(a)(z-a)^{-1}\cdots\partial_{\hat{I}_l}(a)(z-a)^{-1}\right)dz, \tag{8.1}$$

where $k \geq 1$, $(i_1, \ldots, i_k)$ is a tuple of integers with $i_j \in \{1, \ldots, N\}$, and T_k denotes the collection of all partitions of $\{1, \ldots, k\}$, that is, all tuples $(I_1, \ldots, I_l)$ of nonempty, pairwise disjoint subsets of $\{1, \ldots, k\}$, with l varying over $\{1, \ldots, k\}$, such that $\cup_j I_j = \{1, \ldots, k\}$. For a subset $I = \{t_1, \ldots, t_p\}$, with $t_1 < t_2 < \cdots < t_p$, we have denoted by $\hat{I}$ the ordered p-tuple $(i_{t_1}, \ldots, i_{t_p})$.

We just briefly remark, without giving the details, how to carry out the proof if $\mathcal{A}_\infty$ is replaced by $\mathcal{A}_\infty^0$. Suppose that C'_a, C_a are positive constants such that $\|a\|_n \leq C'_a C_a^n$ for all n. Let K be the constant $\frac{1}{2\pi}\int_C |f(z)|\|(z-a)^{-1}\||dz|$, $M = \sup_{z\in C}\{\|(z-a)^{-1}\|\} > 0$, and let P_k denote the set of all tuples $(p_1, \ldots, p_l)$ of integers with $l \geq 1$, $p_i > 0$ for all i and $\sum_i p_i = k$. It is clear that the set P_k has the same cardinality as the set of all tuples of integers $(q_1, \ldots, q_l)$ with $l \geq 1$, $1 \leq q_1 < \cdots < q_l = k$, which has the cardinality 2^{k-1}. Let us also fix a tuple $(i_1, \ldots, i_k)$ and recall the expression for $\partial_{i_1}\cdots\partial_{i_k} f(a)$ given by (8.1). Now, for a fixed choice of $(p_1, \ldots, p_l) \in P_k$, we have

$$\begin{aligned}
&\sum_{(I_1,\cdots,I_l)\in T_k\,:\,|I_j|=p_j \text{ for all } j} \|\partial_{\hat{I}_1}(a)(z-a)^{-1}\cdots\partial_{\hat{I}_l}(a)(z-a)^{-1}\| \\
&\leq \sum_{(I_1,\ldots,I_l)\in T_k\,:\,|I_j|=p_j \text{ for all } j} M^l \|\partial_{\hat{I}_1}(a)\|\cdots\|\partial_{\hat{I}_l}(a)\| \\
&\leq M^l \|a\|_{p_1}\cdots\|a\|_{p_l} \\
&\leq M^l (C'_a)^l C_a^{p_1+\cdots+p_l} \\
&= (MC'_a)^l C_a^k \leq (1+MC'_a)^k C_a^k,
\end{aligned}$$

since $l \leq k$.

Thus, we can estimate the right-hand side of (8.1) by $K((1+MC'_a)C_a)^k|P_k| = \frac{K}{2}(2(1+MC'_a)C_a)^k \leq K\beta^k$, where $\beta = 2(1+MC'_a)C_a$. Since $\|f(a)\|_0 \leq K$, and for fixed $k \geq 1$, the number of tuples $(i_1, \ldots, i_k)$ with each $i_j \in \{1, \ldots, N\}$ is N^k, we get the following estimate:

$$\|f(a)\|_n \leq K\sum_{k=0}^{n}(N\beta)^k = \frac{K}{N\beta - 1}((N\beta)^{n+1} - 1) \leq \frac{KN\beta}{N\beta - 1}(N\beta)^n,$$

which shows that $f(a) \in \mathcal{A}_\infty^0$. □

Corollary 8.1.6 *Let π be a $*$-homomorphism from $\mathcal{A}_\infty$ to $\mathcal{B}(\mathcal{K})$, with $\pi(1) = 1_\mathcal{K}$, where $\mathcal{K}$ is a separable Hilbert space. Then π extends to a $*$-homomorphism of $\mathcal{A}$. When G is compact, we can get the same conclusion if we replace $\mathcal{A}_\infty$ by $\mathcal{A}^0_\infty$.*

Proof:
We give a proof for $\mathcal{A}_\infty$ only. The proof for $\mathcal{A}^0_\infty$ is similar. Let $a \in \mathcal{A}_\infty$ be a self-adjoint element. For $\epsilon > 0$, let $b = (1+\epsilon)\|a\| - a$, which is a positive invertible element of $\mathcal{A}_\infty$. Since $\sigma(b)$ does not contain 0, we can choose a holomorphic function f defined in a neighborhood of $\sigma(b)$ such that $f(b) = \sqrt{b}$. By Lemma 8.1.5, $\sqrt{b} \in \mathcal{A}_\infty$. Now, as π is a $*$-homomorphism on $\mathcal{A}_\infty$, $\pi(b) = \pi((\sqrt{b})^2) = \pi(\sqrt{b})^2 \geq 0$. Therefore, $\pi(a) \leq (1+\epsilon)\|a\|$, and since this is true for arbitrary $\epsilon > 0$, we conclude that $\|\pi(a)\| \leq \|a\|$ for all self-adjoint $a \in \mathcal{A}_\infty$, and $\|\pi(a)\| \leq 2\|a\|$ for all $a \in \mathcal{A}_\infty$. By the norm-density of $\mathcal{A}_\infty$ in $\mathcal{A}$, π extends to a $*$-homomorphism on the whole of $\mathcal{A}$. □

Now we shall introduce some more notation and terminology and prove some preparatory results. If $\mathcal{E}$ is any Banach space with a strongly continuous representation of G given by γ_g, we denote by $\mathcal{E}_\infty$ the intersection of the domains of the generators of different one-parameter subgroups, just as in case of $\mathcal{A}_\infty$. That is,

$$\mathcal{E}_\infty := \cap_{i_1,i_2,\ldots}\mathrm{Dom}(\partial^{\mathcal{E}}_{i_1}\partial^{\mathcal{E}}_{i_2}\cdots\partial^{\mathcal{E}}_{i_k};\, k = 1,2,\ldots).$$

Here, we have denoted by $\partial^{\mathcal{E}}_k$ the generator of the one-parameter group $\gamma_{t\exp(\chi_k)}$. If there is no confusion about $\mathcal{E}$, then we may denote $\partial^{\mathcal{E}}_k$ by just ∂_k. We introduce a family of norms $\| \|_n$ as follows:

$$\|\xi\|_n := \sum_{i_1,i_2,\cdots i_k;k\leq n} \|\partial_{i_1}\partial_{i_2}\cdots\partial_{i_k}(\xi)\|, \quad \xi \in \mathcal{E}_\infty.$$

Clearly, $\mathcal{E}_\infty$ is complete in the locally convex (Fréchet) topology given by this countable family of norms.

For two Banach spaces $\mathcal{E}^i$ with corresponding G-actions γ^i_g $(i = 1, 2)$, we denote by $\mathcal{B}(\mathcal{E}^1_\infty, \mathcal{E}^2_\infty)$ the space of all linear maps S from $\mathcal{E}^1$ to $\mathcal{E}^2$ such that $\mathcal{E}^1_\infty$ is in the domain of S, $S(\mathcal{E}^1_\infty) \subseteq \mathcal{E}^2_\infty$, and S is continuous with respect to the Fréchet topologies of the respective spaces. We call a linear map L from $\mathcal{E}^1$ to $\mathcal{E}^2$ to be *covariant* if $\gamma^1_g(\mathrm{Dom}(L)) \subseteq \mathrm{Dom}(L)$, and $L\gamma^1_g(\xi) = \gamma^2_g L(\xi)$ for all $g \in G, \xi \in \mathrm{Dom}(L)$.

Lemma 8.1.7 *If L from $\mathcal{E}^1$ to $\mathcal{E}^2$ is bounded (in the usual Banach space sense) and covariant in the above sense, then $L \in \mathcal{B}(\mathcal{E}^1_\infty, \mathcal{E}^2_\infty)$.*

Proof:
Let $\partial^{\mathcal{E}^1}_i$ and $\partial^{\mathcal{E}^2}_i$ be respectively the generator of the one parameter subgroup

corresponding to χ_i in $\mathcal{E}^1$ and $\mathcal{E}^2$. From the relation $L\gamma_g^1 = \gamma_g^2 L$ and the boundedness of L, it follows that L maps the domain of $\partial_i^{\mathcal{E}^1}$ into the domain of $\partial_i^{\mathcal{E}^2}$ and $L\partial_i^{\mathcal{E}^1} = \partial_i^{\mathcal{E}^2} L$. By repeated application of this argument it follows that $L\partial_{i_1}^{\mathcal{E}^1} \cdots \partial_{i_k}^{\mathcal{E}^1}(\xi) = \partial_{i_1}^{\mathcal{E}^2} \cdots \partial_{i_k}^{\mathcal{E}^2} L(\xi)$ for all $\xi \in \mathcal{E}^1_\infty$, and thus $\|L\xi\|_n \leq \|L\|\|\xi\|_n$. □

Definition 8.1.8 An element of $\mathcal{B}(\mathcal{E}^1_\infty, \mathcal{E}^2_\infty)$ is called a *smooth map*, and if such a smooth map L satisfies an estimate of the form $\|L\xi\|_n \leq C\|\xi\|_{n+p}$ for all n and for some integer p and a constant C, then we say that L is a *smooth map of order* p (with bound $\leq C$).

From the proof of the Lemma 8.1.7 we observe that any bounded covariant map is smooth of order 0 with the bound $\leq \|L\|$. By a similar reasoning we can prove the following lemma.

Lemma 8.1.9 *Suppose that L is a closed (in the Banach space sense), covariant map from $\mathcal{E}^1$ to $\mathcal{E}^2$. Then L is smooth of the order p for some p if and only if $\mathcal{E}^1_\infty$ is in the domain of L.*

Proof:
Since a smooth map by definition contains $\mathcal{E}^1_\infty$ in its domain, it suffices to prove that $\mathcal{E}^1_\infty \subseteq \mathrm{Dom}(L)$ implies that L is a smooth map of some order. For simplicity of notation, we shall use the same symbol ∂_i for both $\partial_i^{\mathcal{E}^1}$ and $\partial_i^{\mathcal{E}^2}$, and also we use the same symbol γ_g for the G-action on $\mathcal{E}^1$ and $\mathcal{E}^2$. Let L be a map as above. Since L is closed in the Banach space sense, and the Fréchet topology in $\mathcal{E}^1_\infty$ is stronger than its Banach space topology, it follows that L is closed as a map from the Fréchet space $\mathcal{E}^1_\infty$ to the Banach space $\mathcal{E}^2$. By the Closed Graph theorem, it is continuous with respect to the above topologies, since it is defined on the Fréchet space $\mathcal{E}^1_\infty$. By the definition of Fréchet space continuity, there exist some C and p such that $\|L(\xi)\|_0 \leq C\|\xi\|_p$. Now, for any fixed k, let $\gamma_t \equiv \gamma_{\exp(t\chi_k)}$. Since γ_t maps $\mathcal{E}^1_\infty$ into itself and L is covariant, we have $L((\gamma_t(\xi) - \xi)/t) = (\gamma_t(L\xi) - L\xi)/t$. Now, since $(\gamma_t(\xi) - \xi)/t \to \partial_k(\xi)$ as $t \to 0+$ in the Fréchet topology, we have that $L((\gamma_t(\xi) - \xi)/t) = (\gamma_t(L\xi) - L\xi)/t$ converges to $L\partial_k\xi$ in the Banach space topology of $\mathcal{E}^2$, and so by the closedness of ∂_k, $L\xi$ must belong to the domain of ∂_k, with $L\partial_k\xi = \partial_k L\xi$. Repeated use of this line of reasoning proves that $L(\mathcal{E}^1_\infty) \subseteq \mathcal{E}^2_\infty$ and $L(\partial_{i_1} \cdots \partial_{i_k}\xi) = \partial_{i_1} \cdots \partial_{i_k}(L\xi)$ for all $\xi \in \mathcal{E}^1_\infty$. Now, a direct computation enables one to show that L is of order p with bound $\leq C$. □

Let us now consider the special case of Hilbert spaces. Of course, Hilbert spaces are Banach spaces, and our discussion on the Fréchet topology on Banach spaces with group action applies to Hilbert spaces as well. However, we shall make a slightly different choice of the family of norms in case of a Hilbert space, such that each of the norms remain Hilbertian. Given a Hilbert space $\mathcal{H}$ with a unitary representation U_g of G in $\mathcal{H}$, with $d_k^{\mathcal{H}} \equiv d_k$ denoting the self-adjoint generator of the unitary one-parameter group $u_t \equiv U_{\exp(t\chi_k)} = e^{itd_k}$, we introduce a countable family of Hilbertian norms $\| \cdot \|_{2,n}; n = 0, 1, 2, \cdots$ on $\mathcal{H}_\infty$ as follows:

$$\|\xi\|_{2,n}^2 \equiv \sum_{i_1, i_2, \cdots i_k; k \leq n} \|d_{i_1} d_{i_2} \cdots d_{i_k}(\xi)\|^2.$$

It can be observed that $\mathcal{H}_\infty$ is complete in the Fréchet topology given by the above family of norms, and we consider $\mathcal{H}_\infty$ as a Fréchet space with this locally convex topology. We call such a pair $(\mathcal{H}, U_g)$ a *Sobolev–Hilbert space*, and define a smooth map between them in the same way as was done before in the context of Banach spaces. The notion of covariant maps and the notation $\mathcal{B}(\mathcal{H}_\infty, \mathcal{K}_\infty)$ for the Hilbert spaces $\mathcal{H}$ and $\mathcal{K}$ respectively are also introduced similarly.

Remark 8.1.10 *A word of caution about the notation: since a Hilbert space is also a Banach space, there is an ambiguity in the notation $\mathcal{H}_\infty$; it may have been preferable to use a different notation for the Fréchet space corresponding to the Hilbertian norms to distinguish it from the Fréchet space obtained from the family of Banach norms $\{\| \cdot \|_n\}$. However, our convention will be that unless mentioned otherwise, if the Banach space under consideration is a Hilbert space and the group action is unitary, we shall work with the Hilbertian norms.*

Let us note the following analogues of Lemma 8.1.7 and Lemma 8.1.9 in this situation, and remark that the proof remains the same, with the replacement of $\| \cdot \|_n$ by the Hilbertian norm $\| \cdot \|_{2,n}$.

Lemma 8.1.11 *If L from $\mathcal{H}$ to $\mathcal{K}$ is bounded (in the Hilbert space sense) and covariant in the above sense, then $L \in \mathcal{B}(\mathcal{H}_\infty, \mathcal{K}_\infty)$.*

Lemma 8.1.12 *Suppose that L is a closed (in the Hilbert space sense), covariant map from $\mathcal{H}$ to $\mathcal{K}$. Then L is smooth of the order p for some p if and only if $\mathcal{H}_\infty$ is in the domain of L.*

However, Hilbert space framework allows the use of the Spectral theorem of self-adjoint operators, and we can prove more than what we could do for a general Banach space.

Theorem 8.1.13 *Let $(\mathcal{H}, U_g), (\mathcal{K}, V_g)$ be two Sobolev–Hilbert spaces as in earlier discussion, and L be a closed linear map from $\mathcal{H}$ to $\mathcal{K}$. Furthermore, assume that $\mathcal{H}_\infty$ is in the domain of $|L|^2$ and is a core for $|L|^2$, and $LU_g = V_g L$ on $\mathcal{H}_\infty$. Then we have the following conclusions.*

(i) L is a smooth covariant map with some order p and bound $\leq C$ for some C.

(ii) L^* (the densely defined adjoint in the Hilbert space sense) will have $\mathcal{K}_\infty$ in its domain.

(iii) L^* is also a smooth covariant map from $\mathcal{K}_\infty$ to $\mathcal{H}_\infty$; with order p and bound $\leq C$ as in (i).

Proof:
Let the polar decomposition of L be given by $L = W|L|$. We claim that both W and $|L|$ are covariant maps. First we note that $\mathcal{H}_\infty$ is also a core for L (being a core for $|L|^2$) and since U_g is a unitary operator that maps $\mathcal{H}_\infty$ into itself, clearly $\mathcal{H}_\infty$ is a core for LU_g and also for $V_g L$. Thus the relation $LU_g = V_g L$ on $\mathcal{H}_\infty$ implies that the operators LU_g and $V_g L$ have the same domain and they are equal. Now, note that L being closed and V_g being bounded, we have that $(V_g L)^* = L^* V_g^* = L^* V_{g^{-1}}$. Furthermore, since U_g^{-1} maps the core $\mathcal{H}_\infty$ for L into itself, one can verify that $(LU_g)^* = U_g^* L^*$. Thus, we get that $U_g L^* = L^* V_g$ for all g. It then follows that $U_g |L|^2 = |L|^2 U_g$ and hence by the Spectral theorem, U_g and $|L|$ will commute. By Lemma 8.1.12, we get that $|L|(\mathcal{H}_\infty) \subseteq \mathcal{H}_\infty$, and $|L|$ is a smooth covariant map of some order.

Now, if P denotes the projection onto the closure of the range of $|L|$, then P clearly commutes with U_g for all g, hence in particular $U_g \mathrm{Ran}(P)^\perp \subseteq \mathrm{Ran}(P)^\perp$. Thus $WU_g P^\perp = WP^\perp U_g = 0 = V_g W P^\perp$. On the other hand, $V_g WP = WU_g P$, because $V_g W|L| = V_g L = LU_g = W|L|U_g = WU_g|L|$. Hence we have that W is a bounded covariant map, and thus by Lemma 8.1.11, it follows that W^* is covariant too, and in particular $W^*(\mathcal{K}_\infty) \subseteq \mathcal{H}_\infty \subseteq \mathrm{Dom}(|L|)$, so that $\mathcal{K}_\infty \subseteq \mathrm{Dom}(L^*) = \mathrm{Dom}(|L|W^*)$. Moreover, from the fact that W and W^* are smooth maps of order 0 with bound ≤ 1 (as $\|W\| = \|W^*\| = 1$) and $|L|$ is a smooth covariant map of some order p with bound $\leq C$ for some C, clearly both $L = W|L|$ and $L^* = |L|W^*$ are smooth covariant maps of order p and bound $\leq C$, which completes the proof. □

Lemma 8.1.14 *Let $(\mathcal{H}_i, U_g^i), i = 1, 2$ and $(\mathcal{K}_i, V_g^i), i = 1, 2$ be Sobolev–Hilbert spaces and let k be any Hilbert space. Then we can construct the Sobolev–Hilbert spaces $(\mathcal{H}_i \oplus \mathcal{K}_i, U_g^i \oplus V_g^i)$ and $(\mathcal{H}_i \otimes k, U_g^i \otimes I)$ (with the symbols carrying their usual meanings) and if $L \in \mathcal{B}(\mathcal{H}_{1_\infty}, \mathcal{H}_{2_\infty})$, $M \in \mathcal{B}(\mathcal{K}_{1_\infty}, \mathcal{K}_{2_\infty})$, then we have the following:*

(i) $L \oplus M \in \mathcal{B}((\mathcal{H}_1 \oplus \mathcal{K}_1)_\infty, (\mathcal{H}_2 \oplus \mathcal{K}_2)_\infty)$, and

(ii) $(\mathcal{H}_1 \otimes k)_\infty$ is the completion of $\mathcal{H}_{1_\infty} \otimes_{\mathrm{alg}} k$ under the respective Fréchet (Hilbertian) topologies and the map $L \otimes_{\mathrm{alg}} I$ on $\mathcal{H}_{1_\infty} \otimes_{\mathrm{alg}} k$ extends to a smooth map on the respective Fréchet space (we shall denote this smooth

map by $L \otimes I$ or sometimes $\tilde{L}$). Furthermore, if L is of order p with some bound C, so will be $\tilde{L}$.

Proof:
The proof of (*i*) is elementary and is omitted. To prove (*ii*), we fix any orthonormal basis $\{e_l\}$ of k and let $\xi = \sum \xi_l \otimes e_l$ be a vector in the domain of the self-adjoint generator of the one parameter unitary group $u_t \otimes I$, where u_t was introduced after Lemma 8.1.9 and it may be observed that the summation needs to be done over a countable set only. Therefore, without loss of generality, we may assume that the set of l's with ξ_l nonzero is indexed by $1, 2, \ldots$. Since $\sum((u_t(\xi_l) - \xi_l)/t) \otimes e_l$ is Cauchy (in the Hilbert space topology) suppose that $\sum((u_t(\xi_l) - \xi_l)/t) \otimes e_l \to \sum \eta_l \otimes e_l$. Clearly, for each l, $\eta_l = \lim_{t\to 0}((u_t(\xi_l) - \xi_l)/t)$, which implies that $\xi_l \in \mathrm{Dom}(d_k)$ and $id_k\xi_l = \eta_l$. Thus, if $\tilde{d}_k$ denotes the self-adjoint generator of the one parameter unitary group $u_t \otimes I$, then we have proved that the domain of it consists of precisely the vectors $\sum \xi_l \otimes e_l$ such that each $\xi_l \in \mathrm{Dom}(d_k)$ and $\sum \|d_k(\xi_l)\|^2 < \infty$. Repeated use of this argument enables us to prove that $(\mathcal{H}_1 \otimes k)_\infty$ consists of the vectors $\xi = \sum \xi_l \otimes e_l$ with the property that $\xi_l \in \mathcal{H}_{1\infty}$ for each l and for any n, $\|\xi\|_{2,n}^2 \equiv \sum_l \|\xi_l\|_{2,n}^2 < \infty$. From this, it is clear that $\sum_{l=1}^m \xi_l \otimes e_l$ converges (as $m \to \infty$) to ξ in each of the $\|.\|_{2,n}$ norms, that is in the Fréchet topology. The rest of the proof follows by observing that for any $\xi = \sum_{\text{finite}} \xi_l \otimes e_l \in \mathcal{H}_{1,\infty} \otimes_{\text{alg}} k$, $\|\tilde{L}(\xi)\|_{2,n}^2 = \sum \|L\xi_l\|_{2,n}^2$.
□

We end this subsection with a discussion on the natural Fréchet topology on a G–$\mathcal{A}$ Hilbert module (E, γ) (see Definition 4.3.1). In such a case, we can consider the Fréchet space $E_\infty \equiv E_\infty^\gamma$ given by the family of norms $\|\cdot\|_{2,n}$ defined in analogy with the Hilbert space case. In analogy with the terminology used by us in the case of Hilbert spaces, we shall call the above Fréchet space a *Sobolev–Hilbert module*. However, we shall mostly need to consider $G - \mathcal{A}$ Hilbert modules of the form $\mathcal{A} \otimes \mathcal{H}$ for some Hilbert space $\mathcal{H}$, and with G-action of the form $\alpha_g \otimes \upsilon_g$ for some actions α and υ on $\mathcal{A}$ and $\mathcal{H}$ respectively. Let us denote by ∇_i the closure of the map $\partial_i \otimes d_i^{\mathcal{H}}$, $i = 1, 2, \ldots, N$, and for an ordered set $J = \{j_1, \ldots, j_k\}$, let $\nabla_J = \nabla_{j_1} \cdots \nabla_{j_k}$. We then introduce $\|\xi\|_{2,n}$ for elements ξ in the domain $E_\infty := \cap_{J=\{j_1,\ldots,j_k\},\, k\geq 0}\mathrm{Dom}(\nabla_J)$ as follows:

$$\|\xi\|_{2,n}^2 := \sum_J \|\nabla_J(\xi)\|^2,$$

where in the above, J varies over all possible tuples $(j_1, \ldots, j_k)$ such that $j_l \in \{1, \ldots, N\}$ and $k \leq n$. It can be seen that E_∞ is a Fréchet space with the family of norms $\{\|\cdot\|_{2,n}\}$. We shall also consider the Fréchet algebra $\mathcal{B}_\infty$, where $\mathcal{B} = \mathcal{L}(E)$, with the G-action given by $\mathcal{B} \ni X \mapsto \beta_g X \beta_g^{-1}$. It is not difficult to prove that $\mathcal{B}_\infty$ maps E_∞ into itself and we have the following analogue of Lemma 8.1.2.

Lemma 8.1.15 *For $X \in \mathcal{L}(E)_\infty$, $\xi \in E_\infty$, we have*

$$\|X\xi\|_{2,n} \leq \|X\|_n \|\xi\|_{2,n}.$$

The proof is omitted, since it is very similar to the proof of Lemma 8.1.2.

An element ξ of a Hilbert $\mathcal{A}$-module E can be naturally identified with an element L_ξ of $\mathcal{L}(\mathcal{A} \oplus E)$ by defining $L_\xi(a \oplus \eta) = (0 \oplus \xi a)$. It is not difficult to see that the operator norm of L_ξ is same as the norm of ξ, thus $\xi \mapsto L_\xi$ is an isometric map. Now, this can be generalized to a $G - \mathcal{A}$ module (E, β). Indeed, $((\mathcal{A} \oplus E), (\alpha \oplus \beta))$ is a $G - \mathcal{A}$ module, and we consider Fréchet topologies on E as well as on $\mathcal{A} \oplus E$. It can be shown that $\xi \mapsto L_\xi$ is actually a homeomorphism of the Fréchet topologies mentioned above.

The Sobolev–Hilbert spaces defined earlier is clearly a special case of Sobolev–Hilbert modules, with the choice $\mathcal{A} = \mathbb{C}$. Furthermore, we shall take a notational convention similar to the one explained in Remark 8.1.10 in the context of Sobolev–Hilbert spaces.

8.1.2 H–P dilation of a class of symmetric covariant quantum dynamical semigroups

Let τ be a densely defined, semifinite, lower semicontinuous and faithful trace on $\mathcal{A}$. Let $\mathcal{A}_\tau \equiv \{x : \tau(x^*x) < \infty\}$. Thus h can be chosen to be $L^2(\tau)$, and $\mathcal{A}$ is naturally imbedded in $\mathcal{B}(h)$. Recall that $\bar{\mathcal{A}}$ denotes the von Neumann closure of $\mathcal{A}$ with respect to the weak topology inherited from $\mathcal{B}(h)$. Clearly $\mathcal{A}_\tau$ is ultra-weakly dense in $\bar{\mathcal{A}}$. Suppose that $\alpha_g(\mathcal{A}_\tau) \subseteq \mathcal{A}_\tau$ and $\tau(\alpha_g(x^*y)) = \tau(x^*y)$ for $x \in \mathcal{A}_\tau$, $y \in \mathcal{A}$, $g \in G$. By polarization, this is equivalent to the assumption that $\tau(\alpha_g(x^*x)) = \tau(x^*x)$ for $x \in \mathcal{A}_\tau$. This allows one to extend α_g as a unitary linear operator on h, to be denoted by u_g, and clearly $\alpha_g(x) = u_g x u_g^*$ for $x \in \mathcal{A}$. To see this, it is sufficient to verify this relation on the vectors in $\mathcal{A}_\tau$ first, and then extend it to the whole of h using the fact that h is the completion of $\mathcal{A}_\tau$.

Lemma 8.1.16 *$g \mapsto u_g$ is strongly continuous with respect to the Hilbert space topology of h.*

Proof:
Let $\mathcal{A}_1 := \{x \in \mathcal{A} : \tau(|x|) = \tau((x^*x)^{\frac{1}{2}}) < \infty\}$. It is known that $\mathcal{A}_1$ is dense in h in the topology of h. Furthermore, for $x \in \mathcal{A}_\tau$ and $y \in \mathcal{A}_1$, $|\tau((u_g(x) - x)^*y)| \leq \|(u_g(x) - x)^*\|\tau(|y|)$, which proves that $g \mapsto \tau((\alpha_g(x) - x)^*y)$ is continuous, by the strong continuity of α in the norm topology of $\mathcal{A}$. However, by the density of $\mathcal{A}_1$ and $\mathcal{A}_\tau$ in h and the fact that u_g is unitary, we conclude that for fixed $\xi \in h$, $g \mapsto u_g\xi$ is continuous in the weak topology of h, and hence is strongly continuous. □

The above lemma allows us to define $\alpha(f)(\xi) = \int f(g)u_g(\xi)dg \in h$ for $f \in C_c^\infty(G), \xi \in h$. Furthermore, from the expression $\alpha_g(x) = u_g x u_g^*$, it is possible to extend α_g to the whole of $\mathcal{B}(h)$ as a normal automorphism group implemented by the unitary group u_g on h and we shall denote this extended automorphism group also by the same notation.

We have the following Hilbert space analogue of the Theorem 8.1.1.

Theorem 8.1.17 *(i) We denote by d_k the self-adjoint generator of the unitary group $u_{\exp(t\chi_k)}$ on h such that $u_{\exp(t\chi_k)} = e^{itd_k}$, and let*

$$h_\infty \equiv \cap_{i_1,i_2,\cdots}\mathrm{Dom}(d_{i_1}d_{i_2}\cdots d_{i_k}; k = 1, 2, \cdots).$$

Then h_∞ is dense in h.

(ii) If we equip h_∞ with a family of Hilbertian norms $\|.\|_{2,n}; n = 0, 1, 2, \cdots$ given by,

$$\|\xi\|_{2,n}^2 \equiv \sum_{i_1,i_2,\cdots i_k; k\leq n} \|d_{i_1}d_{i_2}\cdots d_{i_k}(\xi)\|^2,$$

then h_∞ is complete with respect to the locally convex topology induced by the (countable) family of norms as defined above.

(iii) The family $\{u_g\}$ maps h_∞ into itself and the map $g \mapsto u_g\xi$ is smooth (C^∞) in the Fréchet topology for fixed $\xi \in h_\infty$.

(iv) Let $\mathcal{A}_{\infty,\tau} = \mathcal{A}_\infty \cap h_\infty$, then $\mathcal{A}_{\infty,\tau}$ is a closed two-sided $*$-ideal in $\mathcal{A}_\infty$ and is dense in $\mathcal{A}$, $\mathcal{A}_\infty$, h and h_∞ with respect to the respective topologies.

Proof:
The proof of (*i*)–(*iii*) are similar to that of the Theorem 8.1.1 and hence is omitted. To prove (iv), we need to note first that the elements of the form $\alpha(f)(\xi) = \int_G f(g)u_g\xi\, dg$, with $f \in C_c^\infty(G)$ and $\xi \in \mathcal{A}_\tau$, are clearly in $\mathcal{A}_{\infty,\tau}$. Let us first consider the density of such elements in h and h_∞. Since the topology of h_∞ is stronger than that of h and since h_∞ is dense in h in the topology of h, it suffices to prove that the set of elements of the above form is dense in h_∞ in the Fréchet topology. For this, we take $\xi \in h_\infty$, and choose a net x_ν of elements from $\mathcal{A}_\tau$ which converges in the topology of the Hilbert space h to ξ, and then it is clear that $\alpha(f)(x_\nu) \to \alpha(f)(\xi)$ for all $f \in C_c^\infty(G)$ in the Fréchet topology of h_∞,

since $d_{i_1} \cdots d_{i_k} \alpha(f)(x_\nu - \xi) = (-i)^k \alpha(\chi_{i_1} \cdots \chi_{i_k} f)(x_\nu - \xi)$ by an integration by parts. Thus, it is enough to show that $\{\alpha(f)(\xi), f \in C_c^\infty(G), \xi \in h_\infty\}$ is dense in h_∞ in the Fréchet topology. For this, we choose a net $f_p \in C_c^\infty(G)$ such that $\int_G f_p dg = 1$ for all p and the support of f_p converges to the singleton set containing the identity element of the group G, and then it is simple to see that $\alpha(f_p)(\xi) \to \xi$ in the Fréchet topology. Finally, the norm-density of $\mathcal{A}_{\infty,\tau}$ in $\mathcal{A}$ and the Fréchet density in $\mathcal{A}_\infty$ will follow by similar arguments. □

Remark 8.1.18 *It may be noted that for* $x \in \mathcal{A}_{\infty,\tau}$, $\delta_{i_1} \cdots \delta_{i_k}(x) = (-1)^k d_{i_1} \cdots d_{i_k}(x) \in \mathcal{A} \cap h$. *This follows from the fact that if* y_p *is a net in* $\mathcal{A} \cap h$ *which converges both in the norm topology of* $\mathcal{A}$ *as well as in the Hilbert space topology of* h, *then the norm-limit belongs to* h *and the two limits must coincide as vectors of* h.

Let us now proceed to the theory of H–P dilation for a class of Q.D.S. on $\mathcal{A}$. Let (T_t) be a Q.D.S. on $\mathcal{A}$ which is symmetric with respect to τ, that is, $\tau(T_t(x)y) = \tau(xT_t(y))$ for all positive $x, y \in \mathcal{A}$, and for all $t \geq 0$. At this point, the reader should recall the discussion in Chapter 3 , where an account of such semigroups from the point of view of Dirichlet forms has been given. As it is mentioned there, T_t can be canonically extended to a normal symmetric Q.D.S. on $\bar{\mathcal{A}}$ as well as to a C_0-semigroup of positive contractions on the Hilbert space h. We shall denote all these semigroups by the same symbol T_t as long as no confusion can arise. Furthermore, we assume that T_t on $\bar{\mathcal{A}}$ is conservative, that is, $T_t(1) = 1$ for all $t \geq 0$.

Let us denote by $\mathcal{L}$ the C^*-generator of T_t on $\mathcal{A}$, and by $\mathcal{L}_2$ the generator of T_t on h. Clearly, $\mathcal{L}_2$ is a negative self-adjoint map on h.

We now make a set of further assumptions.

Assumptions

(A1) T_t is covariant, that is, T_t commutes with α_g for all $t \geq 0, g \in G$.

(A2)$\mathcal{L}$ has $\mathcal{A}_\infty$ in its domain.

(A3) $\mathcal{L}_2$ has h_∞ in its domain.

The next lemma summarizes some of the implications of the assumptions made above for future use.

Lemma 8.1.19 We have the following.

(i) $\mathcal{A}_{\infty,\tau}$ is a core for both $\mathcal{L}$ and $\mathcal{L}_2$,
(ii) $\mathcal{L}(\mathcal{A}_{\infty,\tau}) \subseteq \mathcal{A}_{\infty,\tau}$,
(iii) $\mathcal{L}_2(\mathcal{A}_{\infty,\tau}) \subseteq \mathcal{A}_{\infty,\tau}$.

Proof:
By the Theorem 8.1.17, $\mathcal{A}_{\infty,\tau}$ is dense in $\mathcal{A}$ and h in their respective topologies. The hypothesis of covariance of T_t implies that $\mathcal{A}_{\infty,\tau}$ is invariant under T_t. Furthermore, by (A2)–(A3) $\mathcal{A}_{\infty,\tau}$ is in the domains of $\mathcal{L}$ and $\mathcal{L}_2$. Thus by Theorem 1.9 of [37], one has (i). It follows as in the proof of the Theorem 8.1.17 that $\mathcal{L}(\mathcal{A}_\infty) \subseteq \mathcal{A}_\infty$. Similarly, h_∞ is invariant under T_t and is a core for $\mathcal{L}_2$, and $\mathcal{L}_2(h_\infty) \subseteq h_\infty$. Since $\mathcal{L}$ and $\mathcal{L}_2$ coincide on $\mathcal{A}_{\infty,\tau} = \mathcal{A}_\infty \cap h_\infty$, the conclusions follow. □

It is appropriate to remark here that the assumption (**A3**) is the only hypothesis on the generator of the semigroup which involves the Hilbert space generator. However, in an important special case, namely, when the group G is compact and acts ergodically on the algebra $\mathcal{A}$ (see Chapter 2 for the definition of ergodic action on a C^*-algebra), the assumption (**A3**) follows automatically from the other hypotheses as shown below.

Lemma 8.1.20 *Let G be a compact Lie group acting ergodically on a unital C^*-algebra $\mathcal{A}$. Then $h_\infty = \mathcal{A}_\infty$. Moreover, if a Q.D.S. (T_t) on $\mathcal{A}$ satisfies the assumptions (**A1**) and (**A2**) (where symmetry is assumed with respect to the canonical trace τ on $\mathcal{A}$ described in Proposition 8.1.18), then it will satisfy (**A3**) too.*

Proof:
Let us recall from the Proposition 8.1.18 that for π in $\hat{G}$, the set of irreducible representations of G (which is countable since G is a compact Lie group) with n_π the dimension of the space (also denoted by π) of the representation π, $\{t^\pi_{ij} \in \mathcal{A},\ i = 1, 2, \ldots, n_\pi;\ j = 1, 2, \ldots, m_\pi \leq n_\pi;\ \pi \in \hat{G}\}$ is an orthonormal basis of h. Moreover, let $\{\lambda_\pi, \pi \in \hat{G}\}$ be the set of the eigenvalues, of multiplicity n_π^2, of the bi-invariant Laplacian (say Δ_G) on the compact manifold G (compact Lie group) of dimension N. Then by the Weyl asymptotics of the eigenvalues of the Laplacian (see [113]) we have that

$$\lambda_\pi = O(n_\pi^{\frac{2}{N}}),$$

so that for sufficiently large n,

$$\sum_\pi \frac{n_\pi^3}{\lambda_\pi^{2n}} \leq C \sum_\pi n_\pi^{(3-\frac{4n}{N})} < \infty \text{ (where } C \text{ is a positive constant).}$$

It is clear from the Proposition 8.1.18 that $\partial_k = id_k$ on each of the finite dimensional spaces $\mathcal{H}^\pi_l$ spanned by $\{t^\pi_{il},\ i = 1, \ldots, n_\pi\}$. We also note that the bi-invariant Laplacian Δ_G can be expressed as a linear combination of the form $\sum_{j,k=1}^N p_{jk}\, \chi_j \chi_k$ (where p_{jk} are real constants). Let us denote by $\mathcal{L}_0$ the unbounded operator $\sum_{j,k=1}^N p_{jk}\, d_j d_k$ on h, which coincides with $-\sum_{j,k=1}^N p_{jk}\, \partial_j \partial_k$ on each of the finite dimensional subspaces $\mathcal{H}_l^\pi$. It is also clear that $h_\infty \subseteq \mathrm{Dom}(\mathcal{L}_0^n)$ for all nonnegative integer n. The G-action on the $\mathcal{H}_l^\pi$ induces a homomorphism (say Ψ_l^π) of the universal enveloping algebra of the Lie algebra (see [74]), which sends the generator χ_j to id_j. In particular, $\Psi_l^\pi(\Delta_G) = -\mathcal{L}_0|_{\mathcal{H}_l^\pi}$. However, since the G-action on $\mathcal{H}_l^\pi$ is the irreducible representation π, $\Psi_l^\pi(\Delta_G)$ is nothing but $\lambda_\pi I_{\mathcal{H}_l^\pi}$. Thus we have that

$$\mathcal{L}_0|_{\mathcal{H}_l^\pi} = -\lambda_\pi I_{\mathcal{H}_l^\pi}. \tag{8.2}$$

For proving that $h_\infty = \mathcal{A}_\infty$ as sets, it is sufficient to show that for every $k \in \{1, \ldots, N\}$, one has $h_\infty \subseteq \mathrm{Dom}(\partial_k)$ and $\partial_k(h_\infty) \subseteq h_\infty$. To this end let us consider an arbitrary element $v \in h_\infty$ given by an L^2-convergent series $\sum_{\pi,i,j} c_{\pi,i,j} t_{ij}^\pi$. The fact that v is in $h_\infty \subseteq \mathrm{Dom}(\mathcal{L}_0^n)$ implies that (by (8.2))

$$\sum_{\pi,i,j} \lambda_\pi^{2n} |c_{\pi,i,j}|^2 < \infty$$

for every positive integer n. Since by the Proposition 2.1.18, $\|t_{ij}^\pi\| \leq \sqrt{n_\pi}$, it follows that

$$\sum_{\pi,i,j} |c_{\pi,i,j}| \|t_{ij}^\pi\| \leq \left(\sum_{\pi,i,j} |c_{\pi,i,j}|^2 \lambda_\pi^{2n} \right)^{\frac{1}{2}} \left(\sum_\pi \frac{n_\pi^3}{\lambda_\pi^{2n}} \right)^{\frac{1}{2}} < \infty.$$

This proves that the series $\sum_{\pi,i,j} c_{\pi,i,j} t_{ij}^\pi$ converges in the norm of $\mathcal{A}$, hence $v \in \mathcal{A}$. Since $\hat{G}$ is countable, let us identify $\hat{G}$ with $\mathbb{N}$ without loss of generality for the rest of the proof. Denoting by v_n the element $\sum_{l=1}^n \sum_{i,j} c_{l,i,j} t_{ij}^l$, we observe that v_n converges to v in the norm of $\mathcal{A}$. Moreover, since the finite dimensional space h_n spanned by $\{t_{ij}^l, i = 1, \ldots, n_l,\ j = 1, \ldots, m_l; l = 1, \ldots, n\}$, is invariant under d_k, d_k commutes with the projection P_n on h_n. Thus, $d_k v = \lim_{n \to \infty} P_n d_k v = \lim_{n \to \infty} d_k v_n$. If we write $d_k v_n$ as a sum of the form $\sum_{l=1}^n \sum_{i,j} b_{l,i,j} t_{ij}^l$, then it follows similarly that the (a-priori L^2-convergent) series $d_k v = \sum_{l=1}^\infty \sum_{i,j} b_{l,i,j} t_{ij}^l$ converges in the norm of $\mathcal{A}$. Therefore, $d_k v_n$ converges to $d_k v$ in the norm of $\mathcal{A}$. But d_k coincides with $-i\partial_k$ on h_n, hence $\partial_k v_n = i d_k v_n$, which converges in the norm of $\mathcal{A}$ to $i d_k v$. Thus, both v_n and $\partial_k v_n$ converge in the $\mathcal{A}$-norm, from which it follows by using the fact that ∂_k is closed that v is in the domain of ∂_k and $\partial_k v = i d_k v \in h_\infty$. Moreover, since the trace τ is finite, the Fréchet topology of $\mathcal{A}_\infty$ is stronger than that of h_∞, from which it is clear that the assumption (**A2**) implies (**A3**).

□

By Lemma 8.1.19, the domain $\mathcal{A}_{\infty,\tau}$ can be taken as a candidate of $\mathcal{A}_0$ in Theorem 3.2.30. Applying the Theorems 3.2.30 and 3.2.31 with $\mathcal{A}_0 = \mathcal{A}_{\infty,\tau}$, we obtain the following.

(*i*) A Hilbert space $\mathcal{K}$ equipped with an $\mathcal{A}$–$\mathcal{A}$ bimodule structure, in which the right action is denoted by $(a, \xi) \mapsto \xi a,\ \xi \in \mathcal{K},\ a \in \mathcal{A}$ and the left action by $(a, \xi) \mapsto \pi(a)\xi, \xi \in \mathcal{K}, a \in \mathcal{A}$.

(*ii*) A densely defined closable linear map δ_0 from $\mathcal{A}$ into $\mathcal{K}$ such that $\mathcal{A}_{\infty,\tau} \subseteq \mathrm{Dom}(\delta_0)$, and δ_0 is a bimodule derivation. Moreover, for any fixed $a \in \mathcal{A}_{\infty,\tau}$, the map $\mathcal{A}_{\infty,\tau} \ni b \mapsto \sqrt{2}\delta_0(a)b \in \mathcal{K}$ extends to a unique bounded linear map between the Hilbert spaces h and $\mathcal{K}$, and this bounded map will be denoted by $\delta(a)$.

(*iii*) $\partial\mathcal{L}(a, b, c) \equiv \delta(a)^*\pi(b)\delta(c) = \mathcal{L}(a^*bc) - \mathcal{L}(a^*b)c - a^*\mathcal{L}(bc) + a^*\mathcal{L}(b)c$, for $a, b, c \in \mathcal{A}_{\infty,\tau}$.

(*iv*) $\mathcal{K}$ is the closed linear span of $\{\delta(a)b : a, b \in \mathcal{A}_{\infty,\tau}\}$.

(*v*) π extends to a normal $*$-homomorphism on $\bar{\mathcal{A}}$.

Let $R : h \to \mathcal{K}$ be defined by,

$$\mathrm{Dom}(R) = \mathcal{A}_{\infty,\tau}, \quad Ra \equiv \sqrt{2}\delta_0(a).$$

Then R has a densely defined adjoint R^*, whose domain contains the linear span of the vectors $\delta(a)b,\ a, b \in \mathcal{A}_{\infty,\tau}$, and is given by

$$R^*(\delta(a)b) = a\mathcal{L}(b) - \mathcal{L}(a)b - \mathcal{L}(ab).$$

We denote the closure of R by the same notation again. For $a, b \in \mathcal{A}_{\infty,\tau}$,

$$\left(R^*\pi(a)R - \frac{1}{2}R^*Ra - \frac{1}{2}aR^*R\right)(b) = \mathcal{L}(a)b.$$

Furthermore,

$$\delta(a)b = (Ra - \pi(a)R)b, \text{ for } a, b \in \mathcal{A}_{\infty,\tau},$$

$$\mathcal{L}_2 = -\frac{1}{2}R^*R.$$

The next theorem gives the result on the existence of a unitary H–P dilation for T_t.

Theorem 8.1.21 *There exist a Hilbert space k_1 and a partial isometry $\Sigma : \mathcal{K} \to h \otimes k_0$ (where $k_0 = L^2(G) \otimes k_1$) such that $\pi(a) = \Sigma^*(a \otimes I_{k_0})\Sigma$ and $\tilde{R} \equiv \Sigma R$ is covariant in the sense that $(u_g \otimes v_g)\tilde{R} = \tilde{R}u_g$ on $\mathcal{A}_{\infty,\tau}$ where $v_g = L_g \otimes I_{k_1}$, L_g denoting the left regular representation of G in $L^2(G)$.*

Proof:
First a strongly continuous unitary representation of G is constructed in $\mathcal{K}$, using ideas in the proof of the Theorem 6.7.1 (see also [30]) by setting

$$V_g\delta(a)b := \delta(\alpha_g(a))\alpha_g(b) \text{ for } a, b \in \mathcal{A}_{\infty,\tau} \subseteq \mathrm{Dom}(\mathcal{L}),$$

and extending linearly. One can verify these properties of V_g by using the co-cycle relation (6.10) on $\mathcal{A}_{\infty,\tau}$, the covariance of $\mathcal{L}$ and the invariance of the trace τ, under the action of the group G. The strong continuity of $g \mapsto V_g$ follows by showing its weak continuity on vectors of the form $\sum_i \delta(a_i)b_i$, where $a_i, b_i \in \mathcal{A}_{\infty,\tau}$, and then extending the same to the whole of $\mathcal{K}$ by the density of this set of vectors and the unitarity of V_g. From the definition of V_g and R, it follows that $V_g\delta(a) = \delta(\alpha_g(a))$, or $V_gR = Ru_g$ on $\mathcal{A}_{\infty,\tau}$. Exactly as in (6.11) , we conclude that

$$V_g\pi(a)\delta(b) = \pi(\alpha_g(a))V_g\delta(b), \quad \text{for } a, b \in \mathcal{A}_{\infty,\tau},$$

and by density, we have that $\pi(\alpha_g(a)) = V_g\pi(a)V_g^*$ for all $a \in \mathcal{A}$. Thus π extends to a normal covariant $*$-representation of $\bar{\mathcal{A}}$ in $\mathcal{K}$, hence extends to a normal $*$-representation, say $\bar{\pi}$ of the crossed product von Neumann algebra $\mathcal{A} >\!\!\lhd G$, which is the weak closure of the algebra generated by $\{(a \otimes I_{L^2(G)}), (u_g \otimes L_g), a \in \bar{\mathcal{A}}, g \in G\}$ in $\mathcal{B}(h \otimes L^2(G))$. Thus there is an isometry $\Sigma : \mathcal{K} \to h \otimes L^2(G) \otimes k_1$ (for some k_1) such that $\Sigma^*(X \otimes I_{k_1})\Sigma = \bar{\pi}(X)$, for $X \in \bar{\mathcal{A}} >\!\!\lhd G$. So in particular $\Sigma^*(a \otimes I_{k_0})\Sigma = \pi(a)$, and $\Sigma^*(u_g \otimes v_g)\Sigma = V_g$. The rest of the proof follows from the fact that $\Sigma\Sigma^*$ commute with $u_g \otimes L_g \otimes I_{k_1}$ (by Proposition 2.1.7). □

It is clear that for $a \in \mathcal{A}_{\infty,\tau}$,

$$\mathcal{L}(a) = \tilde{R}^*(a \otimes 1_{k_0})\tilde{R} - \frac{1}{2}\tilde{R}^*\tilde{R}a - \frac{1}{2}a\tilde{R}^*\tilde{R}.$$

This enables us to write down the candidate for the unitary dilation for the Q.D.S. T_t.

Before stating and proving the main theorem concerning H-P dilation, we note that the form-generator is given by

$$\mathcal{B}(h) \ni a \mapsto \langle \tilde{R}u, (a \otimes 1)\tilde{R}v\rangle - \frac{1}{2}\langle au, \tilde{R}^*\tilde{R}v\rangle - \frac{1}{2}\langle \tilde{R}^*\tilde{R}u, av\rangle, \quad u, v \in \mathrm{Dom}(\tilde{R}^*\tilde{R}).$$

By the construction in Subsection 3.2.1, there exists a unique minimal Q.D.S. on $\mathcal{B}(h)$, say $\tilde{T}_t$, such that the predual semigroup of $\tilde{T}_t$, say $\tilde{T}_{*,t}$, has the property that all elements of the form $\rho = (1 + \tilde{R}^*\tilde{R})^{-1}\sigma(1 + \tilde{R}^*\tilde{R})^{-1}$ (for $\sigma \in \mathcal{B}_1(h)$) belongs to the generator of $\tilde{T}_{*,t}$ (say $\tilde{\mathcal{L}}_*$), and

$$\tilde{\mathcal{L}}_*(\rho) = \pi_*(\tilde{R}_1\sigma\tilde{R}_1^*) - \frac{1}{2}\tilde{R}_1^*\tilde{R}_1\sigma - \frac{1}{2}\sigma\tilde{R}_1^*\tilde{R}_1,$$

where $\tilde{R}_1 = \tilde{R}(1+\tilde{R}^*\tilde{R})^{-1}$ and π_* denotes the predual of the normal $*$-representation $a \mapsto (a \otimes 1)$ of $\mathcal{B}(h)$ into $\mathcal{B}(h \otimes k_0)$. Note that for $T \in \mathcal{B}_1(h \otimes k_0)$, $\pi_*(T) = \sum_i T_{ii}$, $T_{ii} \in \mathcal{B}_1(h)$ being the diagonal elements of T expressed in a block-operator form with respect to an orthonormal basis of k_0, and the sum is in the trace-norm.

Lemma 8.1.22 *$\tilde{T}_t$ is conservative.*

Proof:
Let $\tilde{\mathcal{L}}$ denote the generator of $\tilde{T}_t$. We claim that $\mathcal{A}_{\infty,\tau} \subseteq \mathrm{Dom}(\tilde{\mathcal{L}})$ and $\tilde{\mathcal{L}} = \mathcal{L}$ on $\mathcal{A}_{\infty,\tau}$. Fix any $a \in \mathcal{A}_{\infty,\tau}$. Let $\mathcal{D}$ denote the linear span of operators of the form $(1+\tilde{R}^*\tilde{R})^{-1}\sigma(1+\tilde{R}^*\tilde{R})^{-1}$ for $\sigma \in \mathcal{B}_1(h)$. Clearly, for $\rho \in \mathcal{D}$, $\mathrm{tr}(\tilde{\mathcal{L}}(a)\rho) = \mathrm{tr}(a\tilde{\mathcal{L}}_*(\rho)) = \mathrm{tr}(\mathcal{L}(a)\rho)$ (using the explicit forms of $\mathcal{L}$ and $\tilde{\mathcal{L}}$), and since by Lemma 3.2.5, $\mathcal{D}$ is a core for $\tilde{\mathcal{L}}$, we have that $\mathrm{tr}(a\tilde{\mathcal{L}}_*(\rho)) = \mathrm{tr}(\mathcal{L}(a)\rho)$ for all $\rho \in \mathrm{Dom}(\tilde{\mathcal{L}}_*)$ and $a \in \mathcal{A}_{\infty,\tau}$. Now, for $\rho \in \mathrm{Dom}(\tilde{\mathcal{L}}_*)$,

$$\mathrm{tr}\left(\frac{\tilde{T}_t(a)-a}{t}\rho\right) = \mathrm{tr}\left(a\left(\frac{\tilde{T}_{*,t}(\rho)-\rho}{t}\right)\right) = \mathrm{tr}\left(a\tilde{\mathcal{L}}_*\left(t^{-1}\int_0^t \tilde{T}_{*,s}(\rho)ds\right)\right)$$
$$= \mathrm{tr}\left(\mathcal{L}(a)\left(t^{-1}\int_0^t \tilde{T}_{*,s}(\rho)ds\right)\right);$$

and we extend this equality by continuity to all $\rho \in \mathcal{B}_1(h)$. Letting $t \to 0+$, we get that $a \in \mathrm{Dom}(\tilde{\mathcal{L}})$ and $\mathrm{tr}(\tilde{\mathcal{L}}(a)\rho) = \mathrm{tr}(\mathcal{L}(a)\rho)$ for all $\rho \in \mathcal{B}_1(h)$, which implies that a is in $\mathrm{Dom}(\tilde{\mathcal{L}})$ and $\tilde{\mathcal{L}}(a) = \mathcal{L}(a)$. This along with (**A2**) leads to the fact that $\mathcal{L}$ and hence $\tilde{\mathcal{L}}$ leaves $\mathcal{A}_{\infty,\tau}$ invariant and hence $(\lambda-\mathcal{L})^{-1}(a) = (\lambda-\tilde{\mathcal{L}})^{-1}(a)$ for all $\lambda > 0$ and all $a \in \mathcal{A}_{\infty,\tau}$. By using the formula $T_t = \mathrm{s}-\lim_{n\to\infty}(1-(t\mathcal{L})/n)^{-n}$, it follows that $\tilde{T}_t(a) = T_t(a)$ for all $a \in \mathcal{A}_{\infty,\tau}$, and hence by the ultra-weak density of $\tilde{\mathcal{A}}_{\infty,\tau}$ in $\bar{\mathcal{A}}$, T_t and $\tilde{T}_t$ agree on $\bar{\mathcal{A}}$, where we use the same notation for the C^*-semigroup T_t and its canonical normal extension on $\bar{\mathcal{A}}$. In particular $\tilde{T}_t(1) = 1$. □

We note that since the set of smooth complex-valued functions on G with compact support is dense in $L^2(G)$ in the L^2-norm, it is clear that $(k_0)_\infty$ is dense in the Hilbert space k_0. Furthermore, note that k_0 can of course be chosen to be separable since $\bar{\mathcal{A}}$ is σ-finite von Neumann algebra and G is second countable.

Theorem 8.1.23 *The Q.S.D.E.*

$$dU_t = U_t(a^\dagger_{\tilde{R}}(dt) - a_{\tilde{R}}(dt) - \frac{1}{2}\tilde{R}^*\tilde{R}dt); \quad U_0 = I \tag{8.3}$$

on the space $h \otimes \Gamma(L^2(\mathbb{R}_+) \otimes k_0)$ admits a **unitary** *operator- valued solution which implements a H–P dilation for T_t.*

Proof:
Since $\tilde{R}^*\tilde{R} = -2\mathcal{L}_2$, and since $h_\infty \subseteq \mathrm{Dom}(\mathcal{L}_2) \subseteq \mathrm{Dom}(\tilde{R})$, the closed Hilbert space operator $\tilde{R}$ is also continuous as a map from h_∞ to $h \otimes k_0$ with respect to the Fréchet topology of the domain. Thus the relation $\tilde{R}u_g = (u_g \otimes v_g)\tilde{R}$ on $\mathcal{A}_{\infty,\tau}$ extends by continuity to h_∞. That is, $\tilde{R}$ is covariant, and by the assumptions made on $\mathcal{L}_2$ at the beginning of this section it can be seen that the conditions of the Theorem 8.1.13 are satisfied, so that there are C, p such that $\|\tilde{R}w\|_{2,0} \leq C\|w\|_{2,p}$. Moreover, by Theorem 8.1.13, we obtain in particular that $\mathrm{Dom}(\tilde{R}^*)$ contains $(h \otimes k_0)_\infty$. For any vector $\xi \in (k_0)_\infty$, it is clear that $h_\infty \subseteq \mathrm{Dom}(\langle \xi, \tilde{R}\rangle^*)$.

We shall now apply the Theorem 7.2.3 of Chapter 7 to prove the existence and unitarity of solution of the Q.S.D.E. (8.3). To this end, take $\mathcal{D}_0 = \tilde{\mathcal{D}_0} = h_\infty$, $\mathcal{V}_0 = \tilde{\mathcal{V}_0} = k_{0\infty}$ and $Z = \begin{pmatrix} -\frac{1}{2}\tilde{R}^*\tilde{R} & -\tilde{R}^* \\ \tilde{R} & 0 \end{pmatrix}$. Let $G_n = n(n - \mathcal{L}_2)^{-1}$, and

$$Z^{(n)} := \begin{pmatrix} -\frac{1}{2}G_n\tilde{R}^*\tilde{R}G_n & -G_n\tilde{R}^* \\ \tilde{R}G_n & 0 \end{pmatrix}.$$

We shall show that the hypotheses of the Theorem 7.2.3 are satisfied. Clearly, $Z^{(n)*}$ and $Z^{(n)}$ belong to $\mathcal{Z}_c$. Furthermore, G_n is covariant map under the action of G and $\|G_n\| \leq 1$, and therefore is smooth of order 0 with bound ≤ 1, and it maps $\mathcal{D}_0$ into itself. We have that for $w \in \mathcal{D}_0$,

$$\begin{aligned} \|\tilde{R}G_n w\|^2 &= \langle \tilde{R}G_n w, \tilde{R}G_n w\rangle = \langle w, G_n^*(-2\mathcal{L}_2)G_n w\rangle \\ &= \langle w, (-2\mathcal{L}_2)^{\frac{1}{2}}G_n^*G_n(-2\mathcal{L}_2)^{\frac{1}{2}}w\rangle \text{ (since } \mathcal{L}_2, G_n \text{ commute)} \\ &= \|G_n(-2\mathcal{L}_2)^{\frac{1}{2}}w\|^2 \leq \|(-2\mathcal{L}_2)^{\frac{1}{2}}w\|^2. \end{aligned}$$

Therefore, for $\zeta \in \mathcal{V}_0$, $w \in \mathcal{D}_0$, we have that

$$\begin{aligned} \|Z_{\hat{\xi}}^{(n)}w\|^2 &= \|-\frac{1}{2}G_n\tilde{R}^*\tilde{R}G_n w + G_n\tilde{R}^*(w\xi)\|^2 + \|\tilde{R}G_n w\|^2 \\ &\leq 2\|G_n^2(-\mathcal{L}_2)w\|^2 + 2\|G_n\tilde{R}^*(w\xi)\|^2 + \|\tilde{R}G_n w\|^2 \\ &\leq 2\|(-\mathcal{L}_2)w\|^2 + \|(-2\mathcal{L}_2)^{\frac{1}{2}}w\|^2 + 2\|\tilde{R}^*(w\xi)\|^2, \end{aligned}$$

and thus $\sup_{n\geq 1}\|Z_{\hat{\xi}}^{(n)}w\|^2 < \infty$ and the condition (7.6) is verified.

To verify that $\lim_{n\to\infty}\langle \hat{\eta}, Z_{\hat{\xi}}^{(n)}\rangle w = \langle \hat{\eta}, Z_{\hat{\xi}}\rangle w$ for all $w \in \mathcal{D}_0$, $\xi, \eta \in \mathcal{V}_0$, we first prove the following general fact.
If L is a closed linear map from h to h with h_∞ in its domain, so that $\|Lw\|_{2,0} \leq M\|w\|_{2,r}$ for some M and r, then for $w \in h_\infty$, each of the sequences G_nLw, LG_nw and G_nLG_nw converges to Lw as $n \to \infty$. To prove this fact, it

suffices to observe that $G_n w$ belongs to h_∞ and since G_n is covariant and $\text{s} - \lim_{n\to\infty} G_n = I$, we have that

$$\|G_n w - w\|_{2,r}^2 = \sum_{i_1, i_2, \cdots i_k; k \le r} \|(G_n - I)(d_{i_1} d_{i_2} \cdots d_{i_k} w)\|_{2,0}^2 \to 0 \text{ as } n \to \infty.$$

Thus we get

$$\begin{aligned}
&\|G_n L G_n w - L w\|_{2,0} \\
&\le \|G_n L(G_n w - w)\|_{2,0} + \|(G_n - I)Lw\|_{2,0} \\
&\le M\|G_n w - w\|_{2,r} + \|(G_n - I)Lw\|_{2,0},
\end{aligned}$$

which proves that $G_n L G_n w \to Lw$. Similarly, one can show $G_n L w \to Lw$ and $LG_n w \to Lw$.

Using this fact, we observe that

$$\begin{aligned}
\langle \hat{\eta}, Z_{\hat{\xi}}^{(n)} \rangle w \;&=\; -\frac{1}{2} G_n \tilde{R}^* \tilde{R} G_n w - G_n \tilde{R}^*(w\xi) + \langle \eta, \tilde{R} G_n w \rangle \\
\to -\frac{1}{2} \tilde{R}^* \tilde{R} w - \tilde{R}^*(w\xi) + \langle \eta, \tilde{R} w \rangle \;&=\; \langle \hat{\eta}, Z_{\hat{\xi}} \rangle w.
\end{aligned}$$

Similar facts can be proved replacing Z by Z^* and $Z^{(n)}$ by $Z^{(n)^*}$. The conditions (7.13) and (7.17) are also simple to verify. Moreover, we have $\mathcal{L}_0 = \tilde{\mathcal{L}_0}$ and $\beta_\lambda = \tilde{\beta}_\lambda$ in this case. Since h_∞ is a core for $\tilde{R}^* \tilde{R}$ and $\tilde{T}_t$ is conservative, it follows from the Remark 3.2.17 of Chapter 3 that $\beta_\lambda = \{0\}$. This proves that U_t exists and is unitary for all t. Moreover, it is clear from the Q.S.D.E. satisfied by U_t that

$$\langle we(0) U_t (a \otimes I) U_t^* w' e(0) \rangle = \langle w, T_t(a) w' \rangle$$

for all $w, w' \in h$ and $a \in \mathcal{A}$. This completes the proof. □

We conclude this subsection by mentioning a few natural examples of Q.D.S. which satisfy the assumptions (**A1**)–(**A3**).

Example 1

Let $\mathcal{A} = C_0(\mathbb{R}^n)$, $G = \mathbb{R}^n$, with the obvious action of $\mathbb{R}^n$ on $\mathcal{A}$ by translation. The trace τ is given by integration with respect to the Lebesgue measure. We take T_t to be the heat semigroup on $\mathbb{R}^n$, which is given by

$$(T_t f)(x) = \frac{1}{(\sqrt{2\pi t})^n} \int_{\mathbb{R}^n} f(y) \exp\left(-\frac{\sum_i (x_i - y_i)^2}{2t} \right) dy, \quad t > 0;$$

and $T_0 f = f$. It can be verified by simple calculation that T_t is indeed covariant and symmetric. Furthermore, the norm-generator $\mathcal{L}$ of T_t is nothing but the differential operator $\sum_{i=1}^n \frac{\partial^2}{\partial x_i^2}$, from which it is clearly seen that (**A2**) and (**A3**) are satisfied.

Example 2

This is an example from noncommutative geometry (see Chapter 9). Consider the noncommutative $2d$-dimensional plane considered in the Chapter 9, with the notation explained there. We claim that the Q.D.S (T_t) generated by the 'Laplacian' $-\sum_{j=1}^{2d} \delta_j^2$ is covariant with respect to the action ϕ_α of $\mathbb{R}^{2d}$, and it is also symmetric with respect to the canonical trace τ on the noncommutative $2d$-plane. To verify the covariance, we observe (using the notation of Chapter 9) the following:

$$\phi_\alpha(b(f)) = b(f_\alpha),$$

where $\hat{f}_\alpha(x) = e^{i\alpha x}\hat{f}(x)$. Thus,

$$T_t(b(f_\alpha)) = \int_{\mathbb{R}^{2d}} e^{-\frac{t}{2}x^2} e^{i\alpha x}\hat{f}(x)W_x dx = \phi_\alpha(T_t(b(f))).$$

Moreover, we have,

$$\tau(T_t(b(f)^*)b(g)) = \int_{\mathbb{R}^{2d}} e^{-\frac{t}{2}x^2}\overline{\hat{f}(x)}\hat{g}(x)dx = \tau(b(f)^* T_t(b(g))),$$

which proves symmetry. A simple computation shows that the assumptions (**A2**) and (**A3**) hold.

8.1.3 E–H dilation of covariant quantum dynamical semigroups

We now build a theory of E–H dilation under the assumption of compactness of G. However, the existence of a trace and the symmetry condition used in the theory of H–P dilation are no longer required. The background is as in the beginning of the Section 8.1, that is, we have a separable C^*-algebra $\mathcal{A} \subseteq \mathcal{B}(h)$ for some separable Hilbert space h, with the action of a Lie group G on $\mathcal{A}$. We first study the relation of covariance with complete smoothness, and begin with the standard scenario of having a Q.D.S. (T_t) on $\mathcal{A}$ with the generator $\mathcal{L}$. Assume that T_t is conservative, or more specifically, $\mathcal{L}(1) = 0$.

Lemma 8.1.24 *T_t is covariant for every t if and only if $\mathcal{L}$ is covariant.*

Proof:

Suppose that T_t is covariant for all t. Take $a \in \mathrm{Dom}(\mathcal{L})$, $g \in G$. We have for every $t > 0$, $\frac{T_t(\alpha_g(a))-\alpha_g(a)}{t} = \alpha_g\left(\frac{T_t(a)-a}{t}\right)$. Thus,

$$\lim_{t\to 0+} \frac{T_t(\alpha_g(a)) - \alpha_g(a)}{t} = \lim_{t\to 0+} \alpha_g\left(\frac{T_t(a) - a}{t}\right) = \alpha_g(\mathcal{L}(a)).$$

Therefore, $\alpha_g(a) \in \mathrm{Dom}(\mathcal{L})$ and $\mathcal{L}(\alpha_g(a)) = \alpha_g(\mathcal{L}(a))$.

Conversely, if $\mathcal{L}$ is covariant, it is clear the resolvent $(\lambda - \mathcal{L})^{-1}$ is covariant for every $\lambda > 0$ and hence $T_t = \mathrm{s} - \lim_{n\to\infty}(1 - \frac{t}{n}\mathcal{L})^{-n}$ is also covariant. □

We now make the basic assumption of this subsection.

(B1) T_t is covariant with respect to the action of a compact group G.

This assumption allows us to construct the structure maps satisfying the covariance condition.

Theorem 8.1.25 *There exist a Sobolev–Hilbert space (k_0, v_g) (with k_0 separable), maps π, δ defined on $\mathcal{A}_\infty$ satisfying the following.*

(i) π is a $*$-representation from $\mathcal{A}_\infty$ to $\mathcal{A} \otimes \mathcal{B}(k_0)$, with $\pi(1) = 1$.
(ii) $\delta : \mathcal{A}_\infty \to \mathcal{A} \otimes k_0$ is π-derivation.
(iii) $\delta(a)^*\delta(b) = \mathcal{L}(a^*b) - \mathcal{L}(a^*)b - a^*\mathcal{L}(b)$ for all $a, b \in \mathcal{A}_\infty$.
(iv) $\pi(\alpha_g(a)) = (\alpha_g \otimes \beta_g)(\pi(a))$, for all $a \in \mathcal{A}_\infty$ and $g \in G$, where β_g is the G-action on $\mathcal{B}(k_0)$ given by $v_g \cdot v_g^*$.
(v) $\delta(\alpha_g(a)) = (\alpha_g \otimes v_g)(\delta(a))$ for all $a \in \mathcal{A}_\infty$, $g \in G$.

Proof:
We observe that the proof consists of an adaptation of the line of reasoning used in the proof of the Theorem 6.7.1, replacing $\mathcal{A}$ by $\mathcal{A}_\infty$ and noting that $\mathcal{L}(\mathcal{A}_\infty) \subseteq \mathcal{A}_\infty$, which allows the algebraic manipulations made in that proof to be justified in this case as well. □

Theorem 8.1.26 *Let $\hat{k_0} = \mathbb{C} \oplus k_0$, and $\hat{v_g} = 1 \oplus v_g$ be the action of G on $\hat{k_0}$. We define, as in Chapters 5 and 6, the structure matrix $\Theta : \mathcal{A}_\infty \to \mathcal{A} \otimes \mathcal{B}(\hat{k_0})$ as follows:*

$$\Theta(a) = \begin{pmatrix} \mathcal{L}(a) & \delta^\dagger(a) \\ \delta(a) & \sigma(a) \end{pmatrix};$$

where $\delta^\dagger(a) = \delta(a^*)^*$, $\sigma(a) = \pi(a) - a \otimes 1_{k_0}$ for $a \in \mathcal{A}_\infty$. We consider the G-action $\alpha_g \otimes \hat{\beta}_g$ on $\mathcal{A} \otimes \mathcal{B}(\hat{k_0})$ (where $\hat{\beta}_g := \hat{v_g} \cdot \hat{v_g}$), and the corresponding Fréchet algebra $(\mathcal{A} \otimes \mathcal{B}(\hat{k_0}))_\infty$. Then Θ is a covariant smooth map from the Fréchet algebra $\mathcal{A}_\infty$ to the Fréchet algebra $(\mathcal{A} \otimes \mathcal{B}(\hat{k_0}))_\infty$.

Proof:
We already know that $\mathcal{L}$ is smooth covariant, say of order p and with a bound C. Thus, $\|\mathcal{L}(a)\|_n \leq C\|a\|_{n+p}$ for all $a \in \mathcal{A}_\infty$ and $n \geq 0$. Since π is a $*$-homomorphism on $\mathcal{A}_\infty$ with $\pi(1) = 1$, it follows from the Corollary 8.1.6 that π extends to a C^*-homomorphism on $\mathcal{A}$, and it is clearly covariant, thus smooth of order 0. Let us now show that δ (and hence $\delta^\dagger$) is smooth. To this end, we fix any $i \in \{1, \ldots, N\}$ and consider the one parameter subgroup

$g_t = \exp(t\chi_i)$ of G, and denote by α_t^i, γ_t^i the automorphism α_{g_t} and $\alpha_{g_t} \otimes v_{g_t}$, respectively. We claim that for all $a \in \mathcal{A}_\infty$

$$\nabla_i \delta(a) = \delta(\partial_i(a)). \tag{8.4}$$

where ∂_i and ∇_i denote the generator of α_t^i and γ_t^i respectively. To prove the above claim, we observe the following with $a \in \mathcal{A}_\infty$:

$$\begin{aligned}
&\left\langle \frac{\gamma_t^i(\delta(a)) - \delta(a)}{t} - \delta(\partial_i(a)), \frac{\gamma_t^i(\delta(a)) - \delta(a)}{t} - \delta(\partial_i(a)) \right\rangle \\
&= \left\langle \delta\left(\frac{\alpha_t^i(a) - a}{t} - \partial_i(a)\right), \delta\left(\frac{\alpha_t^i(a) - a}{t} - \partial_i(a)\right) \right\rangle \\
&= \mathcal{L}(a_t^* a_t) - \mathcal{L}(a_t^*) a_t - a_t^* \mathcal{L}(a_t),
\end{aligned}$$

where $a_t = (\alpha_t^i(a) - a)/t - \partial_i(a)$. Note that in the above we have used the notation $\langle \cdot, \cdot \rangle$ to denote the natural $\mathcal{A}$-valued inner product of the $G - \mathcal{A}$ module $((\mathcal{A} \otimes k_0), (\alpha_g \otimes v_g))$. Now, $a_t \to 0$ as $t \to 0+$ in the Fréchet topology, so $\|\mathcal{L}(a_t^* a_t)\|_0 \leq C\|a_t^* a_t\|_p \leq C\|a_t\|_p^2 \to 0$, and similarly, $\mathcal{L}(a_t^*) a_t + a_t^* \mathcal{L}(a_t) \to 0$. Thus, $\frac{\gamma_t^i(\delta(a)) - \delta(a)}{t} \to \delta(\partial_i(a))$ as $t \to 0+$ in the norm topology of $\mathcal{A} \otimes k_0$, which proves (8.4).

By repeated application of (8.4), we get $\nabla_{i_1} \cdots \nabla_{i_k} \delta(a) = \delta(\partial_{i_1} \cdots \partial_{i_k}(a))$ for all $a \in \mathcal{A}_\infty$ and $i_1, \ldots, i_k$. Observe also that

$$\begin{aligned}
\|\delta(a)\|^2 = \|\delta(a)^* \delta(a)\| &\leq \|\mathcal{L}(a^* a)\| + 2\|\mathcal{L}(a)\| \|a\| \\
&\leq C\|a\|_p^2 + 2C\|a\|_p \|a\| \leq 3C\|a\|_p^2.
\end{aligned}$$

Thus,

$$\begin{aligned}
\|\delta(a)\|_{2,n}^2 &= \sum_{i_1,\cdots,i_k;\ k \leq n} \|\nabla_{i_1} \cdots \nabla_{i_k} \delta(a)\|^2 \\
&= \sum_{i_1,\cdots,i_k;\ k \leq n} \|\delta(\partial_{i_1} \cdots \partial_{i_k}(a))\|^2 \\
&\leq 3C \sum_{i_1,\cdots,i_k;\ k \leq n} \|\partial_{i_1} \cdots \partial_{i_k}(a)\|_p^2 \\
&= 3C \sum_{i_1,\cdots,i_k;\ k \leq n} \left\{ \sum_{j_1,\cdots j_l;\ l \leq p} \|\partial_{j_1} \cdots \partial_{j_l} \partial_{i_1} \cdots \partial_{i_k}(a)\| \right\}^2 \\
&\leq 3C2^p \sum_{i_1,\cdots i_k;\ k \leq n} \sum_{j_1,\cdots j_l;\ l \leq p} \|\partial_{j_1} \cdots \partial_{j_l} \partial_{i_1} \cdots \partial_{i_k}(a)\|^2 \\
&\leq 3.2^p C\|a\|_{2,n+p}^2.
\end{aligned}$$

This completes the proof of smoothness of δ. □

Now, we recall from Chapter 6 that to construct E–H dilation, one needs to iterate the structure matrix Θ, and make estimates in suitable topology. Thus, it can be seen that we shall need to consider ampliation $\Theta \otimes \mathrm{Id}$ of Θ, and to make sense of such ampliation and also to ensure that it remains a smooth map, we must put some restrictions on $\mathcal{L}$. This motivates the discussion that follows. Consider two Fréchet algebras $\mathcal{B}_\infty$ and $\mathcal{C}_\infty$ with respect to the G-actions β_g and γ_g on the C^*-algebras $\mathcal{B}$, $\mathcal{C}$ respectively. Let $\|\cdot\|_n$ denote the n th norm for both the Fréchet algebras. Now, let us recall the definition of a smooth map of order p from $\mathcal{B}_\infty$ to $\mathcal{C}_\infty$. If L is a smooth map of order p, with a bound C, say, then it is clear that each of the maps $L_N \equiv L \otimes \mathrm{Id}_{M_N}$ is also smooth of the same order, but the bound may vary. This motivates the following definition, which is very similar to the concept of complete boundedness.

Definition 8.1.27 A smooth map L from $\mathcal{B}_\infty$ to $\mathcal{C}_\infty$ is said to be *completely smooth* if there is a nonnegative integer p and a constant $C > 0$ such that for integers $n \geq 0$ and $N \geq 1$, one has that

$$\|(L \otimes \mathrm{Id}_{M_N})(\xi)\|_n \leq C \|\xi\|_{n+p} \quad \text{for all } \xi \in (\mathcal{B} \otimes M_N)_\infty.$$

Such a map L is said to be *completely smooth of order* p.

Before we proceed further, we state and prove a simple result which provides many examples of completely smooth maps.

Theorem 8.1.28 *(i) Composition of two completely smooth maps is again completely smooth.*

(ii) The map $L = \partial_i^\beta : \mathcal{B}_\infty \to \mathcal{B}_\infty$, *where* $\partial_i := \partial_i^\beta$ *denotes the generator of the one-parameter group* $\beta_{\exp(t\chi_i)}$ *(where* $\{\chi_1, \ldots, \chi_N\}$ *is a basis of the Lie algebra of G), is completely smooth of order* 1.

Proof:
The fact (*i*) is a simple consequence of the definition of complete smoothness, hence omitted. To prove (*ii*), we note that with $X = \sum_{j=1}^m b_j \otimes c_j$, where $b_j \in \mathcal{B}_\infty$, $c_j \in M_N(\mathbb{C})$ ($N \geq 1$) and a nonnegative integer n,

$$\begin{aligned}
&\|(L \otimes \mathrm{Id})(X)\|_n \\
&= \sum_{i_1,\cdots,i_k;\ k \leq n} \left\| \sum_{j=1}^m (\partial_{i_1} \otimes \mathrm{Id}) \cdots (\partial_{i_k} \otimes \mathrm{Id})(L(b_j) \otimes c_j) \right\| \\
&= \sum_{i_1,\cdots,i_k; k \leq n} \left\| \sum_j (\partial_{i_1} \otimes \mathrm{Id}) \cdots (\partial_{i_k} \otimes \mathrm{Id})(\partial_i \otimes \mathrm{Id})(b_j \otimes c_j) \right\| \\
&\leq \sum_{i_1,\cdots,i_k,\ k \leq n+1} \left\| \sum_j (\partial_{i_1} \otimes \mathrm{Id}) \cdots (\partial_{i_k} \otimes \mathrm{Id})(b_j \otimes c_j) \right\|
\end{aligned}$$

$$= \sum_{i_1,\cdots,i_k,\ k\leq n+1} \|(\partial_{i_1}\otimes \mathrm{Id})\cdots(\partial_{i_k}\otimes \mathrm{Id})(X)\|$$
$$= \|X\|_{n+1}.$$

This completes the proof. □

Remark 8.1.29 *Combining (i) and (ii) of the Theorem 8.1.28, it is clear that any map of the form $\sum_{i_1,\ldots,i_k} \partial_{i_1}\cdots\partial_{i_k}$, where the sum is over a finite set of indices, is completely smooth. This gives typical examples of completely smooth maps that will occur in our discussion.*

Recall that for a $G-\mathcal{A}$ module (E,γ) with the natural Fréchet topology given by the family of norms $\{\|\cdot\|_{2,n}\}$, we can identify E_∞ with a closed subspace of $(\mathcal{K}(\mathcal{A}\oplus E))_\infty$ by the identification $\xi\mapsto L_\xi$ (where $L_\xi(a\oplus\eta)=0\oplus\xi a$). This allows us to extend the notion of complete smoothness.

Definition 8.1.30 We say that a map from a Fréchet algebra $\mathcal{B}_\infty$ to the Fréchet space E_∞ is *completely smooth* if the map is completely smooth in the sense of Definition 8.1.27 when viewed as a map from $\mathcal{B}_\infty$ into $(\mathcal{K}(\mathcal{A}\oplus E))_\infty$ using the embedding $E_\infty\hookrightarrow(\mathcal{K}(\mathcal{A}\oplus E))_\infty$.

Theorem 8.1.31 *Suppose that L is a smooth map from $\mathcal{B}_\infty$ to $\mathcal{C}_\infty$. Then the following statements are equivalent.*

(i) L is completely smooth of order p.

(ii) For any separable infinite dimensional Hilbert space $\mathcal{H}$, The map $\tilde{L}:=L\otimes_{\mathrm{alg}}\mathrm{Id}$ from $\mathcal{B}_\infty\otimes_{\mathrm{alg}}\mathcal{K}(\mathcal{H})$ to $\mathcal{C}_\infty\otimes_{\mathrm{alg}}\mathcal{K}(\mathcal{H})$ (where $\mathcal{K}(\mathcal{H})$ denotes the algebra of compact operators on $\mathcal{H}$) extends to a smooth map of order p between the Fréchet algebras $\tilde{\mathcal{B}}_\infty\equiv(\mathcal{B}\otimes\mathcal{K}(\mathcal{H}))_\infty^{\beta\otimes\mathrm{id}}$ and $\tilde{\mathcal{C}}_\infty\equiv(\mathcal{C}\otimes\mathcal{K}(\mathcal{H}))_\infty^{\gamma\otimes\mathrm{id}}$.

Proof:
Let us choose and fix an orthonormal basis $\{\xi_1,\xi_2,\ldots\}$ of $\mathcal{H}$ and denote by P_N the projection onto the subspace spanned by $\{\xi_1,\ldots,\xi_N\}$. Clearly, $\mathcal{K}_N\equiv P_N\mathcal{K}(\mathcal{H})P_N$ is isomorphic with $M_N(\mathbb{C})$, and thus $\mathcal{B}\otimes\mathcal{K}_N\cong\mathcal{B}\otimes M_N$. For brevity, we shall denote the Fréchet algebra $(\mathcal{B}\otimes M_N)_\infty$ by $\mathcal{B}_\infty^N$ and by e_N the map $\mathrm{Id}\otimes(P_N\ \cdot\ P_N)$. Since $\beta_g\otimes\mathrm{Id}$ commutes with e_N, it is clear that for any $a\in\tilde{\mathcal{B}}_\infty$, we have $e_N(a)\in\mathcal{B}_\infty^N$ and $e_N(a)\to a$ in the Fréchet topology of $\tilde{\mathcal{B}}_\infty$. Similar conclusions hold for $\tilde{\mathcal{C}}_\infty$. Next assume that (i) holds and let C be a constant such that $\|(L\otimes\mathrm{Id}_{M_N})(\xi)\|_n\leq C\|\xi\|_{n+p}$ for all $\xi\in\mathcal{B}_\infty^N$, $N\geq 1$. For $a\in\tilde{\mathcal{B}}_\infty$, $n\geq 0$ and positive integers $N\geq M$, we have

$$\|\tilde{L}(e_N(a))-\tilde{L}(e_M(a))\|_n$$
$$=\|L_N(e_N(a)-e_M(a))\|_n\ \ (\text{where } L_N=L\otimes\mathrm{Id}_{M_N})$$
$$\leq C\|e_N(a)-e_M(a)\|_{n+p}.$$

From this it follows that $\tilde{L}(e_N(a))$ is a Cauchy sequence in the Fréchet topology of $\tilde{\mathcal{C}}_\infty$, hence converges, and the limit defines $\tilde{L}(a)$. Moreover, it is clear that $\|\tilde{L}(a)\|_n \leq C\|a\|_{n+p}$ for all n, thus (ii) is proved. The converse implication $((ii) \Rightarrow (i))$ is the consequence of the observation that for each N, $\mathcal{B}_\infty^N$ and $\mathcal{C}_\infty^N$ are embedded naturally as topological subalgebra in the Fréchet algebras $\tilde{\mathcal{B}}_\infty$ and $\tilde{\mathcal{C}}_\infty$ respectively. □

Consider the trivial ampliation $(T_t \otimes \mathrm{Id})$ of T_t on each of the algebras $\mathcal{A} \otimes M_N$ and also $\mathcal{A} \otimes \mathcal{K}(\mathcal{H})$ for any fixed separable infinite dimensional $\mathcal{H}$ (for example l^2). We denote by $\mathcal{L}_N$ and $\tilde{\mathcal{L}}$ the generators of $(T_t \otimes \mathrm{Id}_{M_N})$ and $(T_t \otimes \mathrm{Id}_{\mathcal{K}(\mathcal{H})})$, respectively. Note that $\tilde{\mathcal{L}}$ has in its domain at least $\mathcal{A}_\infty \otimes_{\mathrm{alg}} \mathcal{K}(\mathcal{H})$ and is given by $\mathcal{L} \otimes_{\mathrm{alg}} \mathrm{Id}$ on that domain, but it is not known a-priori whether the full domain of $\tilde{\mathcal{L}}$ contains the whole of $\tilde{\mathcal{A}}_\infty \equiv (\mathcal{A} \otimes \mathcal{K}(\mathcal{H}))_\infty$. The following result tells precisely when it does so.

Lemma 8.1.32 *Assume that T_t (equivalently $\mathcal{L}$) is covariant. Then $\mathcal{L}$ is completely smooth of some order if and only if the domain of $\tilde{\mathcal{L}}$ contains $\tilde{\mathcal{A}}_\infty$.*

Proof:
Since $\tilde{\mathcal{L}}$ is the generator of a Q.D.S., it is closed in the norm topology of $\tilde{\mathcal{A}}_\infty$. Thus the result follows from Lemma 8.1.9. □

Corollary 8.1.33 *Assume that T_t is covariant and that its generator $\mathcal{L}$ is completely smooth. Consider any separable Hilbert space $\mathcal{H}'$ with a G-action η_g on $\mathcal{K}(\mathcal{H}')$. Then $\mathcal{L} \otimes_{\mathrm{alg}} \mathrm{Id}_{\mathcal{K}(\mathcal{H}')}$ extends to a smooth covariant map of some order on $(\mathcal{A} \otimes \mathcal{K}(\mathcal{H}'))_\infty^{\alpha\otimes\eta}$.*

Proof:
Consider the Q.D.S. $(T_t \otimes \mathrm{Id})$ on $\mathcal{A} \otimes \mathcal{K}(\mathcal{H}')$, (which is covariant under the action $\alpha \otimes \eta$) with the norm-generator $\hat{\mathcal{L}}$, say. Since this Q.D.S. is covariant with respect to $\alpha \otimes \mathrm{id}$ and $\mathcal{L}$ is completely smooth, we have by Lemma 8.1.32 that the domain of $\hat{\mathcal{L}}$ contains $(\mathcal{A} \otimes \mathcal{K}(\mathcal{H}'))_\infty^{\alpha\otimes\mathrm{id}}$. However, it is clear that $(\mathcal{A} \otimes \mathcal{K}(\mathcal{H}'))_\infty^{\alpha\otimes\eta} \subseteq (\mathcal{A} \otimes \mathcal{K}(\mathcal{H}'))_\infty^{\alpha\otimes\mathrm{id}} \equiv \tilde{\mathcal{A}}_\infty$. So, $(\mathcal{A} \otimes \mathcal{K}(\mathcal{H}'))_\infty^{\alpha\otimes\eta}$ is contained in the domain of $\hat{\mathcal{L}}$ and $\hat{\mathcal{L}}$ is also covariant under $\alpha \otimes \eta$. It follows by applying Lemma 8.1.9 that $\hat{\mathcal{L}}$, which is clearly an extension of $\mathcal{L} \otimes_{\mathrm{alg}} \mathrm{Id}$, is smooth of some order in the Fréchet topology of $(\mathcal{A} \otimes \mathcal{K}(\mathcal{H}'))_\infty^{\alpha\otimes\eta}$. □

We now make the following additional assumption about the generator $\mathcal{L}$ of the Q.D.S., namely:

(**B2**) $\mathcal{L}$ is completely smooth.

We shall prove the existence of an E–H dilation of (T_t) under the assumptions (**B1**) and (**B2**), which we assume for the rest of this section. To construct the E–H dilation, we first prove the existence of covariant and completely smooth structure maps.

Since $\mathcal{L}$ is not only smooth, but completely smooth by our assumption, we can conclude the same for the maps π, δ constructed before, and hence Θ will be completely smooth. We next state this as a theorem, the proof of which is omitted and can be obtained by the same line of reasoning as employed in the proof of the Theorem 8.1.26.

Theorem 8.1.34 *The map Θ defined in Theorem 8.1.26 is completely smooth.*

Let us now consider the Hilbert space $\mathcal{H} = \Gamma^f(\hat{k_0})$, the free Fock space over $\hat{k_0}$, equipped with the natural G-action $w_g := \oplus_{n \geq 0}(\hat{v_g})^{\otimes^n}$, and the action $\eta_g = w_g \cdot w_g^*$ on $\mathcal{A} \otimes \mathcal{K}(\mathcal{H})$. Consider $\hat{\mathcal{H}} = \mathbb{C} \oplus \mathcal{H}$ with trivial G action on $\mathbb{C}$. Since Θ is covariant and completely smooth, the ampliation $\Theta \otimes \mathrm{Id}$, composed with the exchange operator from $\mathcal{B}(\hat{k_0}) \otimes \mathcal{K}(\hat{\mathcal{H}})$ to $\mathcal{K}(\hat{\mathcal{H}}) \otimes \mathcal{B}(\hat{k_0})$, from $\mathcal{A}_\infty \otimes_{\mathrm{alg}} \mathcal{K}(\hat{\mathcal{H}})$ to $\mathcal{A} \otimes \mathcal{K}(\hat{\mathcal{H}}) \otimes \mathcal{B}(\hat{k_0})$ admits a smooth covariant extension on the Fréchet algebra $(\mathcal{A} \otimes \mathcal{K}(\hat{\mathcal{H}}))_\infty^{\alpha \otimes (1 \oplus \eta)}$. Moreover, as the Sobolev–Hilbert module $(\mathcal{A} \otimes \mathcal{H})_\infty^{\alpha \otimes w}$ is embedded as a closed subspace in the above Fréchet algebra, it follows that Θ extends to a smooth covariant map on this Sobolev–Hilbert module, again denoted by Θ, which maps $(\mathcal{A} \otimes \mathcal{H})_\infty^{\alpha \otimes w}$ into $((\mathcal{A} \otimes \mathcal{H}) \otimes \mathcal{B}(\hat{k_0}))_\infty^{\alpha \otimes w \otimes \hat{\beta}}$. For simplicity of notation, we shall denote the Sobolev–Hilbert module $(\mathcal{A} \otimes \mathcal{H})_\infty^{\alpha \otimes w}$ by simply $\mathcal{E}_\infty$ since the actions of the group will be fixed as above. By the complete smoothness of Θ, there exist some positive integer p and positive constant M such that

$$\|\Theta(X)\|_n \leq M \|X\|_{n+p} \tag{8.5}$$

for all $X \in \mathcal{E}_\infty$, $n \geq 0$.

We also note that for any $m \geq 0$, the natural embedding of $(\hat{k_0})^{\otimes^m}$ into $\mathcal{H}$ is covariant, thus it is in fact a topological embedding. Using this, it can be seen that $\mathcal{E}_\infty^m := (\mathcal{A} \otimes \hat{k_0}^{\otimes^m})_\infty^{\alpha \otimes v^m}$ (where ${v_g}^m = {v_g}^{\otimes^m}$) is embedded as a Fréchet subspace in $(\mathcal{A} \otimes \mathcal{H})_\infty$. So, we can consider restriction of Θ (hence of all the components $\delta, \delta^\dagger, \mathcal{L}, \sigma$ also) to each of the Sobolev–Hilbert modules $\mathcal{E}_\infty^m$, and the estimate (8.5) continues to hold with the same M and p for all m. Let us denote by $\mathcal{E}^\vee$ the algebraic direct sum $\oplus_m \mathcal{E}_\infty^m$. Clearly, Θ maps $\mathcal{E}_\infty^m$ to $\mathcal{E}_\infty^{m+1}$, hence $\Theta(\mathcal{E}^\vee) \subseteq \mathcal{E}^\vee$.

Next we want to define integral of the form $\int_0^t Y(s) \circ (a_\delta^\dagger + a_\delta + \Lambda_\sigma + \mathcal{I}_\mathcal{L})(ds)$ for appropriate process $(Y(s))$. It turns out that we can do so when $(Y(s))$ is completely smooth in a suitable sense. The crucial point here is that there is some kind of stability of the domain $\mathcal{A}_\infty$ under the action of the maps $\mathcal{L}, \delta, \delta^\dagger, \sigma$. Indeed, $\mathcal{L}(\mathcal{A}_\infty) \subseteq \mathcal{A}_\infty$, and thus for all $a \in \mathcal{A}_\infty$, $\xi \in (k_0)_\infty$, $\delta(a)$, $\sigma_\xi(a)$ belong to $(\mathcal{A} \otimes k_0)_\infty$. At this point, we recall the notations of Corollary 5.2.7, and take $\mathcal{D}_0 = h$, $\mathcal{V}_0 = (k_0)_\infty$. As before, denote by $\mathcal{V}_t$ the subspace consisting of simple $\mathcal{V}_0$-valued functions in k_t and by $\mathcal{V}$ the space of simple $\mathcal{V}_0$-valued functions.

Lemma 8.1.35 *Let $(Y(s))_{s\geq 0} : \mathcal{A}_\infty \otimes_{\text{alg}} \mathcal{E}(\mathcal{V}) \to \mathcal{A} \otimes \Gamma(k)$ be a map-valued adapted process satisfying the following.*

(i) For every fixed $t \geq 0$ and $f \in \mathcal{V}$, $Y(t)(a \otimes e(f))$ belongs to the Fréchet space $(\mathcal{A} \otimes \Gamma(k))_\infty$ and the map $\mathcal{A}_\infty \ni a \mapsto \Omega_{t,f}(a) := Y(t)(a \otimes e(f)) \in (\mathcal{A} \otimes \Gamma(k))_\infty$ is completely smooth.

(ii) For fixed $X \in \mathcal{E}^m_\infty$ and $f \in \mathcal{V}$, $t \mapsto \widetilde{Y(t)}(X \otimes e(f)) := \tilde{\Omega}_{t,f}(X)$ is continuous, where for any separable Hilbert space $\mathcal{H}'$ and for fixed $t \geq 0$ and $f \in \mathcal{V}$, we have denoted by $\tilde{\Omega}_{t,f}$ the trivial ampliation $(\Omega_{t,f} \otimes \text{Id}_{\mathcal{H}'}) : (\mathcal{A} \otimes \mathcal{H}')_\infty \to (\mathcal{A} \otimes \Gamma \otimes \mathcal{H}')_\infty$. Then $Y(t)$ satisfies the following.

(a) For every fixed $a \in \mathcal{A}_\infty$, $\xi \in \mathcal{V}_0$, the map $s \mapsto Y(s)((\mathcal{L}(a) + \langle \delta(a^*), \xi \rangle) \otimes e(f))$ is strongly continuous.

(b) For every fixed $a \in \mathcal{A}_\infty$ and $f \in \mathcal{V}$, define the families of operators $S_a(s) : h \otimes_{\text{alg}} \mathcal{E}(\mathcal{V}_s) \to h \otimes \Gamma_s \otimes k_0$ and $T_a(s) : h \otimes_{\text{alg}} \mathcal{E}(\mathcal{V}_s) \otimes_{\text{alg}} \mathcal{V}_0 \to h \otimes \Gamma_s \otimes k_0$ as follows:

$$S_a(s)(ue(f_s)) = \widetilde{Y(s)}(\delta(a) \otimes e(f_s))u, \tag{8.6}$$

$$T_a(s)(ue(f_s) \otimes \xi) = \widetilde{Y(s)}(\sigma(a)_\xi \otimes e(f_s))u. \tag{8.7}$$

Then the maps $s \mapsto S_a(s)(ue(f_s))$ and $s \mapsto T_a(s)(ue(f_s)\xi))$ are continuous.

The proof is straightforward and hence is omitted.

Definition 8.1.36 *For a map-valued process $(Y(t))$ satisfying (i) and (ii) of the above lemma, we define $\int_0^t Y(s) \circ (a_\delta^\dagger + \Lambda_\sigma)(ds)(a \otimes e(f))$ for $a \in \mathcal{A}_\infty$ and $f \in \mathcal{V}$ by (5.39) using the Corollary 5.2.7 (ii) with $E = G = I$ and $F = H = 0$. Similarly, we define the integral $\int_0^t Y(s) \circ (a_\delta + \mathcal{I}_\mathcal{L})(ds)(a \otimes e(f))$ by (5.38), which is well-defined by (a).*

Lemma 8.1.37 *If $(Y(t))_{t\geq 0}$ satisfies the conditions of Lemma 8.1.35, then so does the integral $Z(t) = \int_0^t Y(s) \circ (a_\delta^\dagger + a_\delta + \Lambda_\sigma + \mathcal{I}_\mathcal{L})(ds)$. Moreover, $Z(t)$ satisfies the estimate*

$$\begin{aligned}&\|\{Z(t)(a \otimes e(f)\}\|^2 \\ &\leq 2e^t \int_0^t \exp(\|f^s\|^2)(1 + \|f(s)\|^2)\|\hat{Y}(s)(\Theta(a)_{\hat{f}(s)} \otimes e(f_s))\|^2 ds,\end{aligned} \tag{8.8}$$

where $f \in \mathcal{V}$, $a \in \mathcal{A}_\infty$ and $\hat{Y}(s) = Y(s)|_{\mathcal{A}_\infty} \oplus \widetilde{Y(s)}|_{(\mathcal{A}\otimes k_0)_\infty}$.

Proof:
The integral exists by Lemma 8.1.35, and it clearly satisfies the estimate (8.8). It follows from this estimate using the complete smoothness of Θ that the map $a \mapsto Z(t)(a \otimes e(f))$ is completely smooth. The condition (ii) of Lemma 8.1.35 can be verified similarly. □

We are now ready to state and prove the main theorem of this section.

Theorem 8.1.38 *There exists an E–H dilation of* (T_t).

Proof:
The idea of the proof is similar to that of the Theorem 5.4.9, the only difference lying in the use of a family of norms instead of a single norm. We define $J^{(n)}(t)$ iteratively as before, with $J^{(0)} = \text{Id}$ and $J^{(n+1)}(t) = \int_0^t J^{(n)}(s) \circ (a_\delta^\dagger + a_\delta + \Lambda_\sigma + \mathcal{I}_\mathcal{L})(ds)$. However, by Lemma 8.1.37, each $J^{(n)}(t)$ is completely smooth, hence can be extended to the algebra $\mathcal{E}^\vee = \oplus_m \mathcal{E}_\infty^m$, and we denote this extension by the same symbol. We can rewrite the estimate (8.8) in the following form:

$$\begin{aligned} &\|J^{(n+1)}(t)(a \otimes e(f)\|_{2,0}^2 \\ &\quad \leq 2e^t \int_0^t (1 + \|f(s)\|^2) \|\hat{J}^{(n)}(s)(\Theta(a)_{\hat{f}(s)} \otimes e(f_s))\|_{2,0}^2 \, ds. \end{aligned} \tag{8.9}$$

Let us now fix an element $a \in \mathcal{A}_\infty^0$, such that $\|a\|_n \leq C'_a C_a^n$ for some $C'_a, C_a > 0$ and also let $\mathcal{U}$ denote the set of simple $(k_0)_\infty^0$-valued functions on $[0, \infty)$. For $f \in \mathcal{U}$ with $\xi_1, \ldots, \xi_r \in (k_0)_\infty^0$ such that $f(t) \in \{\xi_1, \ldots, \xi_r\}$ for all t, it follows by using the complete smoothness of Θ that each of the maps Θ_{ξ_j} is completely smooth of order p and some bound, say C_j, $j = 1, \ldots, r$. Taking C to be the maximum of $C_1, \ldots, C_r$, we get the following estimate:

$$\|\Theta \hat{f}^{(t_1)}(\cdots(\Theta \hat{f}^{(f_{t_e-1})}(\Theta \hat{f}^{(t_l)}(a)\cdots)\|_{2,0} \leq C^l \|a\|_{pl} \leq C'_a (C C_a^p)^l. \tag{8.10}$$

From (8.9) and (8.10), we obtain

$$\begin{aligned} &\|J^{(n)}(t)(a \otimes e(f))\|_{2,0}^2 \\ &\leq 2e^t (C'_a \|e(f)\|)^2 (C^2 C_a^{2p} B)^n \int_0^t \int_0^{t_{n-1}} \cdots \int_0^{t_1} dt_{n-1} \cdots dt_0 \\ &\leq 2e^t (C'_a \|e(f)\|)^2 (C^2 C_a^{2p} B)^n \frac{t^n}{n!}, \end{aligned}$$

where $B := \text{Max}\{1 + \|\xi_j\|^2,\ j = 1, \ldots, r\}$.

That $J(t)(a \otimes e(f)) := \sum_n J_t^{(n)}(a \otimes e(f))$ converges follows as in the proof of the Theorem 5.4.9. Clearly, $J(t)$ satisfies the equation

$$J(t) = \text{Id} + \int_0^t J(s) \circ (a_\delta^\dagger + a_\delta + \Lambda_\sigma + \mathcal{I}_\mathcal{L})(ds).$$

It also follows similarly that $\langle J_t(a \otimes e(g))v, J_t(y \otimes e(f))u\rangle = \langle ve(g), J_t(a^* b \otimes e(f))u\rangle$ for all $a, b \in \mathcal{A}^0_\infty$ and $f, g \in \mathcal{U}$. Furthermore, $J_t(1 \otimes e(f))u = ue(f)$. For $a \in \mathcal{A}^0_\infty$, we define a linear operator $j_t(a)$ on the dense subspace $h \otimes \mathcal{E}(\mathcal{U})$ spanned by the vectors of the form $ve(g)$ with $v \in h$, $g \in \mathcal{U}$, given by

$$j_t(a)ve(g) := J_t(a \otimes e(g))v.$$

Then $j_t(1) = 1$ since $1 \in \mathcal{A}^0_\infty$ and since by Lemma 8.1.5 a positive invertible element $b \in \mathcal{A}^0_\infty$ admits a positive square root $a \in \mathcal{A}^0_\infty$, we have that

$$\langle \xi, j_t(b)\xi\rangle = \| j_t(a)\xi \|^2 \geq 0,$$

for all $\xi \in h \otimes \mathcal{E}(\mathcal{U})$. From this, it follows that for all self-adjoint $b \in \mathcal{A}^0_\infty$, $\epsilon > 0$ and $\xi \in h \otimes \mathcal{E}(\mathcal{U})$, $\langle \xi, j_t((1+\epsilon)\|b\|1 - b)\xi\rangle \geq 0$. Letting $\epsilon \to 0+$, we get

$$\langle \xi, j_t(b)\xi\rangle \leq \|b\| \|\xi\|^2$$

for all ξ in the dense subspace $h \otimes \mathcal{E}(\mathcal{U})$. Thus, $j_t(b)$ extends to a bounded operator on $h \otimes \Gamma$, denoted by the same symbol, then is further extended for an arbitrary $b \in \mathcal{A}^0_\infty$ by decomposing it into the real and imaginary parts. Clearly, j_t is a $*$-homomorphism from $\mathcal{A}^0_\infty$ to $\mathcal{B}(h \otimes \Gamma)$, hence extends to a $*$-homomorphism on the whole of $\mathcal{A}$ by Corollary 8.1.6.

To complete the proof, it suffices to show that j_t maps $\mathcal{A}^0_\infty$ into $\mathcal{A}'' \otimes \mathcal{B}(\Gamma)$. To this end, we observe that by construction, $J_t(a \otimes e(f))$ is an element of the Hilbert module $\mathcal{A} \otimes \Gamma$ for every $a \in \mathcal{A}^0_\infty$ and $f \in \mathcal{U}$. Thus, for $a' \in \mathcal{A}'$, we have

$$(a' \otimes 1) J_t(a \otimes e(f)) = J_t(a \otimes e(f))a'.$$

Therefore, for every $v \in h$,

$$\begin{aligned}
&(a' \otimes 1) j_t(a)ve(f) \\
&= (a' \otimes 1) J_t(a \otimes e(f))v \\
&= J_t(a \otimes e(f))a'v \\
&= j_t(a)(a' \otimes 1)ve(f).
\end{aligned}$$

Hence $j_t(a)$ commutes with $a' \otimes 1$ for all $a' \in \mathcal{A}'$, so it belongs to $\mathcal{A}'' \otimes \mathcal{B}(\Gamma)$. □

Remark 8.1.39 *From the estimates derived and used in the proof of the Theorem 8.1.38, it can be verified that the map-valued Q.S.D.E.*

$$dJ(t) = J(t) \circ (a^\dagger_\delta + a_\delta + \Lambda_\sigma + \mathcal{I}_\mathcal{L})(dt),$$

with a given initial value of $J(0)$, admits a unique solution, in the sense that any two solutions must agree on $(\mathcal{A} \otimes \Gamma)_\infty$. The proof is by iteration, and *making use of the estimate (8.10) for the completely smooth map Θ. We omit the details of this proof, since it is very similar to the proof of uniqueness of map-valued Q.S.D.E. with bounded coefficients studied in Chapter 5.*

Examples

Let M be a compact oriented Riemannian manifold with the Riemannian volume form dvol (see Chapter 9 for details). Let G denote the group of Riemannian isometries of the manifold M. It is known that G is a compact Lie group acting smoothly on M. We consider the commutative C^*-algebra $\mathcal{A} = C(M)$, and the natural action of G on $\mathcal{A}$ is given by

$$\alpha_g(f)(m) = f(gm),$$

where $G \times M \ni (g, m) \mapsto gm$ denotes the action of the isometry g. Let (T_t) be the heat semigroup on functions, and assume furthermore that the heat-kernel $k_t(m, n)$ is of the following form:

$$k_t(m, n) = \phi_t(d(m, n)),$$

where d denotes the Riemannian distance on M and ϕ_t is a family of smooth functions. Many well-known Riemannian manifolds satisfy this property. In particular, all homogenous spaces of Lie groups and rotationally symmetric Riemannian manifolds are of this kind. Since isometries preserve the distance d, that is, $d(gm, gn) = d(m, n)$ for $g \in G, m, n \in M$, and since therefore dvol is also invariant under the action of G, we have the following:

$$\begin{aligned}
\alpha_g(T_t f)(m) &= \int_M \phi_t(d(gm, x)) f(x) \mathrm{dvol}(x) \\
&= \int_M \phi_t(d(m, g^{-1}x)) f(gg^{-1}x) \mathrm{dvol}(g^{-1}x) \\
&= \int_M \phi_t(d(m, y)) f(gy) \mathrm{dvol}(y) \\
&= T_t(\alpha_g(f))(m).
\end{aligned}$$

This proves that T_t is covariant. However, the complete smoothness of the Laplacian has to be verified case by case. For homogeneous spaces, this can be verified by an application of Theorem 8.1.28 and Remark 8.1.29.

Remark 8.1.40 *The canonical heat semigroup on the noncommutative torus (see Chapter 9 for details) satisfies (**B1**) and (**B2**).*

Remark 8.1.41 *It should be pointed out that compactness of the group G has been used by us just in two places: first, to get the covariant structure maps by* *applying the covariant version of the Kasparov embedding theorem (which is not known for arbitrary noncompact groups), and second, proving that $\mathcal{A}^0_\infty$ is dense in $\mathcal{A}$. However, it may happen in some situation that the (not necessarily covariant) structure maps can be written down explicitly, without appealing to the covariant Kasparov's theorem, and it is also directly verifiable that the structure maps are completely smooth. If, moreover, the density of $\mathcal{A}^0_\infty$ can be proven explicitly in this case, then it is possible to prove the existence of an E–H dilation along the line of reasoning employed to prove the Theorem 8.1.38.*

8.2 Dilation of quantum dynamical semigroups on U.H.F. algebra

We recall the class of Q.D.S. described and constructed in the Subsection 3.2.4. We shall also use the notations of the same subsection.

8.2.1 E–H dilation

In this subsection we want to solve the following Q.S.D.E. in $\mathcal{B}(h) \otimes \mathcal{B}(\Gamma(L^2(\mathbb{R}_+, k_0)))$ where $h = L^2(\mathcal{A}, \mathrm{tr})$ and $k_0 = l^2(\mathbb{Z}^d)$ with a fixed orthonormal basis $\{e_j\}$ of $\mathbb{Z}^d$.

$$dj_t(a) = \sum_{j \in \mathbb{Z}^d} j_t(\delta_j^\dagger(a)) da_j(t) + \sum_{j \in \mathbb{Z}^d} j_t(\delta_j(a)) da_j^\dagger(t) + j_t(\hat{\mathcal{L}}(a)) dt, \quad (8.11)$$

$$j_0(a) = a \otimes 1_\Gamma.$$

It should be pointed out that we are not using our coordinate-free language for quantum stochastic calculus in this section. The reason behind the choice of coordinatized version is two-fold: first, it is somewhat natural in the context of the Q.D.S. considered here, and second, this offers the reader who is not so familiar with quantum stochastic calculus as developed in [100] an opportunity to get accustomed to the conventional language of the subject. Keeping this in mind, we have chosen to explain and expand the steps of the proofs in full details.

Let us at first look at the corresponding H–P equation in the Hilbert space $h \otimes L^2(\Gamma(L^2(\mathbb{R}_+, k_0)))$, given by

$$dU_t = \sum_{j \in \mathbb{Z}^d} \{r_j^* da_j(t) - r_j da_j^\dagger(t) - \frac{1}{2} r_j^* r_j dt\} U_t, \quad (8.12)$$

$$U_0 = I_{h \otimes \Gamma}.$$

However, although each $r_j \in \mathcal{A}$ and hence is in $\mathcal{B}(h)$, the equation (8.12) does not in general admit a solution since

$$\left\langle u, \sum_{j\in\mathbb{Z}^d} r_j^* r_j u \right\rangle = \sum_{j\in\mathbb{Z}^d} \|r_j u\|^2 \text{ for all } u \in h,$$

which is not convergent in general and thus $\sum_{j\in\mathbb{Z}^d} r_j \otimes e_j$ does not define an element in $\mathcal{A}\otimes k_0$. For example, let r be the single-supported element $U^{(k)} \in \mathcal{A}$ for some $k \in \mathbb{Z}^d$ so that $r_j = U^{(k+j)}$ is a unitary for each $j \in \mathbb{Z}^d$ and we have

$$\sum_{j\in\mathbb{Z}^d} \|r_j u\|^2 = \sum_{j\in\mathbb{Z}^d} \|u\|^2 = \infty.$$

However, as we shall see, in many situations there exist E–H flows, even though the corresponding H–P equation (8.12) does not admit a solution.

Remark 8.2.1 *There are some cases when an E–H flow can be seen to be implemented by a solution of a H–P equation. For example, given a self-adjoint* $r \in \mathcal{A}_{\mathrm{loc}}$

$$dV_t = \sum_{k\in\mathbb{Z}^d} V_t(S_k^* da_k(t) - S_k da_k^\dagger(t) - \frac{1}{2} S_k^* S_k dt), \; V_0 = 1,$$

where S_k *is defined by* $S_k(a) = [r_k, a]$ *for* $a \in \mathcal{A}_{\mathrm{loc}} \subseteq h$*, admits a unique unitary solution and*

$$a \mapsto V_t^*(a \otimes 1)V_t$$

gives an E–H dilation for T_t (see [91], [92]).

Let $a, b \in \mathbb{Z}_N$ be fixed and $W = U^a V^b \in M_N(\mathbb{C})$. We consider the following representation of the infinite product group $\mathcal{G}' := \prod_{j\in\mathbb{Z}^d} \mathbb{Z}_N$, given by

$$\mathcal{G}' \ni g \mapsto W_g = \prod_{j\in\mathbb{Z}^d} W^{(j)\alpha_j}, \text{ where } g = (\alpha_j).$$

For any $b \in \mathcal{A},\ b = \sum_{g\in\mathcal{G}} c_g U_g$ and for $n \geq 1$ we define

$$\vartheta_n(b) = \sum_{g\in\mathcal{G}} |c_g|\, |g|^n.$$

Now we consider $r \in \mathcal{A},\ r = \sum_{g\in\mathcal{G}'} c_g W_g$ such that $\sum_{g\in\mathcal{G}'} |c_g|\, |g|^2 < \infty$. It is clear that $\vartheta_1(r) = \sum_{g\in\mathcal{G}'} |c_g|\, |g| < \infty$. We note that any $a \in \mathcal{A}_{\mathrm{loc}}$ can be written as $a = \sum_{w\in\mathcal{G}} c_w U_w$, with complex coefficients c_w satisfying $c_w = 0$ for all w such that $\mathrm{supp}(w) \cap \mathrm{supp}(a)$ is empty. It follows that

$$\vartheta_n(a) = \sum_{w\in\mathcal{G}} |c_w|\, |w|^n < \infty \text{ for } n \geq 1,$$

and it is clear that

$$\vartheta_n(a) \leq |a|^n \sum_{w \in \mathcal{G}} |c_w| \leq c_a^n$$

where $c_a = |a|(1 + \sum_{w \in \mathcal{G}} |c_w|)$. Let us consider the formal Lindbladian $\mathcal{L}$ defined by (3.6) associated with the completely positive map $T(a) = r^* a r$:

$$\mathcal{L} = \sum_{k \in \mathbb{Z}^d} \mathcal{L}_k,$$

where $\mathcal{L}_k a = \tau_k \mathcal{L}_0(\tau_{-k} a)$ with $\mathcal{L}_0 a = \frac{1}{2}\left\{\left[r^*, a\right] r + r^* \left[a, r\right]\right\}$ so that one has

$$\mathcal{L}_k(a) = \frac{1}{2}\{\left[r_k^*, a\right] r_k + r_k^* \left[a, r_k\right]\}.$$

Let us denote these two bounded derivations $\left[r_k^*, .\right]$ and $[., r_k]$ on $\mathcal{A}_{\mathrm{loc}}$ by $\delta_k^\dagger$ and δ_k respectively. Thus,

$$\mathcal{L}(a) = \frac{1}{2} \sum_{k \in \mathbb{Z}^d} \delta_k^\dagger(a) r_k + r_k^* \delta_k(a).$$

For $n \geq 1$, we denote the set of integers $\{1, 2, \ldots, n\}$ by I_n and for $1 \leq p \leq n$, $P = \{l_1, l_2, \ldots, l_p\} \subseteq I_n$ with $l_1 < l_2 < \cdots < l_p$, we define a map from the n-fold Cartesian product of $\mathbb{Z}^d$ to that of p copies of $\mathbb{Z}^d$ by

$$\bar{k}(I_n) = (k_1, k_2, \ldots, k_n) \mapsto \bar{k}(P) := (k_{l_1}, k_{l_2}, \ldots, k_{l_p})$$

and similarly, $\bar{\varepsilon}(P) := (\varepsilon_{l_1}, \varepsilon_{l_2}, \ldots, \varepsilon_{l_p})$ for a vector $\bar{\varepsilon} \equiv \bar{\varepsilon}(I_n) = (\varepsilon_1, \varepsilon_2, \ldots, \varepsilon_n)$ in the n-fold Cartesian product of $\{-1, 0, 1\}$.

For brevity of notations, we write $\bar{\varepsilon}(P) \equiv c \in \{-1, 0, 1\}$ to mean that all $\varepsilon_{l_i} = c$ and denote $\bar{k}(I_n)$ and $\bar{\varepsilon}(I_n)$ by $\bar{k}(n)$ and $\bar{\varepsilon}(n)$ respectively. Setting $\delta_k^\varepsilon = \delta_k^\dagger$, $\mathcal{L}_k$ and δ_k for $\varepsilon = -1, 0$ and 1 respectively, we write $R(\bar{k}) = r_{k_1} r_{k_2}, \ldots, r_{k_p}$ and $\delta(\bar{k}, \bar{\varepsilon}) = \delta_{k_p}^{\varepsilon_p} \cdots \delta_{k_1}^{\varepsilon_1}$ for any $\bar{k} = (k_1, k_2, \ldots, k_p)$ and $\bar{\varepsilon} = (\varepsilon_1, \varepsilon_2, \ldots, \varepsilon_p)$. Now we have the following useful lemma.

Lemma 8.2.2 *Let r, a and constant c_a be as above. Then the following hold.*

(i) For any $n \geq 1$,

$$\sum_{\bar{k}(n)} \|\delta(\bar{k}(n), \bar{\varepsilon}(n))(a)\| \leq (2\vartheta_1(r) c_a)^n \text{ for all } a \in \mathcal{A}_{\mathrm{loc}},$$

where $\bar{\varepsilon}(n)$ is such that $\varepsilon_l \neq 0$, for all $l \in I_n$.

(ii) *For any $n \geq 1$ and $\bar{k}(n)$,*

$$\mathcal{L}_{k_n} \cdots \mathcal{L}_{k_1}(a) = \frac{1}{2^n} \sum_{p=0,1\cdots n} \sum_{P \subseteq I_n : |P|=p} R(\bar{k}(P^c))^* \delta(\bar{k}(n), \bar{\varepsilon}_{(P)}(n))(a) R(\bar{k}(P)),$$

where $\bar{\varepsilon}_{(P)}(n)$ is such that $\bar{\varepsilon}_{(P)}(P) \equiv -1$ and $\bar{\varepsilon}_{(P)}(P^c) \equiv 1$.

(iii) *For any $n \geq 1$, $p \leq n$, and $\bar{\varepsilon}(n)$, such that the number of components of $\bar{\varepsilon}$ equal to 0 is p, we have*

$$\sum_{\bar{k}(n)} \|\delta(\bar{k}(n), \bar{\varepsilon}(n))(a)\| \leq \|r\|^p (2\vartheta_1(r)c_a)^n \leq (1 + \|r\|)^n (2\vartheta_1(r)c_a)^n.$$

(iv) *Let $m_1, m_2 \geq 1$; $a, b \in \mathcal{A}_{\text{loc}}$ and $\bar{\varepsilon}'(m_1), \bar{\varepsilon}''(m_2)$ be two fixed tuples. Then for $n \geq 1$ and $\bar{\varepsilon}(n)$ as in (iii), we have,*

$$\sum_{\bar{k}(n), \bar{k}'(m_1), \bar{k}''(m_2)} \|\delta(\bar{k}(n), \bar{\varepsilon}(n))\{\delta(\bar{k}'(m_1), \bar{\varepsilon}'(m_1))(a) \cdot \delta(\bar{k}''(m_2), \bar{\varepsilon}''(m_2))(b)\}\|$$
$$\leq 2^n (1 + \|r\|)^{2n+m_1+m_2} (2\vartheta_1(r)c_{a,b})^{n+m_1+m_2},$$

where $c_{a,b} = \max\{c_a, c_b\}$.

Proof:

(*i*) As r^* is again of the same form as r, it is enough to observe the following

$$\sum_{k_n,\cdots k_1} \| [r_{k_n}, \cdots [r_{k_1}, a] \cdots] \| \leq (2\vartheta_1(r)c_a)^n \text{ for all } a \in \mathcal{A}_{\text{loc}} .$$

In order to prove this let us expand the left-hand side of the above as

$$\sum_{k_n,\cdots k_1} \sum_{g_n,\cdots g_1 \in \mathcal{G}'; w \in \mathcal{G}} |c_{g_n}| \cdots |c_{g_1}| \, |c_w| \, \left\| \left[\tau_{k_n} W_{g_n}, \cdots \left[\tau_{k_1} W_{g_1}, U_w\right] \cdots \right] \right\| .$$

We note that for any two commuting elements A, B in $\mathcal{A}_{\text{loc}}$, $[A, [B, a]] = [B, [A, a]]$. Thus, for the commutator $\left[\tau_{k_n} W_{g_n}, \cdots \left[\tau_{k_1} W_{g_1}, U_w\right] \cdots \right]$ to be nonzero, it is necessary to have $(\text{supp}(g_i) + k_i) \cap \text{supp}(w) \neq \phi$ for each $i = 1, 2, \ldots, n$. Clearly the number of choices of such $k_i \in \mathbb{Z}^d$ is at most $|g_i| \cdot |w|$. Thus we get,

$$\sum_{k_n,\cdots k_1} \left\| \left[r_{k_n}, \cdots \left[r_{k_1}, a\right] \cdots \right] \right\| \leq \sum_{g_n,\cdots g_1 \in \mathcal{G}'; w \in \mathcal{G}} |c_{g_n}| \cdots |c_{g_1}| |c_w| |g_n| \cdots |g_1| |w|^n 2^n \leq (2\vartheta_1(r)c_a)^n.$$

(*ii*) The proof is by induction. For any $k \in \mathbb{Z}^d$ we have,

$$\mathcal{L}_k(a) = \frac{1}{2} \sum_{k \in \mathbb{Z}^d} \delta_k^\dagger(a) r_k + r_k^* \delta_k(a),$$

so it is trivially true for $n = 1$. Let us assume it to be true for some $m \geq 1$ and for any $k_{m+1} \in \mathbb{Z}^d$ consider $\mathcal{L}_{k_{m+1}} \mathcal{L}_{k_m} \cdots \mathcal{L}_{k_1}(a)$. By applying the statement for $n = m$ we get,

$$\mathcal{L}_{k_{m+1}} \mathcal{L}_{k_m} \cdots \mathcal{L}_{k_1}(a) = \frac{1}{2^{m+1}} \sum_{p=0,1\cdots m} \sum_{P \subseteq I_m : |P|=p}$$
$$[\delta_{k_{m+1}}^* \{R(\bar{k}(P^c))^* \delta(\bar{k}(m), \bar{\varepsilon}_{(P)}(m))(a) R(\bar{k}(P))\} r_{k_{m+1}}$$
$$+ r_{k_{m+1}}^* \delta_{k_{m+1}} \{R(\bar{k}(P^c))^* \delta(\bar{k}(m), \bar{\varepsilon}_{(P)}(m))(a) R(\bar{k}(P))\}].$$

Since the elements r_k are commuting with each other, the above expression becomes

$$\frac{1}{2^{m+1}} \sum_{p=0,1\cdots m} \sum_{P \subseteq I_m : |P|=p}$$
$$[R(\bar{k}(P^c))^* \delta_{k_{m+1}}^* \delta(\bar{k}(m), \bar{\varepsilon}_{(P)}(m))(a) R(\bar{k}(P)) r_{k_{m+1}}$$
$$+ r_{k_{m+1}}^* R(\bar{k}(P^c))^* \delta_{k_{m+1}} \delta(\bar{k}(m), \bar{\varepsilon}_{(P)}(m))(a) R(\bar{k}(P))]$$
$$= \frac{1}{2^{m+1}} \sum_{p=0,1\cdots m+1} \sum_{P \subseteq I_{m+1} : |P|=p} R(\bar{k}(P^c))^* \delta(\bar{k}(m+1),$$
$$\bar{\varepsilon}_{(P)}(m+1))(a) R(\bar{k}(P)).$$

(*iii*) Let $P \subseteq I_n$, $|P| = p$, be such that all the components of $\bar{\varepsilon}(P)$ are equal to 0. By simple application of (ii),

$$\delta(\bar{k}(n), \bar{\varepsilon}(n))(a)$$
$$= \frac{1}{2^p} \sum_{q=0,1\cdots p} \sum_{Q \subseteq P : |Q|=q} R(\bar{k}(P \setminus Q))^* \delta(\bar{k}(n), \bar{\varepsilon}_{(Q,P)}(n))(a) R(\bar{k}(Q)), \tag{8.13}$$

where $\bar{\varepsilon}_{(Q,P)}(n)$ is defined to be the map from the n-fold Cartesian product of $\{-1, 0, 1\}$ to itself, given by $\bar{\varepsilon}(n) \mapsto \bar{\varepsilon}_{(Q,P)}(n)$ such that $\bar{\varepsilon}_{(Q,P)}(Q) \equiv -1$, $\bar{\varepsilon}_{(Q,P)}(P \setminus Q) \equiv 1$ and $\bar{\varepsilon}_{(Q,P)}(I_n \setminus P) = \bar{\varepsilon}(I_n \setminus P)$. Now (*iii*) follows from (*i*).

(*iv*) By (8.13) we have,

$$\sum_{\bar{k}(n), \bar{k}'(m_1), \bar{k}''(m_2)} \|\delta(\bar{k}(n), \bar{\varepsilon}(n))\{\delta(\bar{k}'(m_1), \bar{\varepsilon}'(m_1))(a) \cdot \delta(\bar{k}''(m_2), \bar{\varepsilon}''(m_2))(b)\}\|$$
$$= \frac{1}{2^p} \sum_{\bar{k}(n), \bar{k}'(m_1), \bar{k}''(m_2)} \sum_{q=0,1\cdots p} \sum_{Q \subseteq P : |Q|=q} \Upsilon(\bar{k}(n), \bar{k}'(m_1), \bar{k}''(m_2), q, Q),$$

where

$$\begin{aligned}&\Upsilon(\bar{k}(n), \bar{k}'(m_1), \bar{k}''(m_2), q, Q) \\ &\quad = \|R(\bar{k}(P \setminus Q))^* \delta(\bar{k}(n), \bar{\varepsilon}_{(Q,P)}(n)) \, [\delta(\bar{k}'(m_1), \bar{\varepsilon}'(m_1))(a) \\ &\qquad \cdot \delta(\bar{k}''(m_2), \bar{\varepsilon}''(m_2))(b)] \, R(\bar{k}(Q))\|.\end{aligned}$$

Now applying the Leibnitz rule, $\Upsilon(\bar{k}(n), \bar{k}'(m_1), \bar{k}''(m_2), q, Q)$ can be seen to be less than or equal to

$$\begin{aligned}&\|r\|^p \sum_{l=0}^{n} \sum_{L \subseteq I_n : |L| = l} \|\delta(\bar{k}(L), \bar{\varepsilon}_{(Q,P)}(L)) \delta(\bar{k}'(m_1), \bar{\varepsilon}'(m_1))(a)\| \\ &\qquad \|\delta(\bar{k}(L^c), \bar{\varepsilon}_{(Q,P)}(L^c))[\delta(\bar{k}''(m_2), \bar{\varepsilon}''(m_2))(b)]\|.\end{aligned}$$

Using (iii), we obtain,

$$\begin{aligned}&\sum_{\bar{k}(n), \bar{k}'(m_1), \bar{k}''(m_2)} \|\delta(\bar{k}(n), \bar{\varepsilon}(n))\{\delta(\bar{k}'(m_1), \bar{\varepsilon}'(m_1))(a) \cdot \delta(\bar{k}''(m_2), \bar{\varepsilon}''(m_2))(b)\}\| \\ &\leq \frac{(1 + \|r\|)^n}{2^p} \sum_{q=0}^{p} \frac{p!}{(p-q)!\, q!} \sum_{l=0}^{n} \frac{n!}{(n-l)!\, l!} (1 + \|r\|)^{l+m_1} (2\vartheta_1(r) c_a)^{l+m_1} \, . \\ &\qquad (1 + \|r\|)^{n-l+m_2} (2\vartheta_1(r) c_b)^{n-l+m_2} \\ &\qquad \leq 2^n (1 + \|r\|)^{2n+m_1+m_2} (2\vartheta_1(r) c_{a,b})^{n+m_1+m_2}.\end{aligned}$$

□

Now we are in a position to prove the following result about the existence of an E–H flow for the Q.D.S. T_t associated with the element $r \in \mathcal{A}$, discussed after the Remark 8.2.1.

Theorem 8.2.3 *(i) For $t \geq 0$, there exists a unique solution j_t of the Q.S.D.E.,*

$$dj_t(a) = \sum_{j \in \mathbb{Z}^d} j_t(\delta_j^\dagger(a)) da_j(t) + \sum_{j \in \mathbb{Z}^d} j_t(\delta_j(a)) da_j^\dagger(t) + j_t(\hat{\mathcal{L}}(a)) dt, \tag{8.14}$$

$$j_0(a) = a \otimes 1_\Gamma, \ \textit{for all } a \in \mathcal{A}_{\text{loc}},$$

such that $j_t(1) = 1$, for all $t \geq 0$.

(ii) For $a, b \in \mathcal{A}_{\text{loc}}$ and $u, v \in h \equiv L^2(\mathcal{A}, \tau)$; $f, g \in \mathcal{C}$ (set of all bounded continuous functions in $L^2(\mathbb{R}_+, k_0)$),

$$\langle ue(f), j_t(ab) ve(g) \rangle = \langle j_t(a^*) ue(f), j_t(b) ve(g) \rangle. \tag{8.15}$$

(iii) j_t extends uniquely to a unital C^*-homomorphism from $\mathcal{A}$ into $\mathcal{A}'' \otimes \mathcal{B}(\Gamma)$.

Proof:
We note first that $\mathcal{A}_{\text{loc}}$ is a dense $*$-subalgebra of $\mathcal{A}$.
(*i*) As usual, we solve the Q.S.D.E. by iteration. For $t_0 \geq 0, t \leq t_0$ and $a \in \mathcal{A}_{\text{loc}}$, we set

$$j_t^{(0)}(a) = a \otimes 1_\Gamma \text{ and}$$

$$j_t^{(n)}(a) = a \otimes 1_\Gamma$$
$$+ \int_0^t \sum_{j \in \mathbb{Z}^d} j_s^{(n-1)}(\delta_j^\dagger(a)) da_j(s) + \sum_{j \in \mathbb{Z}^d} j_s^{(n-1)}(\delta_j(a)) da_j^\dagger(s) + j_s^{(n-1)}(\hat{\mathcal{L}}(a)) ds. \tag{8.16}$$

Then for $u \in h$ and $f \in \mathcal{C}$, we can show by induction, that

$$\|\{j_t^{(n)}(a) - j_t^{(n-1)}(a)\}ue(f)\|$$
$$\leq \frac{(tc_f)^{n/2}}{\sqrt{n!}} \|ue(f)\| \sum_{\bar{k}(n)} \sum_{\bar{\varepsilon}(n)} \|\delta(\bar{k}(n), \bar{\varepsilon}(n))(a)\|, \tag{8.17}$$

where $c_f = 3e^{t_0}(1 + \|f\|_\infty^2)$. For $n = 1$, by the estimate (5.18) of Lemma 5.2.5 and the inequality $(a + b + c)^2 \leq 3(a^2 + b^2 + c^2)$ for nonnegative numbers a, b, c, we have

$$\|\{j_t^{(1)}(a) - j_t^{(0)}(a)\}ue(f)\|^2$$
$$= \left\| \left\{ \int_0^t \sum_{j \in \mathbb{Z}^d} \delta_j^\dagger(a) da_j(s) + \sum_{j \in \mathbb{Z}^d} \delta_j(a) da_j^\dagger(s) + \hat{\mathcal{L}}(a) ds \right\} ue(f) \right\|^2$$
$$\leq 3e^{t_0} \|e(f)\|^2 \int_0^t \left\{ \sum_{j \in \mathbb{Z}^d} \|\delta_j^\dagger(a)u\|^2 + \sum_{j \in \mathbb{Z}^d} \|\delta_j(a)u\|^2 + \|\hat{\mathcal{L}}(a)u\|^2 \right\}$$
$$(1 + \|f(s)\|^2) ds$$
$$\leq tc_f \|e(f)\|^2 \left\{ \sum_{j \in \mathbb{Z}^d} (\|\delta_j^\dagger(a)u\| + \|\delta_j(a)u\| + \|\mathcal{L}_j(a)u\|) \right\}^2.$$

Thus (8.17) is true for $n = 1$. Inductively assuming the estimate for some $m \geq 1$, we have by the same argument as above,

$$\|\{j_t^{(m+1)}(a) - j_t^{(m)}(a)\}ue(f)\|^2$$
$$= \left\| \left\{ \int_0^t \sum_{j \in \mathbb{Z}^d} [j_{s_m}^{(m)}(\delta_j^\dagger(a)) - j_{s_m}^{(m-1)}(\delta_j^\dagger(a))] da_j(s_m) \right.\right.$$
$$+ \sum_{j \in \mathbb{Z}^d} [j_{s_m}^{(m)}(\delta_j(a)) - j_{s_m}^{(m-1)}(\delta_j(a))] da_j^\dagger(s_m)$$

$$
\begin{aligned}
&\left. +[j_{s_m}^{(m)}(\hat{\mathcal{L}}(a)) - j_{s_m}^{(m-1)}(\hat{\mathcal{L}}(a))]ds_m \right\} ue(f)\Bigg\|^2 \\
&\leq 3e^{t_0}\int_0^t \left\{ \sum_{j\in\mathbb{Z}^d} \|[j_{s_m}^{(m)}(\delta_j^\dagger(a)) - j_{s_m}^{(m-1)}(\delta_j^\dagger(a))]ue(f)\|^2 \right. \\
&\quad + \sum_{j\in\mathbb{Z}^d} \|[j_{s_m}^{(m)}(\delta_j(a)) - j_{s_m}^{(m-1)}(\delta_j(a))]ue(f)\|^2 \\
&\quad \left. + \|[j_{s_m}^{(m)}(\hat{\mathcal{L}}(a)) - j_{s_m}^{(m-1)}(\hat{\mathcal{L}}(a))]ue(f)\|^2 \right\} (1+\|f(s_m)\|^2)ds_m \\
&\leq c_f \int_0^t \left[\sum_{j\in\mathbb{Z}^d} \|[j_{s_m}^{(m)}(\delta_j{}^\dagger(a)) - j_{s_m}^{(m-1)}(\delta_j^\dagger(a)) \right] ue(f)\| \\
&\quad + \sum_{j\in\mathbb{Z}^d} \|[j_{s_m}^{(m)}(\delta_j(a)) - j_{s_m}^{(m-1)}(\delta_j(a))]ue(f)\| \\
&\quad + \|[j_{s_m}^{(m)}(\hat{\mathcal{L}}(a)) - j_{s_m}^{(m-1)}(\hat{\mathcal{L}}(a))]ue(f)\|]^2 ds_m.
\end{aligned}
$$

Now applying (8.17) for $n = m$, we get the required estimate for $n = m + 1$. This proves the claim.

By the estimate of Lemma 8.2.2 (iii), and noting that there are 3^n ways of choosing $\bar{\varepsilon}(n) = (\epsilon_1, \cdots \epsilon_n)$ from $\{-1, 0, 1\}$, we have

$$
\|\{j_t^{(n)}(a) - j_t^{(n-1)}(a)\}ue(f)\| \leq 3^n \frac{(t_0 c_f)^{n/2}}{\sqrt{n!}} \|ue(f)\|(1+\|r\|)^n(1+2\vartheta_1(r)c_a)^n.
$$

Thus it follows that the sequence $\{j_t^{(n)}(a)ue(f)\}$ is Cauchy. We define $j_t(a)ue(f)$ to be $\lim_{n\to\infty} j_t{}^{(n)}ue(f)$, that is

$$
j_t(a)ue(f) = au\, e(f) + \sum_{n\geq 1}\{j_t^{(n)}(a) - j_t^{(n-1)}(a)\}ue(f) \tag{8.18}
$$

and one has

$$
\|j_t(a)ue(f)\| \leq \|ue(f)\| \left[\|a\| + \sum_{n\geq 1} 3^n \frac{(t_0 c_f)^{n/2}}{\sqrt{n!}} (1+\|r\|)^n (1+2\vartheta_1(r)c_a)^n \right]. \tag{8.19}
$$

To show uniqueness, let us assume that j_t and j_t' be two solutions of (8.14). Setting

$$
q_t(a) = j_t(a) - j_t'(a),
$$

we observe that

$$
\begin{aligned}
dq_t(a) = &\sum_{j\in\mathbb{Z}^d} q_t(\delta_j^\dagger(a))da_j(t) \\
&+ \sum_{j\in\mathbb{Z}^d} q_t(\delta_j(a))da^\dagger{}_j(t) + q_t(\mathcal{L}(a))dt,\ q_0(a) = 0.
\end{aligned}
$$

An exactly similar estimate as used before shows that, for all $n \geq 1$,

$$\|q_t(a)ue(f)\| \leq \frac{(t_0 c_f)^{n/2}}{\sqrt{n!}} \|ue(f)\| \sum_{\bar{k}(n)} \sum_{\bar{\varepsilon}(n)} \|\delta(\bar{k}(n), \bar{\varepsilon}(n))(a)\|.$$

Since by Lemma 8.2.2(iii) the above sum grows as n-th power of t, we conclude that $q_t(a) = 0$ for all $a \in \mathcal{A}_{\text{loc}}$, hence the uniqueness of the solution of (8.14). As $1 \in \mathcal{A}_{\text{loc}}$ with $\mathcal{L}_k(1) = \delta_k^\dagger(1) = \delta_k(1) = 0$ it follows from the Q.S.D.E. (8.14) that $j_t(1) = 1$. *(ii)* For $ue(f), ve(g) \in h \otimes \mathcal{E}(\mathcal{C})$ and $a, b \in \mathcal{A}_{\text{loc}}$, we have, by induction,

$$\langle j_t^{(n)}(a^*)ue(f), ve(g)\rangle = \langle ue(f), j_t^{(n)}(a)ve(g)\rangle.$$

Now as n tends to ∞, we get

$$\langle j_t(a^*)ue(f), ve(g)\rangle = \langle ue(f), j_t(a)ve(g)\rangle.$$

We define

$$\Phi_t(a, b) = \langle ue(f), j_t(ab)ve(g)\rangle - \langle j_t(a^*)ue(f), j_t(b)ve(g)\rangle.$$

Setting $(\zeta_k(l), \eta_k(l)) = (\delta_k, \text{Id}),\ (\text{Id}, \delta_k),\ (\delta_k^\dagger, \text{Id}),\ (\text{Id}, \delta_k^\dagger),\ (\mathcal{L}_k, \text{Id}),\ (\text{Id}, \mathcal{L}_k)$ and $(\delta_k^\dagger, \delta_k)$ for $l = 1, 2, \ldots, 7$ respectively, one has

$$|\Phi_t(a, b)| \leq c_{f,g}^n \sum_{l_n, \cdots l_1} \int_0^t \int_0^{s_{n-1}} \cdots \int_0^{s_1}$$

$$\sum_{k_n, \cdots k_1} |\Phi_{s_1}(\zeta_{k_n}(l_n) \cdots \zeta_{k_1}(l_1)a,\ \eta_{k_n}(l_n) \cdots \eta_{k_1}(l_1)b)| ds_0 \cdots ds_{n-1} \quad \text{for all } n \geq 1, \tag{8.20}$$

where $c_{f,g} = (1 + t_0{}^{1/2})(\|f\|_\infty + \|g\|_\infty)$. By the quantum Itô formula and cocyle properties of structure maps, that is, $\hat{\mathcal{L}}(ab) = a\hat{\mathcal{L}}(b) + \hat{\mathcal{L}}(a)b + \sum_{k \in \mathbb{Z}^d} \delta_k^\dagger(a)\delta_k(b)$, we have

$$\begin{aligned}
&\Phi_t(a, b) \\
&= \int_0^t \sum_k \{\Phi_s(\delta_k(a), b) + \Phi_s(a, \delta_k(b))\} f_k(s) ds \\
&+ \int_0^t \sum_k \{\Phi_s(\delta_k^\dagger(a), b) + \Phi_s(a, \delta_k^\dagger(b))\} \bar{g}_k(s) ds \\
&+ \int_0^t \sum_k \{\Phi_s(\mathcal{L}_k(a), b) + \Phi_s(a, \mathcal{L}_k(b)) + \Phi_s(\delta_k^\dagger(a), \delta_k(b))\} ds,
\end{aligned}$$

which gives the following estimate for $n = 1$:

$$|\Phi_t(a,b)| \le c_{f,g} \sum_{l=1\cdots 7} \int_0^t \sum_k |\Phi_s(\zeta_k(l)(a), \eta_k(l)(b))| ds \,. \tag{8.21}$$

If we now assume (8.20) for some $m \ge 1$, an application of (8.21) gives the required estimate for $n = m + 1$.
At this point we note the following, which can be verified by (8.18), (8.19) and Lemma 8.2.2 (*iv*).

(1) For any n-tuple $(l_1, l_2, \ldots, l_n)$ with $l_i \in \{1, 2, \ldots, 7\}$, we have

$$\sum_{k_n,\ldots k_1} \| j_s(\zeta_{k_n}(l_n) \cdots \zeta_{k_1}(l_1)(a) \cdot \eta_{k_n}(l_n) \cdots \eta_{k_1}(l_1)(b)) ve(g) \|$$

$$\le C_{g,a,b}\{(1 + \|r\|)(1 + 2\vartheta_1(r)c_{a,b})\}^{2n} \|ve(g)\|, \tag{8.22}$$

where for any $g \in C$ we set

$$C_{g,a,b} = 1 + \sum_{m \ge 1} 3^m \frac{(t_0 c_g)^{m/2}}{\sqrt{m!}} \{(1 + \|r\|)(1 + 2\vartheta_1(r)c_{a,b})\}^{2m}.$$

(2) For any $s \le t_0$, $p \le n$ and $\bar{\varepsilon}(p)$,

$$\sum_{\bar{k}(p)} \| j_s\{\delta(\bar{k}(p), \bar{\varepsilon}(p))(b)\} ve(g) \|$$

$$\le C_{g,a,b}\{(1 + \|r\|)(1 + 2\vartheta_1(r)c_{a,b})\}^{n} \|ve(g)\|. \tag{8.23}$$

(3) Since $\vartheta_p(a) = \vartheta_p(a^*)$ and $\{\delta(\bar{k}(p), \bar{\varepsilon}(p))(a)\}^*$ can also be written as $\delta(\bar{k}(p), \bar{\varepsilon}'(p))(a^*)$ for some $\bar{\varepsilon}'(p)$, we have

$$\sum_{\bar{k}(p)} \| j_s\{\delta(\bar{k}(p), \bar{\varepsilon}(p))(a)\}^* ue(f) \|$$

$$\le C_{f,a,b}\{(1 + \|r\|)(1 + 2\vartheta_1(r)c_{a,b})\}^{n} \|ue(f)\|. \tag{8.24}$$

For any fixed n-tuple $(l_1, \ldots, l_n)$, it can be seen from the definition of Φ_s that

$$\begin{aligned}
&\sum_{\bar{k}(n)} |\Phi_s(\zeta_{k_n}(l_n) \cdots \zeta_{k_1}(l_1)a, \eta_{k_n}(l_n) \cdots \eta_{k_1}(l_1)b)| \\
&\le \sum_{k_n,\ldots k_1} \|ue(f)\| \cdot \| j_s(\zeta_{k_n}(l_n) \cdots \zeta_{k_1}(l_1)a \cdot \eta_{k_n}(l_n) \cdots \eta_{k_1}(l_1)b) ve(g) \| \\
&+ \| j_s\{(\zeta_{k_n}(l_n) \cdots \zeta_{k_1}(l_1)(a))^*\} ue(f) \| \cdot \| j_s(\eta_{k_n}(l_n) \cdots \eta_{k_1}(l_1)(b)) ve(g) \|.
\end{aligned}$$

The estimates (8.22), (8.23) and (8.24) yield:

$$\begin{aligned}
&\sum_{\bar{k}(n)} |\Phi_s(\zeta_{k_n}(l_n)\cdots\zeta_{k_1}(l_1)a,\ \eta_{k_n}(l_n)\cdots\eta_{k_1}(l_1)b)| \\
&\quad \leq \{(1+\|r\|)(1+2\vartheta_1(r)c_{a,b})\}^{2n}\|ue(f)\|\cdot\|ve(g)\|(C_{g,a,b}+C_{f,a,b}C_{g,a,b}) \\
&\quad = C\{(1+\|r\|)(1+2\vartheta_1(r)c_{a,b})\}^{2n},
\end{aligned}$$

with $C = \|ue(f)\|\cdot\|ve(g)\|(C_{g,a,b}+C_{f,a,b}C_{g,a,b})$.
Now by (8.20),

$$|\Phi_t(a,b)| \leq C\,\frac{(7\ t_0 c_{f,g})^n}{n!}\{(1+\|r\|)(1+2\vartheta_1(r)c_{a,b})\}^{2n},\quad \text{for all } n \geq 1,$$

which implies $\Phi_t(a,b) = 0$.
(*iii*) Let $\xi = \sum c_j u_j e(f_j)$ be a vector in the algebraic tensor product of h and $\mathcal{E}(\mathcal{C})$. If $b \in \mathcal{A}^+_{\text{loc}}$, b is actually an $N^{|b|}\times N^{|b|}$-dimensional positive matrix and hence it admits a unique square root $\sqrt{b} \in \mathcal{A}^+_{\text{loc}}$. For any $a \in \mathcal{A}^+_{\text{loc}}$, setting $b = \sqrt{\|a\|1-a}$ so that $b \in \mathcal{A}^+_{\text{loc}}$, we get

$$\begin{aligned}
0 \leq \|j_t(b)\xi\|^2 &= \langle j_t(b)\xi,\ j_t(b)\xi\rangle \\
&= \sum \bar{c_i}c_j\langle j_t(b)u_ie(f_i),\ j_t(b)u_je(f_j)\rangle \\
&= \sum \bar{c_i}c_j\langle u_ie(f_i),\ j_t(\|a\|1-a)u_je(f_j)\rangle \text{ (by (ii))} \\
&= \|a\|\|\xi\|^2 - \langle \xi,\ j_t(a)\xi\rangle,
\end{aligned}$$

where we have used the fact that $1 \in \mathcal{A}_{\text{loc}}$ and $j_t(1) = 1$. Now let $a \in \mathcal{A}_{\text{loc}}$ be arbitrary and using part (*ii*) and the above inequality for a^*a, we get that

$$\begin{aligned}
\|j_t(a)\xi\|^2 &= \langle j_t(a)\xi,\ j_t(a)\xi\rangle = \sum \bar{c_i}c_j\langle j_t(a)u_ie(f_i),\ j_t(a)u_je(f_j)\rangle \\
&= \sum \bar{c_i}c_j\langle u_ie(f_i),\ j_t(a^*a)u_je(f_j)\rangle = \langle \xi,\ j_t(a^*a)\xi\rangle \\
&\leq \|a^*a\|\|\xi\|^2 = \|a\|^2\|\xi\|^2 \\
&\text{or } \|j_t(a)\xi\| \leq \|a\|\|\xi\|.
\end{aligned}$$

This inequality obviously extends to all $\xi \in h\otimes\Gamma$. Noting that $j_t(1) = 1$, for all t, we get

$$\|j_t(a)\| \leq \|a\| \text{ and } \|j_t\| = 1.$$

Thus, j_t extends uniquely to a unital C^*-homomorphism satisfying the Q.S.D.E. (8.14) and hence is an E–H flow on $\mathcal{A}$ with T_t as its expectation semigroup. That the range of j_t is in $\mathcal{A}''\otimes\mathcal{B}(\Gamma)$ is clear from the construction of j_t. □

8.2.2 Covariance of the E–H flow

We shall show that the E–H flow $\{j_t\}$ constructed by us in the previous subsection is covariant with respect to the actions τ and λ of $\mathbb{Z}^d$, where λ will be defined shortly.

We observe that

$$\delta_k\ \tau_j = \tau_j\ \delta_{k-j} \text{ and } \delta^\dagger{}_k\ \tau_j = \tau_j \delta^\dagger{}_{k-j}, \quad \text{for all } j, k \in \mathbb{Z}^d, \tag{8.25}$$

and we have the following lemma.

Lemma 8.2.4 *(i)* $\hat{\mathcal{L}}\ \tau_j(a) = \tau_j\ \hat{\mathcal{L}}(a)$ for all $a \in \text{Dom}(\hat{\mathcal{L}})$,
(ii) $T_t\ \tau_j = \tau_j\ T_t$*, that is,* T_t *is covariant.*

Proof:

(i) We note that $\mathcal{C}^1(\mathcal{A})$ is invariant under τ and thus for $a \in \mathcal{C}^1(\mathcal{A})$,

$$\begin{aligned}\mathcal{L}\ (\tau_j(a)) &= \frac{1}{2} \sum_{k \in \mathbb{Z}^d} \delta_k^\dagger(\tau_j(a)) r_k + r_k^* \delta_k(\tau_j(a)) \\ &= \frac{1}{2} \sum_{k \in \mathbb{Z}^d} \tau_j\ \delta_{k-j}^\dagger(a) r_k + r_k{}^* \tau_j\ \delta_{k-j}(a) \quad \text{(by 8.25)} \\ &= \frac{1}{2} \tau_j \left\{ \sum_{k \in \mathbb{Z}^d} \delta_{k-j}^\dagger(a) r_{k-j} + r_{k-j}^* \delta_{k-j}(a) \right\} \\ &= \tau_j(\mathcal{L}(a)).\end{aligned}$$

For $a \in \text{Dom}(\hat{\mathcal{L}})$, we choose a sequence $\{a_n\}$ in $\mathcal{C}^1(\mathcal{A})$ and an element $b \in \mathcal{A}$ such that $b = \hat{\mathcal{L}}(a)$, a_n converges to a and $\mathcal{L}(a_n)$ converges to b. As τ_j is an automorphism for any $j \in \mathbb{Z}^d$, $\tau_j(a_n)$ and $\tau_j\ \mathcal{L}(a_n)$ converge to $\tau_j(a)$ and $\tau_j(b)$ respectively. Since $a_n \in \mathcal{C}^1(\mathcal{A})$ and $\mathcal{L}(\tau_j(a_n)) = \tau_j\ \mathcal{L}(a_n)$, we get

$$\tau_j(a) \in \text{Dom}(\hat{\mathcal{L}}) \text{ and } \hat{\mathcal{L}}\ \tau_j(a) = \tau_j\ \hat{\mathcal{L}}(a).$$

(ii) By (i), for $a \in \text{Dom}(\hat{\mathcal{L}})$ and $0 \le s \le t$ we have

$$\frac{d}{ds} T_s \circ \tau_j \circ T_{t-s}(a) = T_s \circ \hat{\mathcal{L}} \circ \tau_j \circ T_{t-s}(a) - T_s \circ \tau_j \circ \hat{\mathcal{L}} \circ T_{t-s}(a) = 0.$$

This implies that $T_s \circ \tau_j \circ T_{t-s}(a)$ is independent of s for every j and $0 \le s \le t$. Setting $s = 0$ and t respectively and using the fact that T_t is bounded we get $T_t\ \tau_j = \tau_j\ T_t$.

□

We note that $j_t : \mathcal{A} \to \mathcal{A}'' \otimes \mathcal{B}(\Gamma(L^2(\mathbb{R}_+, k_0)))$, where $k_0 = l^2(\mathbb{Z}^d)$ with a canonical basis $\{e_k\}$, as mentioned earlier. We define the canonical bilateral shift s by $s_j e_k = e_{k+j}$, for all $j, k \in \mathbb{Z}^d$ and let $\gamma_j = \Gamma(1 \otimes s_j)$ be the second quantization of $1 \otimes s_j$, that is, $\gamma_j e(\sum f_l(\cdot) e_l) = e(\sum f_l(\cdot) e_{l+j})$. This defines a unitary representation of $\mathbb{Z}^d$ in $\Gamma \equiv \Gamma(L^2(\mathbb{R}_+, k_0))$. We set an action $\sigma = \tau \otimes \lambda$ of $\mathbb{Z}^d$ on $\mathcal{A}'' \otimes \mathcal{B}(\Gamma)$, where $\lambda_j(b) = \gamma_j b \gamma_{-j}$ for all $b \in \mathcal{B}(\Gamma)$.

We recall that $a_k(t)$ is given by

$$a_k(t)e(g) = \int_0^t g_k(s)ds\, e(g),$$

and observe that

$$\begin{aligned}\lambda_j a_k(t)e(g) &= \gamma_j a_k(t)\gamma_{-j}e(g) = \gamma_j a_k(t) e\left(\sum_l \langle g, e_{l+j}\rangle(\cdot)e_l\right)\\ &= \int_0^t \langle g, e_{k+j}\rangle(s)ds\ \gamma_j\left(e\left(\sum_l \langle g, e_{l+j}\rangle(\cdot)e_l\right)\right)\\ &= \int_0^t \langle g, e_{k+j}\rangle(s)ds\ \left(e\left(\sum_l \langle g, e_{l+j}\rangle(\cdot)e_{l+j}\right)\right)\\ &= a_{k+j}(t)e(g).\end{aligned}$$

Since $\langle e(f), \lambda_j a_k(t)e(g)\rangle = \langle \lambda_j a_k^\dagger(t)e(f), e(g)\rangle$, it follows that

$$\lambda_j a_k(t) = a_{k+j}(t) \text{ and } \lambda_j a_k^\dagger(t) = a_{k+j}^\dagger(t). \tag{8.26}$$

Theorem 8.2.5 *The E–H dilation j_t of the Q.D.S. (T_t) is covariant with respect to the actions τ and σ, that is,*

$$\sigma_j j_t \tau_{-j}(a) = j_t(a) \ \text{ for all } a \in \mathcal{A},\ t \geq 0 \text{ and } k \in \mathbb{Z}^d.$$

Proof:
For a fixed $j \in \mathbb{Z}^d$ and for all $t \geq 0$, we set $j_t' = \sigma_j j_t \tau_{-j}$. Using the Q.S.D.E. (8.14), Lemma 8.2.4 and (8.25), (8.26) we have for $a \in \mathcal{A}_{\text{loc}}$,

$$\begin{aligned}&j_t'(a) - j_0'(a)\\ &= \int_0^t \sum_{k\in\mathbb{Z}^d} \sigma_j j_s(\delta_k^\dagger(\tau_{-j}(a)))da_k(s) + \int_0^t \sum_{k\in\mathbb{Z}^d} \sigma_j j_s(\delta_k(\tau_{-j}(a)))da_k^\dagger(s)\\ &\quad + \int_0^t \sigma_j j_s(\hat{\mathcal{L}}(\tau_{-j}(a)))ds\\ &= \int_0^t \sum_{k\in\mathbb{Z}^d} \sigma_j j_s \tau_{-j}(\delta_{k+j}^\dagger(a))da_{k+j}(s) + \int_0^t \sum_{k\in\mathbb{Z}^d} \sigma_j j_s\tau_{-j}(\delta_{k+j}(a))da_{k+j}^\dagger(s)\\ &\quad + \int_0^t \sigma_j j_s \tau_{-j}(\hat{\mathcal{L}}(a))ds\\ &= \int_0^t \sum_{k\in\mathbb{Z}^d} j_s'(\delta_k^\dagger(a))da_k(s) + \int_0^t \sum_{k\in\mathbb{Z}^d} j_s'(\delta_k(a))da_k^\dagger(s) + \int_0^t j_s'(\hat{\mathcal{L}}(a))ds.\end{aligned}$$

Since $j_0'(a) = \sigma_j j_0 \tau_{-j}(a) = \sigma_j(\tau_{-j}(a) \otimes 1_\Gamma) = a \otimes 1_\Gamma = j_0(a)$, it follows from the uniqueness of solution of the Q.S.D.E. (8.14) that $j_t'(a) = j_t(a)$ for all $t \geq 0$ and $a \in \mathcal{A}_{\text{loc}}$. As both j_t' and j_t are bounded maps, we have $j_t' = j_t$.

□

9

Noncommutative geometry and quantum stochastic processes

In this chapter, after a brief review of the basics of differential and Riemannian geometry, we shall discuss some of the fundamental concepts of noncommutative geometry. After that, we shall illustrate with examples how quantum dynamical semigroups arise naturally in the context of classical and noncommutative geometry, and how they carry important information about the underlying classical or noncommutative geometric spaces. These semigroups are essentially the 'heat semigroups' with unbounded generator given by the Laplacian (or some variant of it) on the underlying space; and in the classical case, the dilation of such semigroups naturally involve a suitable Brownian motion on the manifold. While the classical theory of heat semigroup and Brownian motion on a manifold is well established and quite rich, there is not yet any general theory of their counterparts in noncommutative geometry. Neither the theory of quantum stochastic calculus nor noncommutative geometry are at a stage for developing a general theory connecting the two. Instead of a general theory, the present state of both subjects calls for an understanding of various examples available, and this is what we try to do in this chapter. We do so at two levels: first, at the semigroup level, and then at the level of quantum stochastic processes coming from dilation of the semigroups.

9.1 Basics of differential and Riemannian geometry

We presume that the reader is familiar with the basic concepts of differential geometry, including tangent, cotangent, differential forms etc. and at least the definition and elementary properties of Lie groups. Let us very briefly review the concepts of connection, curvature and also of Riemannian geometry. For a comprehensive reference, we recommend references [21], [80] and [123]. Let M be a smooth (C^∞) manifold of dimension n. A locally trivial smooth *fibre*

bundle on M with fibre F (where F is also a smooth manifold) consists of a smooth manifold E with a smooth map $\pi : E \to M$ (called the projection map) satisfying the following condition:
for any $p \in M$, we can find an open neighborhood $U = U_p$ of p and a diffeomorphism $\phi_U : U \to U \times F$ such that $p_1 \circ \phi_U = \pi$, where $p_1 : U \times F \to U$ is the projection onto the first coordinate.
The open cover $\{U_p\}$ will be called a trivializing cover. We say that E is a *smooth vector bundle* if F is a finite dimensional real or complex vector space and $\phi_U \circ \phi_V^{-1} : U \cap V \times F \to U \cap V \times F$ is of the form $(x, \xi) \mapsto (x, T_x^{(U,V)}\xi)$, where $T_x^{(U,V)} : F \to F$ is linear. In fact, it is clear that T_x is invertible, that is, T_x is an element of the Lie group $GL(F)$. This provides an alternative formulation of vector bundle in terms of $GL(F)$-valued maps, called *transition functions* or *cocycles* with suitable properties. More precisely, in this approach, a smooth vector bundle with fibre F is described by an open cover $\{U_\alpha\}$ of M together with smooth maps $h_{\alpha\beta} : U_\alpha \cap U_\beta \to GL(F)$ (for α, β such that $U_\alpha \cap U_\beta$ is nonempty) satisfying

$$h_{\alpha\beta}(x)h_{\beta\gamma}(x) = h_{\alpha\gamma}(x) \quad \text{for all } x \in U_\alpha \cap U_\beta \cap U_\gamma,$$

whenever $U_\alpha \cap U_\beta \cap U_\gamma$ is nonempty. Indeed, given a trivializing cover $\{U_\alpha\}$, we can define $h_{\alpha\beta}(x)$ for $x \in U_\alpha \cap U_\beta$ to be the map $T_x^{(U_\alpha,U_\beta)}$. A (continuous / smooth) section of E is a (continuous / smooth) map $s : M \to E$ such that $s(x) \in E_x := \pi^{-1}(\{x\})$ for all $x \in M$. We denote by $\Gamma(E)$ the space of smooth sections of E.

From now on, we use the word 'vector bundle' or simply 'bundle' to mean a smooth vector bundle on a manifold. Let TM denote the set $\{(m, t) : m \in M, t \in T_m M\}$, where $T_m M$ is the space of tangents at the point m. One can equip TM with a canonical smooth manifold structure, such that the map $\pi : TM \to M$ given by $(m, t) \mapsto m$ is smooth. In fact, given a local chart (U, ϕ) around m, with the associated coordinates $(x_1, \ldots, x_n)$ (where n is the dimension of M), we define a bijection ξ_U from $TU := \{(x, t) : x \in U, t \in T_x M\}$ to $\mathbb{R}^n \times \mathbb{R}^n$ by mapping (x, t) to $(x_1, \ldots, x_n) \times (t_1, \ldots, t_n)$, where t_i's are defined by $t = \sum_{i=1}^n t_i \frac{\partial}{\partial x_i}|_x$. The topological and manifold structure on TM are defined by demanding the above bijections to be diffeomorphisms. Clearly, the maps $\phi_U : \pi^{-1}(U) \to U \times \mathbb{R}^n$ is given by $\phi_U(x, t) = (x, t_1, \ldots, t_n)$. The bundle TM is called the *tangent bundle* of M. The smooth sections of this bundle are called *smooth vector fields* (or simply *vector fields*). Similarly, one can define the *cotangent bundle* $T^*M \equiv \Omega^1(M)$, and more generally, the *bundle of k-forms* $\Omega^k(M)$.

Let us now discuss how one can perform various operations on bundles, for example, taking direct sum and tensor product of bundles. This can be described in the language of cocycles mentioned before. Let E_1 and E_2 be two bundles on M with fibres F_1 and F_2 respectively. We construct a canonical bundle $E = E_1 \oplus E_2$ on M with fibre $F_1 \oplus F_2$ as follows. Let $\{U_\alpha\}$ be a common trivializing cover for E_1 and E_2, which can be obtained by taking a common refinement of trivializing covers for the two bundles. Let $h^{(i)}_{\alpha\beta}(i = 1, 2)$ denote the corresponding cocycles. Then, we define E by the cocycle $h_{\alpha\beta} : U_\alpha \cap U_\beta \to GL(F_1 \oplus F_2)$ by setting

$$h_{\alpha\beta}(x) = h^{(1)}_{\alpha\beta}(x) \oplus h^{(2)}_{\alpha\beta}(x).$$

In a very similar way, one can define $E_1 \otimes E_2$, E_1^*, $E_1 \wedge E_2$, $\wedge^k E_1$ (where $\wedge$ stands for the antisymmetric tensor product). These are vector bundles on M with fibres $F_1 \otimes F_2$, F_1^*, $F_1 \wedge F_2$ and $\wedge^k F_1$ respectively.

A hermitian structure or a metric on a bundle E is an association $x \mapsto \langle \cdot, \cdot \rangle_x$, where $\langle \cdot, \cdot \rangle_x$ is a (real-valued) inner product on the finite dimensional vector space E_x and the association $x \mapsto \langle \cdot, \cdot \rangle_x$ is smooth in the sense that for every smooth section s of E, the map $M \ni x \mapsto \langle s(x), s(x) \rangle_x \in \mathbb{R}$ is smooth. A *Riemannian structure* on M is defined to be a hermitian structure on the bundle TM (or equivalently, on T^*M). Such a structure is completely determined by the locally defined, smooth maps g_{ij} on M, given by

$$g_{ij}(m) := \left\langle \frac{\partial}{\partial x_i}|_m, \frac{\partial}{\partial x_j}|_m \right\rangle_m,$$

where $m \in M$ and $(x_1, \ldots, x_n)$ is a choice of local coordinates defined on an open neighborhood of m. Clearly, the matrix $g(m) := ((g_{ij}(m)))$ is an invertible positive definite matrix for every m. It is customary to denote the (ij)th entry of the inverse of $((g_{ij}(m)))$ by $g^{ij}(m)$. We shall refer to the pair (M, g) as a Riemannian manifold with the metric g.

We now come to the notion of connection on a manifold and the associated curvature. There are various equivalent descriptions of connection, but we shall discuss the one which leads to a natural generalization in the noncommutative context. We restrict ourselves to affine connections only.

Definition 9.1.1 An *affine connection* on a manifold M of dimension n is given by a map $\nabla : \Gamma(TM) \times \Gamma(TM) \to \Gamma(TM)$ satisfying the following :

(*i*) $\nabla_X(Y_1 + Y_2) = \nabla_X(Y_1) + \nabla_X(Y_2)$,
(*ii*) $\nabla_X(fY) = f\nabla_X(Y) + (Xf)Y$,
(*iii*) $\nabla_{fX}(Y) = f\nabla_X(Y)$;

where $X, Y, Y_1, Y_2 \in \Gamma(TM)$, $f \in C^\infty(M)$ and $\nabla_X(Y) := \nabla(X, Y)$.

It can be shown by using the above defining properties that $\nabla_X(Y)$ depends on X locally, that is, if $X = X'$ on an open set U, then $\nabla_X(Y)(m) = \nabla_{X'}(Y)(m)$

for all $m \in U$, and for any Y. Similarly, the dependence on Y is also local. $\nabla_X : \Gamma(TM) \to \Gamma(TM)$ is called the *covariant derivative* in the direction of X, and this is a natural generalization of the notion of directional derivatives in $\mathbb{R}^n$. Using the local property of $\nabla_X(Y)$, we can extend the definition to locally defined vector fields, that is, local sections of TM. Indeed, given locally defined vector fields X and Y defined on open subsets U, V respectively, with $U \cap V$ nonempty, we can define $\nabla_X(Y)$ on $U \cap V$ by setting $\nabla_X(Y)(m) = \nabla_{\tilde{X}}(\tilde{Y})(m)$ for all $m \in U \cap V$, where $\tilde{X}, \tilde{Y} \in \Gamma(TM)$ are arbitrary extensions of X,Y respectively to the whole of M. The local property of ∇ ensures that this does not depend on the choice of the extensions. Thus, one can completely determine an affine connection ∇ by the local vector fields $\nabla_{X_i}(X_j)$, where $X_i = \frac{\partial}{\partial x_i}$, $(x_1, \ldots, x_n)$ is a set of local coordinates. The locally defined, smooth functions Γ^k_{ij} defined as follows are called the Christoffel symbols of the connection ∇:

$$\nabla_{X_i}(X_j) = \sum_k \Gamma^k_{ij} X_k.$$

Given an affine connection ∇ on M, a natural notion of parallel translation along a curve can be introduced, which we briefly describe below. Let $\{\gamma(t)\ t \in J\}$ (J is an interval in $\mathbb{R}$) be a piecewise smooth curve in M and $\{V(t),\ t \in J\}$ be a family of tangent vectors such that $V(t) \in T_{\gamma(t)}M$ for all t, and $t \mapsto (\gamma(t), V(t)) \in TM$ is piecewise smooth. Let I be an open subinterval of J such that $\gamma(I)$ is contained in the domain of a single local chart, say (U, ϕ), with the local coordinates $(x_1, \ldots, x_n)$. Let $\gamma_i(t) = x_i(\gamma(t))$ and $V(t) = \sum_i V_i(t) X_i(\gamma(t))$, where $X_i = \frac{\partial}{\partial x_i}$. Suppose that Y, Z are local vector fields defined on U satisfying $Y(\gamma(t)) = \gamma'(t)$, $Z(\gamma(t)) = V(t)$ for all $t \in I$. Note that $\gamma'(t)$ denotes the tangent vector at $\gamma(t)$ given by $\gamma'(t)f = \frac{d}{dt} f(\gamma(t))$, $f \in C^\infty(M)$. Then one gets that

$$\nabla_Y(Z)(\gamma(t)) = \sum_k f_k(t) X_k(\gamma(t)) \ \text{ for all } t \in I,$$

where $f_k(t) = V_k'(t) + \sum_{ij} \gamma_i'(t) V_j(t) \Gamma^k_{ij}(\gamma(t))$. Thus, in particular, $\nabla_Y(Z)$ $(\gamma(t))$ depends only on the values of Y and Z at $\gamma(t)$, that is, on the functions $\gamma'(t)$ and $V(t)$, hence it makes sense to denote $\nabla_Y(Z)(\gamma(t))$ by $\nabla_{\gamma'(t)}(V(t))$.

Definition 9.1.2 We say that $V(t)$ is *parallel* along γ if $\nabla_{\gamma'(t)}(V(t)) = 0$ for all $t \in J$, or equivalently, the functions $V_k(t), k = 1, \cdots, n$ satisfy the following system of ordinary differential equations:

$$V_k'(t) + \sum_{ij} \gamma_i'(t) V_j(t) \Gamma^k_{ij}(\gamma(t)) = 0; \quad k = 1, \cdots, n. \tag{9.1}$$

It follows from the standard results on the existence and uniqueness of the system of first-order differential equations that given an initial values $V(t_0)$ $(t_0 \in J)$, there exists a unique solution $V(t),\ t \in J$ of the above system of equations (9.1). In other words, given a tangent vector $\xi \in T_{\gamma(t_0)}M$, there exists a unique family $V(t)$ which is parallel along γ and $V(t_0) = \xi$. $V(t)$ is called the parallel translate of ξ along γ from t_0 to t, and is denoted by $\tau_t^{\gamma}(\xi)$. Thus, τ_t^{γ} is a diffeomorphism from $T_{\gamma(0)}M$ to $T_{\gamma(t)}M$. Furthermore, it can be shown that

$$\nabla_X(Y)(m) = \lim_{t \to 0+} \frac{(\tau_t^{\gamma_X})^{-1}(Y(\gamma_X(t)) - Y(m)}{t},$$

where γ_X the integral curve of X passing through m.

The next important concept in this context is that of a geodesic, which we define below.

Definition 9.1.3 A piecewise smooth curve $\gamma(t),\ t \in J$ (J being some interval of $\mathbb{R}$) is called a *geodesic* with respect to the connection ∇ if $\gamma'(t)$ is parallel along γ for all $t \in J$.

It is clear that if we write $\gamma(t) = (\gamma_1(t), \cdots \gamma_n(t))$ in some local coordinates, then the functions $\gamma_i(t)$'s must satisfy the following second-order ordinary differential equations:

$$\gamma_k''(t) + \sum_{ij} \gamma_i'(t)\gamma_j'(t)\Gamma_{ij}^k(\gamma(t)) = 0; \quad k = 1, \cdots, n.$$

It follows from the theory of ordinary differential equations that given $m \in M$ and $v \in T_m M$, there is a geodesic $\gamma(t)$ defined for $t \in (-\epsilon, \epsilon)$ for some $0 < \epsilon \le \infty$, satisfying $\gamma(0) = m$ and $\gamma'(0) = v$. It is unique in the sense that if $\beta(t),\ t \in (-\delta, \delta)$ $(\delta > 0)$ is another geodesic with $\beta(0) = m,\ \beta'(0) = v$, then we must have $\beta(t) = \gamma(t)$ for all $t \in (-\eta, \eta)$, where $\eta = \min(\epsilon, \delta)$. Thus, we can find a maximal interval J containing 0 such that a geodesic $\gamma(t)$ with $\gamma(0) = m,\ \gamma'(0) = v$ can be defined for all values of $t \in J$. This geodesic will be called the geodesic passing through m with the initial tangent v, and will be denoted by $\gamma^{m,v}(t)$. We say that the geodesic is infinitely extendible if $\gamma^{m,v}(t)$ is defined for all $t \in \mathbb{R}$.

There are two important geometric objects associated with a given connection ∇, called the torsion and curvature of the connection, which are defined below.

Definition 9.1.4 The *torsion transformation* associated to the connection ∇, denoted by T^{∇}, is a map from $\Gamma(TM) \times \Gamma(TM)$ to $\Gamma(TM)$, defined by:

$$T_{XY} \equiv T^{\nabla}(X, Y) = \nabla_X(Y) - \nabla_Y(X) - [X, Y].$$

The *curvature transformation* R^∇ is the map from $\Gamma(TM)\times\Gamma(TM)\times\Gamma(TM)$ to $\Gamma(TM)$ given by:

$$R_{XY}(Z) \equiv R^\nabla(X, Y, Z) = \nabla_X\nabla_Y(Z) - \nabla_Y\nabla_X(Z) - \nabla_{[X,Y]}(Z).$$

It is a remarkable fact that any affine connection is completely determined by the set of geodesics and the torsion transformation, and also that given an affine connection ∇ with nonzero torsion, we can always find another connection ∇' which has the same curvature but the associated torsion transformation is identically zero.

Before we take up the discussion of connections on Riemannian manifolds, let us remark that the notion of connection can be generalized from the bundle TM to any vector bundle E on M. Indeed, connection on E can be defined as a map from $\Gamma(TM) \times \Gamma(E)$ to $\Gamma(E)$ satisfying the properties (*i*), (*ii*) and (*iii*) in the Definition 9.1.1, with $Y, Y_1, Y_2 \in \Gamma(TM)$ replaced by elements of $\Gamma(E)$. It is also possible to define parallel translation in E along a curve in M, in a very similar way as done for TM. In fact, given an affine connection ∇ on M, we can construct canonical connections on the bundles $\Omega^k(M)$ for any $k = 0, \ldots, n$, which will also be denoted by ∇. To define the extension of ∇ on $\Omega^k(M)$, we first define ∇_X on $\Omega^0(M) \equiv C^\infty(M)$ by setting $\nabla_X(f) = Xf$. The definition of ∇_X on $\Omega^1(M) = T^*M$ is given by

$$\nabla_X(\omega)(m) = \lim_{t\to 0+} \frac{(\tau_t^{\gamma_X})^*(\omega(\gamma_X(t)) - \omega(m)}{t},$$

where $\omega \in \Gamma(\Omega^1(M))$, γ_X integral curve of X and $(\tau_t^{\gamma_X})^* : T^*_{\gamma_X(t)} \to T^*_m M$ is the dual of the parallel translate $\tau_t^{\gamma_X} : T_m M \to T_{\gamma_X(t)}M$. Similarly, ∇_X on $\Gamma(\Omega^k(M))$ can be defined for $k > 1$.

Let us now consider a Riemannian manifold (M, g). In this case, the tangent space $T_m M$ has an additional structure: it is an inner product space with respect to the inner product $\langle\cdot,\cdot\rangle_m$. Thus, given an affine connection ∇ on M, and a piecewise smooth curve γ, it is natural to demand that the associated parallel translation operators τ_t^γ preserve the inner product, that is, τ_t^γ is an isometry from $T_{\gamma(0)}M$ to $T_{\gamma(t)}M$. This motivates the following definition:

Definition 9.1.5 An affine connection ∇ on the Riemannian manifold M is called a *unitary connection* or *metric connection* if τ_t^γ is an isometry for every t and for every piecewise smooth γ.

It is a fundamental result in Riemannian geometry that there exists a unique affine connection on a given Riemannian manifold (M, g) which is unitary and torsionless (that is, the associated torsion transformation is zero).

Definition 9.1.6 The unique unitary and torsionless affine connection on a Riemannian manifold is called the *Riemannian connection* or the *Levi–Civita connection*. The geodesics with respect to the Levi–Civita connection are called Riemannian geodesics.

Given two points p and q of M, we can define a natural distance between them by using the Riemannian metric g. Indeed, we have a natural notion of length of a piecewise smooth curve $\gamma \equiv \{\gamma(t),\ t \in [a,b]\}$, denoted by $l(\gamma)$, which is defined below:

$$l(\gamma) := \int_a^b \langle \gamma'(t), \gamma'(t) \rangle_{\gamma(t)}^{\frac{1}{2}} dt.$$

We say that the curve $\{\gamma(t), t \in [a,b]\}$ joins p and q if $\gamma(a) = p,\ \gamma(b) = q$. Then, the distance between the points p, q, say $d(p,q)$, is defined by

$$d(p,q) = inf\{l(\gamma) :\ \gamma \text{ piecewise smooth, joins } p \text{ and } q\}.$$

The infimum is taken to be infinity if there is no γ joining p and q. It is a remarkable result that d is indeed a metric on M (called the Riemannian distance) and the topology of M coincides with the topology as a metric space with respect to the metric d. This implies in particular that every Riemannian manifold is metrizable. Furthermore, one can show that Riemannian geodesics are the curves which locally minimize the Riemannian distance on M. To be more precise, for $m \in M$, we can find an open neighborhood U of m such that for every $p, q \in U$, there is a Riemannian geodesic γ_{pq} joining p and q, satisfying $l(\gamma_{pq}) \leq l(\gamma)$ for every γ joining p and q. The manifold is said to be *geodesically complete* if every Riemannian geodesic is infinitely extendible. A connected Riemannian manifold is geodesically complete if and only if any two points can be joined by a Riemannian geodesic. Furthermore, this is equivalent to the completeness of the manifold as a metric space with respect to the Riemannian distance. Examples of geodesically complete (to be called just complete from now on) Riemannian manifold are $\mathbb{R}^n$ with the usual Euclidean Riemannian structure, as well as any compact Riemannian manifold.

We conclude this section with a brief discussion on various measures of curvature of a Riemannian manifold (M, g). Let us choose a system of local coordinates $(x_1, \ldots, x_n)$ and suppose that g is given by a matrix $((g_{ij}))$ of locally defined, smooth maps with respect to the above choice of local coordinates. Let $(X_1, \ldots, X_n)$ be a locally defined vector fields with the property that $(X_1(m), \ldots, X_n(m))$ is an orthonormal basis of $T_m M$ (with respect to the inner product given by g) for all m. Moreover, we denote by ∇ the canonical Levi–Civita connection, and let Γ_{ij}^k be the associated Christoffel symbols.

Definition 9.1.7 The *Ricci curvature* of the Riemannian manifold (M, g) is the two-form $Ric \in \Omega^2(M)$ defined by:

$$Ric(X, Y)(m) := \mathrm{tr}_g(L_{XY}(m)), \ \ X, Y \in \Gamma(TM);$$

where $L \equiv L_{XY}(m) : T_m M \to T_m M$ is the linear map given by its action on the basis $\{X_1(m), \ldots, X_n(m)\}$ as $L(X_i(m)) = R_{X, X_i}(Y)(m)$, and where tr_g denotes the trace on the inner product space $(T_m M, \langle \cdot, \cdot \rangle_m)$.

The curvature transformation $R_{ij} \equiv R_{X_i X_j}$ $\left(\text{where } X_i = \frac{\partial}{\partial x_i}\right)$ can be locally written as

$$R_{ij}(X_l) = \sum_k R^k_{lij} X_k,$$

where R^k_{lij} are locally defined smooth functions. Then one has

$$r_{ij} := Ric(X_i, X_j) = \sum_k R^k_{jik}.$$

Definition 9.1.8 We define the *scalar curvature* of (M, g) to be the smooth function $s : M \to \mathbb{R}$ given by

$$s = \sum_i r_{ii}.$$

Definition 9.1.9 A smooth map $f : M \to N$, where M and N are Riemannian manifolds, is called an *isometry* at m if the map $df|_m : T_m M \to T_{f(m)} N$ preserves the Riemannian metric, that is $\langle df|_m(\xi), df|_m(\eta) \rangle_{f(m)} = \langle \xi, \eta \rangle_m$ for all $\xi, \eta \in T_m M$. We say that M and N are *locally isometric* if given any point $m \in M$ we can find an open neighborhood U of m and an open neighborhood $V \subseteq N$ such that there is a diffeomorphism f from U onto V which is also an isometry at every point of U.

It is a fundamental result of Riemannian geometry that a Riemannian manifold is locally isometric to $\mathbb{R}^n$ (for some n) with the usual Euclidean metric if and only if the Levi–Civita connection is identically zero. Such a Riemannian manifold is called *flat*.

9.2 Heat semigroup and Brownian motion on classical manifolds

Let us assume that the reader is familiar with the basic ideas of algebraic topology, including the de-Rham cohomology of a manifold. However, for the sake of completeness, we quickly review the definitions. Let M be a manifold and $\Omega^k = \Omega^k(M)$ $(k = 0, 1, \cdots, n)$ be the space of smooth k-forms. Set $\Omega^k = \{0\}$

for $k > n$, and let $\Omega = \oplus_k \Omega^k$. The differential d maps Ω^k to Ω^{k+1}, and since $d^2 = 0$, $\mathrm{Ran}(d_k) \subseteq \mathrm{Ker}(d_{k+1})$, where d_k denotes the restriction of d on Ω^k. Thus, we have a complex (Ω, d), and it makes sense to talk of the cohomology of this complex, which is called the *de-Rham cohomology* of the manifold. The kth cohomology group, denoted by $H^k_{dR}(M)$, is given by the following:

$$H^k_{dR}(M) = \mathrm{Ker}(d_k)/\mathrm{Ran}(d_{k-1}), \quad k \geq 1;$$

with $H^0_{dR}(M) := \mathrm{Ker}(d_0) = \{f \in C^\infty(M) : \; df = 0\}$. $H^k_{dR}(M)$ is clearly a vector space over $\mathbb{R}$, and its dimension is called the kth *Betti number*, denoted by $\beta^k \equiv \beta^k(M)$. It is a remarkable and nontrivial fact that β^k's are topological invariants of M, so do not depend on the differentiable structure of M. For a compact M, all the Betti numbers are finite. An explicit example of such a manifold is the n-torus $\mathbb{T}^n$, for which $\beta^k = n!/(k!(n-k)!), \; 0 \leq k \leq n, \beta^k = 0$ for $k > n$. Of course, compactness is not a necessary condition for the finiteness of Betti numbers. For example, $\beta^0(\mathbb{R}^n) = 1$, $\beta^k(\mathbb{R}^n) = 0$ for all $k \geq 1$. The *Euler characteristic* $\chi(M)$ of M is defined by $\chi(M) = \sum_k (-1)^k \beta^k(M)$, whenever all the $\beta^k(M)$ are finite. This is an important topological invariant of the manifold.

Assume from now on that M is a connected and orientable manifold of dimension n. Recall at this point that a manifold of dimension n is said to be *orientable* if there exists an atlas (called an oriented atlas) with the property that on the intersection of the domains of any two local charts of the atlas, the Jacobian of transformation from one set of local coordinates to the other has positive determinant. Let us fix an orientable Riemannian manifold (M, g) with an oriented atlas, with the local coordinates $(x_1, \ldots, x_n)$. We define an n-form $dvol$ locally given by $dvol = \sqrt{det(g)} dx_1 \wedge \cdots \wedge dx_n$ where the Riemannian metric is given by the matrix of (locally defined) smooth maps $g = ((g_{ij}))$ in the local coordinates $(x_1, \ldots, x_n)$. One can verify that $dvol$ does not depend on the choice of local coordinates. This n-form will be referred as the *Riemannian volume form*, (or just *volume form*) and we shall consider integration of compactly supported smooth functions on M with respect to this form. This in a natural fashion extends to a regular Borel measure on M, also denoted by vol. We denote by $L^2(M)$ the Hilbert space obtained by completing the space $\{f \in C^\infty_c(M) : \; \int_M |f|^2 dvol < \infty\}$, with respect to the pre-inner product given by $\langle f_1, f_2 \rangle := \int_M f_1 f_2 dvol$. Similarly, we can consider the complexified version of the above L^2-space, denoted by $L^2_{\mathbb{C}}(M)$.

In an analogous way, we can construct a canonical Hilbert space of forms. The Riemannian metric $\langle \cdot, \cdot \rangle_m$ $(m \in M)$ on $T_m M$ induces an inner product on the vector space $T^*_m M$, and hence also $\wedge^k T^*_m M$, which will again be denoted by $\langle \cdot, \cdot \rangle_m$. This gives a natural pre-inner product on the space of compactly supported k-forms by integrating the compactly supported smooth function

$m \mapsto \langle \omega(m), \eta(m) \rangle_m$ over M. We denote the completion of this space by $\mathcal{H}^k \equiv \mathcal{H}^k(M)$. Let $\mathcal{H} = \oplus_k \mathcal{H}^k$. There is an alternative expression of this inner product. This is given in terms of the *Hodge star operator* $* : \Omega \to \Omega$ which sends $\omega \in \Omega^k$ to $\omega' = *(\omega) \in \Omega^{n-k}$. To give the precise definition of $*$, let us choose an oriented atlas with local charts $(x_1, \ldots, x_n)$, and let $(\omega_1, \ldots, \omega_n)$ be locally defined sections of T^*M such that for each m, $(\omega_1(m), \ldots, \omega_n(m))$ is an orthonormal basis for $T^*_m M$ with respect to the inner product on $T_m M$ induced by the Riemannian metric, and furthermore, $(\omega_1, \ldots, \omega_n)$ has positive orientation in the sense that the transformation from $(dx_1, \ldots dx_n)$ to $(\omega_1, \ldots, \omega_n)$ has positive determinant. To define $*$ on Ω^k, it is sufficient to define it on the basis $(\omega_{i_1} \wedge \cdots \wedge \omega_{i_k})$, where $i_1 < \cdots < i_k$. We define $*(\omega_{i_1} \wedge \cdots \wedge \omega_{i_k})$ to be the $(n-k)$-form $\omega_{j_1} \wedge \cdots \wedge \omega_{j_{n-k}}$, where $j_1 < \cdots < j_{n-k}$, and $\omega_{i_1} \wedge \cdots \wedge \omega_{i_k} \wedge \omega_{j_1} \wedge \cdots \wedge \omega_{j_{n-k}} = dvol = \omega_1 \wedge \cdots \wedge \omega_n$. It can be verified by a simple calculation that for $\alpha, \beta \in \Omega^k$,

$$\langle \alpha, \beta \rangle dvol = \alpha \wedge (*\beta).$$

Now, let us view the de-Rham differential $d : \Omega \to \Omega$ as an unbounded, densely defined operator (again denoted by d) on the Hilbert space $\mathcal{H}$ with the domain Ω. We can verify its closability by showing that its adjoint is densely defined on Ω. Indeed, for $\alpha \in \Omega^k$, $\beta \in \Omega^{k+1}$, we have

$$\langle d\alpha, \beta \rangle = \int_M d\alpha \wedge (*\beta) = (-1)^{k+1} \int_M \alpha \wedge d(*\beta) = (-1)^{nk+1} \langle \alpha, (*d * \beta) \rangle,$$

where we have used the fact that M is without boundary so that $\int_M d(\omega) = 0$ for any $n-1$-form ω, and also the observation that $*^2(\eta) = (-1)^{k(n-k)}\eta$ for all $\eta \in \Omega^k$. Thus, the adjoint d^* of d has $\Omega = \oplus_k \Omega^k$ in its domain, and is given by

$$d^*(\beta) = (-1)^{nk+1} * (d(*\beta))$$

for $\beta \in \Omega^{k+1}$. It is clear that the unbounded operator $(d + d^*)$ is symmetric, and it can be shown that it has a unique self-adjoint extension on $\mathcal{H}$, to be denoted again by $d + d^*$.

Definition 9.2.1 The *Hodge Laplacian* (also called the Laplacian on forms) is the unbounded, positive operator $\Delta = (d + d^*)^2$ on $\mathcal{H}$. Δ_k will denote the restriction of Δ on $\mathcal{H}_k$.
The semigroup of positive contractions on $\mathcal{H}$ generated by $-\Delta$, that is, $(e^{-t\Delta})$, is called the *heat semigroup* on forms. Its restriction on $\mathcal{H}_0$, that is, $(e^{-t\Delta_0})$, is referred to as the heat semigroup.

It is obvious that Δ_k maps $\mathcal{H}_k$ into itself, and since $d^2 = 0$, we have $\Delta = d^*d + dd^*$ on Ω. In particular, for $k = 0$, $\Delta_0 = d^*d = -(*d)^2$. In terms of local coordinates $(x_1, \ldots, x_n)$, Δ_0 has the following expression:

$$\Delta_0 f = -\frac{1}{\sqrt{det(g)}} \sum_{i,j=1}^{n} \frac{\partial}{\partial x_j} \left(g^{ij} \sqrt{det(g)} \frac{\partial}{\partial x_i} f \right),$$

for $f \in C^\infty(M)$, and where $g = ((g_{ij}))$ and $g^{-1} = ((g^{ij}))$. If the manifold is flat, local coordinates can be chosen so that $g = I$, and then $\Delta_0 = -\sum_i \frac{\partial^2}{\partial x_i^2}$. Δ_0 is called the Laplacian on functions, or just Laplacian.

The purpose of introducing the operator Δ is to explore the rich interplay between geometry and analysis. We shall see that this operator contains most of the interesting topological and geometric information about the manifold. The next few results, which we state without proofs (and whose proofs can be found in the references quoted earlier) describe how the Betti numbers are related to the Hodge Laplacian.

Proposition 9.2.2 *Assume that M is compact. Then the Hodge Laplacian Δ on $\mathcal{H}$ has compact resolvents. Thus, the set of eigenvalues of Δ is countable, each having finite multiplicities, and accumulating only at infinity. There exists an orthonormal basis of $\mathcal{H}_k$ consisting of eigenfunctions of $\Delta|_{\mathcal{H}_k}$.*

To connect the topology and geometry of M with the analysis of the above unbounded operators defined at the L^2 level, it is necessary to show that any eigenvector of Δ, which is a-priori defined as an L^2 function only, is actually a smooth function. To achieve this, one needs to consider the heat semigroup for forms, and construct the so-called heat kernel. To motivate the definition of the heat-kernel, let us assume that the semigroup $T_t = (e^{-t\Delta_0})$ acting on $\mathcal{H}_0 = L^2(M)$ is an integral kernel operator in the sense that there exists a family of smooth functions $p_t : M \times M \to \mathbb{R}$, $t \geq 0$, such that $p_t(x, y)$ is also smooth in t for every fixed x, y, and

$$(T_t f)(x) = \int_M p_t(x, y) f(y) dvol(y) \quad \text{for all } f \in L^2(M),\ t > 0. \tag{9.2}$$

Since (T_t) is a C_0-semigroup with the generator $-\Delta_0$, it follows that $p_t(x, y)$ must satisfy the following:

$$\frac{\partial}{\partial t} p_t(x, y) = -\Delta_0^x p_t(x, y), \tag{9.3}$$

$$\lim_{t \to 0+} \int_M p_t(x, y) f(y) dvol(y) = f(x). \tag{9.4}$$

It may be remarked that (9.3) is familiar to physicists; it models the flow of heat on a surface or a solid body, and this is where the name 'heat equation' comes from. Before we proceed further, we may note the important implication of the smoothness of $p_t(x, y)$. It can be shown using the expression (9.2) that even if f is not smooth, $T_t f$ must be a smooth function for any positive t. Since M is compact, it is known [39] that the spectrum of the Laplacian Δ_0 consists of only eigenvalues of finite multiplicities, accumulating at infinity. Suppose now that $f \in L^2(M)$ is an eigenvector of Δ_0 corresponding to an eigenvalue λ. It follows from the Spectral theorem for unbounded operator that $T_t f = e^{-\lambda t} f$, and hence it follows that f must be smooth. In particular, $\mathrm{Ker}(\Delta_0)$ consists of smooth functions only.

We now generalize the above discussion for forms of arbitrary order, and give the definition of heat-kernel.

Definition 9.2.3 For $0 \leq k \leq n$, let E_k be the bundle on $M \times M$ with fibre $\wedge^k(T_x^* M) \otimes \wedge^k(T_y^* M)$ at (x, y). A map $e_t : [0, \infty) \to \Gamma(E_k)$ (where $\Gamma(E_k)$ denotes the space of smooth sections of E_k) is said to be a *heat-kernel for k-forms* if for all $x, y \in M$, the map $t \mapsto e(t, x, y) \equiv e_t(x, y) \in \wedge^k(T_x^* M) \otimes \wedge^k(T_y^* M)$ is smooth, and furthermore, we have the following:

$$\frac{\partial}{\partial t} e(t, x, y) = -\Delta_k^x e(t, x, y), \quad \text{and}$$

$$\lim_{t \to 0+} \int_M \langle e(t, x, y), \omega(y) \rangle_y dvol(y) = \omega(x) \ \text{ for all } \omega \in \Omega^k(M),$$

where Δ_k^x means the action of Δ_k on the form $e(t, \cdot, y)$ holding t, y fixed.

The existence of the heat-kernel is the fundamental result in the analysis of the Laplacian and heat semigroup for forms, which we state below (see [113] for a detailed proof).

Proposition 9.2.4 *For every k, there exists a unique heat-kernel* $e_t^k \equiv e^k(t, \cdot, \cdot)$, *and the semigroup* $e^{-t\Delta}$ *on the Hilbert space* $\mathcal{H}_k$ *is given by the following expression on the dense set* $\Omega^k(M) \subset \mathcal{H}_k$:

$$(e^{-t\Delta}\omega)(x) = \int_M \langle e^k(t, x, y), \omega(y) \rangle dvol(y), \quad \omega \in \Omega^k(M).$$

It follows that $\mathrm{Ker}(\Delta_k) \subset \Omega^k(M)$. For $\omega \in \mathrm{Ker}(\Delta_k)$, we have $(dd^* + d^*d)\omega = 0$, hence $\langle d\omega, d\omega \rangle + \langle d^*\omega, d^*\omega \rangle = 0$, which implies that both $d\omega$ and $d^*\omega$ are 0. In particular, ω is a smooth k-form with $d\omega = 0$, so it makes sense to consider the de-Rham cohomology class $[\omega] \in H_{dR}^k(M)$. This gives a map from the kernel of Δ_k to $H_{dR}^k(M)$ and the next result which we state without proof shows that this is an isomorphism.

Theorem 9.2.5 *[Hodge theorem] The above map* $\omega \mapsto [\omega]$ *is an isomorphism of vector spaces.*

Thus, the Betti number $\beta_k = \dim (H^k_{dR}(M))$ is nothing but the dimension of $\mathrm{Ker}(\Delta_k)$. In other words, the topological invariants β_k can be obtained from the spectral information of the operator Δ. It may be remarked here that even though we restricted our discussion to compact-connected oriented manifolds only, many of the results mentioned in this section indeed extend to a much larger class of manifolds, including several noncompact ones. We shall not, however, discuss any noncompact manifold other than the trivial one, namely $\mathbb{R}^n$.

Coming back to the analysis of Δ, let us consider the operator D^+ obtained by restricting $d + d^*$ on the space of even forms, that is, on $\mathcal{H}^+ := \oplus_k \mathcal{H}_{2k}$. Clearly, $D^+ : \mathcal{H}^+ \to \mathcal{H}^-$, where $\mathcal{H}^- := \oplus_k \mathcal{H}_{2k+1}$. Let $D^- := (D^+)^* : \mathcal{H}^- \to \mathcal{H}^+$. The following result is a consequence of the definition of Euler characteristic $\chi(M)$, the Theorem 9.2.5 and the above observations.

Proposition 9.2.6

$$\begin{aligned}\mathrm{Index}(D^+) &:= \dim (\mathrm{Ker}(D^+)) - \mathrm{codim}\,(\mathrm{Ran}(D^+)) \\ &= \dim (\mathrm{Ker}(D^+)) - \dim (\mathrm{Ker}(D^-)) = \chi(M).\end{aligned}$$

This is one of the simplest index theorems, connecting index of a suitable operator defined using the Riemannian metric of the manifold, to a purely topological invariant. In fact, the operator $d + d^*$ is a prototype of first order elliptic operators on manifold. We do not have the scope to discuss the general theory of such operators and the associated index theorems here, but refer the reader to [113] and [53] for details.

Next we try to understand one concrete example of the heat semigroup. Consider the one-dimensional manifold S^1 and let us compute $e^0(t, x, y) \equiv e(t, x, y)$. In this case, using the standard coordinate $\theta \in (0, 2\pi]$ for S^1, we can write the heat equation as follows:

$$\left(\frac{\partial}{\partial t} - \frac{\partial^2}{\partial \theta^2} \right) e(t, \theta, \theta') = 0. \tag{9.5}$$

Let $e(t, \theta, \theta') = \sum_{n \in Z} a_n(t, \theta') e^{in\theta}$ be the Fourier series expansion. From the equation (9.5), we get $\frac{\partial}{\partial t} a_n + n^2 a_n = 0$ for all n, hence

$$a_n(t, \theta') = C_n(\theta') e^{-tn^2},$$

where C_n is a function of θ'. However, we also require that for every $f \in C^\infty(S^1)$, $\lim_{t \to 0+} \int_0^{2\pi} e(t, \theta, \theta') f(\theta') d\theta' = f(\theta)$ for all θ. That is,

$$\sum_{n \in \mathbb{Z}} e^{in\theta} \int_0^{2\pi} C_n(\theta') f(\theta') d\theta' = f(\theta).$$

From this, it is possible to conclude by elementary calculation that $C_n(\theta') = \frac{1}{\sqrt{2\pi}} e^{-in\theta'}$. Thus

$$e(t, \theta, \theta') = \sum_n \frac{1}{\sqrt{2\pi}} e^{in(\theta-\theta')} e^{-tn^2}.$$

It is remarkable that $e(t, \theta, \theta)$ is a constant multiple of $\sum_n e^{-tn^2}$, which behaves like $t^{-\frac{1}{2}}$ for small t. To make a more precise statement, we note the obvious inequalities

$$\sum_{n=1}^{\infty} e^{-tn^2} \leq \int_0^{\infty} e^{-tx^2} dx \leq \sum_{n=0}^{\infty} e^{-tn^2}.$$

Thus

$$2 \int_0^{\infty} e^{-tx^2} dx - 1 \leq \sum_{n \in \mathbb{Z}} e^{-tn^2} \leq 2 \int_0^{\infty} e^{-tx^2} dx + 1.$$

Since the value of the Gaussian integral $\int_0^{\infty} e^{-tx^2} dx$ is $\sqrt{\pi}/(2\sqrt{t})$, we conclude that $\lim_{t \to 0+} t^{\frac{1}{2}} e(t, \theta, \theta)$ exists and is a constant, independent of θ.

Such an asymptotic behavior of the heat kernel is indeed true in general. Below we formulate in precise mathematical terms what an asymptotic expansion means.

Definition 9.2.7 We say that a function $B : (0, \infty) \to \mathbb{R}$ has an *aysmptotic expansion* $\sum_{k=k_0}^{\infty} b_k t^k$ for $t \to 0+$ if for all $N \geq k_0$, we have

$$\lim_{t \to 0+} \frac{B(t) - \sum_{k=k_0}^{N} b_k t^k}{t^N} = 0.$$

We symbolically write this as

$$B(t) \sim \sum_{k=k_0}^{\infty} b_k t^k.$$

We are now in a position to describe the asymptotic behavior of $e^0(t, x, x) \equiv e(t, x, x)$ for small values of t.

Proposition 9.2.8 *Let $e(t, x, y)$ be the heat kernel for functions for a compact oriented Riemannian manifold (M, g) of dimension d. Then there are smooth functions $u_0, u_1, \cdots$ on M such that*

$$(4\pi t)^{\frac{d}{2}} e(t, x, x) \sim \sum_{k=0}^{\infty} u_k(x) t^k.$$

In particular, $u_0(x) = 1$, $u_1(x) = \frac{1}{6} s(x)$, where $s(x)$ denotes the scalar curvature at x.

Similar asymptotic expansion of the heat kernel $e^k(t, x, x)$ for k-forms can be obtained, which can be found in [113]. Let us now express the pointwise asymptotic expansion of $e(t, x, x)$ as given by Proposition 9.2.8 in terms of trace of the heat semigroup, which admits a direct generalization to the non-commutative situation, as will be seen later. Note that the heat semigroup $T_t = e^{-t\Delta_0}$ on functions is given by the smooth integral kernel $e(t, x, y)$, and clearly (as M is compact) $\int_M e(t, x, x)dvol(x) < \infty$. From this, it can be proved using the theory of integral kernel operators (see Section X.1.4. in [77]) that T_t is trace class with $\mathrm{tr}(T_t) = \int_M e(t, x, x)dvol(x)$. By Proposition 9.2.8, it follows that

$$\mathrm{vol}(M) = \lim_{t \to 0+} t^{d/2}(\mathrm{Tr}(T_t)), \tag{9.6}$$

$$s(M) := \int_M s(x)dvol(x) = \frac{1}{6} \lim_{t \to 0+} t^{d/2-1}\{\mathrm{tr}(T_t) - t^{-d/2}vol(M)\}. \tag{9.7}$$

A few words about the Brownian motion on a Riemannian manifold will be appropriate at this point. We assume that the reader is familiar with the basic concepts of probability theory and stochastic processes, including Markov process and stop time. Recall that given a probability space $(\Sigma, \mathcal{F}, P)$ with a right continuous complete filtration $(\mathcal{F}_t)_{t \geq 0}$ on it, we say that a real-valued stochastic process $(Z_t)_{t \geq 0}$, adapted to the filtration (that is, Z_t is measurable with respect to $\mathcal{F}_t$), is an $(\mathcal{F}_t)$-*martingale* if $E_s(Z_t) = X_s$ for all $0 \leq s \leq t$, where E_s denotes the conditional expectation with respect to $\mathcal{F}_s$. We say that an adapted process (Y_t) is a *local martingale* (with respect to the filtration $(\mathcal{F}_t)$) if there is a sequence (τ_k) of $(\mathcal{F}_t)$-stop times, such that $\tau_k \uparrow \infty$ as $k \to \infty$ almost surely, and for each k, $(Y_{\min(\tau_k, t)})_t$ is an $(\mathcal{F}_t)$-martingale. We are now ready to define a Brownian motion.

Definition 9.2.9 Let (M, g) be a compact Riemannian manifold, Δ_0 be the corresponding Laplacian. An M-valued stochastic process $(X_t)_{t \geq 0}$, defined on a probability space $(\Sigma, \mathcal{F}, P)$ and adapted to a right continuous complete filtration $(\mathcal{F}_t)$, is said to be a *Brownian motion* on M starting at the point $m \in M$ if

(*i*) $X_0 = m$,
(*ii*) for almost all $\omega \in \Sigma$, $t \mapsto X_t(\omega) \in M$ is continuous, and
(*iii*) for every smooth real-valued function f on M there exists a local martingale $(Z_t^f)_{t \geq 0}$ (which depends on f) such that

$$f(X_t) = f(X_0) - \frac{1}{2} \int_0^t \Delta_0(f)(X_s)ds + Z_t^f.$$

There are several equivalent definitions of a Brownian motion on M, and our choice as above is referred as the martingale characterization in the classical theory. This is the one which admits a straightforward generalization in the algebraic language. We do not give the construction of Brownian motion on M, but assume that it exists. Let us denote by (X_t^m) a Brownian motion starting at m. From the above definition it is immediate that $Z_0^f = 0$ for all f, and hence $E_0(Z_t^f) = 0$ for all $t \geq 0$, as (Z_t^f) is a local martingale. Thus, $E_0(f(X_t^m)) = f(m) - \frac{1}{2}\int_0^t (\Delta_0 f)(X_s^m)ds$. It follows from this that $E_0(f(X_t^m)) = (e^{-\frac{t}{2}\Delta_0} f)(m)$. In other words, Brownian motion is nothing but a stochastic dilation of the heat semigroup. Given the importance of the heat semigroup from the geometric point of view, it is quite natural that one can extract many interesting geometric information from a Brownian motion on the manifold. There is a very rich set of stochastic tools to analyze Brownian motions on manifolds, and we refer the reader to [46] for a discussion of such topics.

Our discussion on classical analysis of manifolds cannot be complete without a brief mention of the Dirac operator on spinor bundles. Historically, Dirac operators were introduced to write down a square root of the Laplacian. Recall that the Laplacian defined by us is given by $\sum_{ij} g_{ij} \frac{\partial}{\partial x_i}\frac{\partial}{\partial x_j}$ plus a first-order operator which involves curvature of the Riemannian metric. This first-order term is in fact zero for a flat manifold. For simplicity, let us consider the flat manifold $\mathbb{R}^n$ and look for a first-order differential operator of the following for

$$D = \sum_i \gamma_i \frac{\partial}{\partial x_i},$$

where the γ_i's are some formal symbols, but assumed to be independent of x. We want D^2 to be equal to $\Delta_0 = -\sum_{ij} \frac{\partial^2}{\partial x_i^2}$, which is equivalent to the following commutation relations among the symbols γ_i's:

$$\gamma_i \gamma_j + \gamma_j \gamma_i = -2\delta_{ij},$$

where δ_{ij} denotes the Kronecker delta. For a more general manifold with the Riemannian metric given by (g_{ij}), we need to modify the above relations by replacing δ_{ij} on the right-hand side above by g_{ij}. This is made possible by the next result.

Proposition 9.2.10 *Given a finite dimensional real vector space V with an inner product $\langle \cdot, \cdot \rangle$, there exists a (unique upto isomorphism) algebra* Cl(V) *over $\mathbb{R}$, equipped with an injective linear map j from V to* Cl(V) *such that the algebra generated by $j(V)$ is the whole of* Cl(V), *and for every $v \in V$,*

$$j(v)^2 = -\langle v, v \rangle. \tag{9.8}$$

The algebra Cl(V) is called the Clifford algebra of the inner product space V and by identifying $v \in V$ with its image $j(v) \in$ Cl(V), we consider V as a linear subspace of $\mathrm{Cl}(V)$. Thus, the relation (9.8) can be written as $v^2 = -\langle v, v\rangle$, and from this, it can be deduced that $u.v = -\langle u, v\rangle$ for all $u, v \in V$. It is also not difficult to prove that $\mathrm{Cl}(V)$ as a vector space has the dimension 2^n (where n is the dimension of V), thus isomorphic as a vector space (but not as algebra) with $\oplus_{k=0}^{n} \wedge^k V$, and inherits the natural inner product coming from that of V. We shall need to consider the complexification of $\mathrm{Cl}(V)$, that is, $\mathrm{Cl}(V) \otimes_{\mathbb{R}} \mathbb{C}$, to be denoted by $\mathrm{Cl}_{\mathbb{C}}(V)$. It should also be mentioned here that there is a representation $\kappa_n : \mathrm{Cl}_{\mathbb{C}}(V) \to \mathcal{L}(\Sigma_n)$, where Σ_n denotes $\mathbb{C}^{\frac{n}{2}}$ if n is even, and $\mathbb{C}^{\frac{n-1}{2}} \oplus \mathbb{C}^{\frac{n-1}{2}}$ if n is odd. We refer the reader to [53] for the definition of the representation κ_n and its further properties. Thus, for $v \in V$, $\kappa_n(v)$ is a linear map from Σ_n to Σ_n.

Now, let us consider the compact oriented Riemannian manifold (M, g), and suppose that we can construct a complex vector bundle $E = \mathrm{Cl}_{\mathbb{C}}(M)$ on M with the fibre Σ_n. The construction of such a bundle, called the Clifford bundle or the spinor bundle on M, depends on the existence of the so-called spin structure on M. This is a stronger requirement than orientability, and we refer the reader to [53] for the definition and more details. We do not go into that, but remark that a large class of manifold admits a spin structure, and thus it is possible to talk of the bundle of spinors. Such manifolds are called Riemannian spin manifolds, and let us assume that M is one such. Let S denote the space of smooth sections of this bundle, and define a pre-inner product $\langle \cdot, \cdot\rangle_S$ on S by

$$\langle s_1, s_2\rangle_S := \int_M \langle s_1(x), s_2(x)\rangle dvol(x),$$

for $s_1, s_2 \in S$, where $\langle s_1(x), s_2(x)\rangle$ denotes the usual inner product on Σ_n. The Hilbert space obtained by completing S with respect to the above pre-inner product is denoted by $L^2(S)$ and its elements are called the square integrable spinors. The Levi–Civita connection on M induces a canonical affine connection on the bundle E, which will be denoted by ∇^E. To understand the construction of this connection, it is necessary to discuss in some details the concept of principal and associated bundles and connections on them. We refer to the Appendix in the book [53] for a discussion on this topic.

We are now in a position to give a precise definition of the Dirac operator on M.

Definition 9.2.11 The *Dirac operator* is the self-adjoint extension of the following operator D defined on the space of smooth sections of $E = \mathrm{Cl}_{\mathbb{C}}(M)$:

$$(Ds)(m) = \sum_{i=1}^{n} \kappa_n(X_i(m))(\nabla^E_{X_i} s)(m),$$

where $(X_1, \ldots, X_n)$ are local orthonormal (with respect to the Riemannian metric) vector fields defined in a neighborhood of m, and ∇^E is the canonical connection on E mentioned before. Note that here we have viewed $X_i(m) \in T_m M$ as an element of the Clifford algebra $\mathrm{Cl}_{\mathbb{C}}(T_m M)$, hence $\kappa_n(X_i(m))$ is a map on the fibre of E at m, which is isomorphic with Σ_n. The self-adjoint extension of D is again denoted by the same notation.

We observe that $\mathcal{A} \equiv C^\infty(M)$ acts on S by multiplication, and this action extends to a representation, say π, of the C^*-algebra $C(M)$ on th Hilbert space $L^2(S)$. One can verify by a simple computation that for $f \in C^\infty(M)$, $[D, \pi(f)]$ has a bounded extension. Furthermore, the Laplacian and the Dirac operator D on a compact manifold have compact resolvents. It may be remarked here that a similar definition of spinors and the Dirac operator can be given for a noncompact manifold also. However, the analysis of the Dirac operator will become more delicate, and the resolvents of D will no longer be compact. Nevertheless the commutators of D with compactly supported smooth functions will continue to be bounded. The operator D in fact has a lot of information about the underlying Riemannian metric g, since one can recover the metric from D, as was shown in [28]. As we have seen above, the Dirac operator gives an indirect way of looking at the differential structure on a manifold and it is this route which will form the main path in the road to noncommutative geometry.

9.3 Noncommutative geometry

In this section, we briefly outline the basics of noncommutative geometry. Recall from the previous section that the classical Dirac operator, viewed as an unbounded self-adjoint operator on the Hilbert space $L^2(S)$, has bounded commutators with the multiplication operators by smooth functions. This property has been taken to be the definition of a 'Dirac operator' in Connes' formulation of noncommutative geometry.

Definition 9.3.1 A *spectral triple* or *spectral data* is a triple $(\mathcal{A}, \mathcal{H}, D)$ where $\mathcal{H}$ is a separable Hilbert space, $\mathcal{A}$ is a $*$-subalgebra of $\mathcal{B}(\mathcal{H})$ (not necessarily norm-closed) and D is a self-adjoint (typically unbounded) operator such that for each $a \in \mathcal{A}$, the operator $[D, a]$ admits bounded extension. Such a spectral triple is also called an *odd spectral triple*. If in addition, we have $\gamma \in \mathcal{B}(\mathcal{H})$ satisfying $\gamma = \gamma^* = \gamma^{-1}$, $D\gamma = -\gamma D$ and $[a, \gamma] = 0$ for all $a \in \mathcal{A}$, then we say that the quadruplet $(\mathcal{A}, \mathcal{H}, D, \gamma)$ is an *even spectral triple* or *even spectral data*. The operator D is called the *Dirac operator* corresponding to the spectral triple.

Furthermore, given an abstract $*$-algebra $\mathcal{B}$, an odd (even) spectral triple on $\mathcal{B}$ is an odd (even) spectral triple $(\pi(\mathcal{B}), \mathcal{H}, D)$ (respectively, $(\pi(\mathcal{B}), \mathcal{H}, D, \gamma)$) where $\pi : \mathcal{B} \to \mathcal{B}(\mathcal{H})$ is a $*$-homomorphism.

Since in the classical case, the Dirac operator has compact resolvents if the manifold is compact, we say that the spectral triple is of *compact type* if D has compact resolvents. Most of the examples discussed by us later in this chapter, except the noncommutative plane, are of such type. Let us now briefly explain how the concepts similar to those in de-Rham cohomology can be generalized in the noncommutative situation. We begin with the notion of universal derivation, which the next proposition, stated without proof (and a proof can be found in [28] and [87]), makes precise.

Proposition 9.3.2 *Given an algebra $\mathcal{B}$ over $\mathbb{R}$ or $\mathbb{C}$, there is a (unique upto isomorphism) $\mathcal{B}$-$\mathcal{B}$ bimodule $\Omega^1(\mathcal{B})$ and a derivation $\delta : \mathcal{B} \to \Omega^1(\mathcal{B})$ (that is, $\delta(ab) = \delta(a)b + a\delta(b)$ for all $a, b \in \mathcal{B}$), satisfying the following properties:*

(i) $\Omega^1(\mathcal{B})$ is spanned as a vector space by elements of the form $a\delta(b)$ with $a, b \in \mathcal{B}$; and

(ii) for any $\mathcal{B}$-$\mathcal{B}$ bimodule E and a derivation $d : \mathcal{B} \to E$, there is an $\mathcal{B}$-$\mathcal{B}$ linear map $\eta : \Omega^1(\mathcal{B}) \to E$ such that $d = \eta \circ \delta$.

The bimodule $\Omega^1(\mathcal{B})$ is called the *space of universal* 1*-forms* on $\mathcal{B}$ and δ is called the *universal derivation*.

We can also introduce universal space of higher forms on $\mathcal{B}$, $\Omega^k(\mathcal{B})$, say, for $k = 2, 3, \cdots$, by defining them recursively as follows:

$$\Omega^{k+1}(\mathcal{B}) := \Omega^k(\mathcal{B}) \otimes_{\mathcal{B}} \Omega^1(\mathcal{B}),$$

and also, set $\Omega^0(\mathcal{B}) = \mathcal{B}$. The universal space of forms, denoted by $\Omega(\mathcal{B})$, is defined to be the algebraic direct sum $\oplus_k \Omega^k(\mathcal{B})$, which has an algebra structure by tensor multiplication. This is sometimes referred to as the algebra of universal forms on $\mathcal{B}$. It is straightforward to see that the space $\Omega^k(\mathcal{B})$ is spanned as a vector space by elements of the form $a_0\delta(a_1)\cdots\delta(a_k)$, for $a_0, \ldots, a_k \in \mathcal{B}$. Note that here we have suppressed the tensor product symbol between $\delta(a_i)$ and $\delta(a_{i+1})$, and we shall continue to do so as long as there is no possibility of confusion.

Now, suppose that we are given a spectral triple $(\mathcal{A}, \mathcal{H}, D)$. We have a derivation from $\mathcal{A}$ to the $\mathcal{A} - \mathcal{A}$ bimodule $\mathcal{B}(\mathcal{H})$ given by $a \mapsto [D, a]$. Therefore, by Proposition 9.3.2, we can obtain a bimodule morphism from $\Omega^1(\mathcal{A})$ to $\mathcal{B}(\mathcal{H})$, and in fact, this is also clear from the definition of the algebra of universal forms that this morphism extends to an algebra homomorphism, say π, from $\Omega(\mathcal{A})$ to $\mathcal{B}(\mathcal{H})$, such that $\pi(a) = a$ and $\pi(\delta(a)) = [D, a]$. Then we have the following.

Proposition 9.3.3 *Let J_k be the kernel of the restriction of π on $\Omega^k(\mathcal{A}) \subset \Omega(\mathcal{A})$, and let $\mathcal{I}_k = J_k + \delta(J_{k-1})$ for $k \geq 1$, and $\mathcal{I}_0 = J_0 = \{0\}$. Then*

$\mathcal{I} := \oplus_k \mathcal{I}_k$ is a two-sided ideal in $\Omega(\mathcal{A})$ which is also a graded differential ideal, that is, $\delta(\mathcal{I}_k) \subseteq \mathcal{I}_{k+1}$ for all k.

Using this result, we can construct the quotient algebra $\Omega_D(\mathcal{A}) := (\Omega(\mathcal{A}))/\mathcal{I}$, and in fact we have $\Omega_D(\mathcal{A}) = \oplus_k \Omega_D^k(\mathcal{A})$, where $\Omega_D^k(\mathcal{A}) = (\Omega^k(\mathcal{A}))/\mathcal{I}_k$. We call the space Ω_D^k the space of k-forms corresponding to the spectral triple $(\mathcal{A}, \mathcal{H}, D)$. For $\omega \in \Omega^k(\mathcal{A})$, we shall denote by $[\omega]_D$ the element $p_k(\omega) \in \Omega_D^k(\mathcal{A})$, where p_k is the quotient map from $\Omega^k(\mathcal{A})$ to $\Omega_D^k(\mathcal{A})$. Similar to the classical case, we have a well-defined map $d : \Omega_D(\mathcal{A}) \to \Omega_D(\mathcal{A})$ given by

$$d([a_0\delta(a_1)\cdots\delta(a_k)]_D) := [\delta(a_0)\cdots\delta(a_k)]_D.$$

Clearly, $d^2 = 0$, and thus we have a differential complex $(\Omega_D^k(\mathcal{A}), d)$.

Definition 9.3.4 The cohomology of the complex $(\Omega_D^k(\mathcal{A}), d)$ is called the *Connes–de-Rham cohomology* of the spectral triple.

We have a natural notion of unitary equivalence of two spectral triples.

Definition 9.3.5 We say that two spectral triples $(\pi_1(\mathcal{A}), \mathcal{H}_1, D_1)$ and $(\pi_2(\mathcal{A}), \mathcal{H}_2, D_2)$ on $\mathcal{A}$ are *unitarily equivalent* if there is a unitary operator $U : \mathcal{H}_1 \to \mathcal{H}_2$ such that $D_2 = UD_1U^*$ and $\pi_2(\cdot) = U\pi_1(\cdot)U^*$, where $\pi_j,\ j = 1, 2$ are the representation of $\mathcal{A}$ in $\mathcal{H}_j$, respectively.

One can prove that two unitarily equivalent spectral triples will have same Connes–de-Rham cohomology. However, it turns out that Connes–de-Rham cohomology is not as useful in the noncommutative context as its classical counterpart, namely the de-Rham cohomology. In fact, another cohomology theory, formulated by A. Connes, called the cyclic cohomology, has been found to fit naturally with the concept of spectral triples. Using the so-called Chern character constructed from a spectral triple, and its canonical pairing with the C^*-algebraic K-theory (which is a generalization of the topological K theory), one can obtain many interesting information about the spectral triple. There is a natural notion of nontriviality of a spectral triple using the above-mentioned pairing. However, we do not want to go into such aspects of noncommutative geometry.

We shall end the present section with a discussion of concepts like volume form, scalar curvature, etc. in the context of noncommutative geometry. Before that, we need to discuss the notion of Dixmier trace. Let $\mathcal{H}$ be a separable Hilbert space. A *singular trace* on $\mathcal{B}(\mathcal{H})$ is a functional ϕ defined on some ideal of $\mathcal{B}(\mathcal{H})$ containing all trace class operators, with the trace property, that is, $\phi(ab) = \phi(ba)$, such that ϕ vanishes on the trace class operators. Such a functional can never be normal. The scalar mutiple of the usual trace is the only normal faithful semifinite functional on $\mathcal{B}(\mathcal{H})$ with the trace property. While investigating whether there exists a non-normal functional with the trace property, Dixmier came up with the so-called *Dixmier trace*, which we define now. Consider a positive compact operator T such that its eigenvalues are given by the sequence $\lambda_1 > \lambda_2 > \cdots$, with $\lambda_n \to 0$. Suppose furthermore that the sequence λ_n behaves like $\frac{1}{n}$ in the sense that $\mu_n := \frac{1}{\log(n)} \sum_{k=1}^{n} \lambda_k$ is a bounded sequence, that is , $(\mu_n) \in l^\infty(I\!N)$. For a suitable Banach limit $\mathrm{Lim}_{\,\omega}$ on $l^\infty(I\!N)$ associated with a bounded functional $\omega \in (l^\infty(I\!N))^*$, we define the Dixmier trace of T with respect to $\mathrm{Lim}_{\,\omega}$, denoted by $\mathrm{Tr}_\omega(T)$, by setting $\mathrm{Tr}_\omega(T) = \mathrm{Lim}_{\,\omega}((\mu_n))$. We require certain additional properties of the Banach limit $\mathrm{Lim}_{\,\omega}$ to ensure that Tr_ω has the trace property. In fact, we need $\mathrm{Lim}_{\,\omega}$ to satisfy the following for the above purposes.

(*i*) $\mathrm{Lim}_{\,\omega}((c_n)) \geq 0$ if (c_n) is a sequence with $c_n \geq 0$ for all n.
(*ii*) If $(c_n) \in l^\infty(I\!N)$, such that $\lim_{n\to\infty} c_n = c$, then $\mathrm{Lim}_{\,\omega}((c_n)) = c$.
(*iii*) $\mathrm{Lim}_{\,\omega}((c'_n)) = \mathrm{Lim}_{\,\omega}((c_n))$, where c' denotes the sequence $(c_1, c_1, c_2, c_2, \cdots)$.

We refer the reader to the book [28] for the proof that such a Banach limit exists. Let us just mention one choice without proving that it satisfies the conditions (*i*)–(*iii*). Given a sequence $c \equiv (c_n)$ in $l^\infty(I\!N)$, we consider a function $f_c : [0, \infty) \to \mathbb{R}$ by $f_c(t) = c_n$ if $n - 1 < t \leq n$. Define $M(f_c)$ on $(0, \infty)$ by $M(f_c)(\lambda) = \frac{1}{\log(\lambda)} \int_1^\lambda \frac{f_c(t)}{t} dt$. Then the functional $c \mapsto \limsup_{\lambda\to\infty} M(f_c)(\lambda)$ is one choice of $\mathrm{Lim}_{\,\omega}$.

We say that $T \in \mathcal{K}(\mathcal{H})$ belongs to the Dixmier trace class, or it has finite Dixmier trace if the eigenvalues of $|T| := \sqrt{T^*T}$, given by the decreasing sequence (λ_n), satisfy the following

$$\mu_n \equiv \frac{1}{\log(n)} \sum_{k=1}^{n} \lambda_k = \mathrm{O}(1).$$

We shall denote by $\mathcal{L}^{1,\infty}(\mathcal{H})$ the set of all such operators. For a compact self-adjoint operator $T \in \mathcal{L}^{1,\infty}(\mathcal{H})$ it can be shown that the positive and negative parts of T, denoted by T^+, T^- respectively, also belong to $\mathcal{L}^{1,\infty}(\mathcal{H})$. For a choice of $\mathrm{Lim}_{\,\omega}$ as discussed before, we define the Dixmier trace of T by $\mathrm{Tr}_\omega(T) := \mathrm{Tr}_\omega(T^+) - \mathrm{Tr}_\omega(T^-)$. For a not necessarily self-adjoint compact operator $T = \mathrm{Re}(T) + i\ \mathrm{Im}(T)$, such that $\mathrm{Re}(T)$ and $\mathrm{Im}(T)$ are in $\mathcal{L}^{1,\infty}(\mathcal{H})$, we define

$$\mathrm{Tr}_\omega(T) := \mathrm{Tr}_\omega(\mathrm{Re}(T)) + i\ \mathrm{Tr}_\omega(\mathrm{Im}(T)).$$

The following properties of the Dixmier trace Tr_ω, which are simple consequences of the definition, will be useful later on. The reader is referred to [28] for a proof of this result.

Lemma 9.3.6 *(i) $\mathcal{L}^{1,\infty}(\mathcal{H})$ is a both-sided ideal in $\mathcal{B}(\mathcal{H})$. Furthermore, for $T \in \mathcal{L}^{1,\infty}(\mathcal{H})$, $S \in \mathcal{B}(\mathcal{H})$, one has $\mathrm{Tr}_\omega(TS) = \mathrm{Tr}_\omega(ST)$.*
(ii) If $T \in \mathcal{L}^{1,\infty}(\mathcal{H})$, $S \in \mathcal{K}(\mathcal{H})$ (that is, S is a compact operator), then $\mathrm{Tr}_\omega(TS) = 0$.
(iii) Tr_ω is a positive linear functional on $\mathcal{L}^{1,\infty}$.

It is clear that the Dixmier trace of a given operator may depend on the choice of Lim_ω. However, for a large class of operators the Dixmier trace is actually independent of choice of the Banach limit, and this class is called the class of measurable Dixmier trace class operators.

Let us briefly discuss the relevance of Dixmier trace in the context of geometry. For a (possibly unbounded) self-adjoint operator T with finite dimensional null space, we denote by $\hat{T}$ its restriction on the orthogonal complement of the null space, that is $\hat{T} := TP^\perp$, where P is the projection on $\mathrm{Ker}(T)$. Now, consider an n-dimensional compact Riemannian manifold (M, g) with spin structure, and let D be the corresponding Dirac operator on the Hilbert space $\mathcal{H}$ of square integrable spinors. Since D has compact resolvents, the null space of D is finite dimensional, and we have the following proposition.

Proposition 9.3.7 *The operator $|\hat{D}|^{-p}$ is in $\mathcal{L}^{1,\infty}(\mathcal{H})$ if and only if $p \geq n$. Moreover, $\mathrm{Tr}_\omega(|\hat{D}|^{-n})$ is nonzero and independent of the choice of Lim_ω.*

Thus, the dimension n of the manifold has a purely operator theoretic description, namely,

$$n = \min\{p \geq 0 : |\hat{D}|^{-p} \in \mathcal{L}^{1,\infty}(\mathcal{H})\}.$$

In fact, since $|\hat{D}|^{-p}$ is compact for any positive p, it is clear that $\mathrm{Tr}_\omega(|\hat{D}|^{-p}) = \mathrm{Tr}_\omega(|\hat{D}|^{-n}|\hat{D}|^{-(p-n)}) = 0$ for all $p > n$ (by (ii) of Lemma 9.3.6); that is, n is the unique value of p for which $\mathrm{Tr}_\omega(|\hat{D}|^{-p})$ is finite and nonzero. This can be used as the definition of dimension in the noncommutative context. We make the following definition.

Definition 9.3.8 We say that a spectral triple $(\mathcal{A}, \mathcal{H}, D)$ of compact type has *dimension d* if d is the minimum value of $p \geq 0$ satisfying $|\hat{D}|^{-p}$ belongs to $\mathcal{L}^{1,\infty}(\mathcal{H})$ and $\mathrm{Tr}_\omega(|\hat{D}|^{-d})$ is nonzero.

For a general spectral triple, we cannot of course guarantee existence of such a dimension. Moreover, even if a dimension d in the sense of the Definition 9.3.8 exists, it need not be an integer, though the non-integrality of dimension can be seen even in classical situations like fractals.

Not only the dimension, many other geometric and topological quantities associated with the manifold M can be captured by Dixmier trace of appropriate operators. For example, the volume form is obtained by the following formula:

$$\int_M f dvol = c(d)\mathrm{Tr}_\omega(M_f|\hat{D}|^{-d}),$$

where M_f denotes the operator of multiplication in the Hilbert space of square integrable spinors by $f \in C^\infty(M)$, and $c(d)$ is a universal constant depending only on the dimension d of M. This formula in fact extends to noncompact manifolds. If M is a noncompact d-dimensional manifold with a spin structure, we can still define the spinor bundle and the Dirac operator D, but it will not have compact resolvents any more. However, for compactly supported smooth functions f and any $\lambda > 0$, the operator $M_f(|D|+\lambda)^{-d}$ will be compact and in fact, will have finite Dixmier trace which is equal to $c(d) \int_M f dvol$. Thus, it is natural to define the dimension of a not necessarily compact type spectral triple $(\mathcal{A}, \mathcal{H}, D)$ to be the number d if there exists a dense $*$-subalgebra $\mathcal{A}_0 \subseteq \mathcal{A}$ such that for all $a \in \mathcal{A}_0$, $\lambda > 0$, the operator $a(|D| + \lambda)^{-d}$ has finite Dixmier trace, and $\mathrm{Tr}_\omega(a(|D| + \lambda)^{-d}) \neq 0$ for some a. This will be illustrated in the context of the noncommutative plane.

9.4 Examples

We first outline a general scheme of constructing spectral triples from a C^*-dynamical system, that is, a C^*-algebra equipped with a group action. It should be pointed out that all the examples of noncommutative manifolds considered by us are constructed following this scheme.

Let $(\mathcal{A}, G, \alpha)$ be a C^*-dynamical system with G an n dimensional Lie group, and τ a G-invariant trace on $\mathcal{A}$. Let $\mathcal{A}^\infty$ be the space of smooth vectors, $\mathcal{K} = L^2(\mathcal{A}, \tau) \otimes \mathbb{C}^N$, where $N = 2^{\lfloor n/2 \rfloor}$. Fix any basis $\{X_1, X_2, \ldots X_n\}$ of the Lie algebra of G. Since G acts as a strongly continuous unitary group on $\mathcal{H} = L^2(\mathcal{A}, \tau)$ we can form skew-adjoint operators d_{X_i} on $\mathcal{H}$. Let $D : \mathcal{K} \to \mathcal{K}$ be given by

$$D = \sum_k i d_{X_k} \otimes \gamma_k, \tag{9.9}$$

where $\gamma_1, \ldots \gamma_n$ are self-adjoint matrices in $M_N(\mathbb{C})$ such that $\gamma_k \gamma_j + \gamma_j \gamma_k = -2\delta_{kj}$. Then $(\mathcal{A}^\infty, \mathcal{K}, D)$ is a candidate of a spectral triple, and for such a Dirac operator, $[D, \mathcal{A}^\infty] \subseteq \mathcal{A}^\infty \otimes M_N(\mathbb{C})$. It may be noted that for a nonabelian G, $D^2 \neq -\sum_k d_{X_k}^2$, and whether D has compact resolvents or not has to be verified case by case.

9.4.1 Spectral triples on noncommutative torus

For a fixed θ, an irrational number in $[0, 1]$, consider the C^*-algebra $\mathcal{A}_\theta$ (which appeared briefly in Chapter 3) generated by a pair of unitary symbols subject to the relation:

$$UV = \exp(2\pi i\theta)VU \equiv \lambda VU. \tag{9.10}$$

It is known that $\mathcal{A}_\theta$ is a simple C^*-algebra, that is, it has no two-sided ideal other than the trivial ones, namely $\{0\}$ and $\mathcal{A}_\theta$ itself. For details of the properties of such a C^*-algebra, the reader is referred to [28] and [112]. The algebra has many interesting representations.

(*i*) $\mathcal{H} = L^2(\mathbb{T}^1)$, $\mathbb{T}^1$ is the circle, and for $f \in \mathcal{H}$, $(\pi_1(U)f)(z) = f(\lambda z)$. $(\pi_1(V)f)(z) = zf(z),\ z \in \mathbb{T}^1$.

(*ii*) In the same $\mathcal{H}$, with the roles of U and V reversed: for $f \in \mathcal{H}$, $(\pi_2(V)f)(z) = f(\bar{\lambda}z),\ (\pi_2(U)f)(z) = zf(z),\ z \in \mathbb{T}^1$.

(*iii*) In $\mathcal{H} = L^2(\mathbb{R})$, $(\pi_3(U)f)(x) = f(x+1),\ (\pi_3(V)f)(x) = \lambda^x f(x)$.

While the first two were inequivalent irreducible representations, the ultra-weak closure of the third one is a factor of type II_1 (see also Example 2 in Section 3.2).

There is a natural action of the abelian compact group $\mathbb{T}^2$ (2-torus) on $\mathcal{A}_\theta$ given by

$$\alpha_{(z_1,z_2)}\left(\sum a_{mn}U^m V^n\right) = \sum a_{mn} z_1^m z_2^n U^m V^n,$$

where the sum is over finitely many terms and $\|z_1\| = \|z_2\| = 1$. α extends as a $*$-automorphism on $\mathcal{A}_\theta$ and has two commuting generators d_1 and d_2 which are skew $*$-derivations obtained by extending linearly the rule:

$$\begin{aligned} d_1(U) = U,\ d_1(V) = 0 \\ d_2(U) = 0, d_2(V) = V. \end{aligned} \tag{9.11}$$

Both d_1 and d_2 are clearly well defined on $\mathcal{A}_\theta{}^\infty \equiv \{a \in \mathcal{A}_\theta \mid z \mapsto \alpha_z(a)$ is $C^\infty\} \equiv \{\sum_{m,n\in Z} a_{mn} U_m V^n \mid \sup_{m,n} |m^k n^l a_{mn}| < \infty$ for all $k, l \in N\}$. Clearly, $\mathcal{A}_\theta^\infty$ is a dense $*$-subalgebra of $\mathcal{A}_\theta$. A theorem of Bratteli *et al.* [22] describes all the derivations of $\mathcal{A}_\theta$ which maps $\mathcal{A}_\theta^\infty$ to itself : for almost all θ (Lebesgue), a derivation $\delta : \mathcal{A}_\theta^\infty \to \mathcal{A}_\theta^\infty$ is of the form $\delta = c_1 d_1 + c_2 d_2 + [r, .]$, with $r \in \mathcal{A}_\theta^\infty$, $c_1, c_2 \in \mathbb{C}$. Another important fact about $\mathcal{A}_\theta$ is the existence of a unique faithful trace τ on $\mathcal{A}_\theta$ defined as follows:

$$\tau\left(\sum a_{mn}U^m V^n\right) = a_{00}. \tag{9.12}$$

Then one can consider the Hilbert space $\mathcal{H} = L^2(\mathcal{A}_\theta, \tau)$ (see [96] for an account of noncommutative L^p spaces.) and study the derivations there. It is clear that the family $\{U^m V^n\}_{m,n\in Z}$ constitute a complete orthonormal basis in $\mathcal{H}$. The next simple theorem is stated without proof.

Theorem 9.4.1 *The canonical derivations d_1, d_2 are self-adjoint on their natural domains:* $\mathrm{Dom}(d_1) = \{\sum a_{mn}U^m V^n \mid \sum(1+m^2)|a_{mn}|^2 < \infty\}$ $\mathrm{Dom}(d_2) = \{\sum a_{mn}U^m V^n \mid \sum(1+n^2)|a_{mn}|^2 < \infty\}$. *Furthermore if we denote by $d_r = [r, .]$ with $r \in \mathcal{A}_\theta \subset L^\infty(\mathcal{A}_\theta, \tau)$ acting as left multiplication in $\mathcal{H}$, then $d_r{}^* = d_{r^*} \in \mathcal{B}(\mathcal{H})$*

We define, following Connes, the 'standard' or 'canonical' spectral triple on $\mathcal{A}_\theta^\infty$ by choosing the Hilbert space $\mathcal{H} = L^2(\tau) \oplus L^2(\tau)$, with the $\mathbb{Z}_2$-grading given by $\gamma = \begin{pmatrix} 1 & 0 \\ 0 & -1 \end{pmatrix}$, and the Dirac operator

$$D_0 = \begin{pmatrix} 0 & d_1 + i\, d_2 \\ d_1 - i\, d_2 & 0 \end{pmatrix}.$$

Note that $\mathcal{A}_\theta^\infty$ is embedded as a subalgebra of $\mathcal{B}(\mathcal{H})$ by $a \mapsto \begin{pmatrix} a & 0 \\ 0 & a \end{pmatrix}$. The associated 'Laplacian' is given by $\mathcal{L}_0 = -\frac{1}{2}(d_1^2 + d_2^2)$.

We now construct a family of 'perturbed' spectral triples by perturbing the above canonical triple by a bounded element of $\mathcal{A}_\theta^\infty$ in a suitable sense. To this end, choose and fix $r \in \mathcal{A}_\theta^\infty$. We define a spectral triple with the same Hilbert space, grading and representation of the algebra as before, but with the Dirac operator D_0 replaced by $D = D_r$ given by

$$D_r = D_0 + \begin{pmatrix} 0 & d_r \\ d_r^* & 0 \end{pmatrix}.$$

We claim that this new spectral triple is in general not unitarily equivalent (in the sense described below) with the canonical one. This claim is proven by computing the Connes–de-Rham cohomology groups.

Recall that we say that two spectral triples $(\mathcal{A}_1, \mathcal{H}_1, D_1)$ and $(\mathcal{A}_2, \mathcal{H}_2, D_2)$ are unitarily equivalent if there is a unitary operator $U : \mathcal{H}_1 \to \mathcal{H}_2$ such that $D_2 = U D_1 U^*$ and $\pi_2(\cdot) = U\pi_1(\cdot)U^*$, where π_j, $j = 1, 2$ are the representation of $\mathcal{A}_j$ in $\mathcal{H}_j$ respectively. Now, we want to prove that in general the perturbed spectral triple is not unitarily equivalent to the unperturbed one. Let $\Omega^1(\mathcal{A}_\theta^\infty)$ be the universal space of one-forms ([28]) and π be the representation of $\Omega^1 \equiv \Omega^1(\mathcal{A}_\theta^\infty)$ in $\mathcal{H}$ given by

$$\pi(a) \equiv a \otimes I_{\mathbb{C}^2}, \quad \pi(\delta(a)) = [D, a],$$

where δ is the universal derivation.

Note that $[D, a] = i[\delta_1(a)\gamma_1 + \delta_2(a)\gamma_2]$, where $r_1 = \mathrm{Re}(r)$, $r_2 = \mathrm{Im}(r)$, $\delta_1 = d_1 + d_{r_1}$, $\delta_2 = d_2 + d_{r_2}$, and γ_1, γ_2 are Clifford matrices $\begin{pmatrix} 0 & 1 \\ 1 & 0 \end{pmatrix}$ and $\begin{pmatrix} 0 & i \\ -i & 0 \end{pmatrix}$ respectively.

Theorem 9.4.2 *(i) Let $r = U^m$, then $\Omega^1_D(\mathcal{A}^\infty_\theta) := \pi(\Omega^1) = \mathcal{A}^\infty_\theta \oplus \mathcal{A}^\infty_\theta$.*
(ii) $\Omega^2(\mathcal{A}^\infty_\theta) = 0$ for $r = U^m$.

Proof:
(i) Clearly $\pi(\Omega^1) \subseteq \mathcal{A}^\infty_\theta\gamma_1 + \mathcal{A}^\infty_\theta\gamma_2$. The other inclusion follows from the facts that $\delta_2(U^k) = 0, \delta_1(U^k)$ is invertible, and that $\delta_2(V^l)$ is invertible for sufficiently large l.
(ii) Let $J_1 = \text{Ker}\,\pi|_{\Omega^1}$, $J_2 = \text{Ker}\,\pi|_{\Omega^2}$. Then $J_2 + \delta J_1$ is an ideal, implying that $\pi(\delta J_1) = \pi(J_2 + \delta J_1)$ is a nonzero submodule of $\pi(\Omega^2) \subseteq \mathcal{A}^\infty_\theta \oplus \mathcal{A}^\infty_\theta$. Since $\mathcal{A}^\infty_\theta$ is simple there are two possibilities, namely either $\pi(\delta J_1) \cong \mathcal{A}^\infty_\theta$, or $\pi(\delta J_1) = \mathcal{A}^\infty_\theta \oplus \mathcal{A}^\infty_\theta$. To rule out the first possibility we take a closer look at J_1 and $\pi(\delta J_1)$. $J_1 = \left\{\sum_i a_i\delta(b_i) | \sum_i a_i\delta_1(b_i) = 0, \sum_i a_i\delta_2(b_i) = 0\right\}$. Using the fact that δ_1, δ_2 are derivations we get for $\sum_i a_i\delta(b_i) \in J_1$

$$\sum_i \delta_1(a_i)\delta_2(b_i) = -\sum_i a_i\delta_1(\delta_2(b_i)) \tag{9.13}$$

$$\sum_i \delta_2(a_i)\delta_1(b_i) = -\sum_i a_i\delta_2(\delta_1(b_i)). \tag{9.14}$$

Moreover,

$$\begin{aligned}\pi&\left(\sum_i \delta(a_i)\delta(b_i)\right)\\ &= \sum_i (\delta_1(a_i)\gamma_1 + \delta_2(a_i)\gamma_2)(\delta_1(b_i)\gamma_1 + \delta_2(b_i)\gamma_2)\\ &= \sum_i (\delta_1(a_i)\delta_1(b_i) + \delta_2(a_i)\delta_2(b_i)) + \sum_i (\delta_1(a_i)\delta_2(b_i) - \delta_2(a_i)\delta_1(b_i))\gamma_{12},\end{aligned}$$

where $\gamma_{12} = \gamma_1\gamma_2 = -\gamma_2\gamma_1$. Taking $x = U^{-1}\delta(U) + U\delta(U^{-1}) \in \Omega^1$ it is straightforward to verify that $x \in J_1$ and $\pi(\delta(x)) = -2$. This proves $\mathcal{A}^\infty_\theta \oplus 0 \subseteq \pi(\delta J_1)$. We show that the inclusion is proper by showing the nontriviality of coefficient of γ_{12}. Using (9.13), (9.14) we see that the coefficient of γ_{12} is $\sum a_j[\delta_1, \delta_2](b_j) = \sum -im\ a_j[r_1, b_j]$. As before we can find n_0 such that for $l \geq n_0$, $\delta_2(V^l)$ is invertible. If we now choose $a_1 = I, b_1 = V^{n_0}, a_2 = -\delta_2(V^{n_0})\delta_2(V^l)^{-1}, b_2 = V^l, a_3 = (-a_1\delta_1(b_1) - a_2\delta_2(b_2))U^{-1}, b_3 = U$, then the vanishing of the coefficient of γ_{12} will imply that $[r_1, V^{n_0}] = \delta_2(V^{n_0})$ $\delta_2(V^l)^{-1}[r_1, V^l]$ for all $l \geq n_0$ and we note that while the left-hand side is nonzero and independent of l, the right-hand side converges to 0 as $l \to \infty$ leading to a contradiction. Therefore $\mathcal{A}^\infty_\theta \oplus \mathcal{A}^\infty_\theta = \pi(\delta J_1) \subseteq \pi(\Omega^2) \subseteq \mathcal{A}^\infty_\theta \oplus \mathcal{A}^\infty_\theta$. Hence $\Omega^2_D(\mathcal{A}^\infty_\theta) = \frac{\pi(\Omega^2)}{\pi(\delta J_1)} = 0$. □

Thus we have the following theorem.

Theorem 9.4.3 *The spectral triples $(\mathcal{A}^\infty_\theta, \mathcal{H}, D_0)$ and $(\mathcal{A}^\infty_\theta, \mathcal{H}, D)$ are not unitarily equivalent for $r = U^m$.*

The proof is clear since $\Omega^2_{D_0}(\mathcal{A}^\infty_\theta) = \mathcal{A}^\infty_\theta \neq 0 = \Omega^2_D(\mathcal{A}^\infty_\theta)$.

9.4.2 Noncommutative 2d-dimensional plane

In this subsection we shall discuss the geometry of the simplest kind of non-compact manifolds, namely the Euclidean 2d-dimensional space and its non-commutative counterpart. Let $d \geq 1$ be an integer and let $\mathcal{A}_c \equiv C_0(\mathbb{R}^{2d})$, the (non-unital) C^*-algebra of all complex-valued continuous functions on $\mathbb{R}^{2d}$ which vanish at infinity. Then $\partial_j (j = 1, 2, \ldots, 2d)$, the partial derivative in the jth direction, can be viewed as a densely defined derivation on $\mathcal{A}_c$, with the domain $\mathcal{A}_c^\infty \equiv C_c^\infty(\mathbb{R}^{2d})$, the set of smooth complex valued functions on $\mathbb{R}^{2d}$ having compact support. We consider the Hilbert space $L^2(\mathbb{R}^{2d})$ and naturally imbed $\mathcal{A}_c^\infty$ in it as a dense subspace. Then $i\partial_j$ is a densely defined symmetric linear map on $L^2(\mathbb{R}^{2d})$ with the domain $\mathcal{A}_c^\infty$, and we denote its self-adjoint extension by the same symbol. Also, let $\mathcal{F}$ be the Fourier transform on $L^2(\mathbb{R}^{2d})$ given by

$$\hat{f}(k) \equiv (\mathcal{F}f)(k) = (2\pi)^{-d} \int e^{-ik.x} f(x)dx,$$

and M_φ be the operator of multiplication by the function φ. We set $\widetilde{M_\varphi} = \mathcal{F}^{-1} M_\varphi \mathcal{F}$, thus $i\partial_j = \widetilde{M_{x_j}}$. $\Delta \equiv \widetilde{M}_{\sum x_j^2}$ is the self-adjoint positive operator, called the 2d-dimensional Laplacian. Clearly, the restriction of Δ on $\mathcal{A}_c^\infty$ is the differential operator $-\sum_{j=1}^{2d} \partial_j^2$. Let $h = L^2(\mathbb{R}^d)$ and U_α, V_β be two strongly continuous groups of unitaries in h, given by the following:

$$(U_\alpha f)(t) = f(t + \alpha), \quad (V_\beta f)(t) = e^{it.\beta} f(t), \quad \alpha, \beta, t \in \mathbb{R}^d, \ f \in C_c^\infty(\mathbb{R}^d).$$

Here $t.\beta$ is the usual Euclidean inner product of $\mathbb{R}^d$. It is clear that

$$\begin{aligned} U_\alpha U_{\alpha'} &= U_{\alpha+\alpha'}, \\ V_\beta V_{\beta'} &= V_{\beta+\beta'}, \\ U_\alpha V_\beta &= e^{i\alpha.\beta} V_\beta U_\alpha. \end{aligned} \tag{9.15}$$

For convenience, we define a unitary operator W_x for $x = (\alpha, \beta) \in \mathbb{R}^{2d}$ by

$$W_x = U_\alpha V_\beta e^{-\frac{i}{2}\alpha.\beta},$$

so that the Weyl relation (9.15) is now replaced by $W_x W_y = W_{x+y} e^{\frac{i}{2}p(x,y)}$, where $p(x, y) = x_1.y_2 - x_2.y_1$, for $x = (x_1, x_2)$, $y = (y_1, y_2)$. This is exactly the Segal form of the Weyl relation ([51]). For f such that $\hat{f} \in L^1(\mathbb{R}^{2d})$, we set

$$b(f) = \int_{\mathbb{R}^{2d}} \hat{f}(x) W_x dx \in \mathcal{B}(h). \tag{9.16}$$

Let $\mathcal{A}^\infty$ be the $*$-algebra generated by $\{b(f) \ : \ f \in C_c^\infty(\mathbb{R}^{2d})\}$ and let $\mathcal{A}$ be the C^*-algebra generated by $\mathcal{A}^\infty$ with the norm inherited from $\mathcal{B}(h)$. It is possible to verify using the commutation relation (9.15) that $b(f)b(g) = b(f \odot g)$ and $b(f)^* = b(f^\natural)$, where

$$\widehat{(f \odot g)}(x) = \int \hat{f}(x - x')\hat{g}(x')e^{\frac{i}{2}p(x,x')}dx'; \quad f^\natural(x) = \bar{f}(-x).$$

We define a linear functional τ on $\mathcal{A}^\infty$ by setting $\tau((b(f)) = \hat{f}(0)$ $(= (2\pi)^{-d} \int f(x)dx)$, and verify ([51], page 36) that it is a well-defined faithful trace on $\mathcal{A}^\infty$. It is natural to consider $\mathcal{H} = L^2(\mathcal{A}^\infty, \tau)$ and represent $\mathcal{A}$ in $\mathcal{B}(\mathcal{H})$ by left multiplication. From the definition of τ, it is clear that the map $C_c^\infty(\mathbb{R}^{2d}) \ni f \mapsto b(f) \in \mathcal{A}^\infty \subseteq \mathcal{H}$ extends to a unitary isomorphism from $L^2(\mathbb{R}^{2d})$ onto $\mathcal{H}$ and in the sequel we shall often identify the two.

There is a canonical $2d$-parameter group of automorphism of $\mathcal{A}$ given by $\varphi_\alpha(b(f)) = b(f_\alpha)$, where $\hat{f}_\alpha(x) = e^{i\alpha.x}\hat{f}(x)$, $f \in C_c^\infty(\mathbb{R}^{2d})$, $\alpha \in \mathbb{R}^{2d}$. Clearly, for any fixed $b(f) \in \mathcal{A}^\infty$, $\alpha \mapsto \varphi_\alpha(b(f))$ is smooth, and on differentiating this map at $\alpha = 0$, we get the canonical derivations δ_j, $j = 1, 2, \ldots, 2d$ as $\delta_j(b(f)) = b(\partial_j(f))$ for $f \in C_c^\infty(\mathbb{R}^{2d})$. We shall not notationally distinguish between the derivation δ_j on $\mathcal{A}^\infty$ and its extension to $\mathcal{H}$, and continue to denote by $i\delta_j$ both the derivation on $*$-algebra $\mathcal{A}^\infty$ and the associated self-adjoint operator in $\mathcal{H}$. It is clear that we can write down a candidate of spectral triple using this action of $\mathbb{R}^{2d}$ following the scheme outlined in the beginning of this section. Indeed, we consider a suitable integer N such that the $2d$-dimensional Clifford algebra can be represented in $M_N(\mathbb{C})$, and let $\gamma_1, \ldots, \gamma_{2d}$ be $N \times$N matrices satisfying the Clifford relations $\gamma_i\gamma_j + \gamma_j\gamma_i = 0$ if $i \neq j$, and $-2I$ for $i = j$. We take $\mathcal{H}' = \mathcal{H} \otimes \mathbb{C}^N$, and represent $\mathcal{A}$ in $\mathcal{B}(\mathcal{H}')$ by trivial ampliation $\mathcal{A} \ni a \mapsto a \otimes I_{\mathbb{C}^N}$. We continue to denote this ampliation again by a. Let

$$D = \sum_j i\delta_j \otimes \gamma_j.$$

It is indeed straightforward to verify that $[D, a]$ is bounded for all $a \in \mathcal{A}$. However, this is not a compact type spectral triple, as the resolvent of the associated Dirac operator is not compact.

9.4.3 Spectral triples on quantum Heisenberg manifold

Let G be the Heisenberg group.

$$G = \left\{ \begin{pmatrix} 1 & x & z \\ 0 & 1 & y \\ 0 & 0 & 1 \end{pmatrix} : x, y, z \in \mathbb{R} \right\}$$

For a positive integer c, let H_c be the subgroup of G obtained when x, y, cz are integers. The Heisenberg manifold M_c is the quotient G/H_c. Nonzero Poisson brackets on M_c invariant under left translation by G are parametrized by two real parameters μ, ν with $\mu^2 + \nu^2 \neq 0$ (see [111]). For each positive integer c and real numbers μ, ν, Rieffel constructed a C^*-algebra $A^{c,\hbar}_{\mu,\nu}$ as example of deformation quantization along a Poisson bracket [111]. These algebras have further been studied in [1], [2] and [129]. It was also remarked in [111] that it should be possible to construct example of noncommutative geometry as expounded in [28] in these algebras also.

In what follows, for $x \in \mathbb{R}$, $e(x)$ will denote $e^{2\pi i x}$

Definition 9.4.4 For any positive integer c let S^c denote the space of C^∞ functions $\Phi : \mathbb{R} \times \mathbb{T} \times \mathbb{Z} \to \mathbb{C}$ such that

(a) $\Phi(x + k, y, p) = e(ckpy)\Phi(x, y, p)$ for all $k \in \mathbb{Z}$

(b) for every polynomial P on $\mathbb{Z}$ and every partial differential operator $\widetilde{X} = \frac{\partial^{m+n}}{\partial x^m \partial y^n}$ on $\mathbb{R} \times \mathbb{T}$ the function $P(p)(\widetilde{X}\Phi)(x, y, p)$ is bounded on $K \times \mathbb{Z}$ for any compact subset K of $\mathbb{R} \times \mathbb{T}$.

For fixed real numbers $\hbar$, μ and ν such that $\mu^2 + \nu^2 > 0$, let $\mathcal{A}^\infty_\hbar$ denote S^c with product and involution defined by

$$(\Phi \star \Psi)(x, y, p) = \sum_q \Phi(x - \hbar(q - p)\mu, y - \hbar(q - p)\nu, q) \times \Psi(x - \hbar q\mu, y - \hbar q\nu, p - q) \tag{9.17}$$

$$\Phi^*(x, y, p) = \bar{\Phi}(x, y, -p). \tag{9.18}$$

Consider $\pi : \mathcal{A}^\infty_\hbar \to \mathcal{B}(L^2(\mathbb{R} \times \mathbb{T} \times \mathbb{Z}))$ given by

$$(\pi(\Phi)\xi)(x, y, p) = \sum_q \Phi(x - \hbar(q - 2p)\mu, y - \hbar(q - 2p)\nu, q) \times \xi(x, y, p - q). \tag{9.19}$$

This gives a faithful representation of the involutive algebra $\mathcal{A}^\infty_\hbar$. Let $\mathcal{A}^{c,\hbar}_{\mu,\nu}$ denote that norm closure of $\pi(\mathcal{A}^\infty_\hbar)$. It is called the *Quantum Heisenberg Manifold*.

Let $N_\hbar$ be the weak closure of $\pi(\mathcal{A}^\infty_\hbar)$.

We shall identify $\mathcal{A}^\infty_\hbar$ with $\pi(\mathcal{A}^\infty_\hbar)$ without any mention. Since we are going to work with fixed parameters $c, \mu, \nu, \hbar$ we will drop them altogether and denote $\mathcal{A}^{c,\hbar}_{\mu,\nu}$ simply by $\mathcal{A}_\hbar$ here the subscript remains merely as a reminiscent of Heisenberg only to distinguish it from a general algebra.

We mention the following result without proof.

Proposition 9.4.5 *For $\phi \in S^c$, $(r, s, t) \in \mathbb{R}^3 \equiv G$, (as a topological space)*

$$(L_{(r,s,t)}\phi)(x, y, p) := e(p(t + cs(x - r)))\phi(x - r, y - s, p) \quad (9.20)$$

extends to an ergodic action of the Heisenberg group on $\mathcal{A}^{c,\hbar}_{\mu,\nu}$.

Moreover, $\tau : \mathcal{A}^{\infty}_{\hbar} \to \mathbb{C}$, given by $\tau(\phi) = \int_0^1 \int_{\mathbb{T}} \phi(x, y, 0)dxdy$ extends to a faithful normal tracial state on $N_\hbar$.

It is simple to verify that τ is invariant under the Heisenberg group action. So, the group action can be lifted to $L^2(\mathcal{A}^{\infty}_{\hbar})$. We shall denote the action at the Hilbert space level by the same symbol.

Proposition 9.4.6 *Let V_f, W_k, X_r be the operators defined on $L^2(\mathbb{R} \times \mathbb{T} \times \mathbb{Z})$ by*

$$(V_f\xi)(x, y, p) = f(x, y)\xi(x, y, p),$$
$$(W_k\xi)(x, y, p) = e(-ck(p^2\hbar\nu + py))\xi(x + k, y, p),$$
$$(X_r\xi)(x, y, p) = \xi(x - 2\hbar r\mu, y - 2\hbar r\nu, p + r).$$

Let $T \in \mathcal{B}(L^2(\mathbb{R} \times \mathbb{T} \times \mathbb{Z}))$. Then $T \in N_\hbar$ if and only if T commutes with the operators V_f, W_k, X_r for all $f \in L^\infty(\mathbb{R} \times \mathbb{T})$, $k, r \in \mathbb{Z}$.

A proof of this can be found in [129].

Lemma 9.4.7 *Let $S^c_{\infty,\infty,1}$ be the set of all $\psi : \mathbb{R} \times \mathbb{T} \times \mathbb{Z} \to \mathbb{C}$ such that:*

(i) *ψ is measurable,*
(ii) *$\psi_n = \sup_{x\in\mathbb{R},y\in\mathbb{T}} |\psi(x, y, n)|$ is an l_1 sequence,*
(iii) *$\psi(x + k, y, p) = e(ckyp)\psi(x, y, p)$ for all $k \in \mathbb{Z}$.*

Then for $\phi \in S^c_{\infty,\infty,1}$, $\pi(\phi)$ defined by the same expression as in (9.19) gives a bounded operator on $L^2(\mathbb{R} \times \mathbb{T} \times \mathbb{Z})$.

Proof:
Let $\phi' : \mathbb{Z} \to \mathbb{R}_+$ be given by

$$\phi'(n) = \sup_{x\in\mathbb{R},y\in\mathbb{T}} |\phi(x, y, n)| \equiv \phi_n.$$

Then $|(\pi(\phi)\xi)(x, y, p)| \leq (\phi' \star |\xi(x, y, .)|)(p)$, where $\star$ denotes convolution on $\mathbb{Z}$ and $|\xi(x, y, .)|$ is the function $p \mapsto |\xi(x, y, p)|$. By Young's inequality, $\|(\pi(\phi)\xi)(x, y, .)\|_{l_2} \leq \|\phi' \star |\xi(x, y, .)|\|_{l_2} \leq \|\phi'\|_{l_1}\|\xi(x, y, .)\|_{l_2}$, which shows that $\|\pi(\phi)\| \leq \|\phi\|_{\infty,\infty,1} \equiv \|\phi'\|_{l_1}$. □

Remark 9.4.8 *(i) The product and involution defined by (9.17), (9.18) turn $S^c_{\infty,\infty,1}$ into an involutive algebra.*
(ii) $\phi \mapsto \|\phi\|_{\infty,\infty,1}$ is a $\star$-algebra norm.

Lemma 9.4.9 *The algebra $\pi(S^c_{\infty,\infty,1})$ is a subset of $N_\hbar$.*

Proof:
It follows from the Proposition 9.4.6, which characterizes $N_\hbar$. □

Theorem 9.4.10 *The Hilbert space $L^2(\mathcal{A}^\infty_\hbar, \tau)$ is unitarily isomorphic with $L^2(\mathbb{T} \times \mathbb{T} \times \mathbb{Z}) \cong L^2([0, 1] \times [0, 1] \times \mathbb{Z})$.*

Proof:
For $\phi \in S^c_{\infty,\infty,1}$, define $\Gamma\phi : \mathbb{R} \times \mathbb{T} \times \mathbb{Z} \to \mathbb{C}$ by

$$\Gamma\phi(x, y, p) = \begin{cases} e(-cxyp)\phi(x, y, p) & \text{for } y < 1 \\ \phi(x, y, p) & \text{for } y = 1 \end{cases}.$$

Then $\Gamma\phi(x + k, y, p) = \Gamma\phi(x, y, p)$. So, $\Gamma\phi$ is a map from $\mathbb{T} \times \mathbb{T} \times \mathbb{Z}$ to $\mathbb{C}$.

$$\begin{aligned} \tau(\phi^* \star \phi) &= \int_0^1 \int_{\mathbb{T}} \sum_q |\phi(x - \hbar q\mu, y - \hbar q\nu, -q)|^2 dxdy \\ &= \int_0^1 \int_{\mathbb{T}} \sum_q |\phi(x, y, q)|^2 dxdy \end{aligned}$$

since $|\phi(x+k, y, p)| = |\phi(x, y, p)|$ for all $x \in \mathbb{R}$, $y \in \mathbb{T}$, $k, p \in \mathbb{Z}$. Therefore $\tau(\phi^* \star \phi) = \|\Gamma\phi\|^2$, that is, $\Gamma : L^2(\mathcal{A}^\infty_\hbar, \tau) \to L^2(\mathbb{T} \times \mathbb{T} \times \mathbb{Z})$ is an isometry. To see Γ is a unitary observe the following conditions.

(*i*) $N_\hbar \subseteq L^2(\mathcal{A}^\infty_\hbar, \tau)$, since τ is normal.

(*ii*) $\phi_{m,n,k} = \begin{cases} e(cxyp)e(mx + ny)\delta_{kp}, & \text{for } 0 \le y \le 1 \\ \delta_{kp}e(mx) & \text{for } y = 1 \end{cases}$
is an element of $S^c_{\infty,\infty,1} \subseteq N_\hbar$

(*iii*) $\{\Gamma\phi_{m,n,k}\}_{m,n,k\in\mathbb{Z}}$ is an orthonormal basis of $L^2(\mathbb{T}^2 \times \mathbb{Z})$. □

Remark 9.4.11 $\phi \mapsto \phi|_{[0,1]\times\mathbb{T}\times\mathbb{Z}}$ *gives a unitary isomorphism.*

Corollary 9.4.12 *Let M_{yp} be the multiplication operator by the function yp on $\mathcal{H} = L^2(\mathbb{T} \times \mathbb{T} \times \mathbb{Z})$. If we consider $\mathcal{A}^\infty_\hbar$ as a subalgebra of $\mathcal{B}(\mathcal{H})$ by the left regular representation, then $[M_{yp}, \mathcal{A}^\infty_\hbar] \subseteq \mathcal{B}(\mathcal{H})$*

Proof:
Note that for $\phi \in \mathcal{A}^\infty_\hbar$, $(M_{yp}\phi)(x, y, p) = yp\phi(x, y, p)$ gives an element in $S^c_{\infty,\infty,1}$, hence a bounded operator. Moreover,

$$[M_{yp}, \phi]\psi(x, y, p)$$

$$= \sum_q (yp - (y - \hbar q\nu)(p - q))\phi(x - \hbar(q - p)\mu, y - \hbar(q - p)\nu, q)$$
$$\times \psi(x - \hbar q\mu, y - \hbar q\nu, p - q)$$

$$= \sum_q q(y - \hbar(q - p)\nu)\phi(x - \hbar(q - p)\mu, y - \hbar(q - p)\nu, q)$$
$$\times \psi(x - \hbar q\mu, y - \hbar q\nu, p - q)$$
$$= (M_{yp}(\phi) \star \psi)(x, y, p)$$

for $\psi \in \mathcal{A}_\hbar^\infty$. This completes the proof. □

We now construct a class of spectral triples on the quantum Heisenberg manifold.

Theorem 9.4.13 *For the quantum Heisenberg manifold, if we identify the Lie algebra of Heisenberg group with the Lie algebra of upper triangular matrices, then D as described by (9.9) becomes a self-adjoint operator with compact resolvent for the following choice:*

$$X_1 = \begin{pmatrix} 0 & 1 & 0 \\ 0 & 0 & 0 \\ 0 & 0 & 0 \end{pmatrix}, X_2 = \begin{pmatrix} 0 & 0 & 0 \\ 0 & 0 & 1 \\ 0 & 0 & 0 \end{pmatrix}, X_3 = \begin{pmatrix} 0 & 0 & c\alpha \\ 0 & 0 & 0 \\ 0 & 0 & 0 \end{pmatrix},$$

where $\alpha \in \mathbb{R}$ is greater than one.

Proof:
We define D by setting

$$\mathrm{Dom}(D) := \check{\mathcal{D}} \otimes \mathbb{C}^2, \quad D(f \otimes u) := \sum_{j=1}^{3} i d_j(f) \otimes \sigma_j(u),$$

with

$$\check{\mathcal{D}} := \{f \in L^2([0, 1] \times [0, 1] \times \mathbb{Z}) :$$
$$f(x, 0, p) = f(x, 1, p), f(1, y, p) = e(cpy) f(0, y, p)$$

$pf, \frac{\partial f}{\partial x}, \frac{\partial f}{\partial y} \in L^2\}$, where σ_j's are the spin matrices and d_j are defined by:

$$i\, d_1(f) = -i\frac{\partial f}{\partial x}, \; id_2(f) = -2\pi cpxf(x, y, p) - i\frac{\partial f}{\partial y},$$
$$i\, d_3(f) = -2\pi pc\alpha f(x, y, p).$$

Let $\eta : L^2([0, 1] \times [0, 1] \times \mathbb{Z}) \to L^2([0, 1] \times [0, 1] \times \mathbb{Z})$ be the unitary given by

$$\eta(f)(x, y, p) = \begin{cases} e(-cxyp) f(x, y, p) & \text{for } y < 1 \\ f(x, y, p) & \text{for } y = 1 \end{cases}$$

Let D' be defined by

$$\mathrm{Dom}(D') = (\eta \otimes I_2)\mathrm{Dom}(D), \text{ and } D' = (\eta \times I_2) D (\eta \otimes I_2)^{-1}.$$

Then we have $\mathrm{Dom}(D') = \mathcal{D}' \otimes \mathbb{C}^2$, where

$$\mathcal{D}' := \{f \in L^2([0,1] \times [0,1] \times \mathbb{Z}) : \\ f(x,0,p) = f(x,1,p), f(0,y,p) = f(1,y,p), \frac{\partial f}{\partial x}, \frac{\partial f}{\partial y}, pf \in L^2\},$$

and $D'(f \otimes u) = \sum_{j=1}^{3} i\, d'_j(f) \otimes \sigma_j(u)$ where

$$d'_1(f)(x,y,p) = -2\pi i c y p f(x,y,p) - \frac{\partial f}{\partial x}(x,y,p),$$

$$d'_2(f)(x,y,p) = -\frac{\partial f}{\partial y}(x,y,p), \quad d'_3(f)(x,y,p) = 2\pi i p c \alpha f(x,y,p).$$

On $\mathrm{Dom}(D')$, $D' = T + S$ where S and T are defined as:

$$S = 2\pi c M_{yp} \otimes \sigma_1, \quad T = -i\frac{\partial}{\partial x} \otimes \sigma_1 - i\frac{\partial}{\partial y} \otimes \sigma_2 - 2\pi c \alpha M_p \otimes \sigma_3,$$

$$\text{with } \mathrm{Dom}(T) = \mathrm{Dom}(D') \subseteq \mathrm{Dom}(S),$$

and they are self-adjoint operators on their respective domains. Also note that T has compact resolvent and the conclusion follows from the Kato–Rellich theorem (see [77]), since S is relatively bounded with respect to T with relative bound less than $\frac{1}{\alpha} < 1$. □

Theorem 9.4.14 *Let $\mathcal{H}_1 = L^2(\mathcal{A}_\hbar^\infty, \tau) \otimes \mathbb{C}^2$, and let $\mathcal{A}_\hbar^\infty$ be viewed as a subalgebra of $\mathcal{B}(\mathcal{H})$. Then $(\mathcal{A}_\hbar^\infty, \mathcal{H}_1, D)$ is an odd spectral triple of dimension 3.*

Proof:
We shall use the notation used in the statement and proof of the Theorem 9.4.13. That $(\mathcal{A}_\hbar^\infty, \mathcal{H}_1, D)$ is a spectral triple follows from the Theorem 9.4.13. We only have to show $|D|^{-3} \in \mathcal{L}^{(1,\infty)}$, the ideal of Dixmier trace-class operators. To this end, observe the following.

(*i*) Since T is the Dirac operator on $\mathbb{T}^3$, $\mu_n(T^{-1}|_{\mathrm{Ker}\, T^\perp}) = O(1/n^{1/3})$, μ_n stands for the n-th singular value.
(*ii*) S is relatively bounded with relative bound less than $\frac{1}{\alpha} < 1$, hence $\|S(T+i)^{-1}\| \le \frac{1}{\alpha}$ and $\|(1 + S(T+i)^{-1})^{-1}\| \le \frac{\alpha}{\alpha - 1}$.
(*iii*) $\mu_n(AB) \le \mu_n(A)\|B\|$, for bounded operators A,B.

Applying (*i*),(*ii*),(*iii*) to $(D' + i)^{-1} = (T+i)^{-1}(1 + S(T+i)^{-1})^{-1}$ we get the desired conclusion for D' and hence for D. □

Corollary 9.4.15 *Let T, S, D, D' be as in the proof of Theorem 9.4.13 and $A = (\eta \otimes I_2)^{-1} T (\eta \otimes I_2)$. Then $(\mathcal{A}^{\infty}_{\hbar}, \mathcal{H}_1, A)$ is an odd spectral triple of dimension 3.*

Proof:
We only have to show that $[A, \mathcal{A}^{\infty}_{\hbar}] \subseteq \mathcal{B}(\mathcal{H}_1)$. Let $B = (\eta \otimes I_2)^{-1} S (\eta \otimes I_2)$. Since $\eta \otimes I_2$ commutes with S, we have $B = S$. By Corollary 9.4.12, $[B, \mathcal{A}^{\infty}_{\hbar}] \subseteq \mathcal{B}(\mathcal{H}_1)$. Now we complete the proof by applying the arguments used in the proof of Theorem 9.4.14 along with $D = A + B$ completes the proof. □

Remark 9.4.16 *One can similarly show $(\mathcal{A}^{\infty}_{\hbar}, \mathcal{H}_1, A_t \equiv A + tB)$ forms an odd spectral triple of dimension 3, for $t \in [0, 1]$. It may also be noted that D, A constructed above does depend on α.*

9.5 Asymptotic analysis of heat semigroups and Laplacians

We do not try to build any general theory here; instead, we just analyze the behavior of heat semigroups and Laplacians in some of the examples discussed in the previous section.

9.5.1 Volume form and scalar curvature on noncommutative torus

Weyl asymptotics for $\mathcal{A}_\theta$

For classical compact Riemannian manifold (M, g) of dimension d with metric g, recall how we obtain the volume and integrated scalar curvature from the asymptotics of heat semigroup (see equations (9.6), (9.7)). For the noncommutative d-torus (with d even) one possibility is to define its volume V and integrated scalar curvature s by analogy with their classical counterparts as:

$$V(\mathcal{A}_\theta) \equiv V \equiv \lim_{t \to 0+} t^{d/2} \mathrm{tr} T_t, \tag{9.21}$$

$$s(\mathcal{A}_\theta) \equiv s \equiv \frac{1}{6} \lim_{t \to 0+} t^{d/2-1} [\mathrm{tr} T_t - t^{-d/2} V] \tag{9.22}$$

where the heat semigroup T_t in the classical case is replaced by the semigroup associated with the Laplacians mentioned in the preceding section, $T_t^0 = e^{t\mathcal{L}_0}$ and the perturbed one $T_t = e^{t\mathcal{L}}$, with $\mathcal{L} = -\frac{1}{2}\{(d_1 + d_{r_1})^2 + (d_2 + d_{r_2})^2\}$ on $\mathcal{H} \equiv L^2(\mathcal{A}_\theta, \tau)$. Before we can compute these, we need to study the operators $\mathcal{L}_0$ and $\mathcal{L}$ in $L^2(\tau)$ more carefully. The next theorem summarizes their properties for $d = 2$ and we have denoted by $\mathcal{B}_p$ the Schatten ideals in $\mathcal{B}(\mathcal{H})$ with the respective norms.

Theorem 9.5.1 *(i) $\mathcal{L}_0$ is a negative self-adjoint operator in $\mathcal{H} = L^2(\tau)$ with compact resolvent. In fact $\mathcal{L}_0(U^m V^n) = -\frac{1}{2}(m^2 + n^2)U^m V^n$; $m, n \in \mathbb{Z}$ so that $(\mathcal{L}_0 - z)^{-1} \in \mathcal{B}_p(L^2(\tau))$ for $p > 1$ and $z \in \rho(\mathcal{L}_0)$.*

(ii) If $r_1, r_2 \in \mathcal{A}_\theta^\infty$ and they are self-adjoint, then $\mathcal{L} = \mathcal{L}_0 + B + A$, where $B = -\frac{1}{2}(d_{r_1}^2 + d_{r_2}^2 + d_{d_1(r_1)} + d_{d_2(r_2)})$ is bounded and $A = -d_{r_1}d_1 - d_{r_2}d_2$, so that A is compact relative to $\mathcal{L}_0$ and $\mathcal{L}$ is self-adjoint on Dom$(\mathcal{L}_0)$ *with compact resolvent.*

If $r_1, r_2 \in \mathcal{A}_\theta$, then $-\mathcal{L} = -\mathcal{L}_0 - B - A$ as quadratic form on $(\text{Dom}(-\mathcal{L}_0)^{\frac{1}{2}})$ *and*

$$(-\mathcal{L} + n^2)^{-1} = (-\mathcal{L}_0 + n^2)^{-\frac{1}{2}}(I + Z_n)^{-1}(-\mathcal{L}_0 + n^2)^{-\frac{1}{2}}, \qquad (9.23)$$

where $Z_n = (-\mathcal{L}_0 + n^2)^{-\frac{1}{2}}(B + A)(-\mathcal{L}_0 + n^2)^{-\frac{1}{2}}$, is compact for each n with $B = -\frac{1}{2}(d_{r_1}^2 + d_{r_2}^2)$, $A = -\frac{1}{2}(d_1 d_{r_1} + d_{r_1} d_1 + d_2 d_{r_2} + d_{r_2} d_2)$. This defines $\mathcal{L}$ as a self-adjoint operator on $L^2(\tau)$ with compact resolvent. Furthermore, in both cases of (ii), the difference of resolvents $(\mathcal{L} - z)^{-1} - (\mathcal{L}_0 - z)^{-1}$ is trace class for $z \in \rho(\mathcal{L}) \cap \rho(\mathcal{L}_0)$.

Proof:
The proof of (i) is omitted. For proving (ii), we first verify by straightforward calculation that $\mathcal{L} = \mathcal{L}_0 + B + A$ on $\mathcal{A}_\theta^\infty$ and that $A(-\mathcal{L}_0 + n^2)^{-1}$ is compact for every $n = 1, 2, \ldots$. Therefore

$$(\mathcal{L} - \mathcal{L}_0)(-\mathcal{L}_0 + n^2)^{-1} = (\mathcal{L} - \mathcal{L}_0)(-\mathcal{L}_0 + 1)^{-1}(-\mathcal{L}_0 + 1)(-\mathcal{L}_0 + n^2)^{-1},$$

which goes to 0 in operator norm as $n \to \infty$. By the Kato–Rellich theorem (see, for example, [77]), $\mathcal{L}$ is self-adjoint and since

$$(-\mathcal{L} + n^2)^{-1} = (-\mathcal{L}_0 + n^2)^{-1}[1 + (\mathcal{L}_0 - \mathcal{L})(-\mathcal{L}_0 + n^2)^{-1}]^{-1}$$

for sufficiently large n, one also concludes that $\mathcal{L}$ has compact resolvent. Furthermore, for $z \in \rho(\mathcal{L}) \cap \rho(\mathcal{L}_0)$,

$$(\mathcal{L} - z)^{-1} - (\mathcal{L}_0 - z)^{-1} = (\mathcal{L}_0 - z)^{-1}[1 + (\mathcal{L} - \mathcal{L}_0)(\mathcal{L}_0 - z)^{-1}]^{-1} \times(\mathcal{L}_0 - \mathcal{L})(\mathcal{L}_0 - z)^{-1}.$$

Since $(\mathcal{L} - \mathcal{L}_0)(-\mathcal{L}_0 + n^2)^{-\frac{1}{2}}$ is bounded, $(-\mathcal{L}_0 + n^2)^{-\frac{1}{2}} \in \mathcal{B}_3(L^2(\tau))$ and since $(-\mathcal{L}_0 + z)^{-1} \in \mathcal{B}_{3/2}(L^2(\tau))$, it follows that $(\mathcal{L} - n^2)^{-1} - (\mathcal{L}_0 - n^2)^{-1}$ is trace class for $n = 1, 2, \cdots$ by the Hölder inequality.

When $r_1, r_2 \in \mathcal{A}_\theta$, we cannot write the expression for $\mathcal{L}$ as above on $\mathcal{A}_\theta^\infty$, since r_1, r_2 may not be in the domain of the derivations d_1, d_2. For this reason, we need to define $-\mathcal{L}$ as the sum of quadratic forms and standard results as in [107] can be applied here. From the structure of B and A it is clear that Z_n is compact for each n and hence an identical reasoning as above would help us conclude that $\|Z_n\| \to 0$ as $n \to \infty$. Therefore, $(I + Z_n)^{-1} \in \mathcal{B}(\mathcal{H})$ for sufficiently large n and the right-hand side of (9.23) defines the operator $-\mathcal{L}$ associated with the quadratic form with $\mathrm{Dom}((-\mathcal{L})^{\frac{1}{2}}) = \mathrm{Dom}((-\mathcal{L}_0)^{\frac{1}{2}})$. Clearly,

$$\begin{aligned}(-\mathcal{L}+n^2)^{-1} - (-\mathcal{L}_0+n^2)^{-1} &= -(-\mathcal{L}_0+n^2)^{-\frac{1}{2}}(I+Z_n)^{-1}Z_n(-\mathcal{L}_0+n^2)^{-\frac{1}{2}} \\ &= -(-\mathcal{L}_0+n^2)^{-\frac{1}{2}}(I+Z_n)^{-1}(-\mathcal{L}_0+n^2)^{-\frac{1}{2}}(B+A)(-\mathcal{L}_0+n^2)^{-1}\end{aligned}$$

for sufficiently large n, and since

$$(-\mathcal{L}_0+n^2)^{-\frac{1}{2}} \in \mathcal{B}_3, (-\mathcal{L}_0+n^2)^{-\frac{1}{2}}A(-\mathcal{L}_0+n^2)^{-\frac{1}{2}} \in \mathcal{B}_3,$$

it is clear that $(\mathcal{L} - n^2)^{-1} - (\mathcal{L}_0 - n^2)^{-1}$ is trace class. □

The next theorem studies the effect of the perturbation from $\mathcal{L}_0$ to $\mathcal{L}$ on the volume and the integrated sectional curvature for $\mathcal{A}_\theta$.

Theorem 9.5.2 *(i) The volume V of $\mathcal{A}_\theta (d = 2)$ as defined in (9.21) is invariant under the perturbation from $\mathcal{L}_0$ to $\mathcal{L}$.*

(ii) The integrated scalar curvature for $r \in \mathcal{A}_\theta^\infty$, in general is not invariant under the above perturbation.

Proof:
We need to compute $\mathrm{tr}(e^{t\mathcal{L}} - e^{t\mathcal{L}_0})$. Note that if $r_1, r_2 \in \mathcal{A}_\theta^\infty$, then

$$e^{t\mathcal{L}} - e^{t\mathcal{L}_0} = -\int_0^t e^{(t-s)\mathcal{L}}(\mathcal{L} - \mathcal{L}_0)e^{s\mathcal{L}_0}ds,$$

which on two iterations yields:

$$\begin{aligned}e^{t\mathcal{L}} - e^{t\mathcal{L}_0} &= -\int_0^t e^{(t-s)\mathcal{L}_0}(\mathcal{L}-\mathcal{L}_0)e^{s\mathcal{L}_0}ds + \int_0^t dt_1 e^{(t-t_1)\mathcal{L}_0}(\mathcal{L}-\mathcal{L}_0) \\ &\times \int_0^{t_1} dt_2 e^{(t_1-t_2)\mathcal{L}_0}(\mathcal{L}-\mathcal{L}_0)e^{t_2\mathcal{L}_0} - \int_0^t dt_1 e^{(t-t_1)\mathcal{L}}(\mathcal{L}-\mathcal{L}_0) \\ &\times \int_0^{t_1} dt_2 e^{(t_1-t_2)\mathcal{L}_0}(\mathcal{L}-\mathcal{L}_0)\int_0^{t_2} dt_3 e^{(t_2-t_3)\mathcal{L}_0}(\mathcal{L}-\mathcal{L}_0)e^{t_3\mathcal{L}_0} \\ &\equiv I_1(t) + I_2(t) + I_3(t)(\text{ say }). \end{aligned} \tag{9.24}$$

For estimating the trace norms of these terms, we note that the $\mathcal{B}_p$-norm of $(\mathcal{L} - \mathcal{L}_0)e^{s\mathcal{L}_0}$ is estimated as

$$\|(\mathcal{L} - \mathcal{L}_0)e^{s\mathcal{L}_0}\|_p = \|(B + A)e^{s\mathcal{L}_0}\|_p \le \|B\|\|e^{s\mathcal{L}_0}\|_p +$$

$$c_1(\|d_1 e^{s\mathcal{L}_0}\|_p + \|d_2 e^{s\mathcal{L}_0}\|_p) \le c''(\|e^{s\mathcal{L}_0}\|_p + \|d_2 e^{s\mathcal{L}_0}\|_p)$$
$$\le c'(s^{-p^{-1}} + s^{-p^{-1}-\frac{1}{2}}) \le c\, s^{-p^{-1}-\frac{1}{2}}$$

for constants c, c_1, c', c'' since we are interested only for the region $0 < s \le t \le 1$. Using Hölder inequality for Schatten norms and the fact that

$$\|(\mathcal{L} - n^2)^{-1}\| \le \|(\mathcal{L}_0 - n^2)^{-1}[1 + (\mathcal{L} - \mathcal{L}_0)(\mathcal{L}_0 - n^2)^{-1}]^{-1}\| \le \frac{2}{n^2}$$

for sufficiently large n, We get for the third term in (9.24)

$$\|I_3(t)\|_1 \le 2\int_0^t dt_1 \int_0^{t_1} dt_2 \|(\mathcal{L} - \mathcal{L}_0)e^{(t_1-t_2)\mathcal{L}_0}\|_{p_1} \times$$
$$\int_0^{t_2} dt_3 \|(\mathcal{L} - \mathcal{L}_0)e^{(t_2-t_3)\mathcal{L}_0}\|_{p_2} \|(\mathcal{L} - \mathcal{L}_0)e^{t_3\mathcal{L}_0}\|_{p_3}$$
$$\le c(p_1, p_2, p_3)\int_0^t t_1^{-\frac{1}{2}} dt_1 \to 0$$

as $t \to 0$ where $p_1^{-1} + p_2^{-1} + p_3^{-1} = 1$. A very similar estimate shows that

$$\|I_1(t)\|_1 \le \int_0^t ds \|e^{(t-s)\mathcal{L}_0}\|_{p_1} \|(\mathcal{L} - \mathcal{L}_0)e^{s\mathcal{L}_0}\|_{p_2} \le ct^{-\frac{1}{2}}$$

(with $p_2 > 2$ and $p_1^{-1} + p_2^{-1} = 1$) and

$$\|I_2(t)\|_1 \le \int_0^t dt_1 \|e^{(t-t_1)\mathcal{L}_0}\|_{p_1}$$
$$\times \int_0^{t_1} dt_2 \|(\mathcal{L} - \mathcal{L}_0)e^{(t_1-t_2)\mathcal{L}_0}\|_{p_2} \|(\mathcal{L} - \mathcal{L}_0)e^{t_2\mathcal{L}_0}\|_{p_3} \le c',$$

(with $p_1^{-1} + p_2^{-1} + p_3^{-1} = 1$, in particular the choice $p_1 = p_2 = p_3 = 3$ will do) a constant independent of t. From this it follows that

$$\lim_{t\to 0+} t\, \mathrm{tr}(e^{t\mathcal{L}} - e^{t\mathcal{L}_0}) = 0.$$

Thus the invariance of volume under perturbation is proved.

In the case when $r_1, r_2 \in \mathcal{A}_\theta$, then $\mathcal{L} - \mathcal{L}_0 = B + d_1 B_1 + d_2 B_2 + B_1' d_1 + B_2' d_2$ where B, B_1, B_1', B_2, B_2' are bounded. Therefore a term like $e^{(t-s)\mathcal{L}_0} d_1 B_1 e^{s\mathcal{L}_0} = [e^{s\mathcal{L}_0} B_1^* d_1 e^{(t-s)\mathcal{L}_0}]^*$ admits similar estimates to those above and the same result follows.

(*ii*) From the expression (9.22) for the integrated scalar curvature s, we see that for $d = 2$

$$s(\mathcal{L}) - s(\mathcal{L}_0) = \frac{1}{6}\lim_{t\to 0+} \mathrm{tr}(e^{t\mathcal{L}} - e^{t\mathcal{L}_0}), \tag{9.25}$$

if it exists, and conclude that the contribution to (9.25) from the term $I_3(t)$ vanishes as we have seen in (i). We claim that $\mathrm{tr} I_2(t) \to 0$ as $t \to 0+$. In fact since the integrals in $I_2(t)$ converge in trace norm, we have

$$\mathrm{tr} I_2(t) = \int_0^t dt_1 \int_0^{t_1} dt_2 \mathrm{tr}((\mathcal{L} - \mathcal{L}_0) e^{(t_1 - t_2)\mathcal{L}_0} (\mathcal{L} - \mathcal{L}_0) e^{(t - t_1 + t_2)\mathcal{L}_0}),$$

and by a change of variable we conclude the following:

$$|\mathrm{tr} I_2(t)| \leq t \int_0^t \|(\mathcal{L} - \mathcal{L}_0) e^{s\mathcal{L}_0} (\mathcal{L} - \mathcal{L}_0) e^{(t-s)\mathcal{L}_0}\|_1 ds.$$

For $r \in \mathcal{A}_\theta^\infty$, the perturbation $(\mathcal{L} - \mathcal{L}_0)$ is of the form $b_0 + b_1 d_1 + b_2 d_2$ with $b_i \in \mathcal{B}(\mathcal{H})$ for $i = 0, 1, 2$; and the Hilbert–Schmidt norm estimates are:

$$\begin{aligned} \|(\mathcal{L} - \mathcal{L}_0) e^{s\mathcal{L}_0}\|_2 &\leq \|b_0\| \|e^{s\mathcal{L}_0}\|_2 + \sqrt{2}(\|b_1\| + \|b_2\|) \|(-\mathcal{L}_0)^{\frac{1}{2}} e^{s\mathcal{L}_0}\|_2 \\ &\leq c(s^{-\frac{1}{2}} + s^{-\frac{3}{4}}), \end{aligned}$$

where c is a positive constant.

Therefore

$$|\mathrm{tr} I_2(t)| \leq ct \int_0^t (s^{-\frac{1}{2}} + s^{-\frac{3}{4}})((t-s)^{-\frac{1}{2}} + (t-s)^{-\frac{3}{4}}) ds,$$

which clearly converges to zero as $t \to 0+$.

This leaves only the contribution from the term $I_1(t)$, so that one has

$$6(s(\mathcal{L}) - s(\mathcal{L}_0)) = - \lim_{t \to 0+} t \, \mathrm{tr}((\mathcal{L} - \mathcal{L}_0) e^{t\mathcal{L}_0}).$$

As before we note that $(\mathcal{L} - \mathcal{L}_0)$ contains two kinds of terms:

$$B = -\frac{1}{2}(d_{r_1}^2 + d_{r_2}^2), \quad A = -\frac{1}{2}(d_{r_1} d_1 + d_1 d_{r_1} + d_{r_2} d_2 + d_2 d_{r_2}).$$

The term $\mathrm{tr}(A e^{t\mathcal{L}_0}) = 0$ for all $t > 0$ will follow if we can show that $\mathrm{tr}(d_r d_1 e^{t\mathcal{L}_0}) = 0$ for $r \in \mathcal{A}_\theta^\infty$ and for this we note that

$$\begin{aligned} &\mathrm{tr}(d_r d_1 e^{t\mathcal{L}_0}) \\ &= \sum_{m,n} \langle U^m V^n, d_r d_1 e^{t\mathcal{L}_0}(U^m V^n) \rangle \\ &= \sum_{m,n} m \, exp\left(-\frac{t}{2}(m^2 + n^2)\right) \tau(V^{-n} U^{-m} d_r(U^m V^n)) \\ &= \sum_{m,n} m \, exp\left(-\frac{t}{2}(m^2 + n^2)\right) \tau(V^{-n} U^{-m} r U^m V^n - r) = 0. \end{aligned}$$

Finally we consider the contribution due to B. Thus,

$$s(\mathcal{L}) - s(\mathcal{L}_0) = \frac{1}{12} \lim_{t \to 0+} t \, \mathrm{tr}((d_{r_1}^2 + d_{r_2}^2) e^{t\mathcal{L}_0}), \tag{9.26}$$

if it exists. However since $\{t \operatorname{tr}(({d_{r_1}}^2 + {d_{r_2}}^2)e^{t\mathcal{L}_0})\}$ is bounded as $t \to 0+$, we shall interpret the above limit as a special kind of Banach limit as is done in page 563 of [28], denoting the same by Lim. We thus have

$$\begin{aligned} & s(\mathcal{L}) - s(\mathcal{L}_0) \\ &= \frac{1}{12}\text{Lim } t(({d_{r_1}}^2 + {d_{r_2}}^2)e^{t\mathcal{L}_0}) \qquad (9.27) \\ &= \frac{1}{12}\text{Tr}_\omega(({d_{r_1}}^2 + {d_{r_2}}^2){\hat{\mathcal{L}_0}}^{-1}). \qquad (9.28) \end{aligned}$$

Let us now show that in general, the right-hand side of (9.26) is strictly positive. For example, set $r_1 = (U + U^{-1})$ and $r_2 = 0$. Clearly, $r_1, r_2 \in \mathcal{A}_\theta^\infty$, and

$$\begin{aligned} & 6(s(\mathcal{L}) - s(\mathcal{L}_0)) \\ &= \text{Lim } \frac{t}{2}\sum_{m,n} exp\left(-\frac{t}{2}(m^2+n^2)\right)\langle U^m V^n, d_{r_1}^2(U^m V^n)\rangle \\ &= \text{Lim } \frac{t}{2}\sum_{m,n} exp\left(-\frac{t}{2}(m^2+n^2)\right)\tau\{(1-\lambda^{-n})^2\lambda^{2n}U^2 \\ &\quad +(1-\lambda^n)^2\lambda^{-2n}U^{-2} + (2-\lambda^n-\lambda^{-n})\}] \\ &= \text{Lim } \frac{t}{2}\left(2\sum_{m=1}^{\infty} e^{-m^2t/2}+1\right)\left(8\sum_{n=1}^{\infty}\sin^2(\pi\theta n)e^{-n^2t/2}\right). \end{aligned}$$

Next note that for $0 < t < 2$,

$$\begin{aligned} & \sqrt{t}\sum_{n=1}^{\infty}\sin^2(\pi\theta n)e^{-n^2t/2} \\ &\geq \sqrt{t}\sum_{n=1}^{[\sqrt{2/t}]}\sin^2(\pi\theta n)e^{-n^2t/2} \\ &\geq e^{-1}(\sqrt{2}-\sqrt{t})\sum_{n=1}^{[\sqrt{2/t}]}[\sqrt{2/t}]^{-1}\sin^2\pi(n\theta - [n\theta]) \\ &= e^{-1}(\sqrt{2}-\sqrt{t})E(\sin^2\pi X_t), \end{aligned}$$

where for each $0 < t \leq 2$, X_t is a $[0, 1]$-valued random variable with

$$\text{Probability } (X_t = k\theta - [k\theta]) = \left[\sqrt{2/t}\right]^{-1} \quad \text{for } k = 1, 2, \ldots, \left[\sqrt{\frac{2}{t}}\right],$$

and E is the associated expectation. Since θ is irrational, it is known [65] that as $t \to 0+$, the random variable X_t converges weakly to one with uniform distribution on [0, 1] and therefore

$$\liminf_{t\to 0+}\sqrt{t}\sum_{n=1}^{\infty}\sin^2(\pi\theta n)e^{-n^2t/2} \geq \lim_{t\to 0+}\sqrt{t}\sum_{n=1}^{[\sqrt{2/t}]}\sin^2(\pi\theta n)e^{-n^2t/2}$$

$$\geq \sqrt{2}e^{-1}\int_0^1 \sin^2\pi x dx = (\sqrt{2}e)^{-1}.$$

We also have by ([28], page 563) that

$$\lim_{t\to 0+}\sqrt{t}\sum_{m=1}^{\infty}e^{-m^2t/2} = \frac{\sqrt{\pi}}{\sqrt{2}}.$$

By the general properties of the limiting procedure as given in [28], we have that

$$s(\mathcal{L}) - s(\mathcal{L}_0) \geq \frac{2\sqrt{\pi}}{3e}.$$

□

Remark 9.5.3 *From the expression for $s(\mathcal{L}_0)$, we see that for $d = 2$,*

$$6s(\mathcal{L}_0) = \lim_{t\to 0+}\left(\mathrm{tr}(e^{t\mathcal{L}_0}) - \frac{V}{t}\right).$$

Since the expression for tr $e^{t\mathcal{L}_0}$ *and the volume V are exactly the same as in the case of classical two-torus with its heat semigroup, the integrated scalar curvature for $\mathcal{L}_0$ is the same as in the classical case, which is clearly zero. Therefore $s(\mathcal{L})$ is strictly positive for the case considered here.*

Volume form

As in [28], the volume form $v(a)$ on $\mathcal{A}_\theta$ is the linear functional $v(a) = \frac{1}{2}\mathrm{Tr}_\omega(a|\hat{D}|^{-2}P)$, where P is the projection on Ker $(D)^\perp$. We shall now prove that the above volume form is invariant under the perturbation from $\mathcal{L}_0$ to $\mathcal{L}$.

Lemma 9.5.4 *Let T be a self-adjoint operator with compact resolvent such that $\hat{T}^{-1}$ is in $\mathcal{L}^{1,\infty}(\mathcal{H})$. Then for $a \in \mathcal{A}_\theta$ and every $z \in \rho(T)$, one has* $\mathrm{Tr}_\omega(a\hat{T}^{-1}P) = \mathrm{Tr}_\omega(a(T-z)^{-1})$.

Proof:
Note that $(T-z)^{-1} = (\hat{T}-z)^{-1}P \oplus -z^{-1}P^\perp$ and $P^\perp$ is finite dimensional. Therefore $\mathrm{Tr}_\omega(a(T-z)^{-1}) = \mathrm{Tr}_\omega(PaP(\hat{T}-z)^{-1}P)$. On the other hand,

$$\mathrm{Tr}_\omega(PaP\hat{T}^{-1}P - PaP(\hat{T}-z)^{-1}P) = -z\mathrm{Tr}_\omega(PaP\hat{T}^{-1}(\hat{T}-z)^{-1}P) = 0,$$

by an application of Lemma 9.3.6, since $\hat{T}^{-1}$ is in $\mathcal{L}^{1,\infty}(\mathcal{H})$ and $(\hat{T}-z)^{-1}$ is compact. □

Theorem 9.5.5 *If we set* $v_0(a) = \frac{1}{2}\mathrm{Tr}_\omega(a|\hat{D}_0|^{-2})$ *and* $v(a) = \frac{1}{2}\mathrm{Tr}_\omega(a|\hat{D}|^{-2})$ *for* $a \in \mathcal{A}_\theta$, *then* $v_0(a) = v(a)$.

Proof:

Note that $D^2 = -2\begin{pmatrix} \mathcal{L}_1 & 0 \\ 0 & \mathcal{L}_2 \end{pmatrix}$, where

$$\mathcal{L}_1 = \mathcal{L}_0 + d_r d_{r^*} + (d_1 d_{r^*} + d_r d_1) + i(d_2 d_{r^*} - d_r d_2)$$

and

$$\mathcal{L}_2 = \mathcal{L}_0 + d_{r^*} d_r + (d_1 d_r + d_{r^*} d_1) + i(d_2 d_{r^*} - d_r d_2).$$

Moreover, by Theorem 9.5.1, both $\mathcal{L}_1$ and $\mathcal{L}_2$ have compact resolvents with P_1, P_2 being the projections on $\mathrm{Ker}(\mathcal{L}_1)^\perp$ and $\mathrm{Ker}(\mathcal{L}_2)^\perp$ respectively. Thus, by applying Lemma 9.5.4 with $\mathrm{Im}(z) \neq 0$, we get

$$\begin{aligned} & v(a) \\ &= \mathrm{Tr}_\omega(a(-\hat{\mathcal{L}}_1)^{-1} P_1) + \mathrm{Tr}_\omega(a(-\hat{\mathcal{L}}_2)^{-1} P_2) \\ &= \mathrm{Tr}_\omega(a(-\mathcal{L}_1 - z)^{-1} + a(-\mathcal{L}_2 - z)^{-1}) \\ &= \mathrm{Tr}_\omega(a(-\mathcal{L}_0 - z)^{-1} + a(-\mathcal{L}_0 - z)^{-1}) + \mathrm{Tr}_\omega(a(-\mathcal{L}_1 - z)^{-1} \\ &\quad -a(-\mathcal{L}_0 - z)^{-1}) + \mathrm{Tr}_\omega(a(-\mathcal{L}_2 - z)^{-1} - a(-\mathcal{L}_0 - z)^{-1}) \\ &= v_0(a), \end{aligned}$$

since $(-\mathcal{L}_i - z)^{-1} - (-\mathcal{L}_0 - z)^{-1}$ is trace class for $i = 1, 2$. □

9.5.2 Volume form for noncommutative $2d$-plane

In this subsection, we shall use the notation of Subsection 9.4.2. We fix a positive integer d, and $\mathcal{A}$ be the noncommutative $2d$ dimensional plane, defined and studied in Subsection 9.4.2.

Let us first look at the classical counterpart of the noncommutative $2d$-plane. As a Riemannian manifold, $\mathbb{R}^{2d}$ does not posses too many interesting features; it is a flat manifold and thus there is no nontrivial curvature form. Instead, we shall be interested in obtaining the volume form from the operator-theoretic data associated with the $2d$-dimensional Laplacian Δ. Let $\mathcal{T}_t = e^{-\frac{t}{2}\Delta}$ be the contractive C_0-semigroup generated by Δ, called the heat semigroup on $\mathbb{R}^{2d}$. Unlike compact manifolds, Δ has only absolutely continuous spectrum. However, for any $f \in C_c^\infty(\mathbb{R}^{2d})$ and $\epsilon > 0$, $M_f(\Delta + \epsilon)^{-d}$ has discrete spectrum. Furthermore, we have the following theorems.

Theorem 9.5.6 $M_f \mathcal{T}_t$ *is trace-class and* $\mathrm{tr}(M_f \mathcal{T}_t) = t^{-d} \int f(x)dx$. *Thus, in particular,* $v(f) \equiv \int f(x)dx = t^d \mathrm{tr}(M_f \mathcal{T}_t)$.

Proof:
We have $\mathrm{tr}(M_f \mathcal{T}_t) = \mathrm{tr}(\mathcal{F} M_f \mathcal{F}^{-1} M_{\exp(-\frac{t}{2}\sum x_j^2)})$, and $\mathcal{F} M_f \mathcal{F}^{-1} M_{\exp(-\frac{t}{2}\sum x_j^2)}$ is an integral operator with the kernel $k_t(x, y) = \hat{f}(x - y)\exp(-\frac{t}{2}\sum y_j^2)$. It is continuous in both arguments and $\int |k_t(x, x)|dx < \infty$, we obtain by using a result in [57] (p. 114, Chapter 3) that $M_f \mathcal{T}_t$ is trace class and $\mathrm{tr}(M_f \mathcal{T}_t) = \int k_t(x, x)dx = (2\pi)^d t^{-d} \hat{f}(0) = t^{-d} v(f)$. □

As before, we get an alternative expression for the volume form v in terms of the Dixmier trace.

Theorem 9.5.7 *For $\epsilon > 0$, $M_f(\Delta + \epsilon)^{-d}$ is of Dixmier trace class and its Dixmier trace is equal to $\pi^d v(f)$.*

For convenience, we shall give the proof only in the case $d = 1$. We need the following two lemmas.

Lemma 9.5.8 *If $f, g \in L^p(\mathbb{R}^2)$ for some p with $2 \leq p < \infty$, then $M_f \widetilde{M}_g$ is a compact operator in $L^2(\mathbb{R}^2)$.*

Proof:
This is a consequence of the Hölder and Hausdorff–Young inequalities and a proof can be found in [109]. □

Lemma 9.5.9 *Let S be a square in $\mathbb{R}^2$ and f be a smooth function with* $\mathrm{supp}(f) \subseteq \mathrm{int}(S)$. *Let Δ_S denote the Laplacian on S with the periodic boundary condition. Then* $\mathrm{Tr}_\omega(M_f(\Delta_S + \epsilon)^{-1}) = \pi \int f(x)dx$.

Proof:
This follows from [82] by identifying S with the two-dimensional torus in the natural manner. □

Proof of the theorem:
Note that for $g \in \mathrm{Dom}(\Delta) \subseteq L^2(\mathbb{R}^2)$, we have $fg \in \mathrm{Dom}(\Delta_S)$ and

$$(\Delta_S M_f - M_f \Delta)(g) = (\Delta M_f - M_f \Delta)(g) = Bg,$$

where $B = -M_{\Delta f} + 2i \sum_{j=1}^2 M_{\partial_j(f)} \circ \partial_j$. From this follows the identity

$$M_f(\Delta + \epsilon)^{-1} - (\Delta_S + \epsilon)^{-1} M_f$$
$$= (\Delta_S + \epsilon)^{-1} B(\Delta + \epsilon)^{-1}. \tag{9.29}$$

Now, from the Lemma 9.5.8, it follows that $B(\Delta + \epsilon)^{-1}$ is compact, and since $(\Delta_S + \epsilon)^{-1}$ is of Dixmier trace class (by the Lemma 9.5.9), we conclude that the right-hand side of (9.29) is of Dixmier trace class with the Dixmier trace $= 0$. The theorem follows from the general fact that $\mathrm{Tr}_\omega(xy) = \mathrm{Tr}_\omega(yx)$, if y is of Dixmier trace class and x is bounded (see [28]). □

Similar computation can be done for the noncommutative case. The Laplacian $\mathcal{L}$ generated by the canonical derivation δ_j on $\mathcal{A}^\infty$ is given by

$$\mathcal{L}(b(f)) = -\frac{1}{2}b(\Delta f),\ f \in C_c^\infty(R^{2d}).$$

Since in $L^2(R^{2d})$, $\frac{1}{2}\Delta$ has a natural self-adjoint extension (which we continue to express by the same symbol), $\mathcal{L}$ also has an extension as a negative self-adjoint operator in $\mathcal{H} \cong L^2(R^{2d})$, and we define the heat semigroup for this case as $T_t = e^{t\mathcal{L}}$. A simple computation using (9.16) shows that

$$\mathcal{L}(b(f)) \equiv -\frac{1}{2}b(\Delta(f)) = -\frac{1}{2}\sum_j \{[p_j, [p_j, b(f)]] + [q_j, [q_j, b(f)]]\}, \tag{9.30}$$

where $p_j = -i\partial_j, q_j = M_{x_j}$. The second form of $\mathcal{L}$ will be the one that will appear naturally in the later discussion of the stochastic dilation of T_t.

By analogy we can define the volume form on $\mathcal{A}^\infty$ by setting

$$v(b(f)) := \lim_{t\to 0+} t^d \operatorname{tr}(b(f)T_t).$$

Then we have the following theorem.

Theorem 9.5.10 $v(b(f)) = \int f dx$

Proof:
The kernel $\tilde{K}_t$ of the integral operator $b(f)T_t$ in $\mathcal{H}$ is given as

$$\tilde{K}_t(x, y) = \hat{f}(\underline{x} - \underline{y})e^{-t|y|^2/2}e^{ip(x,y)/2}.$$

As before we note that K_t is continuous in R^{2d} and $\tilde{K}_t(x, x) = k_t(x, x) = \hat{f}(0)e^{-t|x|^2/2}$. Using [57] we get the required result. □

9.6 Quantum Brownian motion on noncommutative manifolds

We recall the martingale definition of Brownian motion on a manifold given earlier in this chapter. To extend the notion of Brownian motion to the noncommutative setting, let us first introduce a notion of operator-valued martingale.

Definition 9.6.1 A *filtration* on a Hilbert space $\mathcal{H}$ is given by an increasing family $(P_t)_{t\geq 0}$ of projections in $\mathcal{H}$ satisfying the following conditions.

(*i*) $P_t \to I$ in the strong operator topology as $t \to \infty$.
(*ii*) For all $t \geq 0$, $\lim_{s\to t+} P_s = P_t$ in the strong operator topology.

Given such a filtration (P_t), we say that a family $(Y_t)_{t\geq 0} \in \mathcal{B}(\mathcal{H})$ is an *operator-valued martingale* with respect to the filtration (P_t), or just that (Y_t) is an (P_t)-martingale, if

$$P_s Y_t P_s = Y_s, \quad \text{for all } s \leq t.$$

One can also introduce the notion of an operator-valued local martingale, For this, a definition of quantum stop time is necessary. We do not discuss this here, and refer the reader to references [14], [15] and [101] for more details.

Now, let M be a compact Riemannian manifold, and for $m \in M$, let $(X_t^m)_{t\geq 0}$ denote M-valued Brownian motion defined on a probability space $(\Omega, \mathcal{F}, P)$, and satisfying $X_0^m = m$. Consider $\mathcal{A} = C(M)$, and $j_t : \mathcal{A} \to \mathcal{A} \otimes L^\infty(P)$ given by

$$j_t(f)(m) := f(X_t^m).$$

We have already seen that j_t is a dilation of the heat semigroup on $C(M)$. Moreover, $j_t(f) - j_0(f) - \frac{1}{2}\int_0^t j_s((\mathcal{L}(f))ds$ is an operator-valued local martingale, where $\mathcal{L}$ denotes the Laplacian. Motivated by this, we take the viewpoint that a quantum Brownian motion on a noncommutative manifold is nothing but an Evans–Hudson dilation of the heat semigroup on the underlying C^*-algebra. To justify the above viewpoint, we observe the following. If $j_t : \mathcal{A} \to \mathcal{A}'' \otimes \mathcal{B}(\Gamma(L^2(\mathbb{R}_+, k_0)))$ is an E–H dilation of a Q.D.S. (T_t) with the generator $\mathcal{L} = \theta_0^0$ on $\mathcal{A}$, satisfying the Q.S.D.E. $dj_t(a) = \sum_{\alpha,\beta} j_t(\theta_\beta^\alpha(a))d\Lambda_\alpha^\beta(t)$ with $j_0(a) = a \otimes I$, for $a \in \mathcal{A}_0 \subseteq \mathcal{A}$ (where $\theta_\beta^\alpha : \mathcal{A}_0 \to \mathcal{A}$), then for $a \in \mathcal{A}_0$, we have

$$j_t(a) - j_0(a) - \int_0^t j_s(\mathcal{L}(a))ds = \sum_{\alpha,\beta:\ (\alpha,\beta)\neq(0,0)} \int_0^t j_s(\theta_\beta^\alpha(a))d\Lambda_\alpha^\beta(s).$$

It follows that $j_t(a) - j_0(a) - \int_0^t j_s(\mathcal{L}(a))ds$ is an (P_t)-martingale, where P_t denotes the projection in $\Gamma(L^2(\mathbb{R}_+, k_0))$ given by the following:

$$P_t e(f) := e(f\chi_{[0,t]}).$$

Let us make some general remarks regarding the possibility of such a dilation in the light of our discussion in Chapter 8. Recall the general scheme of constructing a spectral triple from a given C^*-dynamical system $(\mathcal{A}, G, \alpha)$, where G is a Lie group. Clearly, the heat semigroup associated to such a spectral triple will be covariant. If G is also compact, we can apply the results of Chapter 8. The complete smoothness of the generator of such heat semigroup is verifiable from Lemma 8.1.28 and Remark 8.1.29. However, we prefer to discuss the construction case by case.

A note on the notational convention for the rest of the book: we shall use the generic notation Γ to denote the symmetric Fock space $\Gamma(L^2(\mathbb{R}_+) \otimes \mathbb{C}^n)$

(where n will be clear from the context); and $I\!E$ will denote the vacuum expectation in the Fock space $\Gamma(L^2(\mathbb{R}_+) \otimes \mathbb{C}^n)$. Following [100], we shall identify the n-dimensional standard classical Brownian motion $(w_1(t), \ldots, w_n(t))$ with the n-tuple of commuting self-adjoint operators $((a_1(t) + a^\dagger(t)), \ldots, (a_n(t) + a_n^\dagger(t)))$ acting on Γ. In particular, the classical stochastic integration with respect to w_i $(i = 1, 2, \ldots, n)$ will be understood as the quantum stochastic integration with respect to $da_i(t) + da_i^\dagger(t)$.

9.6.1 Brownian motion on noncommutative torus

From the quantum stochastic point of view, the two 'Laplacians' $\mathcal{L}_0$ and $\mathcal{L}$ are equally good candidates for driving the Brownian motion on the noncommutative manifold $\mathcal{A}_\theta$. Since $\mathcal{L}$ has been obtained by 'perturbing' $\mathcal{L}_0$, it is natural to construct the quantum Brownian motion corresponding to $\mathcal{L}_0$ first, and then from that the Brownian motion corresponding to $\mathcal{L}$ by some kind of perturbation at the quantum stochastic level. We achieve this in the theorem below. We take $n = 2$ and thus $\Gamma = \Gamma(L^2(\mathbb{R}_+) \otimes \mathbb{C}^2)$ here.

Theorem 9.6.2 *(i) There exists a solution of the following quantum stochastic differential equation (Q.S.D.E.) for $x \in \mathcal{A}_\theta{}^\infty$, and j_t^0 extends to a $*$-homomorphism from $\mathcal{A}_\theta$ to $\mathcal{A}_\theta \otimes \mathcal{B}(\Gamma)$.*

$$\begin{aligned} dj_t^0(x) &= j_t^0(id_1(x))dw_1(t) + j_t^0(id_2(x))dw_2(t) + j_t^0(\mathcal{L}_0(x))dt; \\ j_0^0(x) &= x \otimes I. \end{aligned} \tag{9.31}$$

In fact, the solution is given by

$$j_t^0(x) = \alpha_{(\exp 2\pi i w_1(t),\ \exp 2\pi i w_2(t))}(x),$$

where $(w_1, w_2)(t)$ is the standard two-dimensional Brownian motion. Moreover, $I\!E\, j_t^0(x) = e^{t\mathcal{L}_0}(x)$.
(ii) The following Q.S.D.E. in $\mathcal{H} \otimes \Gamma$ has a unique unitary solution:

$$\begin{aligned} dU_t &= \sum_{l=1}^{2} U_t\{i\, j_t{}^0(r_l)da_l{}^\dagger(t) + i\, j_t{}^0(r_l{}^*)da_l(t) - \frac{1}{2} j_t{}^0(r_l{}^* r_l)dt\}, \\ U_0 &= I. \end{aligned} \tag{9.32}$$

Furthermore, setting $j_t(x) = U_t j_t^0(x) U_t^*$, one has the following Q.S.D.E.:

$$dj_t(x) = \sum_{l=1}^{2}\{j_t(i\delta_l(x))da_l{}^\dagger(t) + j_t(i\delta_l{}^\dagger(x))da_l(t)\} + j_t(\mathcal{L}(x))dt, \tag{9.33}$$

and $I\!E\, j_t(x) = e^{t\mathcal{L}}(x)$.

Proof:
The proof of (i) is a straightforward application of the quantum Itô formula, hence the details are omitted. For (ii), we first note that the existence, uniqueness and unitarity of the solution of the Q.S.D.E. (9.32) follow from Theorems 5.3.1 and 5.3.3. The Q.S.D.E. (9.33) satisfied by j_t can be verified by quantum Itô formula, as done in ref [43], for example. □

9.6.2 Noncommutative 2d-dimensional plane

We briefly discuss the stochastic dilation of the heat semigroups on the noncommutative $2d$-planes considered by us. For the classical (or commutative) C^*-algebra of $C_0(\mathbb{R}^{2d})$ the stochastic process associated with the heat semigroup is the well known standard Brownian motion. For the noncommutative C^*-algebra $\mathcal{A}$ we first realize it in $\mathcal{B}(L^2(\mathbb{R}^d))$ by the Stone–von Neumann theorem on the representation of the Weyl relations [51]

$$(U_\alpha f)(x) = f(x+\beta)$$
$$(V_\beta f)(x) = e^{i\alpha.x} f(x). \tag{9.34}$$

Let $q_j, p_j (j = 1, 2 \ldots d)$ be the generators of V_β and U_α respectively, in fact they are the position and momentum operators in the above Schrödinger representation. For simplicity of writing we shall restrict ourselves to the case $d = 1$, and consider the Q.S.D.E. in $L^2(\mathbb{R}) \otimes \Gamma(L^2(\mathbb{R}_+, \mathbb{C}^2))$:

$$dX_t = X_t\left[-ip\, dw_1(t) - \frac{1}{2}p^2 dt - iq\, dw_1(t) - \frac{1}{2}q^2 dt\right], X_0 = I; \tag{9.35}$$

where w_1, w_2 are independent standard Brownian motions, identified as operators on $\Gamma(L^2(\mathbb{R}_+, \mathbb{C}^2))$ as explained before.

Theorem 9.6.3 *(i) The Q.S.D.E. (9.35) has a unitary solution.*
(ii) If we set $j_t(x) = X_t(x \otimes I_t)X_t^*$ then j_t satisfies the Q.S.D.E.:

$$dj_t(x) = j_t(-i[p, x])dw_1(t) + j_t(-i[q, x])dw_2(t) + j_t(\mathcal{L}(x))dt$$

for all $x \in \mathcal{A}^\infty$ and $\mathbb{E}\, j_t(x) = e^{t\mathcal{L}}(x)$ for all $x \in \mathcal{A}$, where $\mathcal{L}$ is given by (9.30).

Proof:
Consider the following Q.S.D.E. in $\Gamma(L^2(\mathbb{R}_+))$ for each $\lambda \in \mathbb{R}$ and for almost all w_1,

$$dW_t^{(\lambda)} = W_t^{(\lambda)}(-i(\lambda + w_1(t))dw_2(t) - \frac{1}{2}(\lambda + w_1(t))^2 dt), W_0^{(\lambda)} = I.$$

It is clear that $W_t^{(\lambda)} = \exp(-i \int_0^t (\lambda + w_1(s)) dw_2(s))$ which is unitary in $\Gamma(L^2(\mathbb{R}_+))$ for fixed λ and w_1. Next we set $W_t = \int_{\mathbb{R}} E^q(d\lambda) \otimes W_t^{(\lambda)}$ which can be clearly seen to be unitary in $L^2(\mathbb{R}) \otimes \Gamma(L^2(\mathbb{R}_+))$ for fixed w_1, where E^q is the spectral measure of the self-adjoint operator q in $L^2(\mathbb{R})$. Writing $X_t = W_t e^{-ipw_1(t)}$ it is clear that X_t is unitary in $L^2(\mathbb{R}) \otimes \Gamma(L^2(\mathbb{R}_+, \mathbb{C}^2))$. A simple calculation using Itô formula derived in Chapter 5 shows that X_t indeed satisfies (9.35).

The second part of the theorem follows from the observation that for fixed w_1 and w_2, X_t^* and $b(f) \otimes I_\Gamma$ with $f \in C_c^\infty(R^2)$ map $\mathcal{S}(\mathbb{R}) \otimes \Gamma(L^2(\mathbb{R}_+, \mathbb{C}^2))$ into itself. It is also straightforward to see that $j_t(x) = X_t x X_t^* = e^{-iqw_2(t)} e^{-ipw_1(t)} x e^{ipw_1(t)} e^{iqw_2(t)} = \phi_{(-w_1(t), -w_2(t))}(x)$. □

9.6.3 Quantum Heisenberg manifolds

We recall the notation introduced in the Subsection 9.4.3, and choose $\Gamma = \Gamma(L^2(\mathbb{R}_+, \mathbb{C}^3))$. Let us define the self-adjoint operators $i\, d_j^{(0)}$, $j = 1, 2, 3$, by setting $d_j^{(0)} = d_j$ for $j = 1, 3$, and taking $d_2^{(0)} = -\frac{\partial}{\partial y}$. Denote by r the unbounded operator on the Hilbert space $\mathcal{H} \equiv L^2([0,1] \times [0,1] \times \mathbb{Z})$ given by

$$r = -2\pi c (M_x \otimes I \otimes M_p),$$

where M_x and M_p denote multiplication operator by x and p respectively. It is clear (see for example [107]) that r is essentially self-adjoint on the product domain $L^2([0,1] \times [0,1]) \otimes_{\mathrm{alg}} \mathrm{Dom}(M_p)$, and we denote the self-adjoint extension again by r. Let us also observe that $id_2 = i\, d_2^{(0)} + r$. Since the unbounded self-adjoint operators $id_2^{(0)}$ and r commute, being nontrivial in the different tensor components, it follows that $\mathrm{Dom}(D)$ (where D is the Dirac operator on the quantum Heisenberg manifold) is a domain of essential self-adjointness for all the operators $i\, d_j^{(0)}$, $i\, d_j$ and r; and we denote the self-adjoint extensions by the same notation. Furthermore, we note that $i\, d_j^{(0)}$'s commute amongst each other. We set

$$u_t = u_t^w := \exp\left(i \sum_{j=1}^{3} d_j^{(0)} w_j(t) \right),$$

where $w(t) \equiv (w_1(t), w_2(t), w_3(t))$ is a three-dimensional standard Brownian motion. Clearly, for almost all w, u_t is unitary, and we also define the homomorphism $\alpha_t : \mathcal{B}(\mathcal{H}) \to \mathcal{B}(\mathcal{H}) \otimes \mathrm{L}^\infty(w)$ by setting for $a \in \mathcal{B}(\mathcal{H})$,

$$\alpha_t(a) = u_t a u_t^*.$$

Next we want to take into account the perturbation of the commuting triplet $(i\, d_1^{(0)}, i\, d_2^{(0)}, i\, d_3^{(0)})$ by $(0, r, 0)$, somewhat along the lines of [43]. But r is unbounded, so the method of [43] cannot be directly applied. However, the following simple observation comes to the rescue. There is a natural 'imprimitivity' between the map $e^{i\beta d_1}$ and M_x, viz.

$$e^{i\beta d_1} M_x e^{-i\beta d_1} = M_{\overline{x+\beta}}, \tag{9.36}$$

where $\overline{a+b} := (a+b)$ mod 1. Thus, u_t leaves the domain of r invariant and

$$\alpha_t(r) = u_t r u_t^* = -2\pi c(M_{\overline{x+w_1(t)}}) \otimes I \otimes M_p,$$

from which it is clear that $\alpha_t(r)$ is a self-adjoint operator with Dom(D) being a domain of its essential self-adjointness for almost all w. Furthermore, observe that $\alpha_t(r)$ and $\alpha_s(r)$ commute for all $s, t \geq 0$. Similarly, we can make sense of $\alpha_t(r^2)$.

Now, let us consider the following Q.S.D.E. with unbounded coefficients in the Hilbert space $\mathcal{H} \otimes \Gamma \equiv \mathcal{H} \otimes \Gamma(L^2(\mathbb{R}_+, \mathbb{C}^3)) \cong \mathcal{H} \otimes L^2(w)$:

$$dU_t = U_t\{i\alpha_t(r)dw_2(t) - \frac{1}{2}\alpha_t(r^2)dt\}, \quad U_0 = I. \tag{9.37}$$

By the results of Chapter 7, the above Q.S.D.E. is seen to have a unitary solution. In fact, we can explicitly write down its solution in terms of classical stochastic integral:

$$U_t = \exp\left(i \int_0^t \alpha_s(r)dw_2(s)\right).$$

Finally, if we set

$$j_t(a) = U_t(\alpha_t(a))U_t^*$$

for $a \in \mathcal{A}^\infty$, it follows that j_t satisfies the equation

$$dj_t(a) = \sum_{j=1}^{3} j_t([i\, d_j, a])dw_j(t) + j_t(\mathcal{L}(a))dt,$$

where

$$\mathcal{L}(a) = \frac{1}{2}\sum_{j=1}^{3}[d_j, [d_j, a]].$$

Thus, j_t satisfies the desired Q.S.D.E.. To conclude that it is indeed an E–H dilation of the Q.D.S. generated by the Laplacian $\mathcal{L}$, that is, j_t is a quantum Brownian motion on the quantum Heisenberg manifold, we have to prove that the range of j_t is in $\mathcal{A}'' \otimes \mathcal{B}(\Gamma)$. However, it is not clear whether α_t maps $\mathcal{A}$ into $\mathcal{A}$, so the explicit definition of j_t as given above does not give information about the range of j_t. To conclude that $j_t(\mathcal{A}) \subseteq \mathcal{A}'' \otimes \mathcal{B}(\Gamma)$, we need to take an indirect route of appealing to the construction of Chapter 8; more precisely to the Theorem 8.1.38 and Remark 8.1.39. The structure maps $[id_j, \cdot]$ and $\mathcal{L}$ appearing in the Q.S.D.E. satisfied by j_t are clearly completely smooth with respect to the action of the Heisenberg group, by Proposition 8.1.28 and Remark 8.1.29. Moreover, the elements $\phi_{m,n,k}$ introduced earlier in this chapter (Subsection 9.4.3) belong to $\mathcal{A}_{\infty}^{0}$, from which it follows that $\mathcal{A}_{\infty}^{0}$ is dense in $\mathcal{A}$. Thus by Remark 8.1.41, we can adapt the arguments of Theorem 8.1.38 to obtain an E–H flow with the above structure maps, which maps $\mathcal{A}$ into $\mathcal{A}'' \otimes \mathcal{B}(\Gamma)$. By Remark 8.1.39, this flow must coincide with j_t on $\mathcal{A}_{\infty}^{0}$, which proves that the range of j_t is a subset of $\mathcal{A}'' \otimes \mathcal{B}(\Gamma)$.

References

1. B. Abadie, 'Vector Bundles' over quantum Heisenberg manifolds. In *Algebraic Methods in Operator Theory*. (Basel: Birkhauser, 1994), pp. 307–15.
2. B. Abadie, Generalized fixed-point algebras of certain actions on crossed products. *Pacific J. Math.* **171**:1 (1995), 1–21.
3. L. Accardi and C. C. Heyde (editors), *'Probability towards 2000.' Papers from the symposium held at Columbia University, New York, October 2–6, 1995*; Lecture Notes in Statistics, Vol. 128. (New York: Springer-Verlag, 1998).
4. L. Accardi, R. Alicki, A. Frigerio and Y. G. Lu, An invitation to the weak coupling and low density limits. In *Quantum Probability and Related Topics*, Vol. VI, ed. L. Accardi. (River Edge, NJ: World Scientific Publishing, 1991).
5. L. Accardi, A. Frigerio and J. T. Lewis, Quantum stochastic processes. *Publ. Res. Inst. Math. Sci.* **18**:1 (1982), 97–133.
6. L. Accardi and S. Kozyrev, On the structure of Markov flows. *Irreversibility, probability and complexity (Les Treilles/Clausthal, 1999); Chaos, Solitons and Fractals* **12**:14–15 (2001), 2639–55.
7. L. Accardi and Y. G. Lu, Wiener noise versus Wigner noise in Quantum Electrodynamics. In *Quantum Probability and Related Topics*, Vol. VIII. (River Edge, NJ: World Scientific Publishing, 1993), pp. 1–18.
8. L. Accardi, Y. G. Lu and I. Volovich, *Quantum Theory and its Stochastic Limit*. (Berlin: Springer-Verlag, 2002).
9. S. Albeverio and R. Hoegh-Krohn, Ergodic actions by compact groups on C^*-algebras. *Math. Z.* **174**:1 (1980), 1–17.
10. S. Albeverio and R. Hoegh-Krohn, Dirichlet forms and Markovian semigroups on C^*-algebras. *Commun. Math. Phys.* **56** (1977), 173–87.
11. S. Albeverio and D. Goswami, A remark on the structure of symmetric quantum dynamical semigroups on von Neumann algebras. *Infin. Dimens. Anal. Quantum Probab. Relat. Top.* **5**:4, (2002), 571–9.
12. W. O. Amrein, J. M. Jauch and K. B. Sinha, *Scattering Theory in Quantum Mechanics*. (Reading, MA: W. A. Benjamin, 1977).
13. W. Arveson, The index of a quantum dynamical semigroup. *J. Funct. Anal.* **146**:2 (1997), 557–88.
14. S. Attal and A. Coquio, Quantum stopping times and quasi-left continuity. *Ann. Inst. Henri Poincaré, Probab. Stat.* **40** (2004), 497–512.
15. S. Attal and K. B. Sinha, Stopping semimartingales on Fock space. In *Quantum Probability Communications*, Vol. IX. (River Edge NJ: World Scientific, 1998), pp. 171–86.
16. V. P. Belavkin, Quantum stochastic calculus and quantum nonlinear filtering. *J. Multivariate Anal.* **42** (1992), 171–202.

17. V. P. Belavkin, Quantum stochastic positive evolutions: characterization, construction, dilation. *Commun. Math. Phys.* **184** (1997), 533–66.
18. B. V. R. Bhat, Cocycles of CCR flows. *Mem. Amer. Math. Soc.* **149**:709 (2001), 114 pp.
19. B. V. R. Bhat and K. R. Parthasarathy, Markov dilations of noncommutative dynamical semigroups and a quantum boundary theory. *Ann. Inst. Henri Poincaré, Probabilités et Statistiqués* **31**:4 (1995), 601–51.
20. B. V. R. Bhat and K. B. Sinha, Examples of unbounded generators leading to nonconservative minimal semigroups. In *Quantum Probability and Related Topics*, Vol. IX, (River Edge, NL: World Scientific, 1994).
21. R. L. Bishop and R. J. Crittenden, *Geometry of Manifolds* (Reprint of the 1964 original). (Providence, RI: AMS Chelsea Publishing, 2001).
22. O. Bratteli, G. A. Eliott and P. E. T. Jorgensen, Decomposition of unbounded derivation into invariant and approximately inner parts. *J. Riene Angew. Math.* **346** (1984), 166–93.
23. O. Brattelli and D. W. Robinson, *Operator Algebras and Quantum Statistical Mechanics I.* (New York: Springer Verlag, 1979).
24. E. Christensen and D. E. Evans, Cohomology of operator algebras and quantum dynamical semigroups. *J. London Math. Soc.* **20** (1979), 358–68.
25. A. M. Chebotarev, *Lectures on Quantum Probability*, Aportaciones Matematicas: Textos [Mathematical Contributions: Texts], Vol. 14. (Mexico: Sociedad Matematica Mexicana, 2000).
26. A. M. Chebotarev, The theory of conservative dynamical semigroups and its applications, Preprint, no. 1 (Moscow: Moscow Institute of Electronic Engineering, 1990).
27. P. R. Chernoff, Note on product formulas for operator semigroups, *J. Funct. Anal.* **2** (1968), 238–42.
28. A. Connes, *Noncommutative Geometry*. (San Diego, CA: Academic Press Inc., 1994).
29. A. M. Chebotarev and F. Fagnola, Sufficient conditions for conservativity of minimal quantum dynamical semigroup. *J. Funct. Anal.* **153**:2 (1998), 382–404.
30. P. S. Chakraborty, D. Goswami and K. B. Sinha, A covariant quantum stochastic dilation theory. In *Stochastics in Finite and Infinite Dimensions*, Trends in Mathematics. (Boston, MA: Birkhäuser Boston, 2001), pp. 89–99.
31. P. S. Chakraborty, D. Goswami and K. B. Sinha, Probability and geometry on some noncommutative manifolds. *J. Operator Theory* **49** (2003), 187–203.
32. F. Cipriani and J.-L. Sauvageot, Derivations as square roots of Dirichlet forms. *J. Funct. Anal.* **201**:1 (2003), 78–120.
33. K. R. Davidson, *C^*-algebras by example*, Fields Institute Monographs, Vol. 6. (Providence, RI: American Mathematical Society, 1996).
34. E. B. Davies, Quantum dynamical semigroups and neutron diffusion equation. *Rep. Math. Phys.* **11** (1977), 169–89.
35. E. B. Davies, *Quantum Theory of Open Systems*. (New York: Academic Press, 1976).
36. E. B. Davies, Generators of dynamical semigroups. *J. Funct. Anal.* **34** (1979), 421–32.

37. E. B. Davies, *One-parameter Semigroups*. (London: Academic Press, 1980).
38. E. B. Davies and J. M. Lindsay, Non-commutative symmetric Markov semigroups. *Math. Z.* **210** (1992), 379–411.
39. E. B. Davies and Y. Safarov (eds), *Spectral Theory and Geometry. Papers from the ICMS Instructional Conference held in Edinburgh, March 30–April 9, 1998*, London Mathematical Society Lecture Note Series, 273. (Cambridge: Cambridge University Press, 1999).
40. J. Dixmier, *von Neumann Algebras*. (New York: North Holland Publishing Company, 1981).
41. R. Durret, *Stochastic Calculus. A Practical Introduction*, Probability and Stochastics Series. (Boca Raton, FL: CRC Press, 1996).
42. M. P. Evans and R. L. Hudson, Multidimensional quantum diffusions. In *Quantum Probability and Applications*, Vol. III, eds. L. Accardi and W. von Waldenfels, Lecture Notes in Mathematics, 1303. (Heidelberg: Springer, 1988).
43. M. P. Evans and R. L. Hudson, Perturbations of quantum diffusions. *J. London Math. Soc.* **41** (1990), 373–84.
44. M. P. Evans, Existence of quantum diffusions, *Probab. Th. Rel. Fields* **81** (1989), 473–83.
45. D. E. Evans and J. T. Lewis, *Dilation of Irreversible Evolutions in Algebraic Quantum theory*, Communication of the Dublin Institute of Advanced Studies, Vol. 24. (Dublin: Dublin Institute of Advanced Studies, 1977).
46. M. Emery, *Stochastic Calculus in Manifolds (with an appendix by P. A. Meyer)*, Universitext. (Berlin: Springer-Verlag, 1989).
47. F. Fagnola, On quantum stochastic differential equations with unbounded coefficients. *Probab. Th. Rel. Fields* **86** (1990), 501–16.
48. F. Fagnola and K. B. Sinha, Quantum flows with unbounded structure maps and finite degrees of freedom. *J. London Math. Soc.* (2) **48**:3 (1993), 537–51.
49. F. Fagnola and S. J. Wills, Solving quantum stochastic differential equations with unbounded coefficients. *J. Funct. Anal.* **198**:2 (2003), 279–310.
50. W. Feller, *An Introduction to Probability Theory and its Applications*, Vol. 2. (New York: John Wiley, 1966).
51. G. B. Folland, *Harmonic Analysis in Phase Space*. (Princeton, NJ: Princeton University Press, 1989).
52. G. W. Ford, J. T. Lewis and R. F. O'Connell, Quantum Langevin equation. *Phys. Rev. A* **37**:11 (1988), 4419–28.
53. T. Friedrich, *Dirac Operators in Riemannian Geometry*, Graduate Studies in Mathematics, Vol. 25. (Providence, RI: American Mathematical Society, 2000).
54. M. Fukushima, *Dirichlet Forms and Markov Processes*. (New York: North-Holland, 1980).
55. L. Garding, Note on continuous representations of Lie groups. *Proc. Natl. Acad. Sci. U.S.A.* **33** (1947), 331–32.
56. J. M. Gracia-Bondia, J. C. Varilly and H. Figuera, *Elements of Noncommutative Geometry*. (Basel: Birkhauser, 2000).
57. I. C. Gohberg and M. G. Krein, *Introduction to the Theory of Nonselfadjoint Operators*, Translations of Mathematical Monographs, Vol. 18. (Providence, RI: American Mathematical Society, 1969).
58. D. Goswami, J. M. Lindsay and S. J. Wills, A continuous stochastic Stinespring's theorem. *Math. Ann.* **319**:4 (2001), 647–73.

59. D. Goswami, J. M. Lindsay, K. B. Sinha and S. J. Wills, Dilations of completely positive flows on a von Neumann algebra. *Pacific J. Math.* **211**:2 (2003), 221–47.
60. D. Goswami, A. Pal and K. B. Sinha, Stochastic dilation of a quantum dynamical semigroup on a separable unital C^* algebra. *Infin. Dimens. Anal. Quantum Probab. Relat. Top.* **3**:1 (2000), 177–84.
61. D. Goswami and K. B. Sinha, Minimal Quantum Dynamical Semigroup on a von Neumann algebra. *Infin. Dimens. Anal. Quantum Probab. Relat. Top.* **2**:2 (1999), 221–39.
62. D. Goswami and K. B. Sinha, Hilbert modules and stochastic dilation of a quantum dynamical semigroup on a von Neumann algebra. *Commun. Math. Phys.* **205**:2 (1999), 377–403.
63. D. Goswami, L. Sahu and K. B. Sinha, Dilation of a class of quantum dynamical semigroups with unbounded generators on UHF algebras. *Ann. Inst. H. Poincaré (Probab. Statist.)* **41**:3 (2005), 505–22.
64. Y. Haga, Crossed products of von Neumann algebras by compact groups. *Tohoku Math. J.* (2) **28**:4 (1976), 511–22.
65. H. Helson, *Harmonic Analysis*, Advanced Book Program. (Reading, MA: Addison-Wesley, 1983).
66. E. Hewitt and K. A. Ross, *Abstract Harmonic Analysis*, Vol. II, Die Grundlehren der mathematischen Wissenschaften, Band 152. (New York–Berlin: Springer-Verlag, 1970).
67. R. Hoegh-Krohn, M. B. Landstad and E. Stormer, Compact ergodic groups of automorphisms. *Ann. Math.* (2) **114**:1 (1981), 75–86.
68. A. S. Holevo, Covariant quantum dynamical semigroups: unbounded generators. In *Irreversibility and Causality (Goslar, 1996)*, Lecture Notes in Physics, vol. 504 (Berlin: Springer, 1998), pp. 67–81.
69. A. S. Holevo, On the structure of covariant dynamical semigroups. *J. Funct. Anal.* **131**:2 (1995), 255–78.
70. R.L. Hudson and K.R. Parthasarathy, Quantum Ito's formula and stochastic Evolutions. *Commun. Math. Phys* **93** (1984), 301–23.
71. R.L. Hudson and K.R. Parthasarathy, Stochastic dilations of uniformly continuous completely positive semigroups. *Acta Applicande Mathematicae* **2** (1984), 353–78.
72. R.L. Hudson and P. Robinson, Quantum diffusions and the noncommutative torus. *Lett. Math. Phys.* **15** (1988) 47–53.
73. R. L. Hudson and P. Shepperson, Stochastic dilations of quantum dynamical semigroups using one dimensional quantum stochastic calculus. In *Quantum Probability and Applications*, V, eds. L. Accardi and W. von Waldenfels. (Singapore: World Scientific, 1988).
74. J.E. Humphreys, *Introduction to Lie Algebras and Representation Theory*, Graduate Texts in Mathematics, Vol. 9, (New York–Berlin: Springer-Verlag, 1972).
75. J. Jacod, *Calcul stochastique et problèmes de martingales*, [Stochastic calculus and martingale problems], Lecture Notes in Mathematics 714, (Berlin: Springer, 1979).
76. R. V. Kadison and J. R. Ringrose, *Fundamentals of the Theory of Operator Algebras*, Vol. I and II. (New York: Academic Press, 1983).
77. T. Kato, *Perturbation Theory for Linear Operators*, (Reprint of the 1980 edition), Classics in Mathematics. (Berlin: Springer-Verlag, 1995).

78. T. Kato, On the semigroups generated by Kolmogoroff's differential equations. *J. Math. Soc. Jap.* **11** (1954), 169–89.
79. B. Kümmerer and H. Maassen, The essentially commutative dilations of dynamical semigroups on M_n. *Commun. Math. Phys.* **109** (1987), 1–22.
80. S. Kobayashi and K. Nomizu, *Foundations of Differential Geometry*, Vol. I and II (reprint of the 1969 original), Wiley Classics Library. (New York: John Wiley & Sons, Inc., 1996).
81. E.C. Lance, *Hilbert C^*-modules: A Toolkit for Operator Algebraists*, London Mathematical Society Lecture Note Series, Vol. 210. (Cambridge: Cambridge University Press, 1995).
82. G. Landi, *An Introduction to Noncommutative Spaces and their Geometries*, Lecture Notes in Physics New Series Mathematical Monographs, Vol. 51. (Berlin: Springer-Verlag, 1997).
83. G. Lindblad, On the generators of quantum dynamical semigroups. *Commun. Math. Phys.* **48** (1976), 119–30.
84. J.M. Lindsay and K.R. Parthasarathy, On the generators of quantum stochastic flows. *J. Funct. Anal.* **156** (1998), 521–49.
85. J.M. Lindsay and S.J. Wills, Existence, positivity and contractivity for quantum stochastic flows with infinite dimensional noise. *Probab. Theory Related Fields* **116**:4 (2000), 505–43.
86. J.M. Lindsay and S.J. Wills, Markovian cocycles on operator algebras adapted to a Fock filtration. *J. Funct. Anal.* **178**:2 (2000), 269–305.
87. J.-L. Loday, *Cyclic Homology*, Appendix E by María O. Ronco. Second edition. Chapter 13 by the author in collaboration with Teimuraz Pirashvili. Grundlehren der Mathematischen Wissenschaften [Fundamental Principles of Mathematical Sciences], 301. (Berlin: Springer-Verlag, 1998).
88. T. Matsui, Markov semigroups on UHF algebras, *Rev. Math. Phys.* **5**:3 (1993), 587–600.
89. P.A. Meyer, *Quantum Probability for Probabilists*, 2nd edn, Lecture Notes in Mathematics 1538. (Heidelberg: Springer Verlag, 1993).
90. J.A. Mingo and W.J. Philips, Equivariant triviality theorems for Hilbert C^*-modules. *Proc. Amer. Math. Soc.* **91**:2 (1984), 225–30.
91. A. Mohari, *Quantum Stochastic Calculus With Infinite Degrees of Freedom*, Ph.D. thesis. Indian Statistical Institute, New Delhi, 1991.
92. A. Mohari, Quantum stochastic differential equations with unbounded coefficients and dilations of Feller's minimal solution. *Sankhyá*, Ser. A, **53** (1991), 255–87.
93. A. Mohari and K.R. Parthasarathy, On a class of generalised Evans–Hudson flows related to classical Markov processes. In *Quantum Probability and Related Topics*, Vol VII. (River Edge, NJ: World Scientific, 1992), pp. 221–49.
94. A. Mohari and K.B. Sinha, Quantum stochastic flows with infinite degrees of freedom and countable state Markov processes. *Sankhyá*, Ser. A, **52** (1990), 43–57.
95. A. Mohari and K.B. Sinha, Stochastic dilations of minimal quantum dynamical semigroup. *Proc. Indian Acad. Sc. (Math. Sc.)*, **103**:3 (1992), 159–73.
96. E. Nelson, Notes on non-commutative integration. *J. Funct. Anal.* **15** (1974), 103–16 .
97. E. Nelson, Analytic vectors. *Ann. Math.* (2) **70**:3 (1959), 572–615.

98. N. E. Wegge-Olsen, *K-theory and C^*-algebras: A Friendly Approach*, Oxford Science Publications. (New York: The Clarendon Press, Oxford University Press, 1993).
99. K. R. Parthasarathy, An additional remark on unitary evolutions in Fock space. In *Séminaire de Probabilités* XXV, Lecture Notes in Mathematics 1485. (Berlin: Springer, 1991), pp. 37–8.
100. K. R. Parthasarathy, *An Introduction to Quantum Stochastic Calculus*, Monographs in Mathematics. (Basel: Birkhäuser Verlag, 1992).
101. K. R. Parthasarathy and K. B. Sinha, Stop times in Fock space stochastic calculus. *Probab. Theory Related Fields* **75** (1987), 317–49.
102. K. R. Parthasarathy and K. B. Sinha, Quantum Markov processes with Christensen–Evans generator in a von Neumann algebra. *Bull. London Math. Soc.* **31**:5 (1999), 616–26.
103. K. R. Parthasarathy and K. B. Sinha, Markov Chains as Evans–Hudson diffusion in Fock space. In *Séminaire de Probabilités* XXIV, eds. J. Azema, P. A. Meyer and M. Yor, Lecture Notes in Mathematics 1426. (Berlin: Springer-Verlag, 1988/89), pp. 326–369.
104. W. Pauli, *Festschrift zum 60 Geburstage A. Sommerfields*. (Leipzig: Horzel, 1928), p. 30.
105. G. K. Pedersen, *C^*-algebras and their Automorphism Groups*, London Mathematical Society Monographs, Vol. 14. (London–New York: Academic Press, Inc. [Harcourt Brace Jovanovich, Publishers], 1979).
106. C. Piron, *Foundations of Quantum Physics*. (Reading, MA: W. A. Benjamin, 1976).
107. M. Reed and B. Simon, *Methods of Modern Mathematical Physics*, Vol. 1. (London–New York: Academic Press, 1978).
108. M. Reed and B. Simon, *Methods of Modern Mathematical Physics*, Vol. 2. (London–New York: Academic Press, 1978).
109. M. Reed and B. Simon, *Methods of Modern Mathematical Physics*, Vol. 3. (London–New York: Academic Press, 1978).
110. M. Reed and B. Simon, *Methods of Modern Mathematical Physics*, Vol. 4. (London–New York: Academic Press, 1978).
111. M. Rieffel, Deformation quantization of Heisenberg manifolds. *Commun. Math. Phys.* **122** (1989), 531–62.
112. M. A. Rieffel, *Noncommutaive Tori—A Case Study of Noncommutative Differentiable Manifolds*, Contemporary Mathematics, Vol. 105. (Providence, RI: American Mathematical Society, 1990).
113. S. Rosenberg, *The Laplacian on a Riemannian Manifold*. (Cambridge: Cambridge University Press, 1997).
114. J.-L. Sauvageot, Tangent bimodule and locality for dissipative operators on C^*-algebras. In *Quantum Probability and Applications* IV, eds. L. Accardi and W. Von Waldenfels, Lecture Notes in Mathematics 1396. (Berlin: Springer-Verlag, 1989), pp. 322–38.
115. Segal, I. E., A non-commutative extension of abstract integration. *Ann. Math.* (2) **57** (1953), 401–57.
116. K. Shiga, Representations of a compact group on a Banach space. *J. Math. Soc. Japan* **7** (1955), 224–48.
117. K. B. Sinha, Quantum dynamical semigroups. *Operator Theory: Advan. Applic.* **70** (1994), 161–9.

118. K. B. Sinha, Quantum mechanics of dissipative systems. *J. Indian Inst. Sci.* **77** (1997), 275–9.
119. K. B. Sinha, Quantum stochastic calculus and applications—a review. In *Probability Towards 2000* (New York, 1995), Lecture Notes in Statistics. 128. (New York: Springer, 1998), pp. 307–18.
120. M. Skeide, Hilbert modules in quantum electrodynamics and quantum probability. *Commun. Math. Phys.* **192** (1998), 569–604.
121. M. Skeide, Generalised matrix C^*-algebras and representations of Hilbert modules. *Math. Proc. R. Ir. Acad.* **100** A:1 (2000), 11–38.
122. M. Skeide, von Neumann modules, intertwiners and self-duality. *J. Operator Theory* **54**:1 (2005), 119–24.
123. M. Spivak, *A Comprehensive Introduction to Differential Geometry*, Vol. I–V. (Wilmington, DE: Publish or Perish, Inc., 1979).
124. V. S. Sunder, *An Invitation to von Neumann Algebras*. (Berlin: Springer, 1987).
125. M. Takesaki, *Theory of Operator Algebras I*. (New York: Springer-Verlag, 1979).
126. L. van Hove, Quantum mechanical perturbations giving rise to a statistical transport equation. *Physica* **21** (1955), 517–40.
127. D. V. Voiculescu, K. J. Dykema and A. Nica, *Free Random Variables. A Noncommutative Probability Approach to Free Products with Applications to Random Matrices, Operator Algebras and Harmonic Analysis on Free Groups*, CRM Monograph Series, 1. (Providence, RI: American Mathematical Society, 1992).
128. J. von Neumann, *The Geometry of Orthogonal Spaces*, Vol. II. (Princeton, NJ: Princeton University Press, 1950).
129. N. Weaver, Sub-Riemannian metrics for quantum Heisenberg manifolds. *J. Operator Theory* **43** (2000), 223–42.
130. K. Yosida, *Functional Analysis*. (Berlin: Springer Verlag, 1974).

Index